Sustainable Materials and Smart Practices

NCSMSP-2021

1st National Conference on Sustainable Materials and Smart Practices, 17th and 18th December 2021, SATHYAMANGALAM, INDIA

BANNARI AMMAN INSTITUTE OF TECHNOLOGY
Stay Ahead

NCSMSP

Editors

M. Vasudevan, V. Preetha, B. Jeyanth

Peer review statement

All papers published in this volume of "Materials Research Proceedings" have been peer reviewed. The process of peer review was initiated and overseen by the above proceedings editors. All reviews were conducted by expert referees in accordance to Materials Research Forum LLC high standards.

Published under License by **Materials Research Forum LLC**
Millersville, PA 17551, USA

Published as part of the proceedings series
Materials Research Proceedings
Volume 23 (2022)

ISSN 2474-3941 (Print)
ISSN 2474-395X (Online)

ISBN 978-1-64490-194-6 (Print)
ISBN 978-1-64490-195-3 (eBook)

Distributed worldwide by

Materials Research Forum LLC
105 Springdale Lane
Millersville, PA 17551
USA
http://www.mrforum.com

Manufactured in the United States of America
10 9 8 7 6 5 4 3 2 1

Table of Contents

Preface

This book reviews sustainable building materials meant for construction works and its smart practices followed in the construction sector. Specific topics include fiber reinforced concrete, composite sections and the use of mineral admixtures.

Technical Review Committee Members

Sustainable Materials and Smart Practices - NCSMSP-2021 Materials Research Forum LLC
Materials Research Proceedings **23** (2022) 1-8 https://doi.org/10.21741/9781644901953-1

Experimental Study on Light Transmitting Concrete Using Plastic Optical Fiber

S. Loganayagan[1,a], A. Mohanraj[1,b], V. Hariharan[1,c], N. Praveen[1,d]

[1]Department of Civil Engineering, Bannari Amman Institute of Technology, Erode 638401, India

[a] loganayagans@bitsathy.ac.in ,[b] mohanraja@bitsathy.ac.in, [c] hariharanv.ce19@bitsathy.ac.in,
[d] praveen.ce19@bitsathy.ac.in

Keywords: Plastic Optical Fiber, Light Transmission, Traditional Concrete, Plastic Fiber

Abstract. The present study aims at manufacturing the concrete specimens by reinforcing optical fiber and scrutinizing it with the traditional concrete. The concrete specimens were subjected to totally different things such as compressive strength tests, light-transmission tests, etc. The compressive strength results obtained for the semi transparent concrete specimens were virtually the same as that of the traditional concrete specimen. The results of the transmission take a look at were satisfactory because the Plastic Optical Fiber retained its potency. Thus, it's evident that the transparency of the concrete structures are often introduced with the insertion of optical fiber while not compromising the strength, that may be a breakthrough to the aspiration of achieving some new feat in modern architecture.

Introduction

Concrete may be an artifact fabricated from a mix of cement, sand, coarse aggregate, pebbles and water, typically outlined as artificial stone. A word thought to be of French origin, concrete is related to engineering achievements bridges, power plants, foundries. Light transmission concrete additionally referred to as semi transparent concrete, may be a Crete based mostly artifact having light-transmissive property [1]. Light-weight transmissive property is especially thanks to uniform distribution of optical fibers throughout Its body. it's additionally notable to be transparent concrete. Translucent stone isn't solely a clear, compact artifact, but also a tool of expression within the hands of artists and architects. Philosophical, visual and material signifiers are all incorporated. Litracon presents the thought of sunshine transmission concrete within the variety of a widely applicable new artifact. It is often used for interior or exterior walls, illuminated pavements or maybe in art or design objects the clear concrete is employed for esthetic application too, by inserting the optical fibers in concrete. Each natural further as artificial light passes through the clear concrete due to optical fibers. Additionally it also makes the concrete as fiber bolstered that is its other advantage[2]. Our study, not solely strained with the ornamental purpose however the impact of fiber application in strength side is additionally taken into thought. Load carrying capacity of clear concrete will increase once fibers are organized in several layers and additionally numerous patterns are often created to create the concrete ornamental.

Materials

The materials used for making light transmitting concrete using plastic optical fiber are discussed below with its characteristics and properties before mixing with various materials.

Cement

Ordinary Portland cement of 43- Grade out there in native market is employed within the investigation[3]. The cement used has been tested for numerous properties as per IS: 4031-1988 and located to be confirmed to varied specifications of IS: 12269-1987 having specific gravity of 3.0. the figure 1 shows the ordinary Portland cement.

Figure.1 Ordinary Portland cement

Manufactured sand (M sand)
Manufactured sand (M sand) is used as a substitute for natural river sand in this study. Particles passing through 4.75 mm sieve are used[4]. Crushed angular aggregates of size 20mm and 10mm showing specific gravity 2.79 and 2.74 respectively were used in the study. Water for curing and mixing confirming to IS 456 was used.

.

Figure.2 Manufactured sand

Coarse Aggregate
The coarse combination is the strongest and porous element of concrete[5]. Presence of coarse combination reduces the drying shrinkage and different dimensional changes occurring on account of movement of wetness. In our investigation we have a tendency to have used the combination passing through 20mm IS Sieve and holding on 12.5mm sieve. The particular gravity of combination was found out to be 2.50

Figure. 3 Coarse Aggregate

Plastic Optical Fiber

Plastic glass fiber is an optical fiber that's created out of chemical compounds, just like glass glass fiber, POF transmits light through the core of the fiber. Its chief advantage over the glass product, different facet being equal, is its hardiness below bending and stretching[6]. Glass fiber utilized in telecommunications is ruled by European Standards EN 60793-2-40-2011.

The idea of using light to send messages has been developed since the eighth century B.C., when the Greeks used fire signals for sending alarms or calls for help. It was only in the mid 1960s that Charles K. Kao determined that glass had a loss of 20db/km, which spurred researchers into exploring methods for making glass more pure[7]. This discovery sparked a revolution in the telecommunication industry as a new industry of processing optical fibers became commercially important. These optical fibers have great light transmission capability.

The typical fibers today are made out of glass or plastic since it is possible to make them thin and long. Also both glass and plastic are transparent at particular Wavelengths, which allow the fiber to guide light efficiently. The fiber is constructed with a core with high index surrounded by a layer of cladding at lower index. The core and cladding can be made out of both plastic and glass. For plastics, the core can be polystyrene or polymethyl methacrylate and the cladding is generally silicone or Teflon for glasses both the cladding and the core are made out of Silica with small amounts of decants such as Boron, Germanium to change its Index. Major differences exist between the two materials when it comes to making the optical fiber. In plastic core fibers they are more flexible and inexpensive compared to glass fibers[8]. They are easier to install and can withstand greater stresses and weigh 60% less than glass fiber. But losses, giving them very limited use in communication applications. Such plastic fibers are practical for short runs such as within buildings. Therefore, due to their restrictive nature glass core fibers are much more widely used because they are capable of transmitting light effectively over large distances.

Materials Research Proceedings **23** (2022) 1-8 https://doi.org/10.21741/9781644901953-1

Figure 4 Plastic Optical Fiber

Characteristics of Plastic Optical Fiber

Generally, fiber protective cover is created of silicon refractive index of 1.46.High index of refraction distinction is maintained between core and cladding[9]. High numerical aperture. Have high mechanical flexibility and low price.Industry-standard IEC (60793-2-40 A 4a.2) step-index fiber includes a core diameter of 1mm.Attenuation loss is concerning I dB/m @ 650 nm. Bandwidth measure is-5 MHz-km @ 650 nm.

Light Theory

The effectiveness of the wire depends on its ability to guide the light ray far distances with as little scattering or absorption of the light as possible. Doing so means that the optical fiber must exhibit total internal reflection within the wire, Thus when considering the propagation of light for an optical fiber the refractive index of the dielectric medium needs to be accounted for. As light rays become incident on an interface between two dielectrics with different index of refractions, refraction occurs between the two mediums. This can be best described by using Snell's Law of Refraction which states,

$$N1\sin\varnothing 1 = N2\sin\varnothing 2,$$

This equation shows that at certain angles partial internal reflection will arise, as well at other angles total internal reflection will occur as shown in following figure

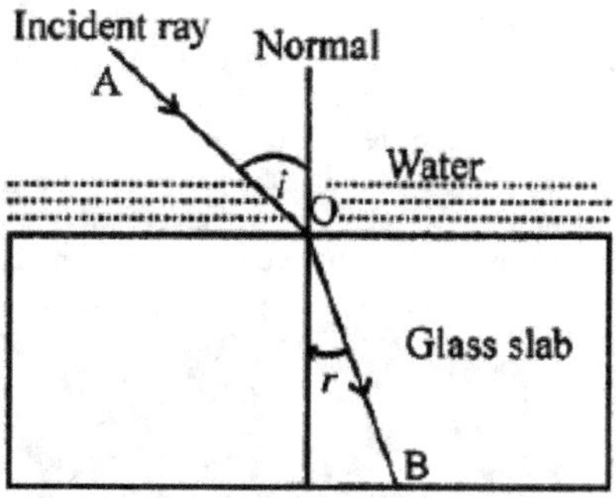

Figure 5 Light reflection diagram

Mix Design for Light Transmitting concrete

Mix designs are often outlined because of the method of choosing appropriate ingredients of concrete and deciding their relative proportions with the object of manufacturing concrete of bound minimum strength and sturdiness as economically as attainable. The main objective is to stipulate the minimum strength and sturdiness[10]. It conjointly reveals the relation between combination and paste. the opposite conditions being equal for possible mixes, the strength of concrete numerous as an inverse function of the water- cement ratio. Since, the standard of water depends upon the number of paste, it's necessary that as very little paste as attainable should be used and therefore the importance of grading.

Indian Standard Recommended Method 456

The Mix style for M20 grade of concrete has been prepared with the design stipulations. Reinforced concrete construction minimum M20 grade of concrete is used as Mix style are often outlined as the method of choosing appropriate ingredients of concrete and deciding their relative proportions with the object of manufacturing concrete of certain minimum strength and sturdiness as economically as attainable. In our investigation we've created M20 grades of concrete. The combination magnitude relation obtained when the combination style as per IS 456 for M20 (1:1.5:3). Further, the poured the concrete within the cube Moulds and completely different samples were created that are as follows

Conventional Concrete of grade M20.

Castings of cubes were performed based on the design mix and action of various specimens of light transmitting concrete. Once the concrete is totally mixed the concrete is poured within the cube, compaction is done by the vibration machine[11]. Concrete cubes were removed from the Moulds when 24 hrs and those were placed into the curing tank. Curing was done for 7, 14 and 28 days for all samples.

Placing of optical fiber

Placing of optical fiber are completely different ways of putting of glass fiber within the concrete are as follows

A. Direct mix
B. Indirect mix
C. Injection methodology

Materials Research Proceedings **23** (2022) 1-8 https://doi.org/10.21741/9781644901953-1

In the concrete block the optical fibers were placed, which have very small diameter, so that a bunch of fiber are prepared. According to the proportion of optical fiber, the number of fibers in particular sizes like 150, 200 fibers for 4 % & 5 % respectively. These bunches are passed transversely from the holes made at opposite face of mould, the bundles of optical fibers are laid such that it extends from face for length of at least 1 cm to make it stable and to avoid sagging during concreting and compaction. In our analysis we have used one among the ways known as the direct methodology. Once the optical were placed within the concrete they ought to be transmit

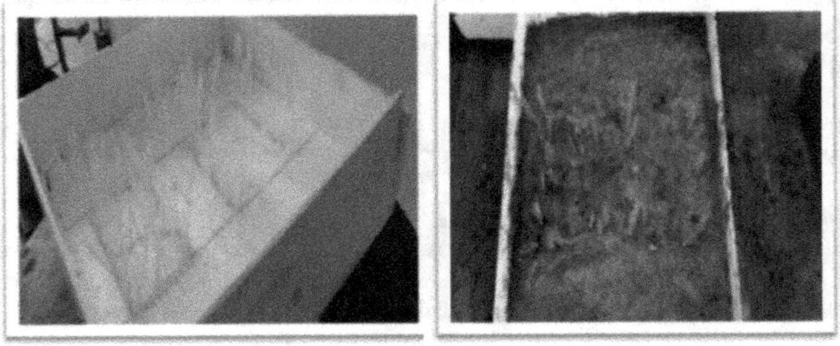

Figure.6 Placing of Optical Fiber in Concrete.

Experimental Test Results

The Light transmitting plastic optical fiber has various experimental setups to categorize the optical fiber and to check the strength of the normal concrete with light transmitting concrete.

Compressive Strength of concrete

It is defined as the Characteristic strength of 150mm size concrete cubes @28 days. As we all know that concrete is a mixture of sand, cement, and aggregate. The strength of the concrete depends upon many factors like individual compressive strength of its constituents (Cement, Sand, aggregate), quality of materials used, air entrainment mix proportions, water-cement ratio, curing methods and temperature effects. Compressive strength gives an idea of the overall strength and above-mentioned factors[12]. Through conducting this test, one can easily judge the concrete strength psi and quality of concrete produced. The table 1 gives the results of cubes tested on different days.

Table.1 Compressive strength Test Results

S.No	Test on Days	Weight of concrete (Kg)	Compressive strength(N/mm2)
1	7	7.646	8.88
2	7	7.646	9.02
3	14	7.852	14.44
4	14	7.852	15.2
5	28	7.969	18
6	28	7.969	19.25

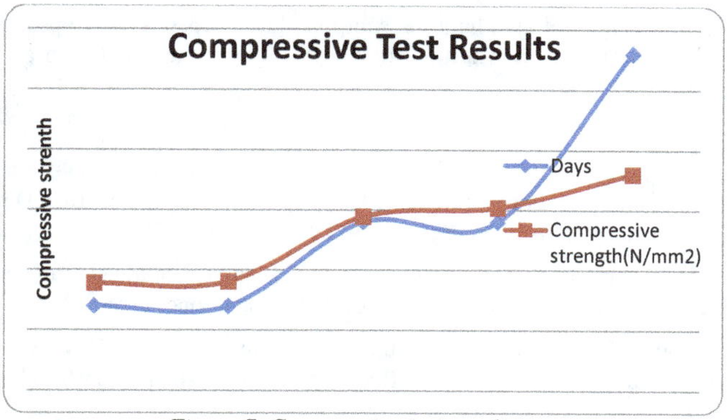

Figure.7. Compressive test results

The above figure explains the compressive strength of the cubes on different days such as 7, 14 & 28. The strength of the cube increases with the number of days as like in the normal concrete. But in this light transmitting concrete has few characteristics of light passing voids or it may be named as path of travel. In addition to this the material properties of concrete were tested. The casted light transmitting concrete was shown in the figure8.

Figure.8 Light transmitting concrete

Conclusion

This study investigates the compressive behavior of light transmitting concrete. The properties of Plastic Optical Fiber (POF) used are investigated. This paper contributes to new alternatives of concrete for property construction. The compressive strength with the quality concrete combine samples was found to be 26.52 N/mm2, and adding POF to the concrete includes a variable impact on the compressive strength, with the best improvement of 34.16 N/mm^2) accomplished with POF concrete with victimization fibers of 1.5 millimeter diameter spaced 10 millimeter apart for a proportion of 1.43% of fibers. It's clear that the compressive strength of the specimens augmented with time. The utmost quantity of sunshine passing through the cubes was 75.53 LUX using fibers of 1.5 millimeter diameter areas 10millimeter apart for a proportion of 1.43% of fibers. Thus the study shows that the transparency of light is feasible in concrete while not poignant in its compressive strength, because the fibers and fiber reinforcement enhance the strength and look.

The transmission of light through light transmitting blocks depends on the percentage of optical fiber used of that surface area. The transmission of light is increased with an increase in the percentage of optical fiber.

References

[1] Altlomate A, Alatshan F, Mashiri F, Jadan M. Experimental study of light-transmitting concrete. International Journal of Sustainable Building Technology and Urban Development. 2016 Oct 1;7(3-4):133-9. https://doi.org/10.1080/2093761X.2016.1237396

[2] Bashbash BF, Hajrus RM, Wafi DF, Alqedra M. Basics of light transmitting concrete. Global Advanced Research Journal of Engineering, Technology and Innovation. 2013;2(3).

[3] Shanmugavadivu PM, Scinduja V, Sarathivelan T, Shudesamithronn CV. An experimental study on light transmitting concrete. Int. J. Res. Eng. Technol. 2014;3(11):160-3. https://doi.org/10.15623/ijret.2014.0323035

[4] Ramana, S. Venkatachalam, K. Vishnuvardhan, M.M. Saravanan, A Study on Fresh and Hardened Properties of Concrete with Partial Replacement of Bottom Ash as a Fine Aggregate, Materials Research Proceedings, Vol. 19, pp 28-35, 2021

[5] Philips, The LED Lighting Revolution: A Summary of the Global Savings Potential, in, Koninklijke Philips Electronics NV. Netherlands, 2012.

[6] Yadav A, Shekhar S, Anand A, Badal A, Zaman B. An investigating study on a new innovative material: transparent concrete. International Journal of Engineering Research and Advanced Development. 2018;4(1):64-72.

[7] He J, Zhou Z, Ou J, Huang M. Study on smart transparent concrete product and its performances. In6thinternational workshop on advanced smart materials and smart structures technology 2011 Jul 25.

[8] Zubia J, Arrue J. Plastic optical fibers: An introduction to their technological processes and applications. Optical fiber technology. 2001 Apr 1;7(2):101-40. https://doi.org/10.1006/ofte.2000.0355

[9] Kaino T. Preparation of plastic optical fibers for near-IR region transmission. Journal of Polymer Science Part A: Polymer Chemistry. 1987 Jan;25(1):37-46. https://doi.org/10.1002/pola.1987.080250105

[10] Sawant AB, Jugdar RV, Sawant SG. Light transmitting concrete by using optical fiber. Int. J. Inventive Eng. Sci. 2014 Dec;3(1):23-8.

[11] Sunil J, Amol K, Chetan K, Pravin T. A study on translucent concrete product and its properties by using optical fibers. International Journal of Modern Engineering Research (IJMER). 2015;5:53-7.

[12] Kister G, Winter D, Gebremichael YM, Leighton J, Badcock RA, Tester PD, Krishnamurthy S, Boyle WJ, Grattan KT, Fernando GF. Methodology and integrity monitoring of foundation concrete piles using Bragg grating optical fibre sensors. Engineering Structures. 2007 Sep 1;29(9):2048-55. https://doi.org/10.1016/j.engstruct.2006.10.021

Sustainable Materials and Smart Practices - NCSMSP-2021 Materials Research Forum LLC
Materials Research Proceedings **23** (2022) 9-15 https://doi.org/10.21741/9781644901953-2

Performance Evaluation of RCC Beam Column Joint with Aramid Fibre

V. Jayanthi [1,a,*], S. Mehala[2,b], P. Rajesh Kumar [2,c]

[1] Bannari Amman Institute of Technology Sathyamangalam, Erode, India

[2] V.S.B College of Engineering Technical Campus,Kinathukadavu,Coimbatore, India

[a]JAYANTHIV@bitsathy.ac.in, [b] mehalacivil2021@gmail.com, [c]professorrajeshkumar2021@gmail.com

Keywords: Column Joint, Aramid Fibre, Moment Resistance Earthquake

Abstract-In reinforced concrete structure beam, the beam column joints are more crucial zones which are most susceptible to failure during earthquake excitation. It is very important to design the joint to dissipate the large amount of energy to the neighbouring elements without loss of strength and ductility. Several retrofitting methods are employed to strengthen the beam column joint. The effective method one among is Fibre Reinforced Polymer (FRP) composites. In this paper, the deflection of beam column joint wrapped with aramid fibre has been calculated and it is compared with Ansys software solution. From the results, the deflection of the joint is decreased when aramid fibre is used.

Introduction

Moment resistance is considered to be challenges part in beam column joint During Earthquake excitation beam column joint behaves to significant influence and response result to entire structure. In general fibers are the principal load-carrying members, while the surrounding matrix keeps them in the desired locations and orientation, acting as a load transfer medium between them, and protects them from environmental damage [1].One of the principal challenges in structural engineering concerns the development of modern design concepts to better protect structures, with the occupants and contents, from the damaging effects of destructive environmental forces including those due to winds, waves and earthquakes. Passive energy dissipation devices, when incorporated into a structure, absorb or consume a portion of the input energy, thereby reducing energy dissipation demand on primary structural members and minimizing possible structural damage[2].FRP-retrofitted structure showed that the shear capacity, ductility, and hysteretic energy dissipation capacity of each story were greatly improved and that the beam-column joint failure mode was prevented and the structure experienced much less damage compared with the original structure[12].When moderate structural strengthening but high resistance for fire is simultaneously sought such as for building structures, the basalt fiber strengthening will be a good alternative methodology among other fiber reinforced polymer (FRP) strengthening systems.[4] On the time of earthquake, the top bar and bottom bar tends to pull in opposite directions, this moment is balanced at the position of beam column joint hence beam column joint has to be reviewed and additional material has to be incorporate to with stand effects of earthquake resistance.. Repairing a damaged beam column joint is difficult when compared with enhancing the strength of concrete. In this literature beam column joint is strengthen by Aramid fibre.

Aramid Fibre
Aramid fibre or aromatic polyamide shown in Figure 1 is a strong synthetic fibre and good resistance to heat. These polymers are widely use in military. It has been first used in vehicle armor and as days passed it came to practice to be used in military helmets, protective gloves, and fireproof suits for fire fighters. These aramid fibres are chemically resistance to organic solvents. It gives good resistance to abrasion and cutting. From the previous studies it is clear that these fibres are with high quality and low weight. These fibres are extremely impervious to effect and scraped area harm. It is Waterproof when added with epoxy resin. It can be utilized as a composite with elastic holding its adaptability. High ductile modulus and low breakage extension joined with great protection from chemicals settle on it the correct decision for various composite basic parts in different applications. To paste fibre fortified polymer composites epoxy resins have been used. Epoxies are high strength, low shrinkage, and excellent adhesion to various substrates, effective electrical insulation, chemical and solvent resistance, low cost, and low toxicity.

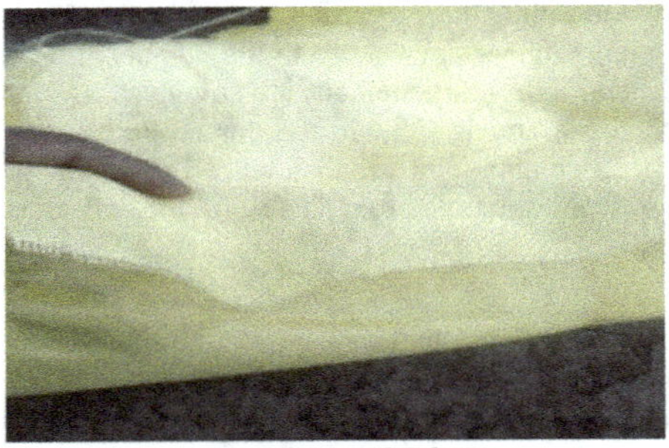

Fig.1. Aramid Fibre

Scope

1. To determine the Structural Behaviours of basalt fibre wrapping at beam column joint.
2. To determine the Durability of basalt fibre wrapping concrete.
3. To prevent the concrete from corrosion by wrapping in beam column joint
4. To prevent the concrete from cracks.

Material Tests

Tests on Cement
The ordinary Portland cement confirming to IS 4031 was used for the preparation of specimens. OPC 53 grade was used.

- Specific gravity of cement , G = 3.134
- Percentage of fineness = 3.6%
- Consistency of the given sample of cement, P = 32%
- Final Setting Time of Cement = 10 hours
- Initial Setting time 30 minutes

Tests on Fine Aggregate
It confirming to zone II of IS 383-1970.Sand is used in the work which has the particle was less than 4.75mm.

- Fineness modulus = 4.83
- Bulk density of fine aggregate in loose state= 1620.64 kg/m^3
- Bulk density of fine aggregate in roddedstate= 1782.46kg/m^3

Tests on Coarse Aggregate
The coarse aggregate particles passing through 20mm and retained on 12.5 mm I.S Sieve used as the natural aggregate which met the grading requirement of IS383-1970.

- Specific gravity of coarse aggregate = 2.6
- Water absorption = 0.56%
- Coarse aggregate Impact Value = 15%
- Fineness modulus = 8.4
- Bulk density of coarse aggregate in loose state = 1553.46 kg/m^3
- Bulk density of coarse aggregate in rodded state= 1693.15kg/m^3

Preparation of Concrete
M20 grade of concrete with slump value of 23mm and flow value of 75% has been used in this analysis. Based on the wrapping arrangement of aramid fibre the beam column joint section were classified as A type, B type and C type and it is shown in Figure 2. The testing arrangement of beam column section is shown in Figure 3.

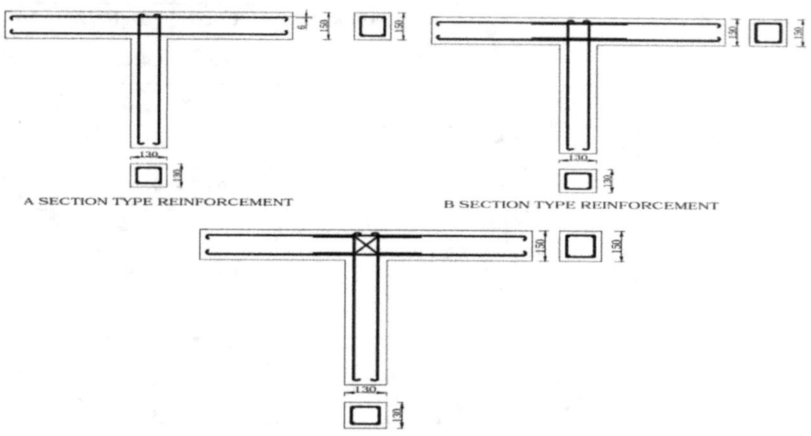

Fig.2. wrapping arrangement at junction

Fig.3. C type reinforcement loading

From the experimental work, it is clearly observed that C type arrangement increases the beam column joint strength. In this C type arrangement the diagonal cross bar at the joint of the specimen provides the better result and the result values were tabulated in Table 1.

Table 1. Test Result

Type of Specimen	Days	Initial Load		Ultimate Load	
		Load (KN)	Deflection (mm)	Load (KN)	Deflection (mm)
A Conventional	7	10	1.6	16	3
	14	6	1.4	18	3.5
	28	8	3.8	18	5.7
B	7	16	6	22	6.6
	14	18	6.3	22	6.7
	28	12	3.4	20	5.4
C	7	18	6.8	22	4.8
	14	18	4.1	24	4.2
	28	14	5	24	4.5

Numerical Analysis

Using Abaqus software, the beam column joint were analysed for Type A, Type B and Type C arrangements. From this the maximum principal stress and maximum principal strain were calculated and it is shown in Figure 3.

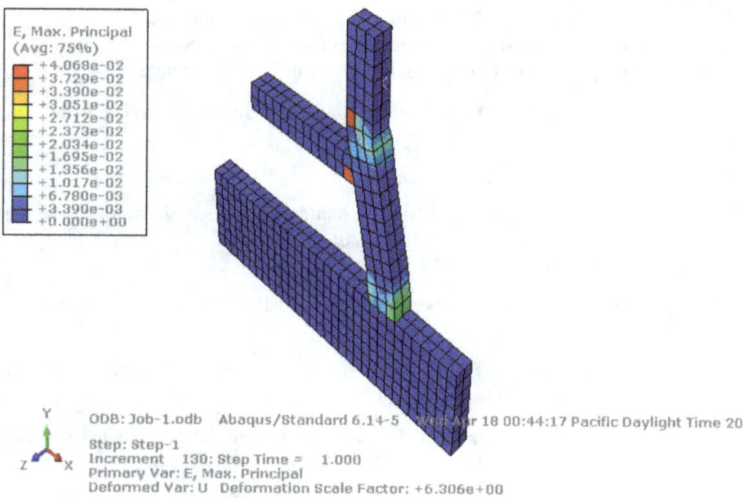

Fig.3 Type C arrangement with aramid fibre

Also the load deflection curve is shown in Figure 4 for Type A and Type B beam column joint.

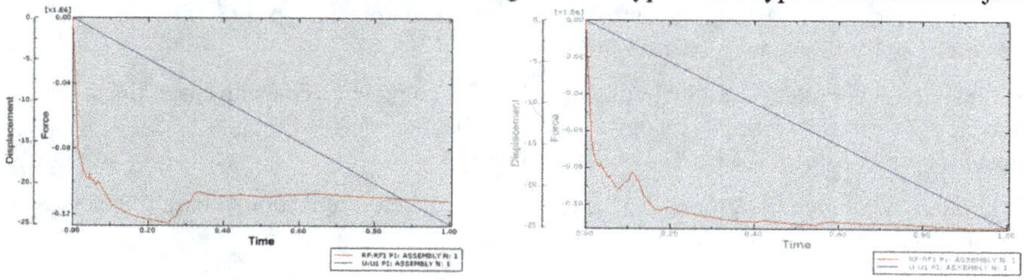

(a) (b)

Fig. 4 (a) Type A (b) Type B

Conclusion

From the above experimental and numerical studies, the performances of beam column joints were concluded.

1. Type C joint arrangement gives 92 % better result than conventional (Type A) arrangement
2. The deflection of Type C is 82.5% lower than Type A arrangement
3. The maximum principal stress is achieved in Type C arrangement
4. The maximum principal strain was experienced in Type C arrangement

The above results shows that Aramid fibre was found to be more effective in beam column joint wrapping.

References

[1] Abdullah, M. M. and Jallo. E. K. (2012). "Mechanical properties of glass fiber reinforced concrete."Al-Rafadain Engineering Journal, Vol.20, No.5, pp.128-135. https://doi.org/10.33899/rengj.2012.61048

[2] T. T. Soong and G. F. Dargush, Passive Energy Dissipation Systems in Structural Engineering, Wiley, 368.

[3] Junji Takagi, Some Properties of glass fiber reinforced concrete, fiber reinforced concrete, ACI SP-44, Detroit, Michigan, 1974, PP. 93-111 .

[4] Adaganti, S. Y. (2014). "Recovery and re-use of Glass fibers for composite reinforcements." Journal of Scientific and Industrial Research, Vol. 73, No. 10, pp. 671-673.

[5] JongsungSim, Park, C & Moon, D 2005, 'Characteristics of basalt fiber as a strengthening material for concrete structures', Composites, Vol. 36, pp.504-512. https://doi.org/10.1016/j.compositesb.2005.02.002

[6] Li, W & Xu, J 2009, 'Mechanical properties of basalt fiber reinforced geopolymeric concrete under impact loading', Materials Science and Engineering, Vol. 505, pp. 178-186. https://doi.org/10.1016/j.msea.2008.11.063

[7] Ludovico, M., Prota, A., &Manfredi, G., 2010, 'Structural upgrade using basalt fibers for concrete confinement', Journal of Composites for Construction, Vol.14, pp.541-552. https://doi.org/10.1061/(ASCE)CC.1943-5614.0000114

[8] Lopresto V., Leone C., &Iorio, D., 2011, 'Mechanical characterization of basalt fiber reinforced plastic', Composites, Vol.42, pp. 717-723. https://doi.org/10.1016/j.compositesb.2011.01.030

[9] Manikandan V., Jappes, W., Kumar, S., &Amuthakkannan, P 2012 'Investigation of the effect of surface modifications on the mechanical properties of basalt fiber reinforced polymer composites', Composites, Vol. 43, pp.812-818. https://doi.org/10.1016/j.compositesb.2011.11.009

[10] Quagliarini, E., Monni, F., Lenci, S., &Bondioli, F., 2012, 'Construction and Building Materials', Vol. 34, pp. 372-380. https://doi.org/10.1016/j.conbuildmat.2012.02.080

[11] Shi, J., Zhu, H., Wu, Z., Seracino, R., & Wu, G., 2013, 'Bondbehavior between basalt fiber–reinforced polymer sheet and concrete substrate under the coupled effects of freeze-thaw cycling and sustained load', Journal of Composites for Construction, Vol. 17, pp.530-542. https://doi.org/10.1061/(ASCE)CC.1943-5614.0000349

[12] Yuan, F., Pan, J., & Leung CKY 2013, 'Flexural behaviors of ECC and concrete / ECC composite beams reinforced with basalt fiber-reinforced polymer', Journal of Composites for Construction, Vol. 17, pp. 591-602. https://doi.org/10.1061/(ASCE)CC.1943-5614.0000381

Sustainable Materials and Smart Practices - NCSMSP-2021 Materials Research Forum LLC
Materials Research Proceedings **23** (2022) 16-24 https://doi.org/10.21741/9781644901953-3

Reuse of Coarse Aggregate as Recycled Aggregate in Concrete

R. Krishnakumar[1,a], M. Krishnagopal[1,b], P. Vishnu[1,c], O.P. Deepak[1,d*]

[1]Department of civil engineering, Bannari amman institute of tech., Erode, Tamilnadu, India

[a]krishnakumar.ce18@bitsathy.ac.in , [b]krishnagopal.ce18@bitsathy.ac.in , [c]vishnu.ce18@bitsathy.ac.in,*[d]deepak@bitsathy.ac.in

Keywords: Recycled Concrete Aggregate (RCA), Slump Test, Compaction, Compression, Ultrasonic Pulse Velocity, Physical and Mechanical Properties

Abstract: In the latest period, the volume of building material has greatly extended, raising social and environmental concerns surrounding waste recycling. Construction and demolition waste is a significant factor in global generated waste. Construction necessarily requires the use of massive quantities of aggregates. When the structure's useful life is up, it will be demolished, and all of the demolished waste will disposed of in landfills. It became increasingly difficult to identify large areas for landfills. Crushed, graded inorganic particles that have been processed from construction materials and demolition debris make up recycled aggregates. Aggregates that have been recycled for use in high-strength structural concrete. In this work, recycled coarse aggregate was employed to replace virgin coarse aggregate. The properties of both fresh and hardened concrete made from 100 % recycled coarse aggregate which has been internally or externally replaced found, and the effects are compared to concrete by applying the virgin coarse aggregate.

Introduction

Due to industrialization, India's urbanization rates are generally high. India's GDP is increasing at a rate of 9% each year. Infrastructure that is built quickly. Development needs a great number of resources, construction materials, land requirements, and the site. Concrete is preferred for huge construction projects. It takes a greater lifespan, lower operating costs, and better performance. Smaller is superior when it comes to achieving a GDP rate. Structures are now being demolished and high rise apartments are now being built. Protection of the environment is a must a factor that is intimately connected to survival the race of humanity Biological influences, for eg.sustainability, natural resource protection play an important role in long production. Construction needs with in modern period due to demolished materials are dumped during modernization on terrain that's not being used anywhere. Situations such as these have an influence on the land's fertility as per a Hindu report, India generates 23.75 billion us dollars online as of March 2007.Each year, massive amounts of demolition waste are produced. As per Central Pollution Control Board report. Concrete compensates 40% of total construction demolition garbage, following by ceramics at 30%, plastics at 5%, wood at 5%, metal at 5%, & unspecified mixes at 10%. As according global insight, expansion in the global construction sector is expected to occur in a 4800 billion $ increase in construction spending in 2013. These statistics show a tremendous growth in construction sector.

Aggregates are probably going to represent 70-80% of the totals utilized in substantial assembling. The coarse total records for 60-67 percent of the aggregate, while the fine total records for 33-half of the aggregate. As indicated by the Fredonia gathering's new exploration, worldwide interest for development totals is relied upon to surpass 26 billion tons by 2012. China is by a wide margin the most dynamic client, representing 25% of all clients, trailed by the United Kingdom (12%) and the United States (10%). (10%). India is likewise among the best ten clients. As far as ecological effect, one ton of regular total produces 0.0046 million tons of carbon dioxide, yet one ton of reused total discharges just 0.0024 million tons. The carbon impression of both regular and reused total might be determined because of the worldwide use of 10 billion metric huge loads of total each year for substantial creation. While contrasted with regular total concrete, the utilization of reused total upgrades drying shrinkage creep and porosity to water while diminishing pressure strength. To the extent that total substitution concerned, it was around 10-30%. Reusing brings down costs (LCC) and Carbon force (LCCO2) by 34-41 % and 23-28 %, as needs be.

Fig 1, Recycled coarse aggregate

Advantage of recycling materials

- Precast and set gutters and kerbs are created with from this materials
- Potential savings: - Concrete is undisturbed, and the new cost in cement is expected to be compensated by the minimal price of Recycled Concrete Aggregate (RCA). By replacement 20% of the cement with fly ash, the alkali silica reaction might be reduced (ASR).
- Eco friendly: - Resources are not extracted, and transportation is minimized. There is also a reduction in the number of land required.
- Save time: Thus, there is no compelling reason to trust that assets will open up. Because of less pounding, fossil fuel byproducts is diminished. As indicated by Dutch standard VBT 1995, all substantial with a trademark strength of 65 MPa can have up to 20% of normal total supplanted with RCA or reused blended totals (RMA) without additional testing.

Disadvantages

- Less quality (for example, compressive strength drops by 10-30%).
- The average time it takes to purchase materials could have an effect on the project's life cycle.
- Land, as well as specialised machinery and equipment, are required (more cost).
- Water absorption is exceptionally high (up to 6 percent).
- It shrinks and creeps more because it cures.

Objectives

- To specify the maximum number of construction which can be performed with the resources available.
- To minimize the environmental effects of waste material.
- To compare the results of various tests performed on recycled and natural aggregates.
- To discover expense solutions such as transportation and excavation.

Methodology

The plain concrete cement (PCC) and the buildup concrete cement (RCC) were bought from different sources. This material is squashed to isolate the totals and diminish their size to minuscule pieces. In the research center, test methods are applied to these isolated totals as per Indian Standard code, and the discoveries are contrasted with regular totals. Reused total assists with decreasing the ecological effect of waste. At the point when transportation and assembling processes are limited, development costs are significantly lower while using a rate.

Testing of Recycled Aggregate

Particle size distribution: For the coarse and fine aggregates which could be used in the concrete, sieve analysis was conducted. Fine aggregate sieve sizes varied from 2.36mm to 75mm for coarse aggregate and 40mm to 4.75mm for coarse aggregate.

Table 1, PHYSICAL PROPERTY OF FINE & COARSE AGGREGATE

AGGREGATE	FINENESS MODULUS	DENSITY (kg/m^3)	SPECIFIC GRAVITY
Fine aggregate	2.77	1752	2.50
Coarse aggregate	7.086	1805.62	2.84
Recycle C.A	7.476	1660.44	2.74

Slump Test: The slump test is used to determine whether or not fresh concrete is practical. The test is simple and affordable. It could be used both in the lab and on the work site. Although the test is simple, it must be performed out together with precaution since any disturbance in the method could result in a substantial fall.

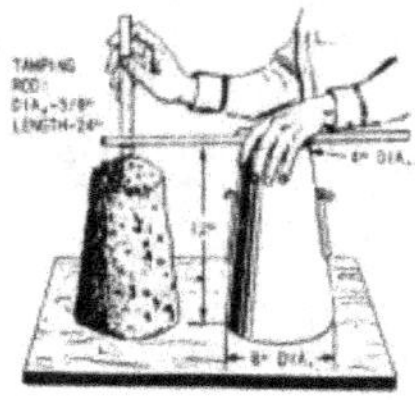

Fig 2, Slump Test

Compacting Factor Test: The compacting factor test, which is far more accurate than the slump test in detecting the workability of fresh concrete, was also used to determine its workability. The compacting factor test, commonly known as the "drop test," measures the weight of fully compacted vs partially compacted concrete.

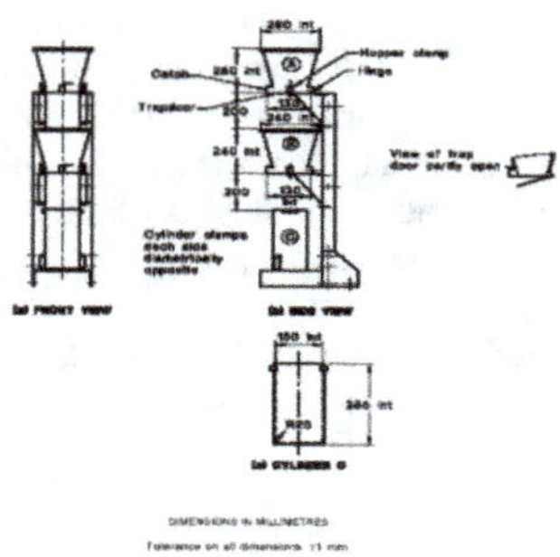

Fig 3, Compacting factor test

Table 2,THE COMPACTING FACTOR FOR MIX CONCRETE

Percentage of Recycled aggregate (w/c=0.55)	0%	20%	40%	60%	80%	100%
Compacting factor	0.839	0.897	0.933	0.875	0.851	0.879

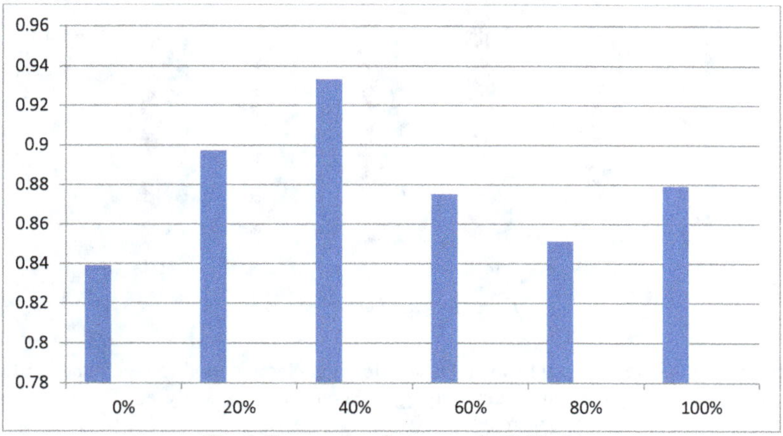

Fig 4, Compaction factor results(0.55w/c)

Non-destructive test

1. **Schmidt's Rebound hammer test**: Schmidt's Rebound Hammer, made in 1948, is perhaps the most broadly utilized surface hardness indicator. It contains a rounded packaging with an

Materials Research Proceedings **23** (2022) 16-24 https://doi.org/10.21741/9781644901953-3

unclogger coasting spring control hammer. The protuberance returns when the unclogger is squeezed against the substantial's surface. It withdraws against the draw of the spring. Therefore, when the sledge hits the substantial, the spring control mass pulls it back, driving the rider to haul along the directing scale. By squeezing a button, the rider can be held set up while the perusing is taken. Each mallet has a special trademark that necessities be altered prior to utilizing on substantial that contains total from a specific provider.

Table 3, RESULTS OF REBOUND HAMMER TEST

Percentage of recycled aggregate	0%	20%	40%	60%	80%	100%
Days	Compressive Strength (MPa)					
3	15.46	20.45	10	15.33	14.3	12.8
7	14.16	15.9	14.76	16.93	12.46	14.16

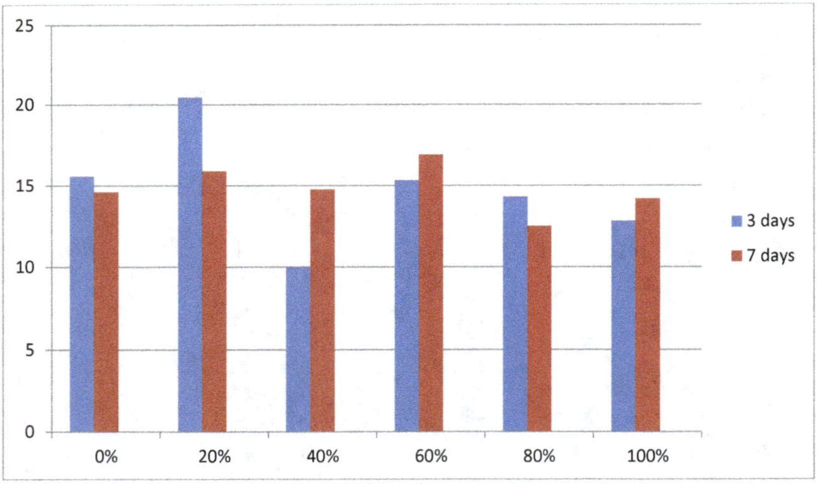

Fig 5, Rebound hammer test

2. **Ultrasonic pulse test**: The ultrasonic pulse velocity method is used to determine the time of travel of an ultrasonic pulse propagating through the concrete to be analyzed. This test is used to assess the internal strength of concrete.

Fig 6, Ultrasonic pulse test

Table 4, ULTRASONIC PULSE TEST RESULTS

Percentage of recycled aggregate	0%	20%		40%	60%	80%	100%
Days		Velocity (m/s)					
3	4110	4210		4190	3860	4020	4120
7	4160	3960		4160	4250	4230	4010
Days		Compressive strength (Mpa)					
3	20.4	19.06		18.0	13.5	16.1	17.3
7	21.7	15.2		19.2	23.6	19.3	19.8

3. **Compressive Test**: The measured maximum resistance of a concrete to axial loading is defined as in its compressive strength. Because compression testing is simple to operate, it is the most common test for hardened concrete specimens. The compression test could be used to assess the strength of concrete specimens with various proportion of recycled aggregate replacement. The compression test utilized specimens that have been 150mmx150mmx150mm in dimension

Table 5, COMPRESSIVE STRENGTH (Mpa) WITH AGE

Percentage of recycled aggregate	0%	20%	40%	60%	80%	100%
Days	Compressive strength (Mpa)					
3	23.11	26.14	25.3	28.96	25.07	21.41
7	29.78	31.11	31.4	33.92	27.63	31.4

Materials Research Proceedings **23** (2022) 16-24 https://doi.org/10.21741/9781644901953-3

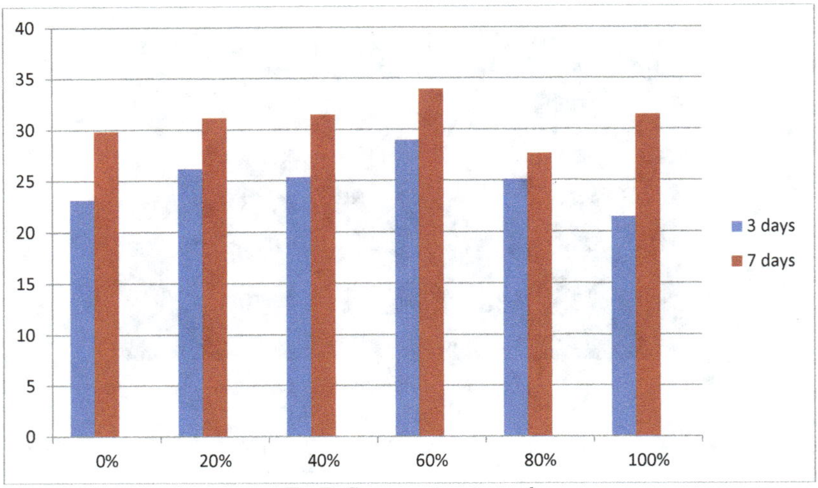

Fig 7, Compressive strength

Results and Discussion

Size distribution: The natural coarse aggregate is coarser than the recycled coarse aggregate. Like a consequence, the fineness modulus of 95 percent recycled coarse aggregate is 7.476, slightly higher than the fineness modulus of natural coarse aggregate, which is 7.086.

Physical and Mechanical Properties: Table 1 shows that the bulk density and specific gravity of recycled coarse aggregate are 1660.44 kg/m3 and 2.74, correspondingly. The figures are lower than the density of natural coarse aggregate, which is 1805.62 kg/m3 and has a density of 2.84. In the recycled coarse aggregate, the original material was bonded with mortar. This is paired with a light and porous mortar. As a result, as compared to natural aggregate, recycled aggregate's specific gravity and density are often lower. Since recycled coarse aggregate has a lower specific gravity, the amount of recycled coarse aggregate used in concrete has reduced. Concrete's mechanical qualities are influenced by aggregate mechanical properties. The mechanical properties of RCA are lower than those of NCA, according to the literature (ECCO 1999; López-Gayarre et al. 2009; Maleev et al. 2010; Poon et al. 2003; Hansen 1992).

Workability: The higher angularity and surface hardness of RCA particles reduce concrete workability and make proper finishing more difficult (Yrjanson 1989). Topçu and Günçan (1995) found that when the proportion of RCA in the concrete increases, the degree of workability decreases. As a result, RCA concrete requires more water to provide the same workability as NCA concrete. As per the outcomes of the workability tests, the greatest workability were obtained in a new concrete mix with 40% RCA replacement.

Compressive strength: As demonstrated, the strength of concrete of RCA concrete is often lower than that of NCA concrete. The strength of concrete of RCA concrete is typically 5 to 10% less than of NCA concrete (ACPA 2009). It can, however, be reduced by up to 25% depending on the RCA's quality. If the RCA is generated from an old concrete source that was originally created with a lesser water cement ratio than the new concrete, it may have similar, if not higher, concrete strength than NCA concrete. Concrete's compressive strength can also be affected by fine RCA. The strength of concrete of RCA concrete, according to Tavakoli and Soroushian (1996), is influenced by RCA's source concrete's coarse aggregate to fine aggregate ratio. The coarser the lower. The higher the mortar-to-fine-aggregate ratio, the more mortar is bonded to coarse RCA particles. As a result, the strength of RCA concrete is diminished. The graph depicts the variation in RCA compressive strength as a function of age. It was discovered that recycled coarse aggregate concrete with such a high proportion of RCA achieved higher force more quickly than normal concrete. When the recycled coarse aggregate is 60 percent replenished, the compressive strength is at its highest. The findings of the Rebound hammer test and Ultrasonic pulse velocity tests agree with the compressive strength of the concretes produced with varied percent substitutions of RCA.

Conclusion

There is a great need to protect our environment and finite natural resources for future generations on a global scale. To assist safeguard public capacity and land area, construction and demolition materials can be recycled. This could help us reduce the need for quarrying and other forms of environmental degradation. Recycled aggregates could benefit a variety of applications, including concrete making. The many elements of recycled aggregates and their application in concrete have been discussed in this study.

Reference

[1] M.Chakradhara Rao "Recycled coarse aggregate and its use in concrete", ICI journal January-march 2011, pp. 27- 40.

[2] Bureau of Indian standards (1970) Indian standard specification for coarse (IS:2863) and fine aggregate (IS:383) from natural sources.

[3] Bureau of Indian standards IS:10262(1982 & 2009) Indian standard recommended guide line for concrete mix.

[4] M.S.SHETTY, Book on concrete technology.

[5] RILEM (1994) Specifications for concrete with recycled aggregates, materials and structures, 27, pp. 557- 559. https://doi.org/10.1007/BF02473217

[6] F. Yuan, L.-Y. Shen, and Q.-M. Li, "Emergy analysis of the recycling options for construction and demolition waste," Waste Management, vol. 31, no. 12, pp. 2503–2511, 2011. https://doi.org/10.1016/j.wasman.2011.07.001

[7] K. P. Verian, W. Ashraf, and Y. Cao, "Properties of recycled concrete aggregate and their influence in new concrete production," Resources, Conservation and Recycling, vol. 133, pp. 30–49, 2018. https://doi.org/10.1016/j.resconrec.2018.02.005

[8] V. B. M. Kumar, H. Ananthan, and K. V. A. Balaji, "Experimental studies on utilization of recycled coarse and fine aggregates in high performance concrete mixes," Alexandria Engineering Journal, vol. 57, no. 3, pp. 1749–1759, 2018. https://doi.org/10.1016/j.aej.2017.05.003

[9] F. Yuan, L.-Y. Shen, and Q.-M. Li, "Emergy analysis of the recycling options for construction and demolition waste," Waste Management, vol. 31, no. 12, pp. 2503–2511, 2011. https://doi.org/10.1016/j.wasman.2011.07.001

[10] R. Neves, A. Silva, J. de Brito, and R. V. Silva, "Statistical modelling of the resistance to chloride penetration in concrete with recycled aggregates," Construction and Building Materials, vol. 182, pp. 550–560, 2018. https://doi.org/10.1016/j.conbuildmat.2018.06.125

[11] M. Buhamad, Effect of Recycled Aggregate on the Chloride Diffusion in Fully Saturated Concrete, Kuwait University, Kuwait City, Kuwait, 2012

[12] J. Wang, H. Yuan, X. Kang, and W. Lu, "Critical success factors for on-site sorting of construction waste: a China study," Resources, Conservation and Recycling, vol. 54, no. 11, pp. 931–936, 2010. https://doi.org/10.1016/j.resconrec.2010.01.012

[13] N. Kartam, N. Al-Mutairi, I. Al-Ghusain, and J. Al-Humoud, "Environmental management of construction and demolition waste in Kuwait," Waste Management, vol. 24, no. 10, pp. 1049–1059, 2004. https://doi.org/10.1016/j.wasman.2004.06.003

Development of Sustainable Bricks by Utilizing Vermicompost as an Alternate Material – A Waste Created Brick

P. Karthigai Priya[1,a,*], S. Vanitha[2,b]

[1]Department of Civil Engineering, School of Environmental and Construction Technology, Kalasalingam Academy of Research and Education, Krishnankoil, Tamil Nadu, India – 626128

[a*]karthigaipriya@klu.ac.in [b]s.vanitha@klu.ac.in

Keywords: Bricks, Vermicompost, Compressive Strength, Water Absorption, Loss on Ignition, Replacement

Abstract. Many different types of wastes are being generated on a daily basis which includes household waste, industrial waste, hazardous waste, construction and demolition waste etc. This generation of waste has increased enormously, and it needs to be considered as a prominent problem, as the waste management and waste disposal is not being done properly. Thus, it will lead to environmental degradation and pollution. On the other hand, over utilization of materials for construction activities leads to its extinguishing. Both scenarios will pose a threat to the environment. So, this study is carried out by utilizing vermicompost as a replacement material for red soil in the manufacturing of clay bricks. Initially, the physical properties of the raw materials are studied and the brick specimens are casted by varying vermicompost at 5, 10, 15 & 20% for red soil and they are tested for compressive strength, water absorption, loss on ignition. Among the replacements done, 10% of vermicompost for red soil gave higher strength to the bricks which is 6.23 N/mm^2 higher than the control specimen. The water absorption for all the trials is found to be less than 20% as per IS 3495 part 2. The loss in weight of the bricks due to ignition is less than 15% for all the specimens manufactured. Therefore, the utilization of vermicompost will reduce the waste disposal considerably and also the cost of the bricks can be minimally reduced as waste is being incorporated in it.

Introduction

Clay is the foremost raw material that is been used for ages in the manufacturing of bricks [1]. Various elements are present in the clay bricks. Even though the clay bricks contain silica in major proportion, the alumina present in it is responsible for imparting the plasticity to the bricks so that they can be moulded in to various shapes and sizes easily [2]. Compressive strength is one of the main properties that is required to be considered in the manufacturing of bricks as it will directly impact the mechanical property of the bricks [7]. On the other hand, waste generation and its disposal has become a major problem nowadays. There are various techniques such as incineration to reduce the waste to landfill. In our country, the waste generated contains a lot of organic matter. So composting is one of the best techniques that can be adopted to reduce the volume of waste that is sent to landfill [8].

In recent times, the motive towards attaining sustainability materials in increasing [9]. Also, the bricks have taken various forms such as fly ash bricks, cement bricks, hollow bricks etc., but still many people stick to the conventional clay bricks and the manufacturing of clay bricks is also not affected due to this change. To achieve sustainability a lot of waste material is being utilized in the construction materials and it is validated through proper researches [10,11]. In this study, an approach towards waste reduction is chosen so that the waste can be effectively reproduced in to newer form to use it for an alternate purpose. So, the waste generated within the university is composted and the by product is used as a replacement material in brick manufacturing. This paper is an attempt to produce bricks by using vermicompost as a replacement material. So, the physical and microstructural properties of materials are analysed and studied first and then then brick specimens are casted and they are tested for its strength and quality.

Materials and methods

Generally, clay and red soil are the materials used in the making of the bricks [1]. The type of the soil used in the manufacturing of bricks will vary from region to region, as the locally available material is preferred in the making of bricks.

Materials

In this study, the locally available clay and red soil is used in the brick making process. In addition to it vermicompost is added as a replacement material for red soil. Vermicompost is a by product obtained from the decomposition of waste using earthworms. The vermicompost used in this study is obtained from the university's vermicompost plant. The waste generated within the campus and the hostel buildings is used to produce the vermicompost.

Methods

Initially, the characteristic properties of the raw materials used in the study are tested. Sieve analysis, plastic limit, liquid limit and specific gravity are the tests done to determine the characteristic properties. The percentage composition of the raw materials is determined by Energy Dispersive X Ray Analysis (EDAX). Also, the microstructure of the raw materials is identified by Scanning Electron Microscopy (SEM) with a Scanning Electron microscope EVO 18 (CARL ZEISS). Then the brick specimens are casted with varying proportions of vermicompost at the rate of 5, 10, 15, 20% for red soil and they are tested for compressive strength, water absorption, loss on ignition. The compressive strength is carried out in a compression testing machine.

Results and discussions

Physical properties

The colour of the clay, red soil and vermicompost as observed is dark brown, red and brown in colour respectively. The specific gravity of the vermicompost and red soil are found to be 1.90 and 2.67 respectively. The plastic limit and liquid limit of the clay is found to be 29.52 and 56% respectively.

Particle size distribution

The gradation of the raw materials namely red soil and vermicompost is performed by means of sieve analysis and the curve is plotted for the both. From the graph, the coefficient of uniformity (Cu) and the coefficient of curvature (Cc) is calculated. The Cc and Cu values of red soil as observed from the graph are 0.61 & 6.1 respectively whereas for compost the values of Cc and Cu are 0.9 and 3 respectively. From the Cu and cc values and as per IS 1498: 1970 [6] both the materials come under poorly graded soil. The particle size distribution graph for red soil vermicompost is shown in Fig. 1 and Fig. 2 respectively. From the graph it can be observed that the shape looks almost similar.

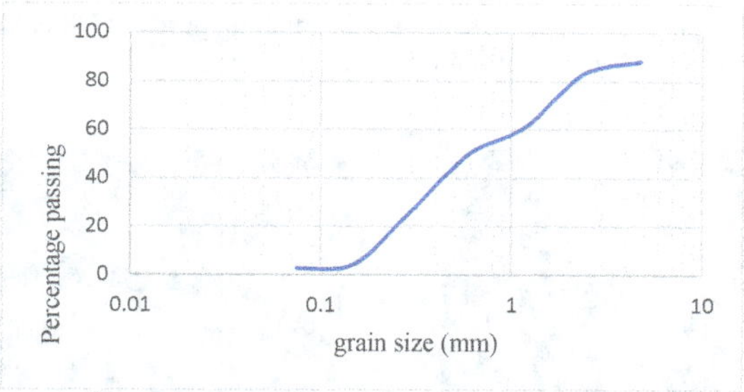

Figure 1. Particle size distribution for red soil

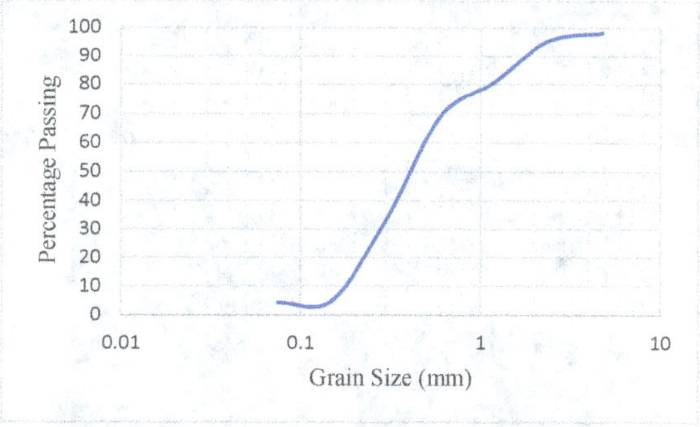

Figure 2. Particle size distribution for vermicompost

Composition

The elemental composition for red soil, vermicompost and clay are represented in Table 1.

Table 1. Composition of raw materials

Element	Weight %		
	Red soil	**Vermicompost**	**Clay**
C	6.7	19.9	38.4
O	49.3	49.4	43.5
Na	3.5	1.7	-
Mg	1.2	-	0.8
Al	10	4.8	3.7
Si	17.5	13.6	8.9
K	0.8	1	0.7
Ca	4.8	4.7	0.9
Ti	0.8	0.6	-
Fe	5.3	3.9	3.2
Cl	-	0.3	-
Mo	-	0.3	-

The main element essential for the making of bricks are silica and alumina and they are present in vermicompost also. The presence of alumina will impart plasticity to the bricks [2]. As alumina is present in vermicompost, it will not affect the brick making process. Other elements such as Fe, Ca, K, O, Na, Ti are also present in both red soil and vermicompost.

Scanning Electron Microscopy

Figure 3. SEM image of red soil

Figure 4. SEM image of vermicompost

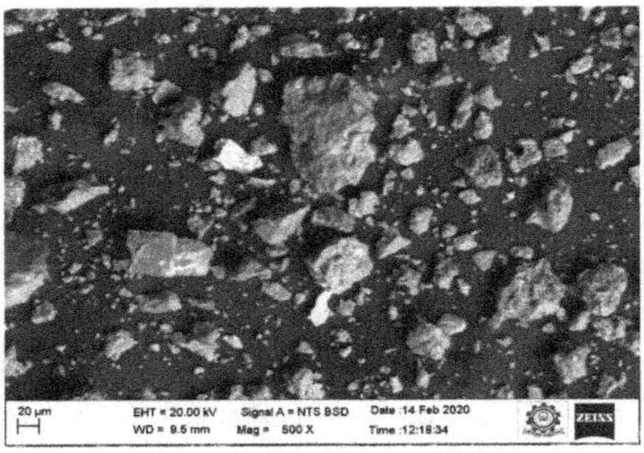

Figure 5. SEM image of clay

The Fig. 3, 4 & 5 shows the SEM image of red soil, vermicompost and clay respectively. From the Fig. 3 & 4 it can be noted that the shape of the particles of red soil is angular with sharp edge and that of the vermicompost is found to be angular with round edges. The black colour spot that is visible in between the particles are the pores present in the material. Also, it can be noticed that the structure of vermicompost is denser than the red soil [3,12]. The particles of vermicompost are packed more closer to each other than the red soil which indicates that the particles of various sizes are present in it. So, vermicompost will fill the pores present in the matrix thus making the entire matrix denser [4]. From Fig. 5, the shape of the particles of clay are found to be spherical and angular in shape. It is evident that the black spot present between the particles of the clay is more thus it increases the porosity. So, the water absorption required for the brick matrix will get higher.

Compressive strength

The load carrying capacity of the bricks are tested by means of compressive strength test and it is carried out in a compression testing machine. The compressive strength of the manufactured brick specimens with varying percentage of compost is shown in Table 2 and the Fig. 6. Shows the pictorial representation of compressive strength for various percentage of compost for red soil in brick making.

Table 2. Compressive strength of brick specimens with vermicompost

S. No	Vermicompost percentage in brick specimens (%)	Compressive strength (N / mm²)
1.	0	5.49
2.	5	5.89
3.	10	6.23
4.	15	4.83
5.	20	4.51

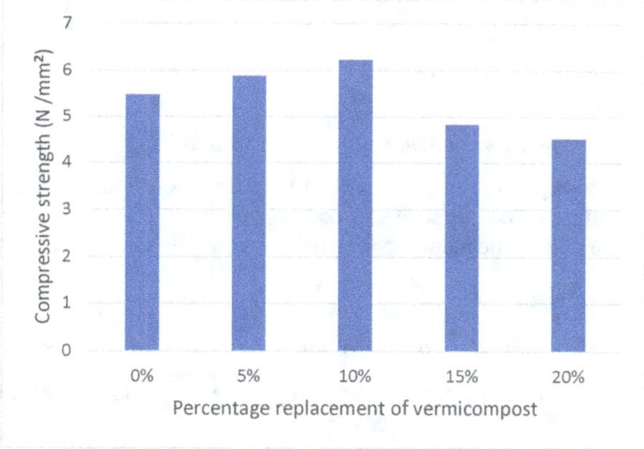

Figure 6. Compressive strength of brick specimens

From the Table 2 and Fig. 6 it can be noted that the compressive strength of the brick specimens with 5% and 10% of vermicompost replacement for red soil increases and it is also found greater than the control specimen. After 10% of replacement the compressive strength of the brick specimens tends to decrease gradually.

Water absorption

Water absorption is the test done to assess the quality of the bricks. The test is carried out in accordance to IS 3495 (II): 1992 [5].

Table 3. Water absorption of brick specimens

S. No	Vermicompost percentage in brick specimens (%)	Water Absorption (%)
1.	0	12
2.	5	10.15
3.	10	9.39
4.	15	9.06
5.	20	9.8

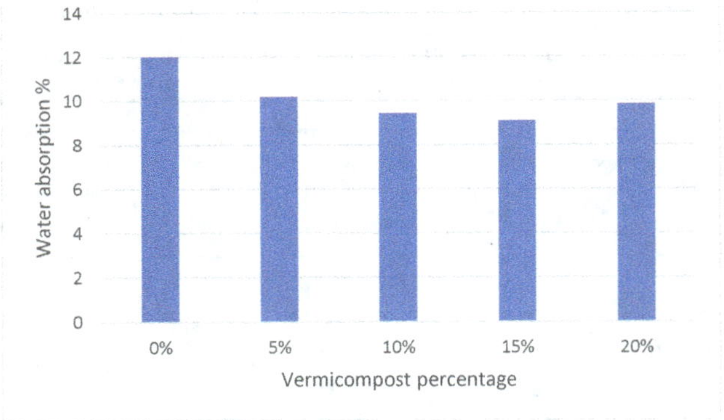

Figure 7. Graph showing water absorption of brick specimens

From the Table 3 and Fig. 4, it can be seen that the water absorption is less than 20% for all the brick specimens manufactured. As per the IS code provision the water absorption of the bricks should not exceed 20%. Hence the manufactured bricks with different percentage of vermicompost possess good quality.

Loss on Ignition

Table 4. Loss on ignition of brick specimens

S. No	Vermicompost percentage in brick specimens (%)	Loss on ignition (%)
1.	0	12.79
2.	5	13.64
3.	10	13.16
4.	15	13.38
5.	20	12.23

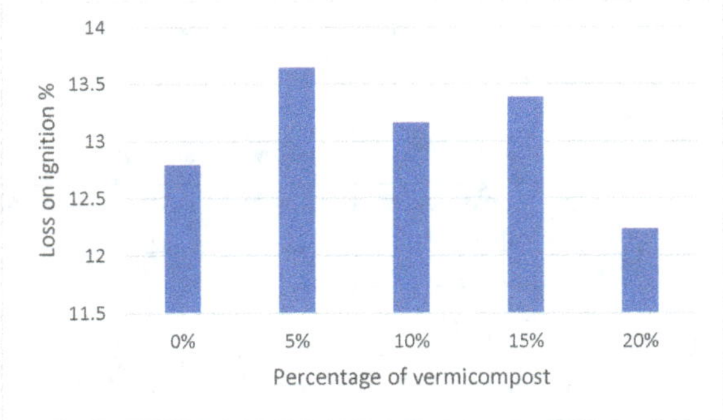

Figure 8. Graph showing loss on ignition

Table 4 shows the loss on ignition values of brick specimens with various percentage of vermicompost replacement. And the Fig. 8 shows the pictorial representation of the Table 4. This test is carried out by keeping the brick specimens in muffle furnace at 600 degrees Celsius for a period of 6 hours. It can be noted that the loss on ignition decreases with increase in percentage of compost.

Conclusion

- Brick specimens were able to cast successfully with varying percentage of vermicompost.
- While speaking about the physical properties, the specific gravity of the vermicompost is found to be lesser than the red soil.
- In the microstructural study carried out for the raw materials, it is found that the vermicompost possess denser microstructure than all other materials.
- The particle size distribution of vermicompost and red soil comes under poorly graded soil and both the curves look alike.
- The compressive strength of the bricks with vermicompost increases up to 10% addition and beyond that the strength tends to decrease. In other words, the bricks with 10% vermicompost possess higher compressive strength than the control specimen.
- The water absorption of all the bricks is found less than 20% as specified in the IS code for water absorption.
- The loss on ignition of the bricks for all the trials is found to be less than 15%.
- Overall, it can be suggested that, bricks can be manufactured effectively by replacing vermicompost up to 10%.
- By this, the waste to landfill can be reduced and bricks can be manufactured sustainably and also the cost of the bricks can be lowered as waste material is added to it.

Acknowledgement

The authors extend their gratitude to K.S. Ajay Venkadesh, K. Harish Kumar, A. Arumugam, B. Hariharan, A. Nithish Kumar for providing their extended support to perform the experimental study. The authors are thankful to M Vijay and S Subbaih Kumar for helping in carrying out the preliminary study.

References

[1] A.R. Wahid, Y.A. Yatmo, K.D. Paramita, More than Just a Material Perfection: Preserved Human-Environment Relationship in Traditional Brick-Making Scenarios, J Phys: Conf. Ser 1655 (2020) 012125. https://doi.org/10.1088/1742-6596/1655/1/012125

[2] M. Ali, S. Room, M.I. Khan, F. Masood, R.A. Memon, R. Khan, A.M. Memon, Assessment of local earthen bricks in perspective of physical and mechanical properties using Geographical Information System in Peshawar, Pakistan. Structures 28 (2020) 2549-2561. https://doi.org/10.1016/j.istruc.2020.10.075

[3] P.K. Priya, S. Vanitha, PL Meyyappan, Characteristic and microstructural study on an alternate material in brick manufacturing, IOP Conf. Ser: Mater Sci Eng 955 (2020) 012038. https://doi.org/10.1088/1757-899X/955/1/012038

[4] J.E. Oti, J.M. Kinuthia, J.J. Bai, Engineering properties of unfired clay masonry bricks. Eng Geology 107(3-4) (2009) 130-9. https://doi.org/10.1016/j.enggeo.2009.05.002

[5] IS 3495: 1992 Part 2 – Determination of water absorption.

[6] IS 1498: 1970 – Classification and identification of soils for general engineering purposes.

[7] M.S. El-Mahllawy, Characteristics of acid resisting bricks made from quarry residues and waste steel slag, Constr Build Mater. 22(8) (2008) 1887-96. https://doi.org/10.1016/j.conbuildmat.2007.04.007

[8] T. Narayana, Municipal solid waste management in India: From waste disposal to recovery of resources. Waste management 29(3) (2009) 1163-6. https://doi.org/10.1016/j.wasman.2008.06.038

[9] X.Y. Wang, L. Sun, C.L. Liu, Present situation of land subsidence and its mechanism in China. Global View of Engineering Geology and the Environment, Proceedings of the International Symposium and 9th Asian Regional Conference of IAEG (AREG2013), Beijing, China (2013) 24-25.

[10] P.L. Paihte, A.C. Lalngaihawma, G. Saini, Recycled Aggregate filled waste plastic bottles as a replacement of bricks, Mater Today: Proceedings 15 (2019) 663-8. https://doi.org/10.1016/j.matpr.2019.04.135

[11] A. Khitab, M.S. Riaz, A. Jalil, R.B. Khan, W. Anwar, R.A. Khan, M.T. Arshad, M.S. Kirgiz, Z. Tariq, S. Tayyab, Manufacturing of Clayey Bricks by Synergistic Use of Waste Brick and Ceramic Powders as Partial Replacement of Clay. Sustainability 18 (2021) 10214.

[12] K.Y. Chiang, K.L. Chien, S.J. Hwang, Study on the characteristics of building bricks produced from reservoir sediment. J Hazardous Mater 159(2-3) (2008) 499-504. https://doi.org/10.3390/su131810214

Sustainable Materials and Smart Practices - NCSMSP-2021 Materials Research Forum LLC
Materials Research Proceedings **23** (2022) 33-44 https://doi.org/10.21741/9781644901953-5

Experimental Study on Practice of Cement Treated Subbase (CTSB) Layer in Flexible Pavement of National Highways in India

S. Loganayagan[1,a], N. Chandra Mohan[1,b], S. Ramakrishnan[2,c], T. Gokulakannan[1,d]

[1]Civil Engineering, Bannari Amman Institute of Technology, Erode 638401, India

[2]Civil Engineering, Sri Krishna college of engineering and technology, Coimbatore 641042, India

[a]loganayaganme@gmail.com, [b]chandramohann@bitsathy.ac.in, [c]ramakrishnans@skcet.ac.in, [d]gokulakannan.ce18@bitsathy.ac.in

Keywords: Cement Foundation, Bituminous Concrete, Ground Floor, Easily Paved Surface

Abstract. Due to the large number of infrastructure projects taking place in rural and urban areas there has been a shortage of building materials. The road industry is looking at ways to improve low-quality materials that are easily accessible for use in road construction. Cement / lime treatment has become an acceptable way to increase soil strength and consistency with moderate proportions, to reduce the number of compounds. The Indian roads congress (IRC) has developed a special edition for the mixed construction of the base / ground floor. There is no design guide currently available for the under the cement base. To overcome this problem, the aim of the current project is to create a chart of the paved area using concrete and limestone on rural and urban roads with small and medium vehicles. It not only saves money but also helps to increase the life cycle of roads. At the base of the road, there are different soils or granite materials available for construction, but they may indicate insufficient structures and lead to significant road stress and reduced life. However, the addition of a stabilizing agent such as cement, asphalt, lime or other non-traditional materials can improve soil properties. Among these various stable materials, cemented materials improve strength and high strength, and demonstrate the excellent performance of the paved system and high durability. Solid foundations can provide inexpensive solutions to many common designs and building conditions. Cement Treated Sub Base (CTSB) is a common method used on road foundations to improve its engineering properties due to the durability of cement where moisture is present and extends the healing time. The bonded base material provides additional strength and support without increasing the overall thickness of the mortar layers. Depending on the needs of the project, CTB increases construction speed, improves the capacity of the pavement structure, or in some cases reduces the full-time project. In addition, a strong foundation reduces deviation due to heavy traffic loads, thereby extending the life of the pavement. CTB base thickness is reduced due to higher carrying capacity compared to granular base thickness.

Materials Research Proceedings **23** (2022) 33-44 https://doi.org/10.21741/9781644901953-5

Introduction

In the flexible pavement of bitumen concrete, the use of Cement Treated Sub – base is the new and modern technological method in India. The main reason for the sub base is that it reduces the project cost when compared to Granular sub-Base courses at Flexible pavement of using Grade 2 at Bituminous base and Bituminous Concrete . It increases the speed of the project and extends the major life time of the flexible pavement. It exhibits and results in good performance, High Durability & High bearing strength.[1-2]Sub base treats and helps to subgrade some addition of stabilizing and traditional agents to improve the quality of the soil. The best new method in Indian civil Technology but it may also fail in high embankments at flexible pavement. Main objective is material properties analysis, mix designing procedures, Trail stretch, construction methodology of cement treated sub-base(CTSB) as per control standards as per IRC SP 89-2010.This report is to Conforms / finalize the Design Mix for cement treated sub-base (CTSB) as per the approved pavement design & IRC – 89-2010.

Note : Cholapuram to Thanjavur (NH45C) by laying this technique for **first time in Tamil Nādu**. Patel Infrastructure Limited, Kumbakonam

Equipments for Lab JMF

- Cube molds. (150 x 150 x 150 mm)
- Sieves (as required)
- Trays, scoop, tape., etc.
- Vibrating Hammer
- Compression testing machine
- CBR testing machine & Molds.
- Proctor Testing equipment.

Acceptance Criteria & QC Checks:

- ✓ Grading shall Satisfies to limit specified in IRC SP 89-2010 Table 8 –Grading III
- ✓ Design JMF Mix should attain the minimum compressive strength of 3 Mpa in 7 days.
- ✓ Properties of stabilizer and stabilizing material shall satisfy the prescribed codal provisions.
- ✓ Confirmatory Test – Durability test prescribed in IRC SP 89-2010

Materials Research Forum LLC
https://doi.org/10.21741/9781644901953-5

Flowchart Mix Design

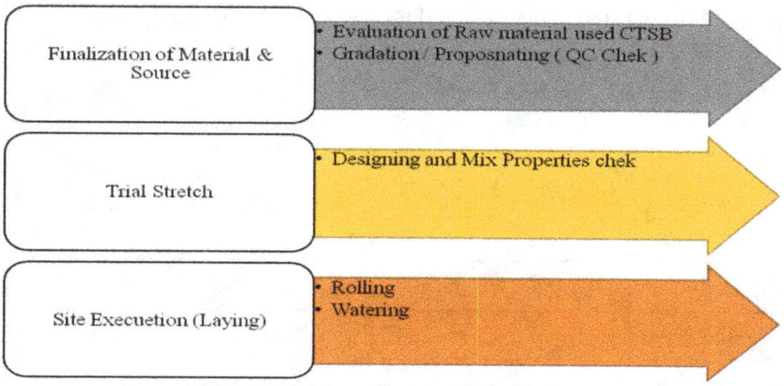

Figure -1 Flow Chart of Mix Design

The chart explains the basic concept from materials collections to site execution with the proportion of mix design based on the principle of MORTH.

Materials & Source

Table -1 *Material Source for Design [3-5]*

S.No	Material Description	Source of Materials
1	CTSB Crushed Material	Ragathampatti Quarry / Crusher
2	40mm Coarse Aggregate	
3	Cement	M/s Dalmia cement , M/s Ultratech Cements, M/s India Cements
4	Water	Borehole @ 138+200 LHS & 162+500 LHS

Evaluation Of Raw Materials Used For CTSB Crushed Materials:

Clean, hard and durable CTSB Crushed Material (53mm down) crushed in 250 TPH Crusher plant (Ragathampatti Quarry / Crusher) has been selected for the Design and Casting mix which is blended with crusher dust to achieve the required grading properties.

Properties of coarse aggregate

Table -2 Properties of Coarse Aggregate

Test Descriptions	Test Result	Specification limit
Aggregate Impact Value	36.22	Max. 35%
10% Fines Value	Test in Progress	Min 50 KN
Organic Content	Test in Progress	Max 2%
Water absorption	0.857	Max. 2%

Materials Research Proceedings **23** (2022) 33-44 https://doi.org/10.21741/9781644901953-5

Fine Aggregate (40MM Aggregate - Crusher Dust)

Clean, hard, durable crusher obtained from crushing of stone boulders crushed by 250 TPH crusher plant (RAGATHAMPATTI) has been selected for the design which complies with the requirement of the MORTH specifications.

Properties of 6MM down. (Crusher Dust)

Table -3 Properties of Fine Aggregate

S.No	Test Descriptions	Test Result	Specifications
1	Liquid Limit	26.6	Max 45%
2	Plasticity Index (PI)	N.P	Max.20%
3	Total SO_4 content	Test in Progress	Max 0.2
4	Organic Content	Test in Progress	Max 2%

Water

Clean water from Bore Hole at 138+200 LHS 162+500 LHS has been selected for the production of design and site mix. The water plays major role in the each and every of construction and the design mix proportions. The properties and characteristics of water are given in table 4.

Table -4 Properties of Water

S.No	DESCRIPTIONS	138+200 LHS	162+500 LHS	Limits
1	Organic	89	32	Max 200
2	Inorganic	165	392	Max 3000
3	Suspended Solids (mg/l)	4.0	10	Max 2000
4	p^H	7.38	7.4	Min 6.0
5	Sulphate ($S0_4$) (mg/l)	10.47	76.3	Max 400
6	Chlorides (Cl) (mg/l)	22.69	195	Max 500
7	Acidity	0.40	0.8	Max 5
8	Alkalinity	7.80	23.2	Max 25

***Note:* *If the Mix is produced in any plant and water used for the mix shall be compliance to IRC SP 89 -2010 limits*

Cement

Cement of Ordinary Portland Cement 53 grade from M/s Dalmia, M/s Ultratech Cements, M/s India Cements compliance the requirement of NORTH & IS 269specification is adopted for the design and Casting of mix.

Properties of Cement

The properties of cement for various manufacturers are listed in the table 5

Table -5 Properties of Cement

Descriptions	M/s Dalmia	M/s Ultratech Cements	M/s India Cements	Limits	Remarks
Consistency	30	29.0	29	-	Physical properties (In-House Testing)
Initial Setting time	190	165	170	Min 30 min	
Final setting Time	305	295	280	Max 600 min	
Compressive Strength a) 3 days b) 7 days c) 28 days	 34.63 46.19 TYC	 34.53 44.36 56.04	 35.65 44.41 60.96	a) 27 MPa b) 37 MPa c) 53 MPa	
Fineness	1.73	1.73	1.69	Max 10%	
Soundness by Le-Chatelier	2.78	2.27	2.50	Max 10%	
Total Loss on Ignition(% by Mass)	1.55	2.30	1.82	Not more than 4%	Chemical properties (MTC)
Insoluble Residue(% by mass)	1.07	1.58	2.30	Not more than 5%	
Ratio of Lime to% Slica,Alumina,Iron oxide	0.93	0.91	0.83	Max 1.02% Min 0.80%	
Ratio of % Alumina to Iron oxide	1.00	0.93	1.04	Not less than 0.66%	
Sulphuric Anhydride	2.94	2.37	2.65	Max 3.5% for C3A≤5% Max 3.0% for C3A≤5%	
Magnesia(% by mass)	0.88	1.26	0.84	Not more than 6%	

Gradation of Aggregate and Proposition Finalization

Individual aggregates were tested in CTSB sieves for blending according to IRC SP 89- 2010 Table 8 Grading III.Gradation used for determination of workability, durability, stability, friction resistance, permeability.

Individual Gradation Results

Description	CTSB Crushed Material	
Source of Material	Ragathampatti Quarry / Crusher	Average of **9 Samples**

Table -6 Individual Gradation Results

IS Sieve Size (mm)	Sample No. 01	Sample No. 02	Sample No. 03	Sample No. 04	Sample No. 05	Sample No. 06	Sample No. 07	Sample No. 08	Sample No. 09	Average of 9 samples
53	100	100	100	100	100	100	100	100	100	100
26.5	89.03	89.33	89.16	89.05	89.09	89.18	89.05	89.13	88.59	89.07
9.5	69.02	68.62	68.70	70.16	69.57	69.37	68.56	69.73	69.53	69.25
4.75	55.75	55.34	56.30	56.76	56.93	55.71	55.16	56.18	56.15	56.03
2.36	43.39	44.21	43.14	44.99	44.61	44.18	44.78	44.99	44.51	44.31
0.425	23.07	23.47	23.97	23.59	23.68	23.86	23.98	23.40	24.18	23.69
0.075	0.62	7.81	7.83	7.95	7.02	7.83	7.26	8.45	8.63	7.04

The individual gradation result is based on the California bearing ratio from each sample of mix design for cement treated bases. sieve size according to **IS code** and **IRC037.**

Description	(40mm Aggregate)	
Source of Material	Ragathampatti Quarry / Crusher	Average of **9 Samples**

Table -6.1 Individual Gradation Results

IS Sieve Size (mm)	Sample No. 01	Sample No. 02	Sample No. 03	Sample No. 04	Sample No. 05	Sample No. 06	Sample No. 07	Sample No. 08	Sample No. 09	Average of 9 samples
53	100	100	100	100	100	100	100	100	100	100
26.5	17.58	19.76	17.95	17.93	17.19	17.92	17.66	19.53	18.32	18.20

Individual gradation of coarse aggregate by sieve analysis by conventional cement treated mix design at guidelines for the design of flexible pavement **IRC-37-2018**in a gradation and size analysis a sample of dry aggregate of known weight is separated through a series of sieves with progressively smaller openings.

Theoretical Blending

Theoretical blending is done on various propositions and the proposition which achieves the values nearer to the mid limit of the limits specified in **IRC SP 89- 2010 Table 8 Grading-** III is selected for the practical job mix and design. The cement treated sub base is protective for flexible pavement.

Table -7 Theoretical Blending Results

I.S SIEVE (mm)	**Average % passing of Individual grading**		**THEORETICAL BLENDING AS BELOW PROPORTION**						**Limits as per IRC 89-2010 Table 8-Grading III**
			BLENDING – 01	BLENDING – 02	BLENDING – 03	BLENDING- 04	BLENDING- 05	BLENDING- 06	
	CTSB	**26.5mm Aggregate**	70%	75%	80%	85%	90%	95%	
		40mm Aggregate	30%	25%	20%	15%	10%	5%	
53	100.00	100.00	100.0	100.0	100.0	100.0	100.0	100.0	**100**
26.5	89.07	18.7	67.8	71.4	74.9	78.4	82.0	85.5	**70-100**
9.5	69.25	0.00	48.5	51.9	55.4	58.9	62.3	65.8	**50-80**
4.75	56.03	0.00	39.2	42.0	44.8	47.6	50.4	53.2	**40-65**
2.36	44.31	0.00	31.0	33.2	35.4	37.7	39.9	42.1	**30-50**
0.425	23.69	0.00	16.6	17.8	19.0	20.1	21.3	22.5	**15-25**
0.075	7.04	0.00	4.9	5.3	5.6	6.0	6.3	6.7	**3-10**

The graph defines that theoretical bending by observing IRC-37-2012 via code book analysis while preparation of mix design at flexible pavement by using sieve gradation.

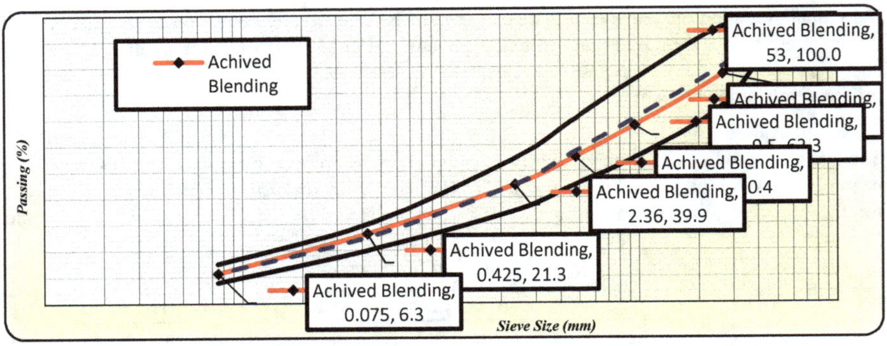

Figure -2 Theoretical Blending Graph

In the graph x and y axis shows the sieve sizes and passing percentage of our sample, The effect of fine percentage on the properties of subbase material.

Practical Blending for CTSB
Cement stabilization is one of the new techniques and recycle road base material and helps a durable lifelong pavement.[6 - 9]

Table -8 Practical Blending of Design

IS Sieve Size (mm)	Proposition No-05: 53mm down -90%; 40mm Aggregate-10%						Limits as per IRC 89-2010 Table 8 - Grading III		
	Sample 1	Sample 2	Sample 3	Sample 4	Sample 5	Average	Lower	Upper	Middle
53.000	100.00	100.00	100.00	100.00	100.00	100.00	100	100	100
26.500	80.70	79.47	81.46	82.83	82.21	81.33	70	100	85
9.500	61.40	60.39	60.37	61.74	61.22	61.02	50	80	65
4.750	48.75	48.08	47.81	49.44	47.58	48.33	40	65	52.5
2.360	37.40	36.93	36.32	37.14	36.12	36.78	30	50	40
0.425	18.74	18.84	17.59	18.81	17.81	18.36	15	25	20
0.075	5.61	5.67	5.28	4.77	5.20	5.31	3	10	7

The practical blending and moisture contents for determining optimum moisture content and maximum dry density are explained in figure 3.

Figure -3 Practical Blending Graph

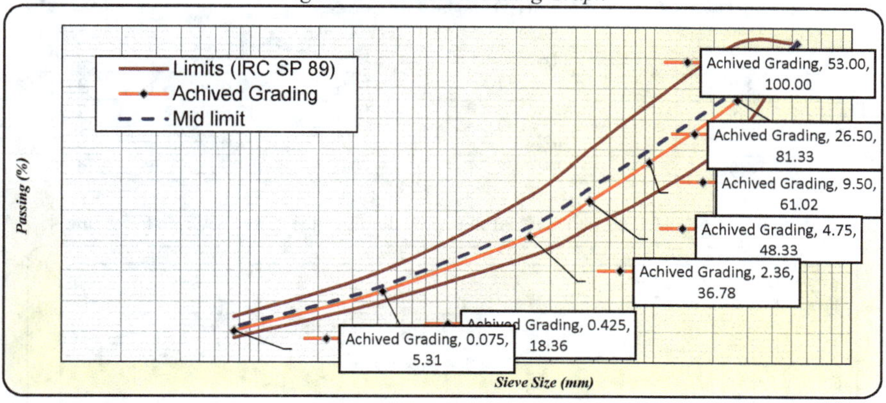

The above graph defines that x axis has sieve size and y axis size passing percentage, the upper and lower line shows the limits of practical blending and the continuous center lines shows that the achieved grading of the mix design. The dotted lines show the mid limit of design .

Design Mix

After finalizing the grading and proposition, CTSB design is conducted in following manner

1. Stabilizing material in the finalized proposition is mixed with cement (2.5%, 3.0%, 3.5%, 4.0%, 4.5%) and optimum moisture content is determined by proctor heavy compaction method is per IS 2720 (Part -8).
2. In accordance with Optimum moisture content results then a set of cubes are casted for (2.5%, 3.0%, 3.5%, 4.0%,4.5%)cement content.
3. Moist curing is done for 7 days and compression test is conducted and strength is obtained for respective cement contents.
4. Graph is plotted for Cement content Vs strength checking and optimum cement content is obtained from the graph.
5. After finalization of Cement content is 7 days strength checking and confirmatory test are conducted to verify the adopted cement content is safe for the mix.[10 - 13]

Determination of Optimum Cement Content For CTSB

The optimum cement content for cement treated sub base has been given in the table.9

Table -9 Design Mix

S.No.	Stabilizer / Cement Content (%)	Stabilizing Material Content (%)		Age Of Test	Average Compressive Strength (Mpa)
		Crushed CTSB Material	40mm Coarse Aggregate		
1	2.5	90%	10%	7 days	2.30
2	3.0	90%	10%	7 days	2.75
3	3.5	90%	10%	7 days	3.20
4	4.0	90%	10%	7 days	3.75
5	4.5	90%	10%	7 days	4.20

The design mix is casted by cube and strength checked by curing as shown in the figure.4.

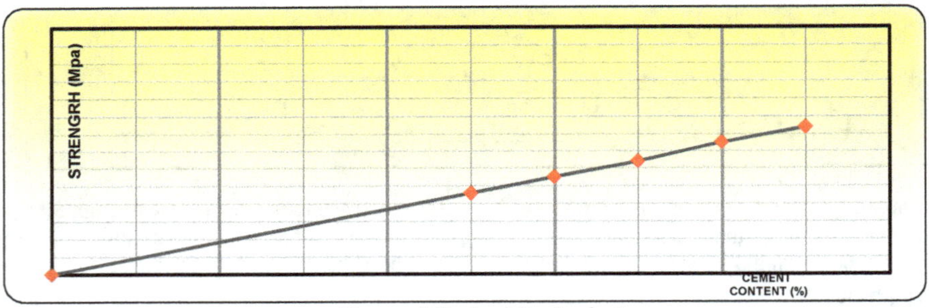

Figure -4 Design Mix Analyzing

The design mix graph defines that cement content has x axis and strength as y axis by using different brand of cement (type-OPC53).[14-15]

Table -10 Cement content for CTSB

S.No.	*Description*	*Results*	
A	Optimum Cement content (From Graph)	3.30	%
B	Considering Factor of Safety of 5% (% of Cement increased by = A*5%)	0.20	%
C	Finalized Cement Content (A+B)	3.50	%

By analyzing the sub base design of flexible pavement optimum cement content used as **4%.**

Confirmatory Tests

A. Confirmatory test set of cubes at Cement content of 3.50% is casted and 7 days compressive strength has been checked with three different sources of cement and found satisfactory on limits of IRC SP 89-2019.

B. Durability Test (Index of resistance to water on strength) is conducted as per IRC SP 89-2010 - Cl.4.7.2-Method -1 and found satisfactory on limits of IRC SP 89-2019.[16]

Table -11 Confirmatory Test

Cement Type / Source	Stabilizer/ Cement Content (%)	Stabilizing Material Content (%)	Confirmatory Test results (MPA) Min 3.0 MPA	Durability Test Results (%) (Min 80%)
OPC 53 Grade / Ultra tech	3.5 %	90% Crushed CTSB Material + 10% 40mm Coarse Aggregate	3.51	86.34
OPC 53 Grade / Dalmia			3.45	87.92
OPC 53 Grade /India Cements			3.25	84.37

As the Job mix Design with 3.50% cement content satisfies all limits specified on IRC SP 89 - 2010 on compressive strength test and durability test. Hence cement content 3.50 % is safer and adopted for laying trails and site execution.

Conclusion

Durable Life of pavement. Thickness of Bitumen Courses will be reduced. Project will be finished at an earlier time. Less traffic problems only faced due to earlier finishing of the project. Mainly reduces the project cost about 18 lakh per/km approx. Uniform distribution of Load in Cement treated service road as compared to conventional road. Aggregate consumption is less for the case of stabilized base compared to that of the conventional method. But in the surface of the high embankment which shows in the metal beam barrier at earthen shoulder which results as a failure of sub base course while high air voids can't rise up in the cement layer it may lead to the crack of the roads. Granular Sub Base will be replaced in the delta region of the project. Depending on the needs of the project, CTB increases construction speed, improves the capacity of the pavement structure, or in some cases reduces the full-time project. Analysis in IRC37-2012 shows that the cement treated sub base naturally produces transverse and longitudinal crack but helps in the thickness of dense bituminous macadam and surface coarse as grade 2.

References

[1] Chummuneerat S, Jitsangiam P, Nikraz H. Shrinkage behavior of cement modified base course materials for Western Australian pavements. InInternational Public Works Conference, 2013, Darwin, Northern Territory, Australia 2013 Aug. https://doi.org/10.7158/C11-714.2013.11.1

[2] Ismail A, Baghini MS, Karim MR, Shokri F, Al-Mansob RA, Firoozi AA, Firoozi AA. Laboratory investigation on the strength characteristics of cement-treated base. InApplied mechanics and materials 2014 (Vol. 507, pp. 353-360). Trans Tech Publications Ltd. https://doi.org/10.4028/www.scientific.net/AMM.507.353

[3] Varma BP, Sirisha KP. Study of Processing and Machinery in Cement Industry. International Journal of Engineering and Innovative Technology. 2013 Nov;3(5):385-93.

[4] Shafeek AM, Salah H, Shehata N, Saddek AB. The impact of cooling water types on the cement clinker properties. Egyptian journal of petroleum. 2018 Sep 1;27(3):277-84. https://doi.org/10.1016/j.ejpe.2017.04.003

[5] Kosmatka SH, Kerkhoff B, Panarese WC. Design and control of concrete mixtures. Skokie, IL: Portland Cement Association; 2002 Jan 1.

[6] Patil VP, Karvekar AV. A Review on a Study of Cement Treated Base and Sub-Base in Flexible Pavement. Technology (IJCIET). 2017 May;8(5).

[7] IRC: 115. Guidelines for Structural Evaluation and Strengthening of Flexible Road Pavements Using Falling Weight Deflectometer (FWD) Technique. New Delhi: The Indian Roads Congress.

[8] Javdanian H, Lee S. Evaluating unconfined compressive strength of cohesive soils stabilized with geopolymer: a computational intelligence approach. Engineering with Computers. 2019;35(1):191-9. https://doi.org/10.1007/s00366-018-0592-8

[9] Little DN, Nair S. NCHRP web-only Document 144: Recommended practice for stabilization of subgrade soils and base materials. Transportation Research Board of the National Academies, Washington, DC. 2009.

[10] Smith T, Eng P, Barnes CL, Eng P, Zupko S. The Use of Engineered Soils in Canada. In Transportation 2014: Past, Present, Future-2014 Conference and Exhibition of the Transportation Association of Canada//Transport 2014: Du passé vers l'avenir-2014 Congrès et Exposition de l'Association des transports du Canada 2014.

[11] Congress IR. Guidelines for the design of flexible pavements. Indian code of practice, IRC. 2001;37.

[12] George KP. Minimizing cracking in cement-treated materials for improved performance 2002.

[13] Kishore K. Mix Design For Concrete Roads As Per IRC: 15-2011.

[14] Bagarre ED. Utilisation des graveleux latéritiques en technique routière. 1990.

[15] Crepin Z, Mohamed G, Ezechiel A, Crespin YA. Optimisation cement content for gravelly lateritic for realization of pavements base in North Benin. 2013(2).

[16] Tapase AB, Ranadive MS. Performance evaluation of flexible pavement using the finite element method. InGeo-China 2016 2016 (pp. 9-17). https://doi.org/10.1061/9780784480090.002

Buckling Behaviour of Cold Formed Steel Column Subjected to Axial Compression Loading

V. Jayanthi[1,a*], B. Mathu Pradeep[1,b], S. Dhanish Ahmed[1,c], M. Prajesh[1,d], S. Abishake[1,e]

[1]Department of Civil Engineering, Bannari Amman Institute of Technology, Sathyamangalam, Tamilnadu, India

[a]jayanthiv@bitsathy.ac.in, [b]mathupradeep.ce20@bitsathy.ac.in, [c]dhanishahmed.ce20@bitsathy.ac.in, [d]prajesh.ce20@bitsathy.ac.in, [e]abishake.ce20@bitsathy.ac.in

Keywords: Cold Formed Steel, Stiffness, Axial Compression Loading, Buckling Load, Abaqus

Abstract. Cold formed steels are durable, reliable and cost effective options for low and mid-rise buildings. Analysing the behaviour of cold formed steel structures is a vital prerequisite for their further large scale applications. The previous studies on cold formed steel structures stated stiffness as a major reason for its application in the construction field. In this paper the cold formed steel structures were analysed for their behaviour when subjected to axial compression loading. The different heights of columns such as 700 mm, 750 mm and 800 mm has been taken for a constant thickness of about 2 mm. The strength and displacement of axial member against loading was investigated and the results were compared with the load displacement curve obtained numerically using Abaqus and CFS software. In this investigation, an attempt is made to study the ultimate load carrying capacity and the corresponding displacement with respect to height and the mode of failure.

Introduction

Generally, construction sites use two different types of steel, classified based on their manufacturing process namely hot rolled steel and cold rolled steel (CFS). Cold formed steel as the name indicates is manufactured at room temperature. It is very efficient and the fabrication cost is lower than that of hot rolled steel. Since the cold formed steel members are formed at room temperature, the material do not experience a thermal stress than that the hot rolled steel does, so it is stronger. In the past, researchers have investigated various buckling modes of commonly used cold-formed steel sections. Computational Analysis of Cold-Formed Steel Columns with Initial Imperfections reported by MarselGarifullin, UdoNackenhorst [1] was based on nonlinear buckling analysis of the cold formed steel C-section columns and load bearing capacity on the imperfections in columns. They have analysed nonlinear buckling &Linear eigenvalue buckling. Thomas H.-K. Kang, Kenneth A. Biggs, and Chris Ramseyer [2] analysed the Buckling Modes of Cold-Formed Steel Columns. In their research, the effect of maximum load on C section was analysed and it is found that the orientation of the column plays a major role. Two different orientations of column were tested to determine the effect of orientation in the failure pattern and [1][2] found the buckling values of CFS.

S. Narayanan and M. Mahendran [3]worked on Ultimate Capacity of Innovative Cold-formed Steel Columns. Experiments and numerical analyses were used to investigate the behaviour of CFS by distortional buckling and ultimate strength was also tested. Different heights of column section were tested on axial compression load and analyzed the cold formed section in Abaqus and

Materials Research Proceedings **23** (2022) 45-49 https://doi.org/10.21741/9781644901953-6

predicted the values. Sreedhar Kalavagunta, Sivakumar Naganathan [4] provided an Experimental Study of Axially Compressed Cold Formed Steel Channel Columns. Axial compressive load was applied on cold formed steel column. Failure at local buckling and distortional buckling were investigated. Cold formed steel uses length and thickness parameters. They have compared numerical and test results with other specification.

Fadhluhartini Muftah [5] calculated the ultimate load of built up cold formed steel column with various lengths of section. The constant size was tested and the ultimate capacity of column was predicted based on the torsional buckling and flexural buckling. The cold formed steel's top and bottom were connected to hot rolled plate with drilled with bolts to study buckling behaviour column and other codes were compared.[6]C. C. Weng conducted compression test on cold formed steel column with two types of columns one stub(small) and the other column was to check the flexural, torsional and flexural-torsional buckling of column. In this investigation compression load capacity of section was predicted with the change in the thickness on section.

Section Properties
The channel section of different lengths has been used in this investigation. The shape of the angle section and the properties of channel section are tabulated in Table 1.

Table 1. Dimensions of the section

Designation H x B X C X T mm	Identification (mm)	DIMENSIONS				Mass kg/m	Area A	CENTRE OF GRAVITY	INERTIA I	RADIOUS OF GUY	
		H mm	B mm	C mm	T mm						
90X40X15 X2	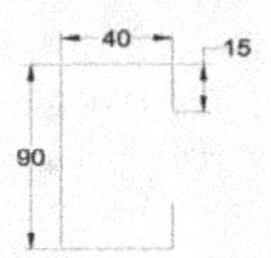	90	40	15	2	2.87	3.65	1.88	13.33	3.23	1.86

Experimental work
Cold formed steel channel section with the specification of IS-811 cold formed light gauges was selected. Design grade of 16 gauge CFS and different heights of 700 mm, 750 mm, 800 mm semi rigid plate connected were on two ends. Then the plate 4mm thick in hot rolled plate was welded with glass weld. Experimental work was done in universal testing machine (UTM) in order to determine axial compression load on specimen shown in Figure 1. The experimental results are shown in Table 2.

Materials Research Forum LLC
https://doi.org/10.21741/9781644901953-6

Figure 1 Experimental setup

Table 2. Experimental results

S NO.	DESCRIPTION	HEIGHT OF SPECIMEN (mm)	FAILURE [BOTTOM TO TOP](mm)	LOAD (KN)	DIPLACEMENT (mm)	COMPRESSIVE STRENGTH (N/mm2)
1	CFS	700	150	32.50	2.40	180.56
2	CFS	750	354	30.10	3	167.22
3	CFS	800	668	30.60	3.50	160.77

Numerical analysis

In this study, the finite element program ABAQUS as well as CFS (Cold Formed Steel component design tool) were used to analyse the columns under axial loading conditions. Initially the size of the member and thickness of the section was given as coordinates. The test results were shown in Figure 2 and the values are compared with the practical work and it is tabulated in Table 3.

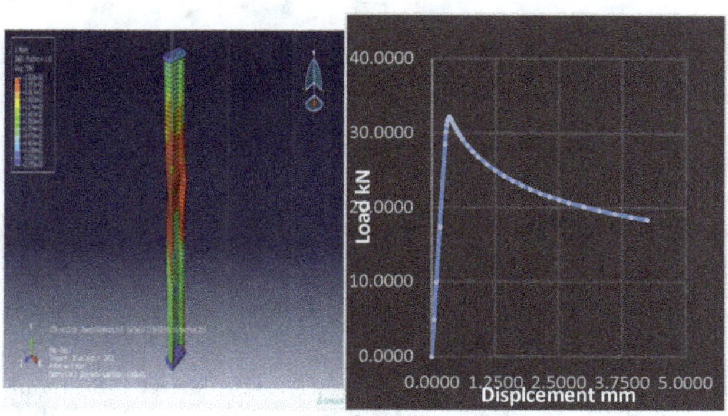

Figure 3 Deformed position

Table 3. Experimental vs Abaqus software results

Specimen No.	Height (mm)	Load (kN) (From Experiments)	Load (kN) (From CFS)	Load (kN) (From Abaqus)
1	CFS 700	32.5	39.4	31.835
2	CFS 750	30.1	40.2	32.2
3	CFS 800	30.6	40.9	31.03

Result and Discussion

This investigation provides the comparison of effective finite model analysis like Abaqus and CFS for the analysis of failure mode of channel section for different heights. Initially the sectional properties of channel section have been reported in this paper. The failure mode of channel section has been predicted using the finite element model and compared with the experimental results and it is shown in Figure 4. The results from the experimental are comparatively lesser than the results from the simulation tests (about 15% to 25%). From the results of this investigation, it is decided to modify the properties of CFS sections to improve the load carrying capacity in future.

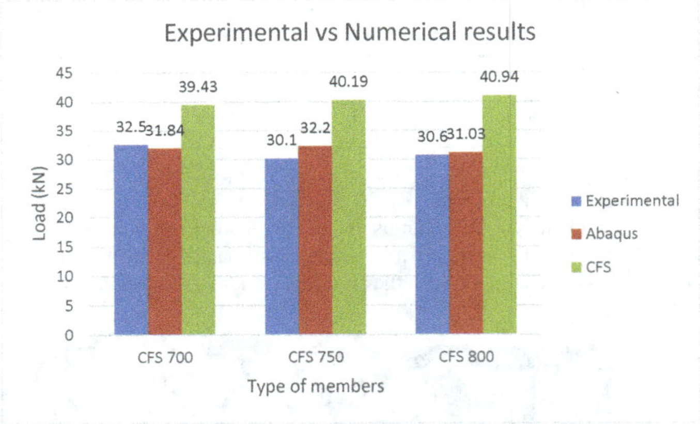

Figure 4. Comparison of Results

References

[1] MarselGarifullin, UdoNackenhorst Computational Analysis of Cold-Formed Steel Columns with Initial Imperfections 2015. https://doi.org/10.1016/j.proeng.2015.08.239

[2] Thomas H.-K. Kang, Kenneth A. Biggs, and Chris Ramseyer Buckling Modes of Cold-Formed Steel ColumnsVol. 5,August 2013

[3] S. Narayanan and M. Mahendran, Ultimate Capacity of Innovative Cold-formed SteelColumnspp. 489-508,2003. https://doi.org/10.1016/S0143-974X(02)00039-1

[4] SreedharKalavagunta,SivakumarNaganathan Experimental Study of Axially Compressed Cold Formed Steel Channel ColumnsVol 6 (4) April 2013. https://doi.org/10.17485/ijst/2013/v6i4.7

[5] MohdSyahrulHisyamMohdSaniultimate load of built up cold formed steel columnvol 9,2014

[6] C. C. Weng and TeomanPekoz, compression test on cold formed steel column 1990. https://doi.org/10.1061/(ASCE)0733-9445(1990)116:5(1230)

[7] MarselGarifullin, UdoNackenhorst Computational Analysis of Cold-Formed Steel Columns with Initial Imperfections 2015. https://doi.org/10.1016/j.proeng.2015.08.239

[8] Thomas H.-K. Kang, Kenneth A. Biggs, and Chris Ramseyer Buckling Modes of Cold-Formed Steel ColumnsVol. 5,August 2013

[9] S. Narayanan and M. Mahendran, Ultimate Capacity of Innovative Cold-formed SteelColumnspp. 489-508,2003. https://doi.org/10.1016/S0143-974X(02)00039-1

[10] SreedharKalavagunta,SivakumarNaganathan Experimental Study of Axially Compressed Cold Formed Steel Channel Columns Vol 6 (4) April 2013. https://doi.org/10.17485/ijst/2013/v6i4.7

[11] MohdSyahrulHisyamMohdSaniultimate load of built up cold formed steel colum, vol v 9,2014

[12] C. C. Weng and TeomanPekoz, compression test on cold formed steel column 1990. https://doi.org/10.1061/(ASCE)0733-9445(1990)116:5(1230)

Experimental Investigation on Partial Replacement of Cement with Metakaoline, Fly Ash and Silica Fume in Concrete

C. Pradeep Kumar[1,a,*], L. Anandhi[2,b]

[1] Karpagam College of Engineering, Coimbatore, Tamilnadu, India

[2] Peri Institute of Technology, Chennai, Tamilnadu, India

[a]pradeeppandian93@gmail.com, [b]anandhicvl@gmail.com

Keywords: Metakaoline, Silica Fume, Fly Ash, RCPT, Acid Test

Abstract. The process of construction begins some thousand years ago and the monuments that we are seeing are the proof for that. The materials that are being used in the construction some thousand years ago are not the same which are being used nowadays. The construction materials get modified and new inventions are taking place due to the continuous research. The scope of this project is to make building materials sustainable and to make them eco-friendly. In the olden days lime was used as a mortar and now it's being replaced by cement. But still research is going on to make better binding materials to make construction reliable. As an initiative, in our project we are going to partially replace cement with metakaoline, fly ash and silica fume simultaneously. We are going to replace cement with each additive in the order of 5%, 7.5%, 10%, 12.5%, and 15%. Therefore, the summation of replacement of all additives will be in the order of 15%, 22.5%, 30%, 37.5%, 45%. The concrete specimens are casted based on the replacement ratios along with the conventional concrete specimens. The specimens are subjected to the various laboratory tests and it was found that a 10% replacement of cement with additive materials to be optimum.

Introduction

The main constituent of construction is concrete. Cement used in concrete is ordinary Portland cement which emits carbon dioxide that causes pollution to the environment. Reduction in the usage of OPC in concrete gains a greatest advantage in the environment. As the nation is focusing more on the infrastructure development the usage of cement in the structure is increasing tremendously. So we are in need of finding an alternative resource that can replace cement and also which does not harm the environment. In this research work we are going to replace the OPC with metakaolin, silica fume and fly ash in different percentage in order to find the optimum percentage of the replacement. The metakaoline is used for greater binder in concrete.

Material Investigation

The materials which are to be used in the research are subjected to basic material test and the results are furnished in the Table 1.

Table 1 – Preliminary test results of aggregates

Fine aggregate [4]		
S.no	**Properties**	**Results**
1.	Specific gravity	2.60
2.	Bulk density	1469 kg/ m^3
3.	Fineness modulus	3.16
4.	Water absorption	0.25 %
Coarse aggregate [4]		
1.	Specific gravity	2.7
2	Water absorption	0.15%
3	Bulk density	1579 kg/ m^3
4	Fineness modulus	4.71
Cement [5]		
1	Specific gravity	3.12
2	Fineness modulus	2.86
3	Soundness of cement	2 mm
4	Initial setting time	45 min
5	Final setting time	10 hours
Metakaoline [2]		
1	Specific gravity	3.4
2	Bulk density	1523 kg/m^3
3	Fineness modulus	2.78
4	Water absorption	0.26 %
Fly ash [8]		
1	Specific gravity	2.18
2	Bulk density	1235 kg/m^3
3	Fineness modulus	2.72
4	Water absorption	0.265 %
Silica fume [8]		
1	Specific gravity	2.3
2	Bulk density	320 kg/m^3
3	Water absorption	0.245 %

Mix design:

M20 grade concrete is used for the research. Following the IS codes the mix design has been arrived and the quantity calculation for different % of replacement are shown in Table 2. [6, 9]

Cement:	FA	:	CA	:	W/C
1 :	1.62	:	2.68	:	0.5

Table 2 – Mix proportions for various % of replacement

Mix (%)	Cement (Kg)	Metakaolin (Kg)	Silica fume (kg)	Fly ash (kg)	Fine aggregate (kg)	Coarse aggregate (kg)
Control	13.16	-	-	-	24.25	48.50
15%	11.21	0.65	0.65	0.65	24.25	48.50
22.5 %	10.19	0.99	0.99	0.99	24.25	48.50
30 %	9.2	1.32	1.32	1.32	24.25	48.50
37.5 %	8.21	1.65	1.65	1.65	24.25	48.50
45 %	8.32	1.97	1.97	1.97	24.25	48.50

During the concrete mix the water content is increased based on water absorption of the additive materials

Experimental Investigation

Mixing was done in a laboratory by hand mixing. Fig.1 and Fig.2 shows the hand mixing of concrete.

Fig.1 Mixing

Fig.2 – Casting

Slump Cone Test

Table 3 represents the slump value of various mixes.

Table 3 – slump cone test results

Grade of concrete	Mix	Slump value
M20 GRADE	CC	85
	M1	74
	M2	78
	M3	80
	M4	81
	M5	77

Acid Test Results

Test Result of HCL

Table 4 represents the test results of acid test. From the test results the concrete containing fly ash and other materials were found to exhibit resistance more than the zero percentage replacement specimens. Fig.3 shows the graphical representation of test results of acid test.

Table 4 – Test result of HCL

S.No	% of replacement	Loss in weight (grams)
1	Control	280
2	15%	211
3	22.5 %	170
4	30 %	162
5	37.5 %	247
6	45 %	253

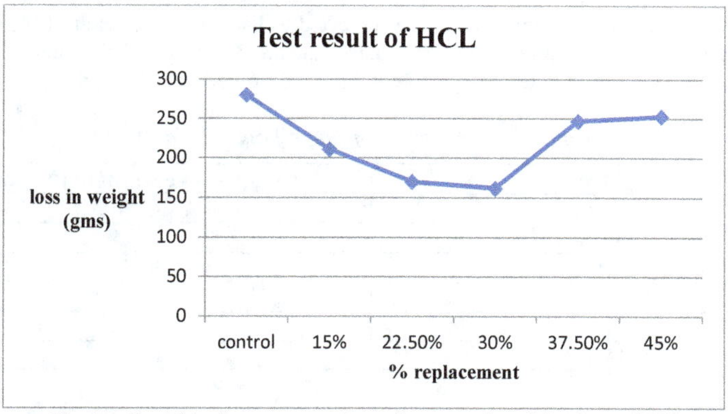

Fig.3 – Graphical representation of test results of Acid test (HCl)

Test Result of H_2SO_4

Table 5 shows the test results of sulphuric acid test. From the table, the replaced cement concrete shows higher resistance than that of normal cement concrete up to 37.5%. And the resistance of 45% is higher but it is less than normal cement concrete. The graphical representation of the results of the sulphuric acid test is shown in figure 4.

Table 5 - Test results of sulphuric acid test

S.No	% of replacement	Loss in weight (grams)
1	Control	540
2	15%	492
3	22.5 %	459
4	30 %	441
5	37.5 %	510
6	45 %	523

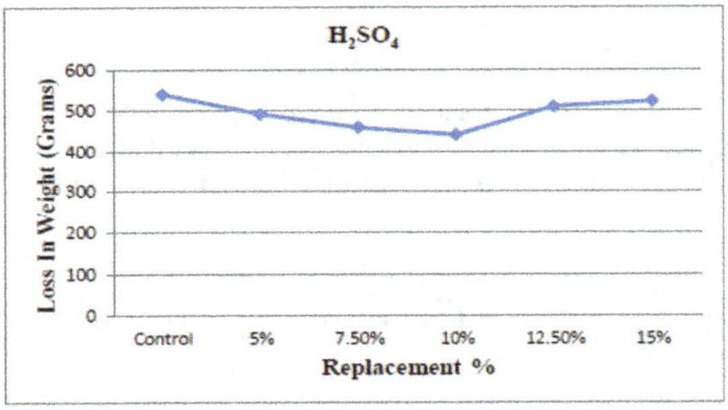

Fig.4 – Graphical representation of sulphuric acid test

Mechanical Properties

Compressive Strength Test Results

Once the curing period ends the concrete are subjected to hardened concrete test. As a first step the concrete is subjected to compressive strength test and are tabulated in Table 6. Fig.5 represents the graphical representation of test results of compressive strength test. [5, 7, 9]

Table 6 - Compressive Strength Value (28days)

Mix	Compressive strength (N/mm^2)
Control	25.76
15%	26.25
22.5 %	26.88
30 %	27.41
37.5 %	24.32
45 %	19.28

Materials Research Forum LLC
https://doi.org/10.21741/9781644901953-7

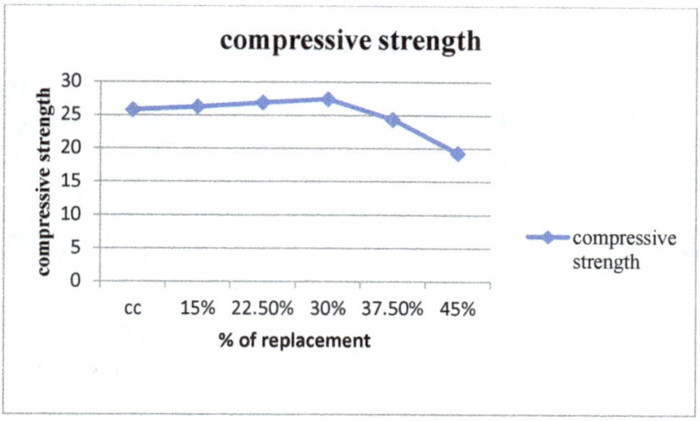

Fig.5 –compressive strength results after 28 days

Split Tensile Strength Test Result

Split tensile strength test was conducted in universal testing machine and test results are listed in Table 7. Fig.6 shows the split tensile strength test.Fig.7 shows the graphical representation of split tensile strength test results. [5, 7, 9]

Table 7 - split tensile strength test (28 days)

Mix	Split tensile strength (N/mm^2)
Control	2.28
15%	2.30
22.5 %	2.33
30 %	2.35
37.5 %	2.21
45 %	1.97

Fig.6 - split tensile strength test

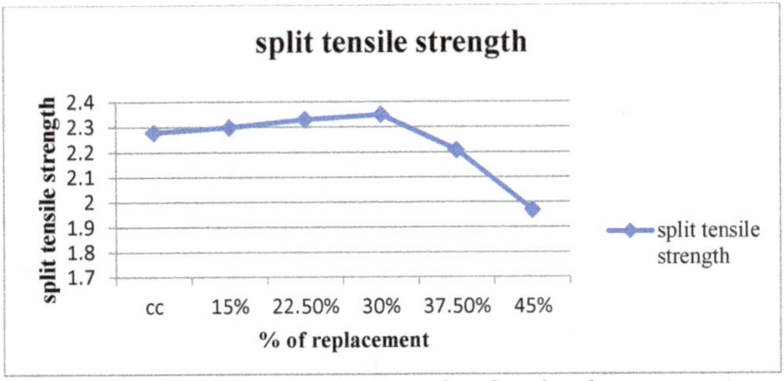

Fig.7 – Split tensile strength test result – Graphical representation

Modulus of Elasticity Test Results

Table 8 shows the test results of modulus of elasticity test.Fig.8 shows the experimental set up of modulus of elasticity test. [5]

Table 8 - Modulus of Elasticity test results

Mix	Young's Modulus (N/mm2)
Control	22360
15%	22516
22.5 %	23000
30 %	23162
37.5 %	22986
45 %	22788

Fig.8 – Modulus of elasticity test set up

Permeability Test Results

Permeability test was carried out and the test results have been furnished in the Table 9. Fig.9 shows the graphical representation of test results of permeability test. [8 ,9]

Table 9 - Charge Passed In RCPT

Grade of concrete	% of replacement of metakaolin, silica	Charge passed (coulomb)	Chloride permeability as per ASTM C1202
M20	Control	1158	High
M20	15%	652	moderate
M20	22.5 %	281	Very low
M20	30 %	272	Very low
M20	37.5 %	298	low
M20	45 %	307	low

The rapid chloride ion permeability measured at 28 day is illustrated. We can see that the total charge passed through the cement replaced concrete decreases when the percentage of additives increases.

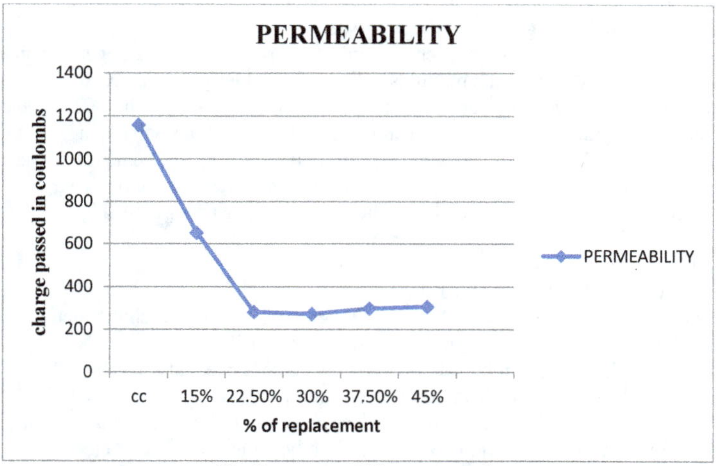

Fig.9 – permeability test results

Results & Discussions

1. From the acid test it has been observed that the loss in weight is more for the conventional concrete and for the 37.5% and 45% replacement of cement with additives when compared to the 15%, 22.5% and 30% replacement of cement with additives. The reason behind this is, as the alkaline content is low it neutralizes the acid and thereby the loss in weight is less

for the lower replacement percentage when compared to the higher replacement percentage of additives. [12]
2. From the compressive strength test we can find that the strength is increasing for the concrete with the replacement of cement by additives in the range of 15%, 22.5% and 30%. It is because the voids have been filled by the additive materials. Therefore the strength gets increased when the grain size of the additive materials is well graded with aggregate.
3. On further adding the additive materials in the range of 37.5% and 45% the strength decreases since the grain size of the additive materials become uniformly graded which leads to the decrease in strength. [1, 10]
4. From split tensile strength test it has been found that the strength gets increased upto 30% replacement of cement with additives. On mixing with water, the additives develop a homogenous layer with the aggregates because it acts as filler materials as well as binding materials [1, 3, 6]. Hence it gives better split tensile strength than the conventional concrete. If the percentage of replacement is being increased the minor cracks in the concrete gets developed which leads to the failure of the concrete or in other words reduction in the tensile strength. [11]
5. Modulus of elasticity will be greater for the concrete which is having well graded aggregates. The well graded aggregates in concrete can be achieved due to the addition of additives which reduces the voids present in the concrete. Therefore the replacement of cement with 30% of additives will results in better modulus of elasticity values as the voids in the concrete is being filled by the additives. On increasing the percentage of replacement the well graded aggregates will change into uniformly graded aggregates which leads to increase in voids thereby the modulus of elasticity values gets decreased.
6. From RCPT test we can found that the permeability is low for the 30% replacement of cement with additives. Grain size of additives is less when compared to aggregates thereby we are getting the low permeability. On partially replacing the cement with additives the voids present in the concrete is getting reduced. On increasing the percentage of additives the permeability gets increased due to the uniform grain size. [3, 10, 11]

Conclusion

The following conclusions were arrived from the research

1. The basic test results were conducted on the materials from which the material properties were found out which is used in the mix design calculation.
2. Various test has been conducted on the hardened concrete and it has been found that 30% replacement of cement with additives is optimum
3. From acid test it has been found that the loss in weight of the 30% replaced concrete is less when compared to conventional concrete. So it becomes dense concrete.
4. Form compressive strength test it has been found that the 30% replaced concrete has 6% more strength than the conventional concrete.
5. From split tensile strength test it has been found that the 30% replaced concrete has 3% more strength than the conventional concrete.
6. On conducting RCPT test we can found that the permeability for the 30% replacement of cement with additives shows better result when compared to all other percentage of replacement.
7. From the above results we can conclude that the 30% replacement of cement with additives shows better permeability, durability and strength properties. So we can replace cement

with metakaolin, silica fume and fly ash up to 30% without affecting durability and strength properties of the concrete.

8. Finally we suggest that this concrete can be utilized as a sustainable construction material.[11,12]

References

[1] Abideng Hawa, Danupon Tonnayopas, Woraphot Prachasaree(2014), "Performance Evaluation Of Metakaolin Based Geopolymer Containing Parawood Ash And Oil Palm Ash Blends", materials science,vol:20, pg: 1292–1320. https://doi.org/10.5755/j01.ms.20.3.4543

[2] Aboulayta,M.Riahib, M.Ouazzanitouhami(2017), "Properties Of Metakaolin Based Geopolymer Incorporating Calcium Carbonate", Advanced powder technology, Vol-28,pg:2393-2401. https://doi.org/10.1016/j.apt.2017.06.022

[3] Dinakar, Pradosh .K, Sahoo, G.Sriram, (2013), "Effect of Metakaolin Content on the Properties of High Strength Concrete", International Journal of Concrete Structures and Materials, Vol-7, pg: 215-223. https://doi.org/10.1007/s40069-013-0045-0

[4] IS 383 - 1970 'Indian Standard Specification for Coarse and Fine Aggregate from Natural Sources for Concrete' 11th reprint, September 1993, Bureau of Indian Standards, New Delhi - 110 002.

[5] IS - 516 - 1959 'Indian Standard Methods of Test for Strength of Concrete', 13th reprint, June 1989, Bureau of Indian Standards, New Delhi - 110 002.

[6] IS 10262 - 2009 'Indian Standard Recommended Guidelines for concrete Mix Design', Bureau of Indian Standards, New Delhi - 110 002.

[7] IS 5816 - 1970 'Indian Standard Methods of tests for Splitting Tensile Strength of Concrete Cylinders' 3rd Print, October 1985, Indian Standard Institute New Delhi - 110 002.

[8] J.M.Khatib, J.J.Hibbert,(2005), "Selected Engineering Properties Of Concrete Incorporating Slag And Metakaolin", Construction And Building Materials , vol-19, pg: 460-472. https://doi.org/10.1016/j.conbuildmat.2004.07.017

[9] Mukundlahotia,Pratiknarangb,(2017), "Mix Design Factors And Strength Prediction Of Metakaolin-Based Geopolymer",Ceramics international,vol-43,pg:11433-11441. https://doi.org/10.1016/j.ceramint.2017.06.006

[10] Raphaelle Pouhet, (2015), "Formulation And Durability Of Metakaolin-Based Geopolymers", universitepaul Sabatier- Toulouse, vol-3, pg: 129-315

[11] Pouhet, M. Cyr (2016) "Formulation And Performance Of Metakaolin Geopolymer Concretes", construction and building materials, vol:120,pg:130-150. https://doi.org/10.1016/j.conbuildmat.2016.05.061

[12] Zhongheshui, Wei Chen,Chunhuashen.R, (2003),Enhancing Microstructure And Durability Of Concrete From Ground Granulated Blast Furnace Slag And Metakaolin As Cement Replacement Material, Journal of Material Research And Technology, vol-2, pg:52-59. https://doi.org/10.1016/j.jmrt.2013.03.010

Feasibility Study on Utilization of Compost as a Partial Replacement of Fine Aggregate in Conventional Concrete

Vanitha Sankararajan[1,a,*], Karthigai Priya Pandiaraj[2,b]

[1]Department of Civil Engineering, Kalasalingam Academy of Research and, Tamil Education, Krishnankovil Nadu, India

[2]Department of Civil Engineering, Kalasalingam Academy of Research and Education, Krishnankovil, Tamil Nadu, India

[a]s.vanitha@klu.ac.in, [b]karthigaipriya@klu.ac.in

Keywords: Solid Waste, Compost, Concrete, Compressive Strength, Loss of Compressive Strength

Abstract. Solid waste management is the biggest challenge due to population growth. Lack of awareness about onsite processing of solid waste implies production of huge quantity of solid waste which is unmanageable. In this study an attempt is made to utilize the degraded solid waste called compost which is collected from municipal site and is used as a filler material in concrete. Compost is replaced for fine aggregate with 5%, 10%, 15% and 20% in conventional concrete. The compressive strength results at the age of 28 days are 30.5, 19.8, 16.8 and 18.2 N/mm2 respectively. Result indicates 5% replacement of compost is effective in concrete. The concrete specimen made with compost (CVC) and conventional concrete (CC) are exposed to an elevated temperature of 200°C, 400°C, 600°C and 800°C respectively. Result indicates the conventional concrete shows 5.3%, 6.1%, 15% and 24% compressive strength loss whereas 5% compost replaced specimen shows 10%, 11 %, 23.2% and 58.2% respectively. At an elevated temperature, both conventional concrete and compost-based concrete lost its strength. The reason could be thermal incompatibility between the aggregates and concrete. However, the loss of compressive strength is almost double in CVC than CC. The possible reason could be volatile compounds may evaporate in CVC at high temperature. The weight lost is observed in CC varies from 2.4 % to 4.5 % whereas CVC shows 6% to 7% from 200°C to 800°C. Weight lost is 3% more in CVC specimen than CC specimen.

Introduction

Concrete is considered as composite material which consists of cement, fine aggregate, coarse aggregate and water and it is used all over the world. Every material in concrete plays a vital role in making structure strong. For example cement binds all the material together and brings hardened structure. Coarse and fine aggregate fill the pores in concrete. Particularly fine aggregate fills fine pores in concrete thus making structure impermeable. Coarse and fine aggregate occupies 70% of total volume in concrete [1]. Common river sand is used as fine aggregate in concrete. Over exploitation of river sand affects natural resources, river bed and its eco system. Currently manufactured sand (M sand) is a quarry dust replaces river sand which is obtained from crushing of stone. If the same condition persists, there is a demand for M sand may occur in future.

Effective solid waste management is major important problem in India. Lack of responsibility is major reason for production of more volume [2]. Improper disposal leads dangerous problem such as unhygienic condition, stagnation of rain water, mosquito breeding, attractive to rats, odour problem, spreading of diseases, affecting cattle and ground water contamination. A priority is to move from waste dump to waste management system [3]. Most of the time liners are damaged in waste dumpsite and affect ground water nearby. Sustainable and maximum resource extraction from waste is an urgent requirement in current scenario for protecting the environment [2,4]. It is an urgent need to utilize the solid waste for useful purposes continuously for solving these issues. In recent past, the waste material from one industry can be used as a raw material for other industry. For instance, bagasse waste from sugar industry is used as a raw material in thermal power plant industry, paper and pulp industry, agricultural fertilizer etc. Similarly fly ash from thermal power plant is used as a raw material for replacement of cement in construction industry, soil amendment in agricultural sector.

Many of the researches are moving towards utilization of waste from industries and agro waste for building construction [1,5]. Many researchers used rice husk ash, bagasse ash, groundnut shell ash for replacing cement [6,7,8]. Many researches had undergone their research by replacing fine aggregate in concrete [1,5]. For instance, the waste product from ceramic industry is used as a fine aggregate in building construction [9]. [10] used bio deposited recycled mortar aggregate in building construction. [11] used construction and demolition waste as fine aggregate in building construction. Some researchers attempted to utilize agro waste for building construction. For instance [12] used ground shell ash for sand. Ganion 2014 used partially replaced saw dust as fine aggregate in concrete. Table 1 shows different waste materials used for replacing fine aggregate.

Table1 Waste material used for fine aggregate in building construction

Author	Material Replaced	Remarks
[14]	Fine Aggregate - Partial replacement of waste glass powder	In this work fine aggregate is partially replaced with glass powder with 2 and 2.5%. The compressive strength is linearly increases upto 2 % and then decreases.
[15]	Fine Aggregate -Partial replacement of ceramic waste	Ceramic waste is replaced for fine aggregate with 20%, 25%, and 30% in concrete. Compressive strength at 20% replacement is high.
[5]	Fine Aggregate -Partial replacement of agro - waste.	In this review article different waste such as bagasse ash, sawdust ash, and oyster shell, ground nut shell, rice husk ash, tobacco waste and giant reed is replaced for fine aggregate and optimal replacement is given.
[16]	Fine Aggregate -Partial replacement of paper sludge.	Her paper sludge is replaced for fine aggregate. The optimal proportion is 5% to 10% respectively
[13]	Fine Aggregate - Partial replacement of Saw dust	In this study saw dust is used as fine aggregate. It is recommended to use in interior walls of the buildings. The sawdust reduces 56% of the total cost.
[17]	Fine Aggregate - Partial replacement Cork	Cork is used for replacing fine aggregate and cork has low thermal resistance and good sound absorption.

[18]	Fine Aggregate - Partial replacement Groundnut shell	Ground nut shell is replaced at 0, 5, 15, 25, 50 and 75% with fine aggregate.
[19]	Fine Aggregate - Partial replacement of Compost	In this compost is replaced for portland cement

According to planning commission report 2014, in the year 2011, the number compost unit is 162 and vermi compost unit is 24 in number in Tamilnadu and it ranks first all over India in municipal solid waste processing facilities [2]. In this study an attempt is made to utilize homogeneous compost obtained from municipal solid waste is used for fine aggregate in building construction. The different percentage such as 5%, 10%, 15% and 20% replacement of compost with fine aggregate is made and compressive strength study is made. The concrete is exposed to elevated temperature mainly because the replaced material is organic in nature.The behavior of compost based concrete specimen at an elevated temperature is necessary for this study. The weight loss and compressive strength loss in concrete cube at an elevated temperature is studied.

Methodology

Collection of materials

The materials namely cement, fine aggregate, coarse aggregate, compost and water is used in this study. The basic properties are tested in the laboratory. The specific gravity of the cement is 2.22. The fine aggregate used here is manufactured sand (M Sand). It is obtained from local stone crushing factory. The compost is obtained from local solid waste decomposing plant. Coarse aggregate is obtained from stone crushing factory. The size of the stone used here is passing 22.5mm and retaining in 20 mm sieve. Table 2 shows the physical properties of materials used in the study.

Table 2 Characteristics properties of raw materials used for this study

S.No	Test	Coarse aggregate	Fine aggregate	Compost
1	Specific gravity	3.116	2.33	1.76
2	Fineness Modulus	2.985	4.56	1.77
3	Water absorption (%)	0.81	4.35	2.26

Mix Design

Mix design is an important procedure to attain the maximum strength. Based on the value obtained from the material property tests the mix design was calculated as per IS codal provision of 10262 for M30 grade concrete. Table 3 shows the mix proportion of replacement of compost in conventional concrete.

Mixing and casting

Improper mixing leads to low strength and less durable concrete. First the dry materials like cement, fine aggregate, coarse aggregate and compost mixed thoroughly the water has been added and mixed for another 2-3 minutes. Like mixing, casting also an important, the concrete should be placed into the mould before the initial setting. The concrete cube size of 100mm × 100mm × 100 mm are casted to find out the compressive strength tests.

Table 3 Mix ratio of conventional concrete with compost

Percentage of compost %	Cement (kg/m ^{3})	Fine aggregate (kg/m ^{3})	Coarse aggregate (kg/m ^{3})	Compost (kg/m^3)
0%	478.95	529.52	1101.65	0
5%	478.95	503.07	1101.65	26.45
10%	478.95	476.62	1101.65	52.9
15%	478.95	454.17	1101.65	75.35
20%	478.95	423.72	1101.65	105.8

Curing

Curing is supplying water to the concrete for a certain period at different interval. Curing influences the hydration process which improves the strength. There are different methods of curing can be adopted based on the requirement. Water curing is adopted in this work. For normal concrete the water curing is done for 28 days and the tests carried out after 28 days of curing period.

Experiment for hardened concrete

The compressive strength of the specimen is tested by Compression Testing Machine. To find the temperature resistance, the cubes are kept in oven at different temperature (200°C, 400°C, 600°C and 800°C) with duration of 2 hours. After that the specimen is kept at room temperature to cool down and their compressive strength and weight loss is tested.

Results and discussion

Effect of percentage replacement on compressive strength

Table 4 shows the compressive strength of CC and CVC specimen. It is found that the compressive strength concrete is 30.5 N/mm^2 at 5% replacement of compost whereas control specimen shows 29 N/mm^2. It is almost 2 N/mm^2 more than control specimen. The reason could be compost may fill the minute pores which is present between fine aggregate and coarse aggregates. However 10%, 15% and 20% CVC specimen shows 19.8, 16.8 and 18.2 N/mm^2 respectively. The possible reason could be there is minimum binding between aggregates; compost and cement may decrease the compressive strength.[20] found the cementitious material like metakaolin, ground granulated blast furnace slag (GGBS), fly ash and silica fume can increase the binding between composite materials in concrete.

Table 4 Compressive strength results

S.No	% Replacement	Compressive strength (N/mm^2)
1	0	29.0
2	5	30.5
3	10	19.8
4	15	16.8
5	20	18.2

Effect of compressive strength loss and weight loss at an elevated temperature

To improve the serviceability of building, the building material should be tested whether it resists fire during accidents. It is decided that compost based specimen and conventional concrete specimen are subjected to an elevated temperature. The CC specimen and 5% replaced CVC specimen which gives maximum compressive strength are subjected to an elevated temperature of 200 °C, 400 °C, 600°C and 800°C temperature is shown in Table 5 and Fig.1. It is found that the compressive strength reduction at 200 °C in control specimen and CVC specimen are 5.3% and 10% respectively. The compressive strength reduction at 400°C in CC and CVC specimen are 6.11% and 11% respectively. The compressive strength reduction at 600°C in CC and CVC are 15% and 23.2% respectively. The compressive strength reduction at 800 °C in CC and CVC specimen are 24% and 58.2% respectively. It is found that at all higher temperature the compressive strength reduction is obtained in both CC and CVC specimen. Arioz O (2007) observed the strength of the concrete decreased when the temperature is increased. The reason could be difference in thermal expansion between aggregates and binder, decomposition of cement paste, degradation and phase decomposition of aggregate [22,23]. At higher temperature the physical, chemical and mechanical properties of concrete is affected [22,23]. At higher temperature the bond between aggregates and cement may broken and loss of compressive strength may occur. Demerial B 2010 found above 600°C degeneration of CSH (Calcium Silicate Hydrate) gel increases amount of air voids and formation of more pore structure in the specimen thereby decreases the transmission of sound and the value of ultrasonic pulse velocity while experiment is conducted.

It is noted that the compressive strength lost is more in CVC specimen than CC specimen. The loss is almost double in CVC specimen than CC particularly at 200 °C and 800°C. In compost based specimen, at an elevated temperature of 800° C, the volatile organic compounds may evaporate thereby increases more spaces. Another reason could be the weight lost in more in CVC specimen than CC specimen causing decrease in strength (Table 6).

Table 5. Reduction of compressive strength at an elevated temperature

Temperature (°C)	Loss of compressive strength (%)	
	CC	CVC
200	5.30	10.00
400	6.11	11.00
600	15.00	23.20
800	24.00	58.20

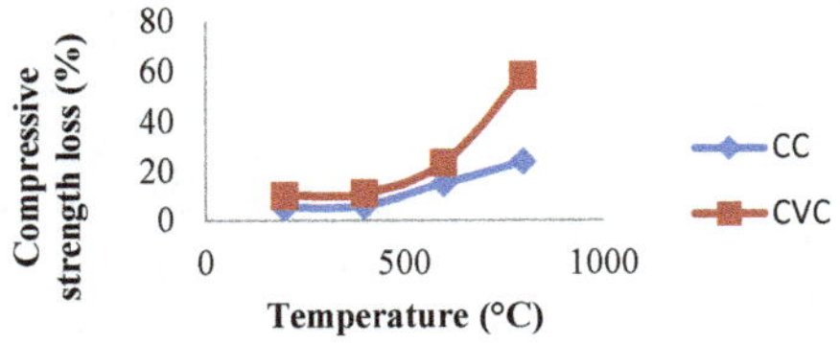

Fig.1. Loss of compressive strength in CC and CVC specimen

Study of weight lost at an elevated temperature
Table 6 and Figure 2 shows the weight lost in CC and CVC specimen at different temperature. From table 6, the weight lost of CC specimen at 200°C is 2.45%. After that, the CC specimen is exposed to 400°C, 600°C and 800°C the weight lost is 3.09, 3.3 and 4.55% respectively. The CVC specimen gives 6.11 % weight lost at 200°C. After exposed to 400°C, 600°C and 800°C the weight lost is 6.5, 6.5 and 7.1% respectively. There is no abrupt change is weight lost is observed beyond 200°C. From Fig. 2, the weight loss is more in CVC specimen than CC specimen. In CVC specimen the weight lost is almost 3% higher than CC specimen at all temperature. [22] found when the concrete is subjected to an elevated temperature; the weight is significantly reduced in conventional concrete after 800°C.

Table 6.Weight loss at an elevated temperature

Temperature (°C)	Weight loss (%)	
	CC	CVC
200	2.45	6.1
400	3.09	6.5
600	3.3	6.5
800	4.55	7.1

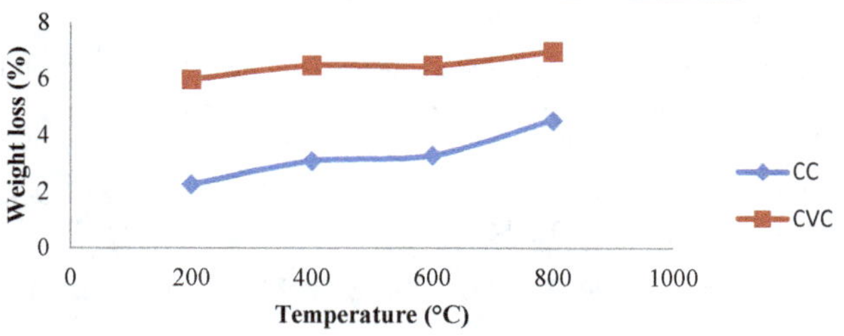

Fig.2 Weight loss in CC and CVC specimen

Tuffail M and Gentcturk B 2017 found free water will expel at 105°C. The dehydration of hydrous calcium aluminium sulfate mineral will occur at the temperature of 80 to 150°C. At 150°C to 170°C, gypsm will be degraded. The degradation of CSH will takes place beyond 400°C. The second phase of CSH decomposes at 600°C to 800 °C forming β-C_2S. The breakdown of CSH is fully occurring at the temperature of 900°C. The dangerous and crucial temperature for concrete is from 400°C to 900°C. In the particular temperature range, the concrete almost lost its strength.

Conclusion
In this study the following conclusion are arrived

1. The compressive strength of concrete and compost based specimen are 29 N/mm^{2}and 30.5 N/mm^2 respectively. Result shows 5% replacement compost in concrete shows encouraging results.
2. At higher temperature the decrease of compressive strength in CC and CVC specimen is observed. Also in CVC specimen, the loss is double than CC. This may be due to escaping

of volatile organic compounds at higher temperature. Another reason could be thermal effect create cracks in concrete by breaking the bond between composite material in concrete thus results reduction in compressive strength. To increase the binding properties between aggregate, cement and compost, pozzolanic material like GGBS, Fly ash, Metakaolin or silica fume can be added for improving the strength.

3. The weight lost of CC and CVC occurred at an elevated temperature. Weight loss is 3% more in CVC than CC.

Acknowledgement
The authors thank Muthulingam G, Arikara Velan N, Deepak V, Dhinesh Kumar N, Sachin Sabariraj for conducting experimental studies.

References

[1] M.K. Dash, S.K. Patro, A.K. Rath, Sustainable use of industrial waste as partial replacement of fine aggregate for preparation of concrete-A review, Int. J. Sustain. Built Environ. 5 (2016) 484-516. https://doi.org/10.1016/j.ijsbe.2016.04.006

[2] S.Kumar, S.R. Smith, G.Fowler, C.Velis, S.J. Kumar, S.Arya, Rena, R.Kumar, C. Cheesman, Challenging opportunities associated with waste management in India, R. Soc. Open Sci. 4 (2021) 1-7. https://doi.org/10.1098/rsos.160764

[3] A. Pappu, M.Saxena, S.R.Asolekhar, Solid waste generation in India and their recycling potential in building materials, Build Environ, 42 6 (2007) 2311-2320. https://doi.org/10.1016/j.buildenv.2006.04.015

[4] Y. Sharifi, M. Houshiar, B.Aghebati, Front. Struct. Civ. Eng. 7 4 (2013) 419-428. https://doi.org/10.1007/s11709-013-0224-8

[5] J.K Prusty S.K Patro S.S Basarkar, Concrete using Agro-waste as fine aggregate for sustainable built environment, 5 (2016) 312-333. https://doi.org/10.1016/j.ijsbe.2016.06.003

[6] G.C.Cordeiro, R.D. Toledo Filho, L.M. Tavares, E.M.R. Fairbairn, Pozzolanic activity and filler effect of sugar cane bagasse ash in Portland cement and lime mortars, Cem Concr Compos. 30 (2008), 410-418. https://doi.org/10.1016/j.cemconcomp.2008.01.001

[7] T.C. Nwofor, S.Sule, Stability of groundnut shell ash (GSA)/ordinary portland cement (OPC) concrete in Nigeria, Advances in Applied Science Research, 3 4 (2012) 2283-2287.

[8] I.O. Obilade, Use of rice husk ash as a partial replacement for cement in concrete, International Journal of Engineering and Applied Sciences, 5 04 (2014), 1-6.

[9] R.Ravindra, A.B.G. Kumar, P Lingaraj, Effect of Size of Crushed Ceramic Tiles as Fine Aggregates in Cement Composites, Recent Trends in Civil Engineering & Technology, 5 3 (2013) 46-53.

[10] Y. Zhu, Q. Li, P. Xu, X. Wang S. Kou, Properties of Concrete Prepared with Recycled Aggregates Treated by Bio-Deposition Adding Oxygen Release Compound, Materials (Basel). 12 13 (2019), 1-13. https://doi.org/10.3390/ma12132147

[11] N.Kartam, N.A. Mutairi, I.A. Ghusain, J.A. Humoud, Environmental management of construction and demolition waste in Kuwait, Waste Manage 24 2004, 1049-1059. https://doi.org/10.1016/j.wasman.2004.06.003

[12] H Mahmoud, Z A Belel, C Nwakaire, Groundnut shell ash as a partial replacement of cement in sandcrete blocks production Int. J. Sustain. Dev. International Journal of Development and Sustainability, 1 3, (2012), 1026-1032.

[13] T.U. Ganiron, Effect of sawdust as fine aggregate in concrete mixture for building construction. Int. J. Adv. Sci. 2014 63 73-82. https://doi.org/10.14257/ijast.2014.63.07

[14] A. Dongre R. Kandoori, Performance of SCC with partial replacement of fine aggregate with waste glass powder, 33rd National Convention of Architectural Engineers and National seminar on "Architectural Engineering Aspect For sustainable Building Envelopes, The Institution of Engineers and Indian Association of Structural Engineers, Institution of Engineers, Khairatabad, Hyderabad (2017).

[15] T.Muralidharan, V.Aravind, B.Logesh, C. Manikandan, Partial replacement of fine aggregate in concrete using ceramic waste, SSRG International Journal of Civil Engineering (SSRG-IJCE) - Special Issue ICETSST - April 2018, 42-48.

[16] S. Ronaldo Gallardo, Q.Mary Ann, Adajar, Structural performance of concrete with paper sludge as fine aggregates partial replacement enhanced with admixtures, IOSR Journal of Engineering 3 11 (2006) 6-15.

[17] D.K. Panesar, B.Shindman The mechanical, transport and thermal properties of mortar and concrete containing waste cork, Cem Concr Compos. 34 (2012) 982-992. https://doi.org/10.1016/j.cemconcomp.2012.06.003

[18] B.H.Sad, Y.D. Amartey, S.Bako. An Investigation into the use of groundnut shell as fine aggregate replacement. Niger J. Technol. 32 1 (2013) 54-60.

[19] S.Muthulakshmi, S.G. Uma G.Hemalatha, Feasibility study of compost as partial replacement of fine aggregate in concrete. Mater. Today, March 2021. https://doi.org/10.1016/j.matpr.2021.02.020

[20] H.D. Saleem, T.Elsokary, Effect of elevated temperature on physic mechanical properties of blended cement concrete, Constr Build Mater. (2011), 25 2, 1009-1017. https://doi.org/10.1016/j.conbuildmat.2010.06.078

[21] O.Arioz, Effect of elevated temperature on properties of concrete, Fire Safety Journal, 42 8 (2007) 516-522. https://doi.org/10.1016/j.firesaf.2007.01.003

[22] M.Tuffail, K.Shahzada, B.Gencturk, J.Wei (2017), 'Effect of elevated temperature on mechanical properties of lime stone, quartzite, granite concrete, Int. J. Concr. Struct. Mater 11 1 (2017) 17-28. https://doi.org/10.1007/s40069-016-0175-2

[23] B.Demerial, O.Kelestemur, Effect of elevated temperature on the mechanical properties of concrete produced with finely ground pumice with silica fume, Fire Saf J. 45 (2010) 385-391. https://doi.org/10.1016/j.firesaf.2010.08.002

Materials Research Proceedings **23** (2022) 68-75 https://doi.org/10.21741/9781644901953-9

Evaluation of Mechanical Properties on Light Weight Concrete by using Silica Fume with M-Sand

V.M. Gnanasundar[1,*], T.S. Palanisamy[2], M. Refak Afrith[3] and B. Pradeepkumar[3]

[1]Bannari Amman Institute of Technology, Erode, Tamil Nadu, India

[2]National Institute of Technology- Karnataka, India

[3]Bannari Amman Institute of Technology, Erode, India

* gnanasundar@bitsathy.ac.in

Keywords: Light Wight Concrete, Density, Silica Fume, M-Sand, Roof

Abstract. Light weight Concrete (LWC) is that the building material employed in the development of building utilization the most recent technology to cut back the self-weight of building. Silica fume is added for achieve strength. Silica fume is added in the percentage of 0%, 10%, 15% and 20%. Light weight Concrete prepared by exploitation light weight combinations (pumice stone) or volcanic stone or silicon oxide. Admixture metal powder as associate in nursing air-entraining agent to the conventional combine concrete. Light weight concrete is restricted to sure functions compared to traditional concrete. However, the introduction of light weight concrete offers additional different to the development business, that presently focuses on natural resources. Light weight concrete plays a distinguished role in reducing the density and to extend the thermal insulation. The density of light weight concrete varies from 1440 to 1840kg/m^3. By exploitation the sunshine weight concrete it minimizes the earthquake or any environment impact. Generally, light weight concrete has wonderful thermal and sound insulating properties, an honest hearth rating, non-combustible and offers price savings through construction speed and simple handling. Then light weight concrete is great for rooftop deck fixes, support profiles, raised floor chunks, and floor deck overlays. Light weight concrete has a lower temperature move rate than typical weight concrete, bringing about better protection. The principal advantage of lightweight cement is that it is incredibly fast and comparatively easy in construction. Light weight combination particle strength varies with kind and strength. In some cases, compressive strength may be exaggerated by commutation a part of the fine light weight combination with smart quality natural sand.

Introduction

Lightweight concrete (LWC) has a density with less than 2000 kg/m3 and a compressive strength of more than 30 N/mm2. The problem in creating LWC is to lower the density while retaining strength and avoiding cost increases. Adding various types of lighter aggregates to the matrix is a standard approach to reduce the density of concrete. The result of the assembling of basic silicon or silicon compounds in electric circular segment heaters is silica seethe, which is a truly fine shapeless silica. Silica smolder, otherwise called consolidated silica, volatized silica, miniature silica, or essentially silica dust, is a dim hued powdered that is like Type 1 ASTM concrete [1]. Silica rage (SF) is a notable pozzolan with cementitious abilities that builds the mechanical properties of substantial plans. It likewise assists with expanding the toughness of cement and ensures it against inserted steel erosion in threatening circumstances [2].

Different examinations have been led on the use of silica rage in substantial blends. As silica smolder is added to light weight concrete, it improves its solidarity. The impact of silica smolder in light weight concrete on long haul concrete compressive strength was investigated [4]. They saw that during the hydration cycle, silica seethe concrete is delicate to temperature varieties, making the substantial glue solidify. The pace of solidarity is eased back when the temperature goes beneath the ideal. At the point when the temperature rises, notwithstanding, the pace of solidarity gain speeds up. Not set in stone that for improved strength, the ideal silica smolder fixation has all the earmarks of being somewhere in the range of 15 and 20%. Adding silica smoke to substantial examples improved the modulus of crack/flexural rigidity, separating flexibility and compressive strength [6&7]. The proportion of water to cementitious in the substantial blend impacts the ascent in strength. How much silica smolder in the substantial blend, then again, impacts its helpfulness. At 28 days, silica smolder replacement in the scope of 10to20 wt. percent of concrete raised compressive strength by 11.65–18.90 percent when contrasted with control substantial examples. As per past study, the best flexural rigidity, parting elasticity, and compressive strength accomplished at 7and28 days old enough with 10to16 wt.

Revealed that with a 12 percent relocation of silica smolder, the best pace of compressive strength upgrade was up to around 8.1 percent, 12%, and 12.8 percent at 7, 28, and 90 days old enough, separately, when differentiated to a control substantial combination [8]. Pozzolanic minerals are a wellspring of nebulous silica in substantial blends. The ascent in strength is expected to the pozzolanic collaboration of calcium hydroxide with indistinct silica. Fine pozzolan particles in the blend fill in as nucleation destinations for the precipitation of hydration items, bringing about homogenous glues. Tricalcium silicate cooperates with water to shape calcium silicate hydrate gel and calcium hydroxide during the concrete hydration response. The unreacted week period of calcium hydroxide created by concrete hydration cooperates with undefined silica from pozzolanic materials to deliver additional calcium silicate hydrate gel, which is the vital strength-contributing stage, expanding the mechanical and sturdiness of the construction properties of concrete [9].

The exhibition of mineral admixtures and their effect on substantial quality were assessed. Mineral admixtures, for example, fly debris and silica rage increment the exhibition of cement in most of occasions, as indicated by the consequences of the exploratory tests. Silica seethe, then again, is great for both the short-and long haul properties of cement. Silica rage advances the sturdiness of cement by expanding its protection from compound assaults and the soluble base silica response.

Different advantages of utilizing silica smolder incorporate further developed consistency, explicit gravity, and air entrainment in mortar and concrete glue. Creep and shrinkage, youthful modulus, and dampness relocation are currently totally further developed when silica smolder is added to high-strength concrete.

Experimental work

Material properties

In this experimental investigation, light weight concrete, silica rage, coarse aggregate, fine aggregate, and water. The actual properties of regular Portland concrete as agreeing ASTM C 150 Type I were explored. The fine totals used in this assessment were from Lawrence and had a fine modulus and explicit gravity of 2.65and2.65, individually.

Table 2 records the actual properties of fine totals. Figure 1 portrays the degree bend for minuscule particles utilized in substantial blends. Igneous has been utilized as coarse total, with a most extreme size going through a 20 mm screen size and a particular gravity of 2.68. The physicochemical attributes of coarse totals are recorded in Table 2. Imporient Chemicals Pakistan outfitted the silica seethe. A gravimetric substance investigation of silica smolder found that over 90% of the silica was in a formless condition.

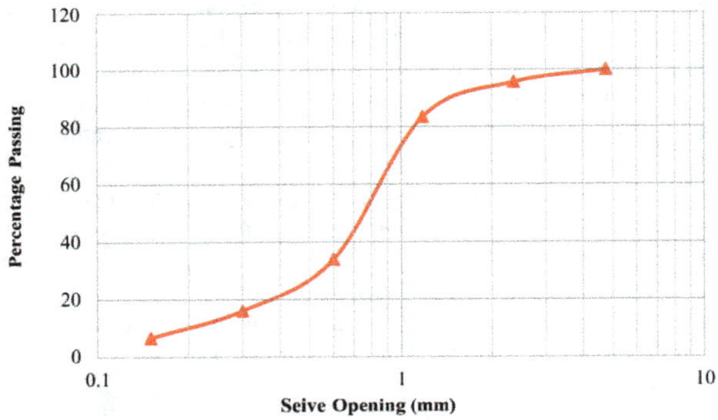

***Fig1.**Gradation curve for fine aggregates*

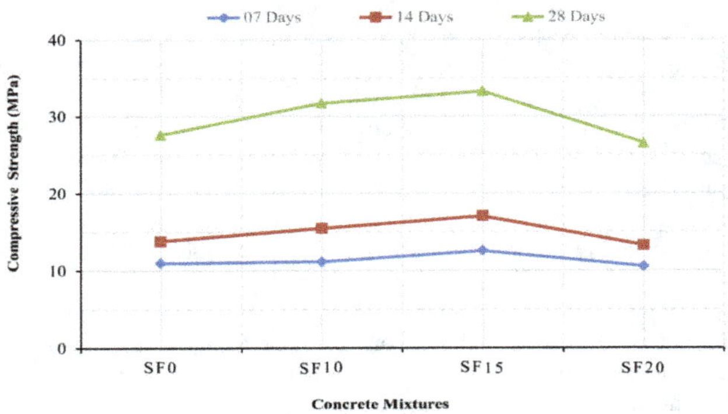

***Fig 2.**Result of compressive strength test at 7,14 and 28-day test*

Materials Research Forum LLC
https://doi.org/10.21741/9781644901953-9

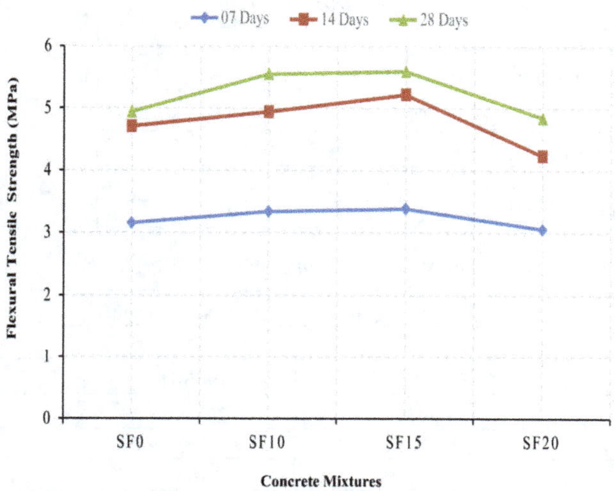

Fig 3.*Results of flexural tensile strength test at 7, 14 and 28 days of age.*

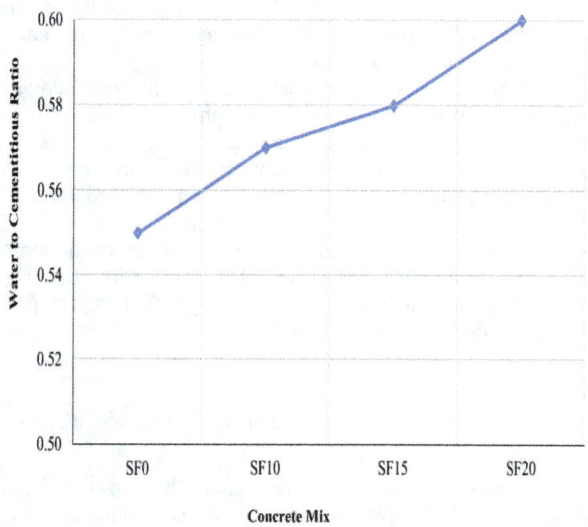

SF10 SF15 SF20

Fig 4.*Water to cementitious ratio of various concrete mixtures.*

Table 1. *Physical properties of cement.*

Property	Cement
Specific Gravity	3.10
Specific Surface Area (m^2/kg)	340
Normal Consistency (%)	29.30
Initial Setting Time (minute)	38
Final Setting Time (minute)	289

Table 2. *Physical properties of fine aggregates and coarse aggregates.*

Property	Fine aggregate	Coarse aggregate
Fineness modulus	2.55	2.832
Dry rodded units	2512.98	1946.15
Specific gravity	2.605	2.608
Absorption (%)	1.507	0.63
Water content (%)	2.02	0.183

Specimens and mixture details

A substantial example's compressive strength and modulus of break were estimated. Every example then, at that point, subbed with 0, 10, 15, or 20% silica rage by weight of concrete. Seven particular periods of examples were assessed: seven, fourteen, and 28 days. The objective control by compressive strength of the substantial blend was determined at 30 MPa following 28 days with a water to fastener proportion of 0.56 and a blend plan proportion of 1: 1.6: 3.5. All examples needed to have a hang of somewhere in the range of 25 and 50 mm. The test examples were restored in water until the day of the test, and the materials were relieved in water until the day of the test. An aggregate of three estimations were gathered for each test. Mean compressive strength was assessed utilizing 150 x 300 mm barrel shaped substantial examples (6 x 12 in). The ASTM C 39 testing strategy was utilized to quantify the compressive strength. Prismatic cement footers with estimations of 100 x 510 mm (4 x 20 in.) were fabricated to appraise the modulus of crack in accordance with ASTM C 78. The substantial mixes are itemized in Table 4. The prefixes SF0, SF10, SF15, and SF20 alludes to substantial examples with 0, 10, 15, and 20 weight percent silica smolder, individually.

Results and discussion

Compressive strength test and modulus of flexural tensile strength test are as discussed below.

Effect on compressive strength

The consequences of the compressive strength test are graphically shown in Figure 2. The inclusion of 10% and 15% silica smolder in concrete helpfully affects the substantial compressive strength, as outlined in Fig.2. Nonetheless, a substantial combination containing 15% silica rage had the increment in compressive strength at all ages. 10.2 percent silica rage by weight of concrete expanded strength by 1.5 percent, 12%, and 15 percent above control examples at 7, 14, and 28 days old enough, separately. At 7, 14, and 28 days old enough, the expansion of 15% silica smolder brought about 14.7 percent, 23.7 percent, and 20.9 percent strength improvement, individually, when contrasted with control examples. The compressive strength of substantial examples was decreased at all ages when additional silica seethe was added, i.e., over 15%. The expansion of silica rage creates heat. By expanding the volume of C-S-H gel shaped, cement's compressive strength can be improved. By joining with free calcium hydroxide present in the combination made because of cementitious materials, silica smolder, which is wealthy in shapeless silica, contributes in the lengthening of the concrete hydration process. A creation and improve of C-S-H gel is shaped because of the connection between silica from mineral admixtures and calcium hydroxide, which contributes essentially to concrete compressive strength. Notwithstanding, in light of the fact that there is just a specific measure of calcium hydroxide in the combination to respond with silica, and there isn't sufficient calcium carbonate in the blend to proceed with the response, the adequacy of adding silica smolder diminishes, and its inclusion just further develops the pressing thickness barely. The expansion 15% silica smoke to substantial blends by weight of concrete expands the compressive strength [11].

Effect on flexural tensile strength

The aftereffects of the flexural elasticity/modulus of crack test agree with the consequences of the pressure strength since modulus of break is straightforwardly associated with compressive strength. Figure 3 portrays the aftereffects of the flexural elasticity test. The steady means 15% silica rage showed the most noteworthy flexural rigidity at all ages. At 7, 14, and 28 days old enough, the expansion of 10% silica seethe helped flexural rigidity by 6.5 percent, 5.5 percent, and 12 percent, individually, when contrasted with control examples. The expansion of 15% silica seethe raised flexural rigidity by 7.5 percent, 11%, and 13 percent at 7, 14, and 28 days old enough, likewise, when contrasted with the control example. Interestingly, the expansion of 20% silica rage brought about a decrease in break modulus.

Table 4.*Mixture proportion for various concrete mix*

Mixture ID	Cement (Kg/m^3)	Silica fume (Kg/m^3)	Sand (Kg/m^3)	Coarse aggregate (Kg/m^3)	Water (Kg/m^3)	Slump (Kg/m^3)	W/C (Kg/m^3)
SF0	35	0	576	1236	194	50	0.55
SF10	31	35	576	1236	194	41	0.57
SF15	29	53	576	1236	194	38	0.58
SF20	28	70	576	1236	194	38	0.60

Effect on water to cementitious ratio

The presence of silica rage upgraded the interest for water in the substantial combination when contrasted with the control example. To acquire the steady droop range, an additional a volume of water decrease specialist was applied dependent on the prerequisite for silica rage. The graphical connection here between W/C proportion and rising water utilization is displayed in Figure 4.

Conclusions

Conclusions are drawn from this experimental research study

1. The light weight concrete with silica fume to concrete improves its compressive strength and flexure tensile strength at all ages.

2. Information on the mechanical properties of high-strength lightweight concrete up to 30 MPa with a corresponding density of 2000 kg/m3 is presented
3. When 15% silica rage was blended in with light weight concrete, compressive strength and flexural elasticity were most elevated at all ages. At the point when 20% silica seethe was added, compressive and flexural elasticity dropped at all ages since there wasn't sufficient calcium hydroxide in the combination to keep the pozzolanic interaction going.
4. Compared to early age strength growth, silica fume aided later age strength gain.
5. In comparison to the control concrete mixture, specimens containing silica fume had a higher water requirement.
6. Silica fumes has a higher fine than concrete and a larger surface area, so the regularity increases exponentially.
7. When compared to our mix the light in weight and compressive strength percentage was found to be increase

References

[1] ACI 116R. Cements and Concrete Terminologies. American Concrete Institute (2005).

[2] Siddique, R. & M.I. Khan. Supplementary Cementing Material. Springer Heidelberg (2011). https://doi.org/10.1007/978-3-642-17866-5

[3] ACI Committee 234. Guide for the use of silica fume in concrete (ACI 234R). ACI Materials Journal 92(4): 437- 440 (1995). https://doi.org/10.14359/978

[4] Aitcin, P.C. & P. Laplante. Long-term compressive strength of silica fume concrete. Journal of Materials in Civil Engineering 2(3): 164-170 (1990). https://doi.org/10.1061/(ASCE)0899-1561(1990)2:3(164)

[5] Newmann, J. & B.S. Choo. Advanced Concrete Technology Constituent Materials. Elsevier Linacre House, Jordan Hill, Oxford (2003).

[6] Katkhuda, H., B. Hanayneh, & N. Shatarat. Influence of Silica Fume on High Strength Lightweight Concrete. World Academy of Science, Engineering and Technology 3(10): 757-764 (2009).

[7] Bhanja, S. & B. Sengupta. Influence of silica fume on the tensile strength of concrete. Cement and Concrete Research 35(4): 743-747 (2005). https://doi.org/10.1016/j.cemconres.2004.05.024

[8] Ganesh, B.K. & P.V.S. Prakash. Efficiency of silica fume in concrete. Cement & Concrete Research 25(6): 1273-1283 (1995). https://doi.org/10.1016/0008-8846(95)00120-2

[9] Detwiler, R.J. & P.K. Mehta. Chemical and Physical Effects of Silica Fume on the Mechanical Behaviour of Concrete. ACI Materials Journal. 86(6): 609-614 (1989). https://doi.org/10.14359/2281

[10] Khedr, S.A. & M.N. Abou-Zeid. Characteristics of Silica-Fume Concrete. Journal of Materials in Civil Engineering, ASCE 6(3): 357-375 (1994). https://doi.org/10.1061/(ASCE)0899-1561(1994)6:3(357)

[11] Bhanja, S. & B. Sengupta. Optimum Silica Fume Content and its Mode of Action on Concrete. ACI Materials Journal 100(5): 407-412 (2003). https://doi.org/10.14359/12816

[12] Ajileye, V.F. Investigations on M. S. (S. F.) As Partial Cement Replacement in Concrete. Global Journal of Researches in Engineering Civil and Structural Engineering 12(1): (2012).

[13] Amudhavalli, N., & K.J. Mathew. Effect of silica fume on strength and durability parameters of concrete, International Journal of Engineering Sciences & Emerging Technologies 3(1): 28-35 (2012).

[14] Sabale, V.D., M.D. Borgave, & S.D. Shinde. Study the Effect of Addition of Silica Fume on Properties of High Strength Concrete. International Journal of Engineering Research & Technology 3(1): 267-270 (2014).

Mechanical Properties of Fiber Reinforced Concrete by using Sisal Fiber with M-Sand as Fine Aggregate

V.M. Gnanasundar[1,*], T.S. Palanisamy[2], G.S. Thirugnanam[3], V. Preetha[1]

[1]Bannari Amman Institute of Technology, Erode, Tamil Nadu, India

[2]National Institute of Technology- Karnataka, India

[3]Builders Engineering College, Erode, Tamil Nadu, India

*gnanasundar@bitsathy.ac.in

Keywords: Sisal Fiber, Mechanical Properties, Sisal Strands

Abstract. Conventional concrete has a low tensile strength, constrained ductility and little protection from crack propagation. The inward miniaturized scale of cracks, prompting weak disappointment of concrete. Investigations have been carried out in many countries on various mechanical properties, physical performance and durability of cement-based matrices reinforced with naturally occurring fibers including sisal, coconut, jute, bamboo, and wood fibers. Raised natural mindfulness and an expanding worry with an unnatural weather change have invigorated the search for materials that can supplant traditional engineered fiber. Characteristic fiber, for example, sisal strands show up as one of the great options since they are accessible in sinewy structure and can be separated from plant leaves, stalk, and products of the soil at exceptionally low expenses. In this work, the impact of sisal strands on the quality of cement for M25 evaluation has been examined by shifting the level of filaments in concrete. Fiber substance were shifted by 0.05%, 0.10%, 0.15%, 0.20%, 0.25%, 0.30%, 0.35% and 0.40% by volume of cement. Cubes, Cylinder and Prism were cast to assess the Compressive, Split Tensile and Flexural Strength test. Every one of the samples was tested for a time of 28 days curing. The results of fiber reinforced concrete for 28 days curing with a varied percentage of fiber were studied and it has been found that there is significant strength improvement with addition of sisal fiber in concrete.

Introduction

Concrete is weak in tension and has a brittle character. The concept of using fibre to improve the characteristics of construction materials is practiced from olden days. The quality of concrete is determined by its mechanical properties as well as its ability to resist deterioration. The strength of the concrete is most important characteristics, as it has strong relationship with quality. Also the stress-strain behaviour of concrete depends upon the properties of the material with which the concrete is made and the loading parameters.

Fibre Reinforced Concrete (FRC) is concrete containing fibrous material which increases its structural integrity. It contains short discrete fibre that are uniformly distributed and randomly oriented. Fibre include steel fibre, glass fibre, synthetic fibre and natural fibre. The use of various fibre which lead varying properties to the concrete. In addition, the character of fibre reinforced concrete changes with varying concretes, fibre materials, geometries, distribution, orientation, and densities

Sisal fibre is derived from the leaves of the plant. It is usually obtained by machine decortications in which the leaf is crushed between rollers and then mechanically scraped. The fibre is then washed and dried by mechanical or natural means. The dried fibre represents only

4% of the total weight of the leaf. Once it is dried the fibre is mechanically double brushed. Fibre is subsequently cleaned by brushing. Dry Fibre are machine combed and sorted into various grades, largely on the basis of the previous in-field separation of leaves into size groups. Mathur (2005) were studied on the building materials such as laminates/panels, doors, roofing sheets, shuttering and dough moulding compound prepared by using natural fibre composites. In this study, the deflection in the plate was 1.95 mm at upon initial loading and 2.57mm after 7 days. This shows good improvement than jute fibres.

Flavio et al.,(2009) investigated the Cracking mechanisms in durable sisal fibre reinforced cement composites. In this paper the sisal fibre with 3% reinforcement as used. A matrix with partial cement replacement by 30% of metakaolin and 20% of calcined waste crushed clay brick was used in order to improve the durability aspects. The post crack modulus decreases at the range compared to the elastic response, and the cracks are arrested within the tensile region which leads to high mechanical performance and energy absorption capacity. Funga (2009) experimented an investigation on the processing of sisal fibre reinforced polypropylene composites. In this paper, a pre-impregnation technique has been introduced for the injection moulding of sisal fibre reinforced polypropylene (PP/SF) composites. Young's modulus by about150%, and the tensile strength was increased by about 10%. Specimens injection moulded using this temperature profile will be called HT (for high temperature) specimens. The reinforcement efficiency for LT specimens is slightly higher than HT samples. Romildo et.al.,(2010) experimented the Physical and mechanical properties of durable sisal fibre cement composites.

The ultimate bending in hot-water immersion is 3.8 times higher and the toughness 42.4 times higher than pc composites. Reis(2012) reported that the fracture mechanics approach of Sisal fibre polymer mortar composites. In this paper, the untreated sisal Fibre and with different surface treatment as reinforcing agent in polymer mortars were made and its fracture propertieswere calculated. Untreated and surface treated sisal Fibre when used as reinforcement contributes significantly to improve the fracture properties both energy and toughness of epoxyand unsaturated polyester polymer mortars.

Materials and methodology

In the present study, sisal fibres were obtained from the Tokyo fibres, Ltd, Coimbatore, India. The fibre were chopped in to 12mm size with a diameter of 300µm.

Cement: Ordinary Portland cement (OPC 53) grade conforming to IS 12269-1987 was used.The specific gravity of cement is 3.115.

Fine Aggregate: River sand was as fine aggregate in the study. It was properly graded to give the minimum voids ratio and shall be free from deleterious materials like clay, silt content and chloride contamination.

Table 1. Properties of river sand

Test for Fine Aggregate	Relevant Code	Results
Specific gravity	IS:2386 -1963 (Part I)	2.65
Bulk density	IS:2386 -1963 (Part III)	1721.32 kg/m^3
Fineness Modulus	IS:383-1970	2.39

Sisal fiber: Sisal fibre is one of the most widely used natural fibres and is very easily cultivated. It has short renewal times and grows wild in the hedges of fields and railway tracks. Nearly 4.5 million tons of sisal fibre is produced every year throughout the world. Tanzania and Brazil are the two main producing countries. Sisal fibre is a hard fibre extracted from the leaves of the sisal plant (Agave sisalana). Though native to tropical and sub-tropical North and South America, sisal plant is now widely grown in tropical countries of Africa, the West Indies and the Far East. Sisal fibres are extracted from the leaves. A sisal plant produces about 200±250 leaves and each leaf contains 1000±1200 fibre bundles which are composed of 4% fibre, 0.75% cuticle, 8%dry matter and 87.25% water. So normally a leaf weighing about 600 g will yield about 3% by weight of fibre with each leaf containing about 1000 fibres.

Table 2 . Chemical Composition of Sisal Fibre

Parameter		Percentage
Cellulose		65%
Hemicelluloses		12%
Lignin		9.9%
Waxes		2%
Total		100%

Casting and Curing of fibre reinforced concrete

In this work, the cube, cylinder and prism were casted with and without addition of fibres. The fibres percentages are added by volume of concrete. Each three specimens were casted for various fibre percentage added in concrete. The beams are casted for optimum percentage based on the test results from various fibre percentage. The casted specimens were demoulded on the next day and kept in water for 28 days.

Testing for fresh and hardened concrete

The fresh and hardened concrete with and without sisal fibre were tested by various methods in the following.

Slump cone test

The concrete slump test is an empirical test that measures the workability of fresh concrete. It measures the consistency of the concrete in that specific batch. This test is performed to check the consistency of freshly made concrete.

Compaction factor test

Compacting factor of fresh concrete is done to determine the workability of fresh concrete by compacting factor test as per IS: 1199 – 1959. The apparatus used is Compacting factor apparatus.

Table 3. Slump and Compaction Value

S.No	Mix Identification Fibre%	Slump value (mm)	Compaction factorvalue
1	0	72	0.80
2	0.05	70	0.82
3	0.10	70	0.82
4	0.15	70	0.82
5	0.20	68	0.82
6	0.25	68	0.82

Compressive strength test

The test was conducted as per IS 516-1959. The cubes of standard size 150 X 150mm X 150mm were used to find the compressive strength of concrete. Specimens were placed on the bearing surface of CTM, of capacity 200 tones without eccentricity and a uniform rate of loading of 140 kg/cm per minute was applied till the failure of the cube. The Compressive strength test is shown in Fig-1

Split tensile strength test

This test was conducted ad per 5816-1970. The cylinders of standard size 150mm diameter and 300mm height was placed on the CTM with capacity 200tonnes, with the diameter horizontal. At the top and bottom two strips of wood where placed to avoid the crushing of concrete specimen at the points where the bearing surface of the compression testing machine and the cylinder meets. The Split tensile strength test is shown in Fig-2.

Flexural strength test

SFRC beams of size 100 X 100 X 500mm were tested using a servo controlled UTM as per the procedure given in ASTM-78.

Results and discussions

Test results of compressive strength, split tensile strength, flexural strength test.

Compressive strength

Irrespective of type of mix, all the specimens showed increased compression values as age of concrete increases and maximum values are reported for 28 days aging. In all curing ages, the maximum compressive strength is achieved for M3 specimens and this increase is almost 16.5% and 13% greater than that of reference specimen and sisal reinforced specimen without polymer. Also, it is noted that M1 specimen shows greater strength next to M3, which expresses the superior performance of latex in terms compressive strength than specimens prepared without latex addition. Moreover, the strength attainment of latex polymer modified concrete was found to be quick when compared with concrete without latex. Thus from the results, it is noticed that the 28 days compressive strengths of latex concrete ranges from 1.15-1.19 times and concrete without latex is 1.85-1.96 times that of 7 days strength. Therefore, reference concrete and sisal fibre concrete modified with latex polymer is suitable for repair works and other works which requires earlier strength attainment.

Split tensile strength

As that of compressive strength, split tensile strength also increases with increase of age of specimens. Also, the entire specimen shows greater strength when compared with reference specimen irrespective of the ages of concrete. From that, the concrete prepared with latex and sisal presented the maximum split tensile strength and the improvement of strength of 28 days specimens are compared and is almost 36% higher than that of reference specimen (M0) and 15.5% than unmodified sisal fibre concrete (M2). Also the concrete modified with latex (M1) shows strength increment upto 13% than M0 but 4% decrement than sisal fibre concrete (M2). Inall respects, sisal fibre reinforced concrete with and without latex shows better performance under split tensile test.

Flexural strength

Under all ages of curing, concrete specimens offered maximum strength for specimens prepared with latex and sisal fibre. Also, as in previous cases, increase of flexural strength is observed with increase of curing period. The maximum strength attainment is achieved by M3 specimens under all curing ages and the 28 days cured specimens are compared here. Therefore, the M3 concrete showed 28% and 12% higher strength than the reference and M2 concrete specimens. The strength variation under flexure follows the same trend as in case of split tensile strength. Next to M3 concrete M2 shows better performance and is almost 15% and 2% greater than M0 and M1 respectively. From this, it is evident that the incorporation of sisal fibre imparts greater strength under flexural loading.

Table 4 : Strength Results

Mix ID	**Compressive strength (N/mm^2)**			**Split tensile strength (N/mm^2)**			**Flexural strength (N/mm^2)**		
	7 days	14 days	28 days	7 days	14 days	28 days	7 days	14 days	28 days
M0	21.76	37.62	42.70	2.19	2.44	2.63	4.36	5.06	5.12
M1	38.5	41.71	45.96	2.34	2.68	2.98	4.91	5.23	5.77
M2	23.08	39.19	44.14	2.78	3.07	3.10	5.29	5.62	5.88
M3	43.15	46.99	49.74	2.97	3.27	3.58	5.72	5.88	6.56

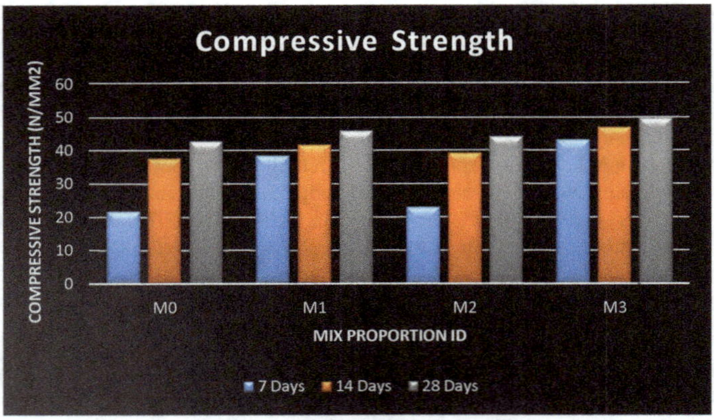

Fig: 1 Compressive Strength results

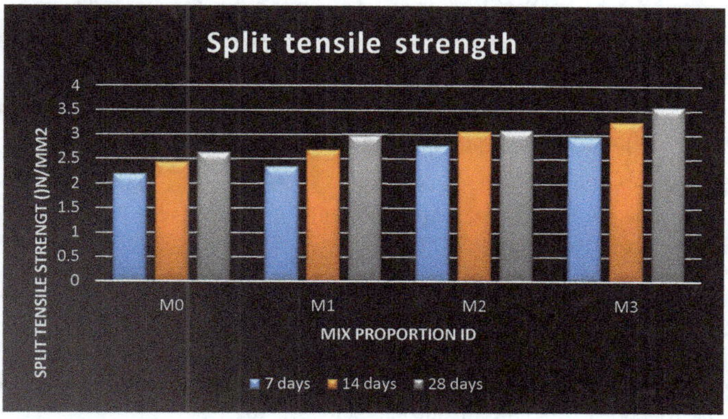

Fig: 2 Split Tensile Strength results

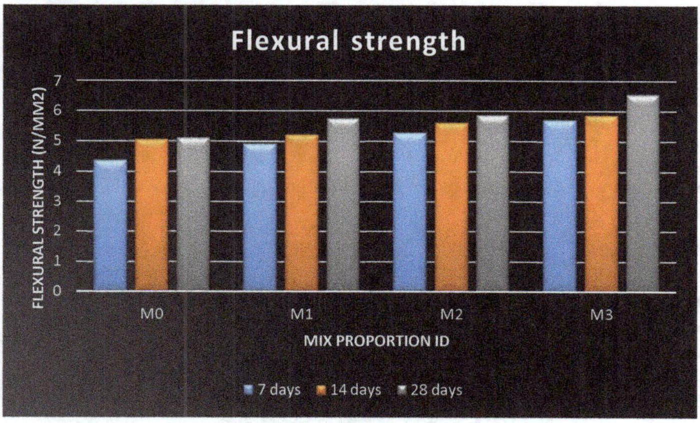

Fig: 3. Flexural Strength results

Conclusion

Based on the test results

The conclusion of the project work is

- The optimum fibre content is at 1.0% fibre content
- The compressive strength, split tensile strength and flexural strength of fibre reinforced concrete shows higher value than conventional concrete.
- The design of sisal fibre reinforced concrete proceeds from knowledge of basic properties under tensile, compression and flexural force, coupled with estimates

of behaviour under secondary loading effects such as creep, thermal and moisture movement.

References

[1] Mathur.V.K. "Composite materials from local resources", Construction and Building Materials 20 (2006) 470-477. https://doi.org/10.1016/j.conbuildmat.2005.01.031

[2] Di Bella.G, Fiore.V , Galtieri.G , Borsellino.C , Valenza.A, "Effects of natural Fibre reinforcement in lime plasters", Construction and Building Materials 58 (2014) 159-165. https://doi.org/10.1016/j.conbuildmat.2014.02.026

[3] L. K. Aggarwal, Cement & Concrete Composites, Bagasse reinforced cement composites, (2005) Vol. 17, pp 107-112. https://doi.org/10.1016/0958-9465(95)00008-Z

[4] G. Ramakrishna and T. Sundararajan, Impact strength of a few natural fibre reinforced cement mortar slabs: a comparative study, Cement & Concrete Composites (2005) , Vol. 27 pp. 547-553. https://doi.org/10.1016/j.cemconcomp.2004.09.006

[5] A. Kriker, G. Debicki, A. Bali, M. M. Khenfer, and M. Chabannet, Mechanical properties of date palm fibres and concrete reinforced with date palm fibres in hot-dry climate, Cement & Concrete Composites (2005) Vol. 27 pp. 554-564. https://doi.org/10.1016/j.cemconcomp.2004.09.015

[6] IS3025 (Part II)- 1983, Indian Standard methods of sampling and test (Physical and Chemical) for water and waste water (part ii) ph value (first revision).

[7] IS 12269: 2013, Indian Standard ordinary portland cement, 53 grade specification (first revision).

[8] IS 383 - 1970, specification for coarse and fine aggregates from natural sources for concrete (second revision)

[9] A. Kriker, G. Debicki, A. Bali, M. M. Khenfer, and M. Chabannet, Mechanical properties of date palm fibres and concrete reinforced with date palm fibres in hot-dry climate, Cement & Concrete Composites, (2005), Vol. 27 pp. 554-564,. https://doi.org/10.1016/j.cemconcomp.2004.09.015

[10] Didier Snoeck and Nele De Belie, Mechanical and self-healing properties of cementitious composites reinforced with flax and cottonised flax, and compared with polyvinyl alcohol fibres, Biosystems Engineering (2012), Vol. 111, pp. 325-335. https://doi.org/10.1016/j.biosystemseng.2011.12.005

[11] G. Ramakrishna and T. Sundararajan, Long-Term Strength and Durability Evaluation of Sisal Fibre Composites Part-I: Cement Mortar Composites, International Journal Of Civil Engineering And Technology (IJCIET), Volume 4, Issue 1, January- February (2013), pp.76-86.

Mechanical Properties of Concrete Using Oyster Shells as Coarse Aggregate

P. Kanaka[1,a,*], S. Southamirajan[1,b], S. Elango[1,c], K. Samruda[1,d], R.A. Akshaya[1,e]

[1]Department of Civil Engineering in Kongunadu College of Engineering and Technology, Trichy, Tamilnadu, India

[*a]kanakasivam@gmail.com, bsouthamirajan@gmail.com,cs.elango@ksrce.ac.in, [d]ksamruda@gmail.com, [e]akshayasaspa23@gmail.com

Keywords: Coarse Aggregate, Fine Aggregate, Oyster Shell, Industrial Waste, Cement Paste

Abstract. In order to investigate the recycling possibilities of coarse oyster shell aggregate, the chemical compound of oyster shell and reactivity of oyster shell with cement paste was examined. Specifically, the mechanical characteristics of fresh concrete and hardened concrete were quantified in terms of oyster shell substitution rates. Test results indicate that there is no interaction between oyster shell and cement paste and that concrete's workability decreases as the substitution rate for oyster shell increases. The decrease was about 20% at a substitution rate of 30%.

Introduction

Concrete is one of the most important materials in developing nations, as the construction industry is the backbone of all other industries. As the most consumed building material, concrete production depends on cement, sand, fine aggregates, and coarse aggregates. As the consumption of concrete grows, the demand for coarse aggregate increases in countries with high infrastructure growth. The mechanical properties of oyster shell make it a suitable substitute for coarse aggregate. Sieved shells can be used for both the waste industry and construction industry. In order to increase the strength of concrete, oyster shell will be added in different variations of 20% and 30% to M20 grade of concrete. Aggregates are inert filter materials of either the coarse or fine variety. Several studies suggesting that Indian oyster shell can be used as a construction material have recently been conducted. Greater usage of oyster shells will cause now no longer most effective saving such creation fabric however additionally assists in fixing the hassle of disposal of this waste products. So, the maximum economical, ecological and boom the convenience of labor creation of the shape is crucial within side the gift economy.So the characteristic of concrete has come into the sector.

As modern engineering practices become extra traumatic, there's a corresponding need for unique forms of cloth with novel residences. Such materials are called composite cloth as it's miles a aggregate of various substances. As current engineering practices grow to be extra traumatic, there's a corresponding need for special sorts of cloth with novel houses.Scientists, engineers and technologists are constantly on the attempting to find substances, which can act as substitute for conventional materials or which own such homes as could allow new design and innovations resulting in to financial system, in order that a shape can be built economically. Many attempts were combination of or extra materials. Such substances are called composite cloth as it's far a combination of different substances.

Materials Testing Program

That allows you to observe the conduct of concrete produced with locally available oyster shell and also recognize the pressure –pressure conduct of a light weight concrete a total number of cube 6 and beam 6 mixes were tried. Six mixes of concrete with 20%,30% extent of oyster shell. And ultimate of coarse aggregate and first-class combination a complete wide variety of 12 specimens became prepared. In all the mixes the equal type of high-quality aggregate i.e. easy river sand has been used.

Oyster Shell

The domestically to be had crushed oyster shell has been used as a rough aggregate on this research. The maximum mixture length of 20mm is used. The experiments are executed to locate the residences of oyster shell i.e,sieve evaluation, particular gravity, density and water absorbtion.The take a look at has been accomplished as according to the process given in IS code.

Figure 1 Oyster Shell

Table 1 Specific Gravity of Fine Aggregate

Sample Number	W1 in kg	W2 in kg	W3 in kg	W4 in kg	Specific Gravity
1	0.632	1.120	1.802	1.512	2.4
2	0.632	1.128	1.746	1.512	2.0
3	0.632	1.122	1.838	1.512	2.8
				Average	2.4

Table 2 Specific Gravity of Coarse Aggregate

Sample Number	W1 in kg	W2 in kg	W3 in kg	W4 in kg	Specific Gravity
1	0.632	1.110	1.702	1.512	2.6
2	0.632	0.978	1.746	1.512	3.0
3	0.632	1.132	1.738	1.512	2.7
				Average	2.6

Table 3 Specific Gravity of Oyster Shell

Sample Number	W1 in kg	W2 in kg	W3 in kg	W4 in kg	Specific Gravity
1	0.632	1.714	1.960	1.512	1.68
2	0.632	1.888	2.060	1.512	1.75
3	0.632	1.980	2.306	1.512	2.37
				Average	1.93

Table 4 Sieve Analysis of Oyster Shell

IS Sieve size	Weight retained (gm)	% weight retained	Cumulative % of weight retained	Cumulative % of passing
80mm	0	0	0	100
40mm	98	98	4.9	95.1
20mm	1590	1688	84.5	15.6
10mm	305	1993	99.75	0.35
4.75mm	7	2000	100	0
2.36mm	0	-	100	-
1.16mm	0	-	100	-
600 mic	0	-	100	-
300 mic	0	-	100	-
150 mic	0	-	100	-
Total	2000	-	822.95	-
Fineness modulus of coarse aggregate = 7.32				

Compressive Strength Testing

Table 5 Compressive Strength of Concrete at 7 Days Curing

Specimen no.	Load (kN)	Compressive strength (N/mm^2) 7 days curing	Type of specimen
1	385	17.11	Conventional concrete
2	380	16.88	Oyster shell concrete 20% replacement
3	382	16.97	Oyster shell concrete 30% replacement

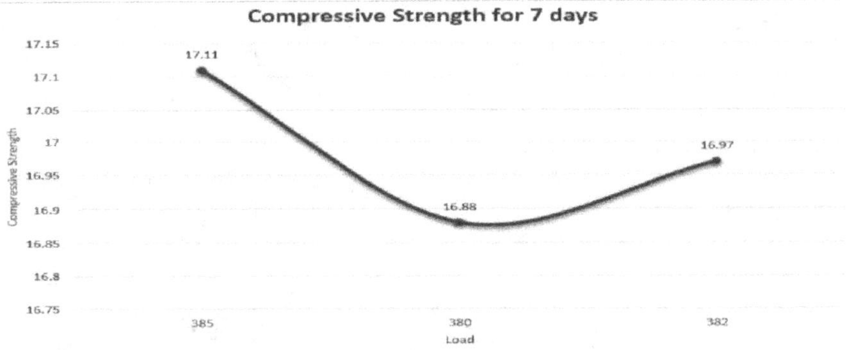

Figure 2 Compressive Strength Graph For 7 Days of Curing

Table 6 Compressive Strength of Concrete at 28 Days of Curing

Specimen no.	Load (kN)	Compressive strength (N/mm^2) 28 days curing	Type of specimen
1	619	27.51	Conventional concrete
2	588	26.13	Oyster shell concrete 20% replacement
3	610	27.11	Oyster shell concrete 30% replacement

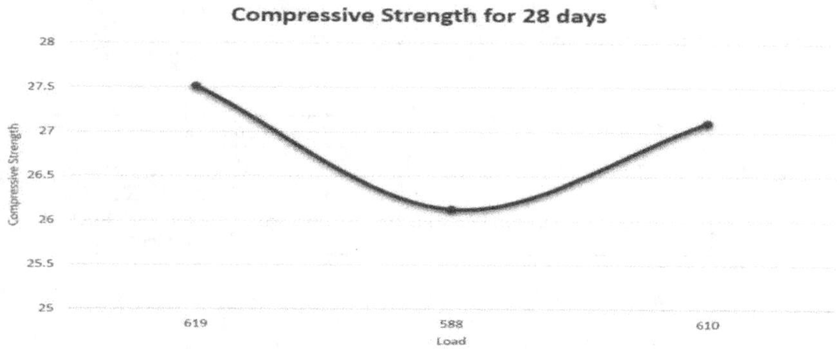

Figure 3 Compressive Strength Graph For 28 Days of Curing

Table 7 Flexural Strength of Concrete at 7 Days Curing

Specimen no.	Load (kN)	Flexural strength (N/mm^2) 7 days curing	Type of specimen
1	5.16	2.58	Conventional concrete
2	4.72	2.36	Oyster shell concrete 20% replacement
3	4.20	2.1	Oyster shell concrete 30% replacement

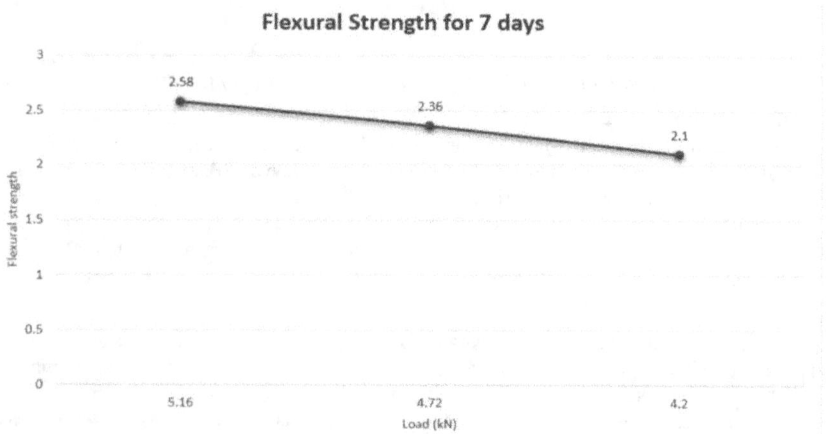

Figure 4 Flexural Strength Graph For 7 Days of Curing

Table 8 Flexural Strength of Concrete at 28 Days Curing

Specimen no.	Load (kN)	Flexural strength (N/mm^2)28 days curing	Type of specimen
1	8.32	4.16	Conventional concrete
2	7.46	3.82	Oyster shell concrete 20% replacement
3	7.40	3.71	Oyster shell concrete 30% replacement

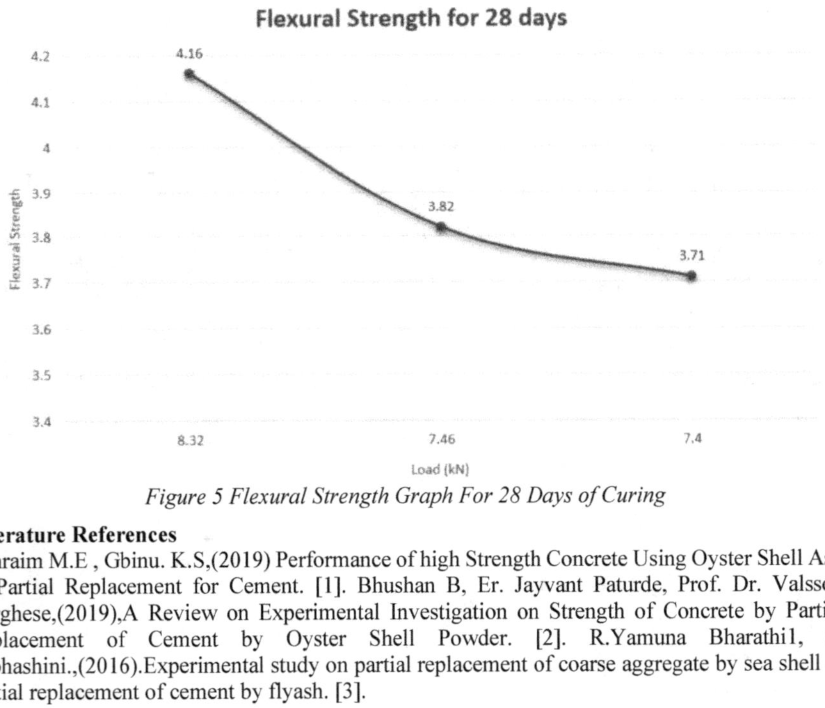

Figure 5 Flexural Strength Graph For 28 Days of Curing

Literature References

Ephraim M.E , Gbinu. K.S,(2019) Performance of high Strength Concrete Using Oyster Shell Ash as Partial Replacement for Cement. [1]. Bhushan B, Er. Jayvant Paturde, Prof. Dr. Valsson Verghese,(2019),A Review on Experimental Investigation on Strength of Concrete by Partial Replacement of Cement by Oyster Shell Powder. [2]. R.Yamuna Bharathi1, S. Subhashini.,(2016).Experimental study on partial replacement of coarse aggregate by sea shell & partial replacement of cement by flyash. [3].

Summary

It has been determined that by using growing the oyster shell content material the electricity values tend to increase. in this project we attempted to replace the coarse combination partially via oyster shell (20%, & 30%) respectively to increase the energy of concrete. but the strength is equal with the traditional concrete only at 20% replacement of mixture via oyster shell as a result increasing the amount of oyster shell past 20% reduces the compressive and flexural too. these project paves a way for defensive our grave which is one of the renewable electricity supply and make a 20% cut off in construction field. The power is progressively decreasing at 30% replacement of oyster shell. So we finish that the coarse mixture changed with oyster shell at 20% in concrete is suitable for creation. Furthermore it reduces the development fee by way of lowering the cost of coarse combination and it also reduces the environmental pollutants because of oyster shell.

References

[1] Ephraim M.E, Gbinu. K.S,(2019) Performance of high Strength Concrete Using Oyster Shell Ash as Partial Replacement for Cement.

[2] Bhushan B, Er. Jayvant Paturde, Prof. Dr. Valsson Verghese,(2019),A Review on Experimental Investigation on Strength of Concrete by Partial Replacement of Cement by Oyster Shell Powder.

[3] Abinaya, S.Prasanna venkatesh ,(2016),An Effect on Oyster Shell Powder's Mechanical Properties in Self Compacting Concrete.

[4] R.Yamuna Bharathi1, S. Subhashini.,(2016).Experimental study on partial replacement of coarse aggregate by sea shell & partial replacement of cement by flyash.

[5] K.P.Ravikumar, M.Karuppiah, I.ArunRaj (2012) Experimental Investigation of Concrete in Partial Replacement of Coarse Aggregate and Cement by Sea Shell and Prosopis Juliflora Ash.

[6] Hai-Yan CHEN, Leo Gu, Zhen-Min .Effects of Crushed Oyster Shell on Strength and Durability of Marine Concrete Containing Fly Ash and Blastfurnace Slag.

Materials Research Proceedings **23** (2022) 90-96 | https://doi.org/10.21741/9781644901953-12

Performance Evaluation of High-Performance Concrete with Steel Fiber

M.P. Thiyaneswaran[1], C. Sowmiyashree[1,a], P. Thiruvenkatesh[1,b], A. Sushanth[1,c]

Civil department, Kongu Engineering Collage, Erode, TN, India

[a]sowmiyashreec.18civil@kongu.edu, [b]thiruvenkateshp.18civil@kongu.edu, [c]sushantha.18civil@kongu.edu

Abstract. Cement concrete is a composite and widely used construction material. The motive of this analysis is based on examination of the use of steel fiber in high performance concrete to enhance the mechanical properties of concrete. The utilization of steel fiber strengthen the concrete by resisting the tensile cracking, enhance the mechanical properties, durability and the serviceability of structure. But higher the utilization of steel fiber can decrease the workability of concrete. In such case, fly ash is used in a range of 20-30% to enhance the workability. Also, the use of fly ash can reduce the greenhouse gas emissions. Thus it creates a sustainable environment. Tests were conducted by preparing the mixes and were compared with the normal conventional concrete properties.

Keywords: Steel Fiber, High Performance Concrete, Fly Ash

Introduction

Concrete is one of the most important, highly utilized building material in the world. Concrete is a mixed material which contains cement, water, aggregates(gravel, sand). When these materials are mixed together, they form a workable paste which then gets hardened over a period of time. Though concrete is an important material, it has it's own consequences. Concrete is a brittle material because of its low tensile strength. The use of plain concrete may also lead to the formation of cracking, spalling, scaling etc. In order to overcome these defects, different types of concrete have been introduced. One such concrete is the HPC.

HPC is generally a quasi brittle material. Utilization of HPC was came into practice in the early 80's. Based on data collected, the life circle of high performance concrete was up to 500 years. High strength, volume stability, low absorption, low heat of hydration, low porosity, good workability and high durability are the features of HPC. With these characteristics, when steel fiber is added to high performance concrete, it is more delicate to loading rate when compared to standard concrete because it's strength gain is greater for the expanding strain rate.

The purpose of adding steel fiber to the concrete is to enhance the durability and toughness of concrete and mortar. Fibers utilized for concrete helps in reducing the shrinkage cracks, increase the overall strength, increasing the energy absorption and decreases the spalling at rise in temperatures. Percentage of steel fiber in concrete should be limited to about 2% of volume because of the reduction in workability. Flexural strength is found to be increased as amount of steel fiber is greater. Compressive and tensile strength also increases with the optimum percentage of steel fiber. Higher percentage addition of steel fiber leads to congestion of fiber which causes balling effect and reduces the bonding with the concrete. It also reduces workability of concrete.

Fly ash is a pozzolan containing siliceous and aluminous materials. This material forms cement when it comes in contact with water. Fly ash in concrete enhance the strength as well as the segregation of concrete and makes it much easier to pump. Concrete which is created with fly ash tends to be more stable and exhibits a higher grade of overall strength than concrete made with conventional cement products. Fly ash enhances workability of plastic concrete, strength , durability of hardened concrete.

OBJECTIVE:

- To determine the mechanical properties of concrete when steel fibers are utilized in concrete.
- To find the viability of using fly ash as a substitute of cement in concrete production.

Overview

Jacek Katzer [1] describes the most interesting and important applications of steel fiber reinforced concrete used worldwide. Over a period of four decades, SFRC has been continuously developed. Because of this, today SFRC is a commercially available and viable construction material. Sivakumar [2] explains experimental results of tests carried on different properties of different fiber reinforced concrete at the low volume portions of fibers with 0.5%. A major importance of these findings is that the steel fibers present in concrete can be partially replaced with the non-metallic fibers but the ductility should not be compromised. Abhijit Wadekar et. al [3] describes that the high strength concrete is slightly a brittle material so that the steel fibers are added to increase its ductility. From this study it is indicated that utilization of steel fibers can enhance split tensile strength of HSFRC as compared with compressive and flexural strength. Abdul Ghaffer et. al [4] explains addition of steel fibers in structural concrete to increase its mechanical properties. The steel fiber in concrete improve the ductility with higher fiber content and the crack widths are less than in plain cement concrete. But workability gets reduced when % of fiber is higher. Soek-Joon Jang et. al [5] explains the results of hooked end steel fibers on mechanical properties of HPC. It also investigates the feasibility of using steel fibers to simplify the critical complicated reinforcement detailing under high shear stress. Khaled Marar et. al [6] discusses the direct shear performance of simple cement and fiber reinforced concrete. It shows that, when the volume portion of fibers improved (0 to 1.5%), the shear strength also improved about 2 times that of simple cement concrete correspondingly. Jagadeesh Kamboj [7] discusses the effects of different types of steel fibers on ductility and mechanical properties of the cement concrete. It is observed that the flexural strength rises with increase of fiber content but up to optimum value and compressive strength also increases but the workability decreases. Gunduz et. al [8] addresses the effects on type of hooked end fibers on HPC by experimental research. Flexural strength and tensile strength tests were conducted. The results shows that these were significant effects on post cracking behavior and fracture energy. The type of hooked end does not affect the fresh concrete properties, however the slump values get decreases. Sivakumar Anandan et. al [9] investigated the mechanical properties of different types of fibers substituted with high strength concrete. The tests were conducted using displacement control to get post peak force displacement relationship. The results showed the effect on reinforcing efficiency on different steel fiber reinforced concrete. Ingrid Lande Larsen et. al [10] explains the effects of fiber reinforcement regarding tensile strength and compressive strength of UPC. The effect depends on the percentage of fiber, type and hybrid combinations. The use of these fiber combinations has the potential to rise compressive, tensile strength of concrete. Ramadoss et. al [11] uses water cement proportions of 0.3 and 0.4, replacement of silica fume at 10% and 15% and crimped steel fibers with aspect ratio 80. Steel fiber addition to silica fume concrete increase toughness, modulus of rupture and resist cracking. Yanxia Ye et. al [12] addresses, the adding of steel strands to concrete have different effects on toughness and mechanical properties. Micro(M), corrugated(C), and end hooked (H) are the steel fibers used. From these types, Micro(M) steel fiber has the significant effect. Jae-Jin Kim et. al [13] explains that the average and equivalent bond strength could be increased using different fibers in HPC. Umberto De Maio et. al [14] addresses about failure analysis of model joined with entrenched truss model to simulate the concrete steel rebase contact. These models are used for calculating load carrying capacity of steel bar-reinforcing concrete with nanoplatelets. Wonsik Shin et. al [15] investigates the effects of preformed micro crack properties and immersion duration in 3.5% NaCl solution on erosion degree of steel strands in HPC. Hamdy Shehab El-Din et. al [16] discusses the effects of various types of steel strands on the mechanical performance of UHPC. Addition of steel fibers has various effects on crack designs, postpone the crack presence and imprison the crack growth in test specimen. Ajimon Thomos [17] describes the behavior of fiber reinforced concrete high performance and high performance external beam column junctions exposed to repeated loading. The trial outcomes showed that ultimate weight of steel fiber reinforced high performance concrete external beam column joins were 15.3% greater that HPC beam column joins. Ayyappa et. al [18] discusses the optimum strength of HPC specimen with or without fiber. The test results shows that the concrete with 30% silica fume and 3% of glass and steel fibers gains the optimum strength than the other proportions. Calik Ozyildirim [19] explains that the steel fibers in UHPC provides tensile volume across cracks, ensuing in high shear volume in twisting associates. UHPC beams with steel strands do not need shear reinforcement. Valcachaela Veera Mnikanta Srikal et. al [20] attempted to find out strength related test like compressive, split tensile and flexural strength using steel fibers with volume of 0.5% and 1%. The steel fiber usage improves concretes mechanical properties such as toughness in tension and durability. Jun Yang et. al [21] aims to determine the compression performance of UHPFRC for thin-walled arch unit plan and a evaluation was made with conventional concrete. It describes that the wall width of 50mm with the stirrups efficiently controlled the instability disappointment of thin walled compression column. Abbas et. al [22] examined that adding of steel fiber varies the failure method of ultra-high performance concrete samples from whole injury or the abrupt blast to some ductile performance. These concrete samples exhibits a high flexural strength properties depending on its mix design. Rui Liu et. al [23] describes that adding steel strands to concrete dramatically increases flexural strength. Steel strands on concrete is uniformly distributed in the matrix in front of the cracking, steel fiber and cement matrix bonds together, as the function of concrete by force. Kiranbala Devi [24] identifies the primary function of fibers is to modify micro and macro cracks and infection crack at their origin and inhibit crack growth. Macdonald [25] observed that the fiber reinforced concrete is considered as a new and challenging composite material. Steel fibers are available in different sixes and lengths depending on aggregate sixes and design requirements. Hamid Behbahani [26] describes that the steel fibers are used to restrict the plastic and drying shrinkage in concrete. Steel fibers can be used as secondary reinforcement used with conventional steel bars or prestressing strand as main reinforcement. Concrete with steel fibers has shown significant improvement in flexural strength. Ajay Dahake [27] aims to find the result of addition of fibers to concrete and determine their outcome on the performance and mechanical properties of reinforced concrete beams. The toughness depends on fiber volume fraction, silica fume content, and fiber aspect ratio. Patodi [28] addresses the performance of concrete with unlike kinds of hybrid fibers used in the concrete mixture. Samples were prepared with and without fly ash and varying volume portion of fibers from 0 to 1%. Best improvement in mechanical properties is concrete having 0.3% re corn and 0.7% of steel fiber. Yuanxun Zheng et. al [29] addresses that the expansion of various kind of steel strands effectively affect the mechanical properties and sturdiness of HLAC. The steel strands utilized are Micro(M), end-snared (H) and corrugated(C). Miniature steel strands have the best supporting exhibition on the mechanical properties and strength is additionally awesome with Miniature steel filaments. Pramod Kawde et. al [30] clarifies that, the regular cement doesn't fulfill the need of the world on account of its low rigidity restricted malleability and less obstruction the breaking. It has been demonstrated that diverse sort of fiber included explicit rate to concrete works on the mechanical properties, solidness and workableness of construction. The outcome showed that the expansion of steel fiber builds a definitive strength and pliability.

Future scope

- In future, first the suitable mix design for the project is prepared.
- The percentage of steel fibers are taken as 0.5%, 1.0%, 1.25%, 1.5%, 1.75%, 2.0%.
- Physical properties of the steel strand reinforced high performance concrete are examined.
- Tests namely compressive, tensile , flexural strengths, durability test were carried out on the specimen. At the end, results obtained from the specimen were observed and further discussed.

Conclusion

The utilization of traditional cement can prompt break development, spalling, pop outs and so forth, High execution concrete is a profoundly fragile material and flops abruptly. The weak method of disappointment is changed by expansion of steel strands in high performance concrete, in to a more bendable one. Concrete has low rigidity, low sturdiness and low explicit strength. Concrete accomplishes indicated compressive strength in 28 days subsequent to projecting and relieving. Original capacity requires appropriate encompassing temperature controlled longer than a month. Pressure face of cement has a cover to ensure the building up steel bars. In the event that elasticity at outrageous fiber surpasses the pliable limit of cement, it breaks and prompts erosion of fortifications. It was seen that steel fiber supported cement works on the substantial flexibility, its post-breaking load conveying limit. Steel strands of various rates 0.5%, 1.0%, 1.25%, 1.5%, 1.75%, 2.0% were added to the substantial to lead the different tests. In reference with the gone through papers, it has been gotten that the utilization of steel filaments at pace of 1.25% has invigorated an ideal. Our task is to add fly ash (partial substitution of concrete) and steel strands with the elite cement. Then, at that point, the tests were directed to notice the flexural, split pliable and compressive strength. The utilization of fly debris works on the manageability of the climate.

References

[1] Jacek Katzer 2006 ,'Steel fibers and Steel fiber reinforced concrete', The Pacific journal of Science and Technology, Vol.7, pp.53-58. http://www.akamaiuniversity.us/PJST7_1_53.pdf

[2] Sivakumar 2011,'Influence of hybrid fibres on the post crack performance of high strength concrete', Journal of Civil Engineering and Construction Technology, Vol.2, pp.147-159. http://academicjournals.org/article/article1379515702_Sivakumar.pdf

[3] Dr.Abhijit P. Wadekar, Prof. Rahul D .Pandit 2014,' study of different types of fibers use in high strength fiber reinforced concrete', International journal of Innovative Research in Advanced Engineering, Vol.1, Issue 9, pp.225-230.

[4] Abdul Ghaffar, Amit S. Chavhan, Dr. R.S Tatwawadi 2014, 'steel fibre reinforced concrete', International journal of Engineering Trends and Technology, Vol.9, pp.791-797. https://doi.org/10.14445/22315381/IJETT-V9P349

[5] Seok-Joon Jang, Dae-Hyun Kang, Kyung-Lim Ahn 2015, 'Feasibillity Of Using High Performance Steel Fiber Reinforced Concrete For Simplifying Reinforcement Details Of Critical members', International journal of Polymer Science. https://doi.org/10.1155/2015/850562

[6] Khaled Marar, Ozgur Eren, Hooman Roughani 2017, 'The Influence Of Amount And Aspect Ratio Of Fibres On Shear Behaviour Of Steel Fiber Reinforced Concrete', KSCE Journal of Civil Engineering, Vol.21, pp.1393-1399. https://doi.org/10.1007/s12205-016-0787-2

[7] Jagadeesh Kamboj 2016, 'Effect of addition of different types of steel fibres on the mechanical aspects of concrete', Chandigarh University,Punjab. http://www.iaeme.com/MasterAdmin/Journal_uploads/IJCIET/VOLUME_7_ISSUE_5/IJCIET_07_05_005.pdf

[8] Y.Gunduz , E.Taskan, Y.Sahin 2016, 'Using hooked end fibres on high performance steel fibre reinforced concrete', WIT Transactions on The Build Environment, Vol.166, pp.265-276. https://www.witpress.com/Secure/elibrary/papers/HPSM16/HPSM16024FU1.pdf

[9] Sivakumar Anandan, Saiful Islam, Roohul Abad Khan 2019, 'Effect of steel fibre profile on the fracture characteristics of steel fibre reinforced concrete beams', Journal of Engineering Research, Vol.7, pp.105-124. https://kuwaitjournals.org/jer/index.php/JER/article/view/6073/287

[10] Ingrid Lande Larsen, Rein Terje Thorstenseen 2020, 'The Influence of steel fibres on compressive and tensile strength of ultra high performance concrete', Construction and Building Materials. https://doi.org/10.1016/j.conbuildmat.2020.119459

[11] P. Ramadoss, K. Nagamani 2013, 'Impact Characteristics Of High Performance Steel Fiber Reinforced Concrete Under Repeated Dynamic Loading',international journal of Civil Engineering, Vol.12, pp.513-520. http://ijce.iust.ac.ir/browse.php?a_id=872&sid=1&slc_lang=fa

[12] Yanxia Ye ,Jilei Liu 2020, 'Experimental study of high strength steel fiber light weight aggregate concrete on mechanical properties and toughness index', Advances in Materials Science and Engineering. https://doi.org/10.1155/2020/5915034

[13] Jae-jin Kim, Yun-sik Jang , Doo-yeol Yoo 2020, 'Enhancing the tensile performance of ultra high performance concrete through mouse novel curvilinear steel fibres',Hongang unweis Republic of Korea. https://doi.org/10.1016/j.jmrt.2020.05.072

[14] Umberto De Maio, Nicholas Fantuzzi, Fabrizio Greco 2020, 'Failure analysis of ultra high performance fiber – reinforced concrete structures enhanced with nanomaterials by using a diffuse cohesive interface approach', Nanomaterials, Vol.10. https://doi.org/10.3390/nano10091792

[15] Wonsik Shin, Doo-yeol yoo 2020, 'Influence of steel fibre corroded through multiple microcracks on the tensile behaviour of a ultra high performance concrete', Construction and Building Materials, Vol.259. https://doi.org/10.1016/j.conbuildmat.2020.120428

[16] Hamdy K.Shehab El-din, Heba A.Mohamed, Mahmoud Abd El-Hak Khater and Sayed Ahmed 2016,'Effect of steel fibers on behaviour of ultra high performance concrete', First International Symposium. https://www.extension.iastate.edu/registration/events/UHPCPapers/UHPC_ID11.pdf

[17] Ajimon Thomas 2014, 'Effect Of Steel Fiber Reinforced High Performance Concrete Exterior Beam Column Joint Subjected To Cyclic Loading', International journal of Engineering Research and Technology, Vol.3, Issue 9, pp.1409-1414. https://citeseerx.ist.psu.edu/viewdoc/download?doi=10.1.1.537.4205&rep=rep1&type=pdf

[18] Ayyappan, R. Kumar 2017,'Mechanical and durability properties of high performance concrete using hybrid fibres', International journal of Scientific Development and Research, Vol.2, Issue 6, pp.122-133. http://www.ijsdr.org/papers/IJSDR1706015.pdf

[19] Celik Ozyildirim 2011, 'Evaluation of ultra high performance fiber reinforced concrete',Virginia center for Transportation Innovation and Research. https://rosap.ntl.bts.gov/view/dot/37801

[20] Vakacharla Veera Mnikanta Srikar , G. Kalyan 2018, 'Performance of concrete with adding of steel fiber', International journal of Engineering and Research Technology, Vol.3, pp.290-308. http://www.ijesrt.com/issues%20pdf%20file/Archive-2018/March-2018/34.pdf

[21] Jun Yang , Hong Zhang 2019, 'Structural Behaviour Of Ultra High Performance Fiber Reinforced Concrete Thin Walled Arch Subjected To Asymmetric Load', Hindawi. https://doi.org/10.1155/2019/9276839

[22] S. Abbas , M.A. Sallem 2016, 'Ultra High Performance Concrete: Mechanical Performance,durability,sustainability and Implimentation Challenge',International journal of Concrete structures and Materials, Vol.10, pp.271-295. https://ijcsm.springeropen.com/articles/10.1007/s40069-016-0157-4

[23] Rui Liu, Yulong Liu 2016, 'Steel Fibre Reinforced Concrete And Its Application Performance', International journal of Multidisplinary Research & Development, Vol.3, Issue 6, pp.341-343. http://www.allsubjectjournal.com/download/2359/3-6-129-578.pdf

[24] Dr.Th.Kiranbala Devi 2013, 'Effect Of Steel Fibre In Reinforced Concrete', International journal of Engineering Research and Technology, Vol.2, Issue 10, pp.2906-2913. https://www.ijert.org/research/effects-of-steel-fibres-in-reinforced-concrete-IJERTV2IS101024.pdf

[25] C.N.Macdonald 2000, 'Plastic And Steel Fibre- Reinforced Concrete Application', Transportation research record. https://onlinepubs.trb.org/Onlinepubs/trr/1984/1003/1003-001.pdf

[26] Hamid Behbahani 2011, 'Steel Fibre Reinforced Concrete', University Putra, Malaysia. https://www.researchgate.net/publication/266174465_Steel_Fiber_Reinforced_Concrete_A_Review

[27] Ajay Dahake 2016, 'Effect Of Steel Fibres On Strength Of Concrete', Journal of Engineering, Science and Management Education, Vol.9, pp.45-51. https://www.researchgate.net/publication/304776998_Effect_of_Steel_Fibers_on_Strength_of_Concrete

[28] S.C. Patodi , C.V. Kulkarni 2012, 'Performance Evaluation Of Hybrid Fibre Reinforced Concrete Matrix' , International journal of Engineering and Applications, Vol.2, Issue 5,

pp.1856-1863.
https://citeseerx.ist.psu.edu/viewdoc/download?doi=10.1.1.637.4713&rep=rep1&type=pdf

[29] Yuanxun Zheng, Xiaolong Wuand Guangxian He 2018, 'Mechanical Properties Of Steel Fibre Reinforced Concrete By Vibrating Mixing Technology' ,School Of Water Conservancy And Environment, Zhengzhou University. https://doi.org/10.1155/2018/9025715

[30] Pramod Kawde , Abhijit Liavudital 2017, 'Steel Fiber Reinforced Concrete' , Impersial College Of Engineering And Research, Vol.3, pp.130-133.
http://www.ijesrt.com/issues%20pdf%20file/Archive-2017/January-2017/17.pdf

Comparative Study on the Behavior of Fiber Reinforced Concrete

S. Ramakrishnan[1,a,*], S. Loganayagan[2,b], N. Chandramohan[2,c] and K. Gowthambalaji[1,d]

[1]Department of Civil Engineering, Sri Krishna College of Engineering and Technology

[2]Department of Civil Engineering, Bannari Amman Institute of technology

[a]ramakrishnans@skcet.ac.in, [b]loganayagans@bitsathy.ac.in, [c]chandramohann@bitsathy.ac.in, [d]18eucv015@skcet.ac.in

Keywords: Fiber Reinforced Concrete, Steel Fiber, Glass Fiber, Carbon Fiber, Compressive Strength, Tensile Strength, Flexural Strength

Abstract. Next to water, concrete is the most consumed material in the world. In the construction industries, concrete is a basic material used for high compressive strength, durable, fire resistant but has low tensile strength. This experimental study aimed to investigation the compressive, tensile and flexural strength of the concrete reinforced with three different fibers. Comparative study has been made between metallic: steel fibers and nonmetallic: glass and carbon fiber reinforced concrete. Fibers were used in concrete with fractions of 0%, 0.5%, 1%, 1.5%, 2% and 2.5% by volume of cement in M20 grade of concrete. In this paper, the behavior of cube, cylinder and prism specimen of fiber reinforced concrete (FRC) were deliberated. Addition of fiber in concrete were increased the basic mechanical properties of concrete increases. The steel fiber reinforced concrete attains higher compressive, flexural and tensile strength than concrete with carbon fiber and glass fiber. Carbon fibered concrete attained higher flexural and tensile strength than glass fibered concrete.

Introduction

Fiber reinforced concrete (FRC) has received with the considerable attention recently due to high toughness increases, then the addition of fiber in the concrete was provide the high tensile and flexural strength. In general, fibers used in concretes are steel fiber, glass fibers, carbon fibers [1]. The addition of steel fibres to cement-based composites is predicted to improve their performance [3]. Steel fiber is a key benefits to include good tensile and flexural strength, resistance of spitting, permeability and resistance of impact. It's a good way to boost the mortar's hardness and plastic shrinkage cracking of resistance [2]. Steel fibre of concrete are being concrete are added to improve its mechanical performance and strength of the tensile. Steel fibre are in the bridging cracks that form in concrete. Steel fibres also strengthen the concrete's toughness [4]. Carbon fiber with good work and compaction, tensile and compressive strength of the concrete contain carbon to increase the content. Also he mentioned that a higher carbon content causes lower strength of compressive, believes is due to mix's workability failing to perform properly and resulting in more empty spaces [5]. Carbon fibre increased density of dry and decreased absorption of water, according to findings. Furthermore, the addition to improve the carbon fibre the concrete's tensile strength and compressive strength, as well as its crack resistance [6]. During the hardening stage of concrete, cracks appear. Carbon fibre reinforced concrete was created as a result of new technology for employing carbon fibres to overcome issues material of cement based which associate, which have low tensile strength and fracture toughness is poor of cementations composite [7]. Various studies on reinforcement of concrete with reasonable quantity addition in the concrete of glass fibres demonstrate improved compression, flexural, and toughness performance; however, the improvement is dependent on type of the glass fibre employed in mixture. Furthermore, glass fibres aid in the preventing the formation of micro-cracks in the concrete contact, resulting in a robust composite with higher crack resistance and ductility [9]. Glass fiber is lesser expensive than the carbon fibre, and the glass fibre offer unique properties such as being undetectable on surface are finished being safe, handle easy to use. When glass fibres are added to the mix, there is no need for change the cement water ratio [8]. The fibres have a channel so the matrix helps to transfer the load between the fibres and protects them from chemical exposure and ambient clamminess by keeping the load-carrying members intact in the required positions and orientation. The influence of fibres in matrix that generates an increase in the matrix's impact strength as well as increases in tension, because concrete is weak in tension. Glass fibres have a high modulus of elasticity and tensile strength [10]. Although prior researchers have researched and separated these concretes, no previous comparisons have been performed using the mix proportions as same, fraction volume of fiber processing the settings, or mechanical testing conditions.

Research Significance

This paper we are going to provide the comparative study on the behavior of compressive, tensile and flexural strength of fiber reinforced concrete and the cost of fiber, that the concrete was containing the various types of the steel fiber, carbon fiber, and glass fiber short fibers. In fiber volume of fraction and length of the fiber then same mixing procedure were subjected. Compare and highlight the superior properties of different fiber reinforced concrete

Material Properties

Cement, Fine aggregate and Coarse aggregate

In this experimental investigation, cement ordinary Portland cement (PPC) (43Grade) used with a 3.18 of specific gravity. 254 min and 48 min was the final and initial setting time of the respective cement. M sand used as a fine aggregate for this study. M sand was confirming as a zone II with 2.45 of specific gravity, 2% of water absorption and 3.18 of fine modulus. Crushed stone particles are larger than 4.75 mm are referred to as coarse aggregates. Crushed stone from a nearby licensed quarry was used in this experiment.20mm coarse aggregate is a maximum size and specific gravity and fine modulus are 2.85 and 7.1 respectively.

Fibers

In this experimental investigation steel fibers, glass and carbon fibers were studied. The properties of the fibers are given in Table 1.

Table 1: *Properties of the Fibers*

Fibers	Steel fiber	Glass fiber	Carbon fiber
Length, [mm]	5	5	5
Diameter, [µm]	60	10	36
Density, [g/cm3]	7.9	2.4	0.95
Modulus, [GPa]	200	72	117.3
Tensile strength, [MPa]	970	1700	2588

Mix proportion

The proportion of mixture was done according to the Indian standard recommend (refer with: IS 10262-2009). Concrete is one of the most commonly utilized construction material. Concrete is a mixture of cement, sand, coarse aggregate, and water. By substation pozzolana (fly ash, silica fume, rice husk ash crushed granulated blast furnace slag) for cement, the cost of concrete is reduced [11]. The binder content in the mix was 383kg/m^3 of PPC cement, then the fine aggregate was taken for mixture is 672 kg/m^3 and the coarse aggregate are taken for mixture is 1100kg/m^3. M20 grade of concrete for mix proportion was1:1.75:2.87 by water cement ratio with weight of 0.5was kept as content. The variable percentage of steel, glass and carbon fiber from 0 to 2.5% of cement in volume batching with the incremental of 0.5.Cement, coarse and fine aggregate they properly mixed together to achieved homogenous before water was added. Totally 5 minute was mixing time of the mixture, the sample was casted, kept for 24 hour before demodulating. Until the day of testing, the sample was place in the curing tank. In the specimens, the compressive strength was containing fiber were cured in the tank for 7 days, and for compressive, tensile and flexural strength, the specimens were cured for 28 days.

Result and Discussion

Compressive strength

The compressive strength of the concrete arrived by crushing the cube of size 150mm at a period of seventh day and twenty eighth day. The compressive strength was the ratio of cross sectional area to ultimate load of the concrete. The average values of the three specimen of fiber reinforced concrete (FRC) (refer with table 2). In seventh day of curing, the average compression strength of the steel fiber reinforced concrete (SFRC) has higher than the glass fiber reinforced concrete (GFRC) as well as carbon fiber reinforced concrete (CFRC). The glass fiber reinforced concrete (GFRC) has a higher than the average compression strength carbon fiber reinforced concrete (CFRC). The graphical representation of average compressive of (FRC) at seventh day (refer with: Figure 1). In twenty eight days of curing, average compression strength of the steel fiber reinforced concrète (SFRC) higher than the glass fiber reinforced concrete (GFRC) as well as carbon fiber reinforced concrete (CFRC). Glass fiber reinforced concrete (GFRC) has higher average compression strength than carbon fiber reinforced concrete (CRFC). The average compressive strength of FRC at twenty eighth day (refer with: Figure 2).

Table 2: *Average Compression strength of FRC*

Sl.No.	% of Fiber	Average Compression strength [MPa]					
		SFRC		GFRC		CFRC	
		7th	28th	7th	28th	7th	28th
1	0.50	25.32	34.69	24.83	34.01	24.29	33.67
2	1.00	26.08	35.72	25.14	34.77	24.94	34.07
3	1.50	26.22	36.26	25.48	35.28	24.96	34.58
4	2.00	26.88	36.99	25.84	35.53	25.04	34.82
5	2.50	27.44	37.63	25.50	35.72	25.38	35.01

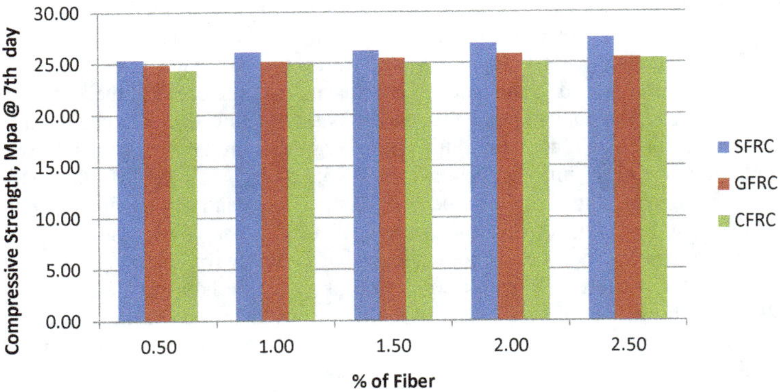

***Figure 1**: Average compressive strength of FRC at 7days*

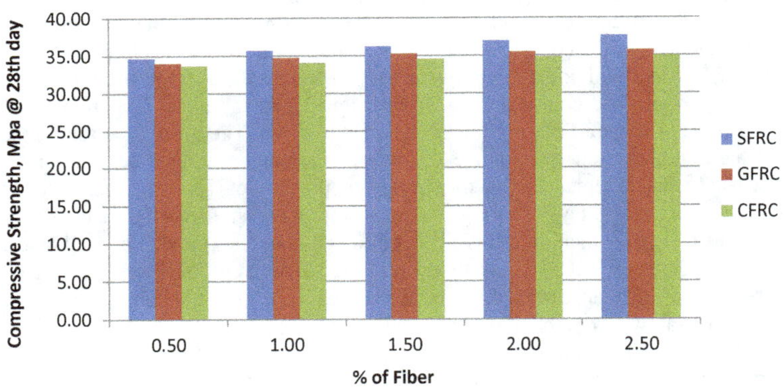

***Figure2**: Average compressive strength of FRC at 28 days*

The percentage of the fiber increase, compression strength of the Fiber reinforced concrete (FRC) increases. Incremental of compression strength of; SFRC was varied from 0 to 16 percentage, GFRC was varied from 0 to 10 percentage and CFRC was varied from 0 to 8 percentage for variable percentage of fiber in concrete. Graphical representation of incremental of compression strength is given. (refer with figure 3)

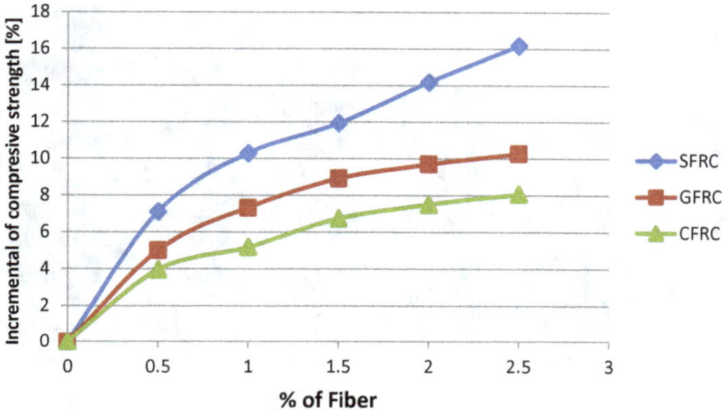

***Figure3**: Incremental of Compression Strength of FRC*

Split tensile strength

The cylinder was used to forecast tensile strength. A cylinder having a crushing strength, with a diameter of 150mm and a depth or height of 300mm is used to define and determine the concrete's strength. The specimen is placed lengthwise in the cylinder and the force is applied by machine. Tested the cylinder of steel fiber, glass fiber, carbon fiber the tensile strength was represented in the MPa. The cylinder's cross section area is 235 sq. cm. Finally, the tensile strength was the ratio of cross sectional area to ultimate load of the concrete. (refer with IS 516-1959) was used to record the highest load was applied to the cylinder. The cylinders ware put through tensile testing, and the tensile strength ware computed. The average values of three specimen of fiber reinforced concrete (FRC) (refer with Table: 3). In 28 days, the average tensile strength of the steel fiber reinforced concrete (SFRC) higher than the carbon fiber reinforced concrete (CFRC) as well as glass fiber reinforced concrete (GFRC). Carbon fiber reinforced concrete (CFRC) has a higher average Tensile strength than the glass fiber reinforced concrete. The average split tensile strength at 28 days (refer with: figure: 4)

***Table 3:** Average split tensile strength and Flexural strength at 28th day.*

Sl.No	% of Fiber	Average split tensile strength, [MPa]			Average flexural strength, [MPa]		
		SFRC	GFRC	CFC	SFRC	GFRC	CFRC
1	0.5	2.57	2.52	2.54	4.71	4.56	4.64
2	1	3.13	2.86	3.03	5.56	4.87	5.22
3	1.5	3.62	3.11	3.34	6.16	5.39	5.77
4	2	3.87	3.38	3.64	6.65	5.82	6.23
5	2.5	4.20	3.46	3.92	7.27	5.911	6.59

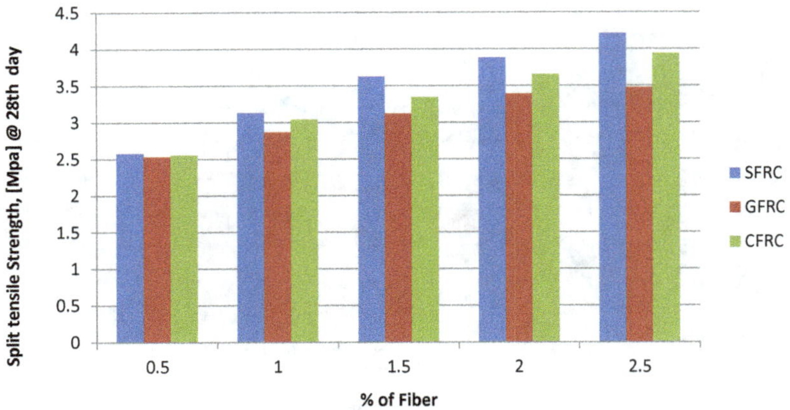

Figure 4: *Average Tensile strength FRC at 28 days*

Incremental of split tensile strength of; SFRC was varied from 0 to 80 percentage, GFRC was varied from 0 to 48 percentage and CFRC was varied from 0 to 68 percentage for variable percentage of fiber in concrete. Graphical representation of incremental of split tensile strength is given (refer with: figure 5).

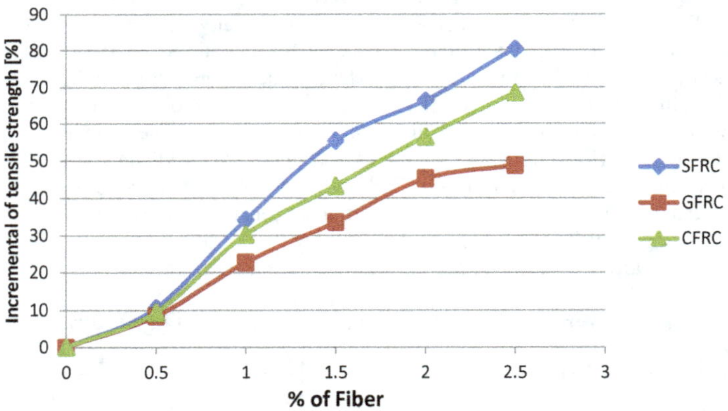

Figure 5: *Incremental of Tensile Strength of FRC*

Flexural strength

Flexural strength was predicted by using prism in the dimensions of (100mmx100mmx500mm). The flexural behavior of materials subjected to simple beam loading using the flexural method. With some materials, it's also named as a transverse beam test. Each increment of loading, maximum strain and maximum fiber stress are determined. The test result of flexural strength prism of steel fiber, glass fiber, carbon fiber are (refer with Table 3). The flexural strength was represented in the [MPa].The cross section area of the prism is 225sq.cm. (Refer with: IS 516-

1959) was used to record the highest load applied to the specimen. In 28th day of the average flexural strength of the steel fiber reinforced concrete (SFRC) comparatively greater than the carbon fiber reinforced concrete (CFRC) as well as glass fiber reinforced concrete (GFRC) (refer with: Figure 6). Carbon fiber reinforced concrete (CFRC) has a higher average flexural strength than glass fiber reinforced concrete (GFRC)

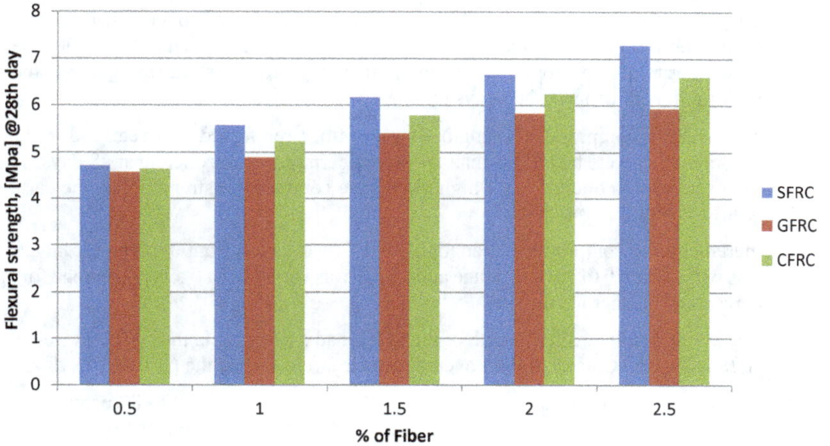

***Figure 6**: Average Flexural strength of FRC at 28 days*

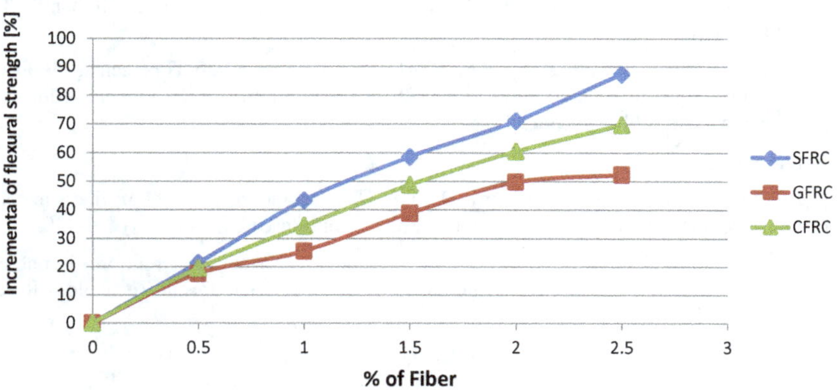

***Figure7:** Incremental of flexural strength of FRC*

The percentage of the fiber increase, flexural strength of the fiber reinforced concrete (FRC) also increases. Incremental of flexural strength of; SFRC was varied from 0 to 87 percentage, GFRC was varied from 0 to 52 percentage and CFRC was varied from 0 to 69 percentage for variable percentage of fiber in concrete. The figure 7 shown the incremental of flexural strength of FRC.

Conclusion

In the fiber reinforced concrete has a component like steel glass and carbon with the specimen of cube, cylinder and prism respectively

1. The average compressive strength of the steel fiber reinforced concrete had higher than the glass fiber reinforced concrete as well as carbon fiber reinforced concrete at 7 days. The glass fiber reinforced concrete has a highest average compressive strength than the carbon fiber reinforced concrete at 7 days
2. The average compressive strength of glass fiber reinforced concrete and carbon fiber reinforced concrete had a less than the steel fiber reinforced concrete at 28 days. The glass fiber reinforced concrete has a highest average compressive strength than the carbon fiber reinforced concrete at 28 days
3. Incremental of compression strength of; SFRC was varied from 0 to 16 percentage, GFRC was varied from 0 to 10 percentage and CFRC was varied from 0 to 8 percentage for variable percentage of fiber in concrete.
4. The average split tensile strength of the SFRC had a higher than the CFRC as well as GFRC at 28 days. CFRC has a highest average tensile strength than the GFRC at 28 days.
5. Incremental of split tensile strength of; SFRC was varied from 0 to 80 percentage, GFRC was varied from 0 to 48 percentage and CFRC was varied from 0 to 68 percentage for variable percentage of fiber in concrete.
6. The average flexural strength of the SFRC had a higher than the CFRC as well as GFRC at twenty eight days of curing. CFRC has a highest average flexural strength than the GFRC at 28 days
7. Incremental of flexural strength of; SFRC was varied from 0 to 87 percentage, GFRC was varied from 0 to 52 percentage and CFRC was varied from 0 to 69 percentage for variable percentage of fiber in concrete.

Reference

[1] C. Selin Ravikumar, V. Ramasamy, Dr. T.S. Thandavamoorthy, Effect of fibers in concrete composites, International Journal of Applied Engineering Research. 10 (2005) 419-430.

[2] S. Vikrant, Vairgade, S. Kavita, Kene, Introduction to steel fiber reinforced concrete on engineering performance of concrete, International Journal of Scientific and Technology Research. 1 (2012) 139-141.

[3] Seong Cheol Lee, Joung-Hwan Oh, Jae-Yeol Cho, Compressive Behaviour of Fiber Reinforced Concrete with End hooked Steel Fibres, Materials.8 (2015) 1442-1458. https://doi.org/10.3390/ma8041442

[4] Ankit Agrawal, Tarit Jain, Sarang Agarwal, Compressive Strength Testing of Steel Fiber Reiforced Concrete in Different Curing Regimes, International Journal of Engineering Research and Technology. 3 (2010) 1-5.

[5] Mohamad Atiyeh and Ertug Aydin, Carbon-Fiber Enriched Cement-Based Composites for Better Sustainability, Materials. 13 (2020) 1-16. https://doi.org/10.3390/ma13081899

[6] Md. Saifiuddin, George abdel-sayed, Nataliya hearn, Absorption and Strength Properties of Short Carbon Fiber Reinforced Mortar Composite, Buildings. 11 (2021) 1-16. https://doi.org/10.3390/buildings11070300

[7] Md. Saifiuddin, George abdel-sayed, Nataliya hearn Effect of Pitch-Based Short Carbon Fibers on the Workability, Unitweight and Aircontent of Mortar Composite, Fibers. 6 (2018) 1-15. https://doi.org/10.3390/fib6030063

[8] S. Hemalath, A.Leema Rose, An Experimental Study on Glass Fiber Reinforced Concrete, International Research Journal of Engineering and Technology. 3 (2016) 2285-2289.

[9] J.D.Chaitanya kumar, G.V.S. Abhilash, P.Khasim Khan, G.Manikanta sai, V.Taraka ram, Exprimenthal Studies on Glass Fiber Concrete, American Journal of Engineering Research. 5 (2016) 100-104.

[10] Mustafa Sakti Wala, Anisha Mire, Amit Kumar Dewangan, Strength Characteristics of Glass Fiber Reinforced concrete, International Journal of Engineering Research and Technology, 8 (2019) 824-828.

[11] S.Ramakrishnan, G.Velrajkumar, S.Ranjith, Behavior of Cement Rice Husk Ash Concrete for Pavement, International Journal of Emerging Trends in Engineering and Development, 1 (2014) 31-41.

[12] IS 516-1959, Indian standard method of tests for strength of concrete

[13] IS 383-1970 Indian standard specification for coarse and fine aggregates from natural sources for concrete

[14] IS 8112- 2013 Indian standard ordinary portland cement, 43 grade specification

[15] IS 10262- 2009 Indian standard concrete mix proportioning - guidelines

A Review on Smart and Eco-Friendly Building Material

T. Jeevetha[1,a,*], S. Vijayashanthy[1,b], J. Dharanidharan[1,c], B. Bala Murali Krishna[1,d]

[1]Department of Civil Engineering, Kongu Engineering College, Erode, Tamil Nadu, India-638060

[a]jeevethat.civil@kongu.edu,[b]vijayashanthi.civil@kongu.edu,[c]dharanidharanj.19civil@kongu.edu, [d]balamuralikrishnab.19civil@konu.edu

Keywords: Self-Cleaning, De-Air Polluting Concept, Nanotechnology, Photo-Catalytic Treatment

Abstract. Due to industrialization and man-made activities in and around the world, Day by day pollution levels are drastically increasing. This being said, building materials in the present context can be produced using novel techniques by applying advanced pollution reduction coatings/strategies. The self-cleaning and de-air polluting concept is emphasized in concrete structures using nanotechnology. Photo-catalytic treatment will be a promising approach as it alleviates structural and aesthetic damages on building materials. Finally, the risk exposure due to pollutants can be computed for better treatment of the same.

Introduction

The Photo-catalytic Titanium dioxide (TiO_2) was widely utilized as a white pigment in the textile industry in the past [1], where the Discover under which they react is consumed as supporting accouterments a part of debasement of maquillages and fabrics. The very first scientific invention on this place started in 1929 [2], providing an important active role of Titanium Dioxide vanishing of paints. Wherein, it was found that the active oxygen is present on the surface of TiO_2, and this is also identified as the cause for bleaching in the dyes and paints and fabrics when they are photo-exposed or irradiated to UV in 1930 [3], but only after the alternate half of XX century, the proposition behind the miscellaneous print-catalysis in presence of essence oxides was observed, and in the 1970s attestations for the print-catalytic exertion of TiO_2 was published [4, 5]. And after all the efforts that were put into research, one important theory was found that TiO_2 has a self-cleaning effect on itself [6]. So, only after this was found then have put in a lot of effort and then tried to include it in building materials with the integration on the surrounding environment like pure and clean air which includes reduction of maintenance cost [7-9]. So then, many buildings imparted this TiO_2 as a photo-catalytic and the self-cleaning of TiO_2, and one such building is in Tokyo, and it was started in 2002 And then emerged other buildings as the Italian kiosk at the Milano Expo 2015 Universal Exposition was among the first constructions that include tone-drawing window spectacles.

Titanium dioxide was widely used in past years starting from self-cleaning aspects [10-14] or it roads [15-19] that are combined with the anti-pollution effects [20-23] that are used for the preservation purpose in the architectural heritage that involves Stones in it [23-28]. When the mechanism is deeply studied and the materials are improved concerning the photo-activity their longevity and the photo-activated effects are treated only marginally. This is especially true when it comes to the materials that are used in building purposes when it comes to outdoor usage, where many other factors come into role, such as they are being exposed to wind, rain, sunlight and these may lead to the deactivation of the TiO_2 component that is present of to the complete degradation of the materials utilizing the erosion mechanism [11, 12, 29-32].

In recent times, when it is analyzed in the aspect of the TiO_2's presence, it is referred to as the cool pigment [33-36]. The fact here is, Cool surfaces mostly won't get heated when they are exposed to the sun, since they are based on the rate of the reflected coincident the rate of radiated thermal radiation is comparable to that of a black body due to solar radiation and strong thermal emittance. So, the builders will usually prefer the cool materials for roofing to minimize the heat in the buildings and also to reduce the peak power needs and the cooling energy needs and also diminish the influence of the local climate and also the heat that is being released to the surrounding is also reduced [37-41]. But these benefits will be affected by the weather in the surroundings and the soil that is present in that area [41-44], it is identified from the State of California, that the roofs except for the commercial buildings will have the solar reflectance of 0.63 [45] after three years of construction as a minimum reflectance value. And it was discovered recently that if the Anatase is added to the material then the drop in the solar reflectance will be minimum even after the exposure to the wind and sun after two solid years will be 0.19 instead of 0.26 [35]. The degradation of photo-catalytic NOx, which is responsible for the nitric acid conformation and also impacts the optic parcels of the nano-patches present in the TiO_2 material, is described as the degradation of photo-catalytic NOx, which is responsible for the nitric acid conformation and also impacts the optic parcels of the nano-patches present in the TiO_2 material, it occurs in an alternate portion of the near-infrared wavelength range between 1500 [29-35].

The main aim of this article is to insist on the change in the outlook of the material which consists of the high solar reflectance in the self-cleaning TiO_2 material. A recent study on this states that TiO_2 NPS are modified to enhance certain optical properties once before getting added to the building material as in acrylic paint. Two acids are involved in this and they are dilute nitric acid and sulfuric acid. And here the admixtures of the modified powders to the mortars and paints and other construction materials, then there's a final process of maquillages neutralization, which is likewise estimated, to check whether the revision attained is sustained in the annulled maquillages. Also, the makeup that has TiO_2 in its anatomized on its optic parcels and print-catalytic performance.

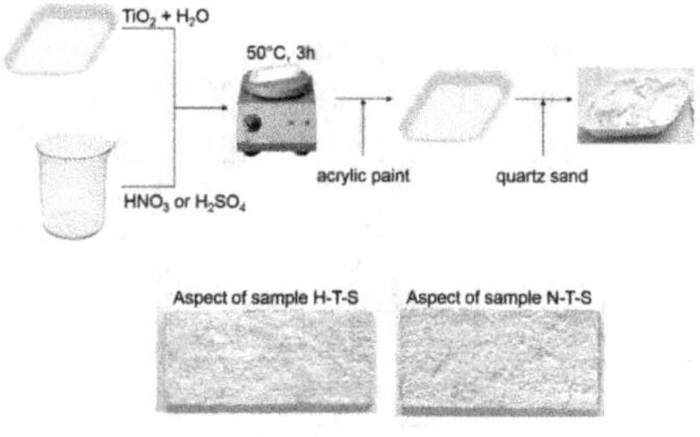

Fig.1. *Samples preparation and sample pictures*

Materials and Methods

The Preparation of sample and Characterization discussed below is common procedure of various paper in the concept of Self-cleaning and De-air polluting by Photo-catalytic treatment.

Preparation of Sample

The preparation of TiO_2 samples is summarized below. The photo-catalyst, which is being used to prepare and to study paint samples, is AEROXIDE®TiO2P25 by Evonik Industries. The Anatase and rutile combinations are about 80-20 with chastity is greater than 99.5 and a specific face area is 35 – 65m2/g. 1 g of nano-particles were first dispersed in distilled water of 4 g, preliminarily to the objectification into the makeup, which produces suspense of TiO_2 and distilled water [14]. Despite the use of ultrasonic treatment, the product will still have agglomerated nano-particles and have a faint milky look. Adulteration of nitric and sulfuric acids is expected. (HNO3 weight is 0.5 and H2SO4 is 0.1) and the result will be added to the aqueous Titanium Dioxide suspension to performance The NPs will be acid-treated, which will have only favorable effects on their reflectance of NIR [35]. A fresh result will be obtained, containing 10% TiO_2 by weight. Before combining the NPS-acid with the cosmetics, the admixture will be faded to an electric plate at around 50 °C to decrease the position of water, providing the NPS-acid a total contact time of 3h and a final TiO_2 concentration of 33.3. (TiO_2 solvent in a ratio 12). The admixture will also be nullified by adding it to a stoichiometric volume of NaOH waterless result later, after the time of commerce and before the objectification of it into the composition. The specimen markers are displayed in Table 1 as a function of the nano-flyspeck treatment. In each composition, at least two samples will be used for optic dimension and the other two for print-catalytic studies.

These print-catalytic maquillages are substantially used in the constructions for producing the coatings that are tone-cleaning, and then a makeup in white color is chosen to help the TiO_2 nano-patches. The same adulterated Nitric acid was first tested, 0.5 by weight, to the makeup and to corroborate that whether the objectification of acid together with the nano- patches could be mischievous to the makeup that was used, and There was no reaction, and uniform mixing of makeup and acid was eventually achieved. Sulfuric acid was used to prove this once more. Following that, it was anticipated that the acid in the NPS would have no deleterious effect on the composition.

Table 1: *Samples and their composition*

Label	TiO2 particles pre-immersion	TiO2 content (% by weight)	Paint content (% by weight)	Sand content (% by weight)	Neutralization with NaOH
Ref	-	-	33.3	66.7	-
H-T	In DI H_2O	20	80	-	-
H-T-S		6.67	26.7	66.7	-
N-T	In HNO_3 0.5 M	20	80	-	-
N-T-S		6.67	26.7	66.7	-
N-TN-S		6.67	26.7	66.7	Yes
S-T	In H_2SO_4 0.1 M	20	80	-	-
S-T-S		6.67	26.7	66.7	-
S-TN-S		6.67	26.7	66.7	Yes

In the dimensional stabilization, fine quartz sand of 0.06–0.1 mm is utilized to produce uniform and crack-free samples of cosmetics containing Titanium Dioxide nanoparticles, as necessary for reflectance optics. In the sample, however, uniformity and nebulosity are essential for reducing light from passing through the sample. The best rate of Titanium Dioxide, water (when mixed), cosmetics, and the beach was found to be 12410, with each sample containing 6.67 percent Titanium Dioxide. This could be compared to commercially available coatings, which typically contain TiO_2 in quantities ranging from 1 to 10 by weight. Beach with limitations of 30 mm, 60 mm, and 3 mm would be included in the final samples.

Characterization

The crystallinity of P25 NPs was evaluated again using X-ray diffraction spectrometry after they were both treated with acid and further annulled. Cu Ka radiation was employed with Philips PW 1830 equipment (applied pressure was 40kV, deg/min/scan/rate). TEM investigations using a Philips CM200 FEG were used to define the phrase "flyspeck morphology" (200 kV applied pressure, electron surge-length 0.251,0.19 nm resolution). Additionally, the samples will be prepped before examination by applying one drop of greasepaint waterless suspense to a bobby grid carpeted with unformed carbon and leaving it to dry.

The samples will be allowed to dry for at least 14 days before beginning the optical measurements to reach an equilibrium quantum of water that will be left out of the material. This is a critical step in preventing dimension crimes caused by variable quantities of water in the samples, as water has a significant impact on reflectance across the light diapason. A Perkin Elmer Lambda 950 spectrometer with a 150 mm integration sphere has been used to perform UV Vis-NIR reflectance measurements between 300 and 2500 nm with a spectral resolution of 5 nm. The tear in the Vis range was set at 2 nm between 300 and 860 nm, and the rest of the exam will be conducted in servo mode with a 15-mm area between 1500 and 2500 nm. Three measurements are obtained on the corridor that will not lap for each set of samples. For conditions of clear sky and air, mass equal to 1[46], the broadband values were also determined using solar spectrum irradiance. The broadband reflectance will be between 1500 and 2500 nm, and the solar reflectance qs and qn2 are independent.

The samples with the print-catalytic exertion of H-T and H-T-S, such as TiO_2 fusions that will stay non-treated with makeup only or with both makeup and beach, were measured and compared to determine the influence of adding beach on print exertion. Following that, a comprehensive characterization of completed samples such as cosmetics, beaches, and NPS will be conducted. rhodamine B – an azo color with a magenta tinge – was chosen to check print-catalytic exertion because it is widely accepted and will be used to test the effectiveness of print-catalytic accouterments that are representatives of organic adulterants, as attested by several independent exploration groups [14, 29, 31, 47-50].

The samples will also be bathed in a waterless solution containing color with a concentration of 10-5 M for 3 hours, dried, and then subjected to artificial light with a UV intensity of 1 mW cm 2 for 4 hours, simulating a solar diapason (Osram Vitalux beacon, 300 W). All samples were also characterized in duplicate, with one sample per material type collecting measurements at the same time. A Konica Minolta radiometer UM-10 was used to set UV irradiation at 365 nm at the same intensity in each test spot. The color was measured before and after the irradiation to use a reflectance spectrophotometry system with a Konica Minolta CM-2600d spectrophotometer. The software Spectra magic NX was then used to convert reflectance data into color coordinates in the color space. The color coordinates in the color space CIE Lab, as defined by the International Commission, were then converted using the software Spectra magic NX [51] The decline in the * (red) match of color, which signifies a loss of red intensity and results in abrasion of the color, was also linked to Rhodamine B declination.

Conclusion

According to previous research, the efficiency of building materials must be increased to extend the building's life. In this context, advanced doped materials based on nanotechnology and requiring extensive preparation will have greater market potential in the materials industry. In the construction era, the use of synthetic building materials aided by nano-particles will result in increased efficiency, followed by pollution prevention strategies. To reduce noxious emissions from construction materials, buildings and other infrastructure development authorities should be advised.

References

[1] A. Khataee, G.A. Mansoori, Nanostructured Titanium Dioxide Materials, World Scientific Publishing Co., Singapore, 2011. https://doi.org/10.1142/8325

[2] E. Keidel, Die Beeinflus sung der lich tech the it von teer far blacken durch titan weiss (Influence of titanium white on the fastness to light of coal-tardays), Farben-Ztg. 34 (1929) Pg: 1242-1243.

[3] C.F. Goodeve, J.A. Kitchener, Photosensitisation by titanium dioxide, Trans. Faraday Soc. 34 (1938) Pg: 570-579. https://doi.org/10.1039/tf9383400570

[4] W. Doerfler, K. Hauffe, Heterogeneous photocatalysis I. The influence of oxidizing and reducing gases on the electrical conductivity of dark and illuminated zinc oxide surfaces, J. Catal. 3 (1964) Pg: 156-170. https://doi.org/10.1016/0021-9517(64)90123-X

[5] A. Fujishima, K. Honda, Electrochemical photolysis of water at a semiconductor electrode, Nature 238 (1972) Pg: 37-38. https://doi.org/10.1038/238037a0

6] R. Wang, K. Hashimoto, A. Fujishima, M. Chikuni, E. Kojima, A. Kitamura, M. Shimohigoshi, T. Watanabe, Light-induced amphiphilic surfaces, Nature 388(1997) Pg: 431-432. https://doi.org/10.1038/41233

[7] M.V. Diamanti, M. Pedeferri, Concrete, mortar and plaster using titanium dioxide nanoparticles: applications in pollution control, self-cleaning and photo sterilization, in Nanotechnol. Eco-Effic. Const, Elsevier, 2013, Pg: 299-326. https://doi.org/10.1533/9780857098832.3.299

[8] A. Fujishima, X. Zhang, D. Tryk, Heterogeneous photocatalysis: from water photolysis to applications in environmental cleanup, Int. J. Hydrogen Energy 32 (2007) Pg: 2664-2672. https://doi.org/10.1016/j.ijhydene.2006.09.009

[9] K. Nakata, A. Fujishima, TiO2 photocatalysis: design and applications, J. Photochem. Photobiol. C Photochem. Rev. 13 (2012) Pg: 169-189. https://doi.org/10.1016/j.jphotochemrev.2012.06.001

[10] A. Folli, C. Pade, T.B. Hansen, T. De Marco, D.E. Macphee, TiO2 photocatalysis in cementitious systems: insights into self-cleaning and de air pollution chemistry, Cem. Concr. Res. 42 (2012) Pg: 539-548. https://doi.org/10.1016/j.cemconres.2011.12.001

[11] L. Graziani, E. Quagliarini, F. Bondioli, M. D'Orazio, Durability of self-cleaning TiO2 coatings on fired clay brick façades: Effects of UV exposure and wet & dry cycles, Build. Environ. 71 (2014) Pg: 193-203. https://doi.org/10.1016/j.buildenv.2013.10.005

[12] M. Baudys, J. Kry' sa, M. Zlámal, A. Mills, Weathering tests of photocatalytic facade paints containing ZnO and TiO2, Chem. Eng. J. 261 (2015) Pg: 83-87. https://doi.org/10.1016/j.cej.2014.03.112

[13] S. Banerjee, D.D. Dionysiou, S.C. Pillai, Self-cleaning applications of TiO2 by photo-induced hydrophilicity and photocatalysis, Appl. Catal. B Environ. 176-177 (2015) Pg: 396-428. https://doi.org/10.1016/j.apcatb.2015.03.058

[14] M.-Z. Guo, A. Maury-Ramirez, C.S. Poon, Self-cleaning ability of titanium dioxide clear paint coated architectural mortar and its potential in field application, J. Clean. Prod. 112 (Part 4, 2016) Pg: 3583-3588. https://doi.org/10.1016/j.jclepro.2015.10.079

[15] M.M. Hassan, Quantification of the environmental benefits of ultrafine/nano titanium dioxide photocatalyst coatings for concrete pavement using hybrid life-cycle assessment, J. Infrastruct. Syst. 16 (2010) Pg: 160-166. https://doi.org/10.1061/(ASCE)IS.1943-555X.0000016

[16] C.S. Poon, E. Cheung, NO removal efficiency of photocatalytic paving blocks prepared with recycled materials, Constr. Build. Mater. 21 (2007) Pg: 1746-1753. https://doi.org/10.1016/j.conbuildmat.2006.05.018

[17] M.M. Ballari, H.J.H. Brouwers, Full-scale demonstration of air-purifying pavement, J. Hazard. Mater. 254-255 (2013) Pg: 406-414. https://doi.org/10.1016/j.jhazmat.2013.02.012

[18] C. Toro, B.T. Jobson, L. Haselbach, S. Shen, S.H. Chung, Photoactive roadways: Determination of CO, NO and VOC uptake coefficients and photo labile side product yields on TiO2 treated asphalt and concrete, Atmos. Environ. 139(2016) Pg: 37-45. https://doi.org/10.1016/j.atmosenv.2016.05.007

[19] E. Boonen, A. Beeldens, Photocatalytic roads: from lab tests to real scale applications, Eur. Transp. Res. Rev. 5 (2013) Pg: 79-89. https://doi.org/10.1007/s12544-012-0085-6

[20] J. Chen, C.-S. Poon, Photocatalytic activity of titanium dioxide modified concrete materials Influence of utilizing recycled glass cullets as aggregates, J. Environ. Manage.90 (2009) Pg: 3436-3442. https://doi.org/10.1016/j.jenvman.2009.05.029

[21] M.-Z. Guo, A. Maury-Ramirez, C.S. Poon, Photocatalytic activities of titanium dioxide incorporated architectural mortars: effects of weathering and activation light, Build. Environ. 94 (Part 1) (2015) Pg: 395-402. https://doi.org/10.1016/j.buildenv.2015.08.027

[22] C. George, A. Beeldens, F. Barmpas, J.-F. Doussin, G. Manganelli, H. Herrmann, J.Kleffmann, A. Mellouki, Impact of photocatalytic remediation of pollutants on urban air quality, Front. Environ. Sci. Eng.10 (2016). https://doi.org/10.1007/s11783-016-0834-1

[23] N. Bengtsson, M. Castellote, Photocatalytic activity for NO degradation by construction materials: parametric study and multivariable correlations, J.Adv. Oxid. Technol. 13 (2010) Pg: 341-349. https://doi.org/10.1515/jaots-2010-0311

[24] P. Munafo, G.B. Goffredo, E. Quagliarini, TiO2-based Nanocoatings for preserving architectural stone surfaces: an overview, Constr. Build. Mater.84 (2015) Pg: 201-218. https://doi.org/10.1016/j.conbuildmat.2015.02.083

[25] D. Colangiuli, A. Calia, N. Bianco, Novel multifunctional coatings with photocatalytic and hydrophobic properties for the preservation of the stone building heritage, Constr. Build. Mater. 93 (2015) Pg: 189-196. https://doi.org/10.1016/j.conbuildmat.2015.05.100

[26] E. Quagliarini, F. Bondioli, G.B. Goffredo, A. Licciulli, P. Munafò, Smart surfaces for architectural heritage: preliminary results about the application of TiO2-based coatings on travertine, J. Cult. Herit. 13 (2012) Pg: 204-209. https://doi.org/10.1016/j.culher.2011.10.002

[27] E. Quagliarini, F. Bondioli, G.B. Goffredo, A. Licciulli, P. Munafò, Self-cleaning materials on architectural heritage: compatibility of photo-induced hydrophilicity of TiO2 coatings on stone surfaces, J. Cult. Hermit. 14 (2013) Pg: 1-7. https://doi.org/10.1016/j.culher.2012.02.006

[28] L. Pinho, M. Rojas, M.J. Mosquera, Ag-SiO2-TiO2 nanocomposite coatings with enhanced photoactivity for self-cleaning application on building materials, Appl. Catal. B Environ. 178 (2015) Pg: 144-154. https://doi.org/10.1016/j.apcatb.2014.10.002

[29] M.V. Diamanti, R. Paolini, M. Rossini, A.B. Aslan, M. Zinzi, T. Poli, M.P. Pedeferri, Long term self-cleaning and photocatalytic performance of anatase added mortars exposed to the urban environment, Constr. Build. Mater. 96 (2015) Pg: 270-278. https://doi.org/10.1016/j.conbuildmat.2015.08.028

[30] M.V. Diamanti, M. Ormellese, M. Pedeferri, Characterization of photocatalytic and super hydrophilic properties of mortars containing titanium dioxide, Cem.Concr. Res. 38 (2008) Pg: 1349-1353. https://doi.org/10.1016/j.cemconres.2008.07.003

[31] A. Calia, M. Lettieri, M. Masieri, Durability assessment of nanostructured TiO2 coatings applied on limestones to enhance building surface with self-cleaning ability, Build. Environ. 110 (2016) Pg: 1-10. https://doi.org/10.1016/j.buildenv.2016.09.030

[32] D. Osborn, M. Hassan, S. Asadi, J. White, Durability quantification of TiO2 surface coating on concrete and asphalt pavements, J. Mater. Civ. Eng. 26(2013) Pg: 331-337. https://doi.org/10.1061/(ASCE)MT.1943-5533.0000816

[33] A.P. Werle, M.L. de Souza, K. Loh, R. Ando, V.M. John, The performance of a self-cleaning cool cementitious surface, Energy Build. 114 (2016) Pg: 200-205. https://doi.org/10.1016/j.enbuild.2015.06.025

[34] R. Sharma, R. Shaw, S. Tiwari, S. Tiwari, Nano-titania decorated fly ash as a self-cleaning antibacterial cool pigment, ACS Sustain. Chem. Eng.3 (2015) Pg: 2796-2803. https://doi.org/10.1021/acssuschemeng.5b00692

[35] R. Paolini, M. Sleiman, M. Pedeferri, M.V. Diamanti, TiO2 alterations with natural aging: unveiling the role of nitric acid on NIR reflectance, Sol. Energy Mater. Sol. Cells. 157 (2016) Pg: 791-797. https://doi.org/10.1016/j.solmat.2016.07.052

[36] S. Wang, J. Zhang, L. Liu, F. Yang, Y. Zhang, Evaluation of cooling property of high-density polyethylene (HDPE)/titanium dioxide (TiO2) composites after accelerated ultraviolet (UV) irradiation, Sol. Energy Mater. Sol. Cells. 143(2015) Pg: 120-127. https://doi.org/10.1016/j.solmat.2015.06.032

[37] M. Santamouris, Cooling the buildings - past, present and future, Energy Build.128 (2016) Pg: 617-638. https://doi.org/10.1016/j.enbuild.2016.07.034

[38] H. Akbari, C. Cartalis, D. Kolokotsa, A. Muscio, A.L. Pisello, F. Rossi, M.Santamouris, A. Synnefa, N.H. Wong, M. Zinzi, Local climate change, and urban heat island mitigation techniques - the state of the art, J. Civ. Eng. Manage. 22(2016) Pg: 1-16. https://doi.org/10.3846/13923730.2015.1111934

[39] R. Levinson, P. Berdahl, A. Asefawberhe, H. Akbari, Effects of soiling and cleaning on the reflectance and solar heat gain of a light-colored roofing membrane, Atmos. Environ.39 (2005) Pg: 7807-7824. https://doi.org/10.1016/j.atmosenv.2005.08.037

[40] H. Akbari, R. Levinson, L. Rainer, Monitoring the energy-use effects of cool roofs on California commercial buildings, Energy Build. 37 (2005) Pg: 1007-1016. https://doi.org/10.1016/j.enbuild.2004.11.013

[41] R. Paolini, M. Zinzi, T. Poli, E. Carnielo, A.G. Mainini, Effect of aging on the solar spectral reflectance of roofing membranes: natural exposure in Roma and Milano and the impact on the energy needs of commercial buildings, Energy Build. 84 (2014) Pg: 333-343. https://doi.org/10.1016/j.enbuild.2014.08.008

[42] M. Sleiman, G. Ban-Weiss, H.E. Gilbert, D. François, P. Berdahl, T.W.Kirchstetter, H. Destaillats, R. Levinson, Soiling of building envelope surfaces and its effect on solar reflectance--Part I: analysis of roofing product databases, Sol. Energy Mater. Sol. Cells.95 (2011) Pg: 3385-3399. https://doi.org/10.1016/j.solmat.2011.08.002

[43] P. Berdahl, H. Akbari, R. Levinson, W.A. Miller, Weathering of roofing materials- an overview, Constr. Build. Mater. 22 (2008) 423-433. https://doi.org/10.1016/j.conbuildmat.2006.10.015

[44] P. Berdahl, H. Akbari, L.S. Rose, Aging of reflective roofs: soot deposition, Appl. Opt. 41 (2002) Pg: 2355-2360. https://doi.org/10.1364/AO.41.002355

[45] California Energy Commission, 2016 Building Energy Efficiency Standards for Residential and Nonresidential Buildings - Title 24, Part 6, (2015).

[46] R. Levinson, H. Akbari, P. Berdahl, Measuring solar reflectance-Part I: defining a metric that accurately predicts solar heat gain, Sol. Energy. 84 (2010) Pg: 1717-1744. https://doi.org/10.1016/j.solener.2010.04.018

[47] I.K. Konstantinou, T.A. Albanis, TiO2-assisted photocatalytic degradation of azo dyes in aqueous solution: kinetic and mechanistic investigations, Appl. Catal. B Environ. 49 (2004) Pg: 1-14. https://doi.org/10.1016/j.apcatb.2003.11.010

[48] J. Zhang, S. Yan, L. Fu, F. Wang, M. Yuan, G. Luo, Q. Xu, X. Wang, C. Li, Photocatalytic degradation of rhodamine B on anatase, rutile, and brookite TiO2, Chin. J. Catal. 32 (2011) Pg: 983-991. https://doi.org/10.1016/S1872-2067(10)60222-7

[49] T. Wu, G. Liu, J. Zhao, H. Hidaka, N. Serpone, Photo assisted degradation of dye pollutants. V. self-photosensitized oxidative transformation of rhodamine B under visible light irradiation in aqueous TiO2 dispersions, J. Phys. Chem. B 102 (1998) Pg: 5845-5851. https://doi.org/10.1021/jp980922c

[50] M.V. Diamanti, F. Lollini, M.P. Pedeferri, L. Bertolini, Mutual interactions between carbonation and titanium dioxide photo activity in concrete, Build. Environ. 62 (2013) Pg: 174-181. https://doi.org/10.1016/j.buildenv.2013.01.023

[51] R.W.G. Hunt, The specification of color appearance. I. Concepts and terms, Color Res. Appl. 2 (1977) Pg: 55-68. https://doi.org/10.1002/col.5080020202

An Experimental Study on the Impact of Incorporating Glass on the Photocatalytic Capacity of Concrete Blocks

S. Karthikeyan[1,a], S. Deepiha[2,b], K. Dinesh[1,c], F. Chriswin[1,d]

[1]Department of Civil Engineering, Kongu Engineering College, Perundurai, Tamil Nadu, India

[2]Department of Civil Engineering, Sri Sairam Engineering College, Chennai, Tamil Nadu, India

[a] karthistruct@gmail.com, [b] deepihasomasundaram18@gmail.com, [c] dineshempire03@gmail.com, [d] fchriswin@gmail.com

Keywords: Concrete, Light Transmission, Optical Fibre, Glass Aggregate

Abstract. Today's world is more concerned about the preservation of natural sources. Consumption of less amount of natural resources in construction activities is equally important as energy saving after construction. To combine both the aspects, one such thing tried in this work is translucent concrete. As everyone knows, concrete is termed as traditional material for a building. The important property hidden in this concrete is light emission. When light hits the concrete, some amount light gets transmitted to the other side of concrete. This characteristic is achievable when optical fibers are included in concrete. Optical fiber is a medium where it is associated with the transmission of information as light pulses along the fiber. In addition, the use of glass as a coarse aggregate is also studied and it also plays a major role in strength of the concrete. Concrete blocks containing glass particles as a partial replacement for coarse aggregates are casted and tested for mechanical properties and light transmission. Results shows the blocks transfer light in an efficient manner without major changes in the strength aspects.

Introduction

Translucent Concrete is a new class of concrete that focus on the conveyance of light. By the help of optical fiber, the transmission of light is practicable. Optical fiber is a medium and a technology accompanied with the transversion of particulars as light impulse along a glass or plastic strand or fiber. The core part of optical fiber is of high refractive index, which emerges the transit of light through it without much loss in potential energy of light. In recent times, the generation of waste glass is increased. So, it was used in this study. The fragmentary substitution of glass is done which results in durable and eco-friendly outcome.

Ahmed M Tahwia et al. [1] conducted research demonstrating that transparent self-compacting concrete may provide a wide range of light transmittance while maintaining good mechanical qualities. According to the findings of Yat Chhoy Wong[2], 10% recycled plastic waste and recycled crushed glass aggregates of the concrete design mix have the potential to be incorporated into concrete footpaths as a sustainable material, resulting in a significant reduction of plastic and glass wastes ending up in landfills. According to Pengwei Guo[3,] the use of glass particles can enhance the microstructure and reduce the permeability of concrete, enhancing sulphate attack resistance and freeze-thaw resistance. In modern architecture, Ping Liang noted how daylight plays an important role in both biological and psychological needs [4].

Nicholas [5] highlighted how using recycled waste glass to replace up to 25% of natural coarse particles in concrete did not have a major impact on its abrasion resistance. According to Zhuan Shen [6, the natural lighting coefficient in the room improved by 100 percent, the light was uniformly raised by approximately 50 percent, and the artificial lighting device's running time was

Materials Research Proceedings **23** (2022) 114-121 https://doi.org/10.21741/9781644901953-15

reduced from 72 percent to 41 percent. Shing Mei [7] discussed how light transmitting concrete (LTC) might minimise light energy consumption and carbon emissions, promoting green building construction, particularly in cities. With the addition of more optical fibres, the optical power of light-transmitting concrete increased nonlinearly. According to Danial Navabi[8,] LTC does not transmit infrared radiation as well as visible ones.

In order to achieve the standard living criterion, Shing Mei Chiew [9] suggests that, in addition to examining light transmittance attributes, simulation modelling of LTC should focus more on thermal comfort and energy savings. According to Nadeem Gulzar Shahmir[10], translucent cement can be used in green structures as a vitality effective and efficient source that has acceptable flexural quality, compressive quality, and light trans mitting qualities when used fine totals.

Experimental Investigation

Materials Used

Cement

Ordinary Portland Cement (OPC) of Grade 53 was employed in this research work. Specific gravity, fineness modulus, initial setting time and final setting time. And their values are 3.15, 6, 38 mins and 585 mins respectively.

Water

The most vital and least costly component of concrete is water. For the experimental studies and curing, ordinary potable water available in the laboratory was used. The specific gravity of water is ideal i.e., 1.

Fine Aggregate

Locally available river sand which passes through 2.5 mm sieve is used as as fine aggregate in the present study. The values for specific gravity and fineness modulus are 2.70 and 2.50 respectively. Water absorption value is 1.3%.

Coarse Aggregate

The work was done with angular coarse aggregate. The aggregate seemed to have a nominal size of 20mm. The values for specific gravity and fineness modulus are 2.80 and 7 respectively. Water absorption value is 1.1%.

Optical Fiber

The optical fiber is collected from the market. The optical fiber is inserted into the wooden mould for about 10 cm, for the light transmission purpose. The diameter and aspect ratio are 0.75 mm and 133.33 respectively.

Figure 1 Wooden Mould with Optical fiber

Glass
The glass we have used is taken from old glass bottles. They are broken down into pieces and sieved using the sieve sets 20 mm, 16 mm and 12 mm. The broken glass particles which were passed through 20 mm sieve and retained on 16 mm and 12 mm sieve.

Figure 2 Broken Glass Aggregate

Superplasticizer
Superplasticizers, often known as water reducers, have a wide spectrum of water reduction properties. In this work, Master Gelenium 430 is used.

Mix Design
IS 10262 – 2019 is used to calculate the mix percentage for M30 grade. The mix proportion obtained is 1:1.74:2.64.

Tests Conducted

Compressive Strength Test For Blocks
The block size employed was 200*100*100 mm in dimension. After wiping away the surface moisture from the specimen, the mould was examined in a saturated state. At the age of 28 days, cubes were examined using a compression testing equipment for this study. The block is placed on the testing equipment and the equipment is switched on. When the block attains a ultimate load the block gets broken or gets cracked and the load transmission is dropped off. The specimens were casted and tested as per IS 516:1959 The blocks tested in compressive testing machine setup is shown in Figure 3.

Figure 3 Block during Compression Test

Water Absorption Test for Blocks

For water absorption test, the block is weighed soon after demolding. This weight is termed as dry weight of the block. For wet weight of the block, the block weight is taken after 28 days of curing. And then the results were made.

Results and discussions

Compressive strength test results

The specimens were tested as per IS 516:1959. The blocks were tested in compressive testing machine as shown in Figure 4. The compressive strength results were tabulated in Table 1.

Figure 4 Compression Test Setup

Table 2 Compression Test Results

Specimen	**% of glass**	**Average compressive strength N/mm²(28 days)**
Conventional concrete	10%	31
Translucent Concrete (Block 1)	0%	32
Translucent Concrete (Block 2)	5%	35
Translucent Concrete (Block 3)	10%	33

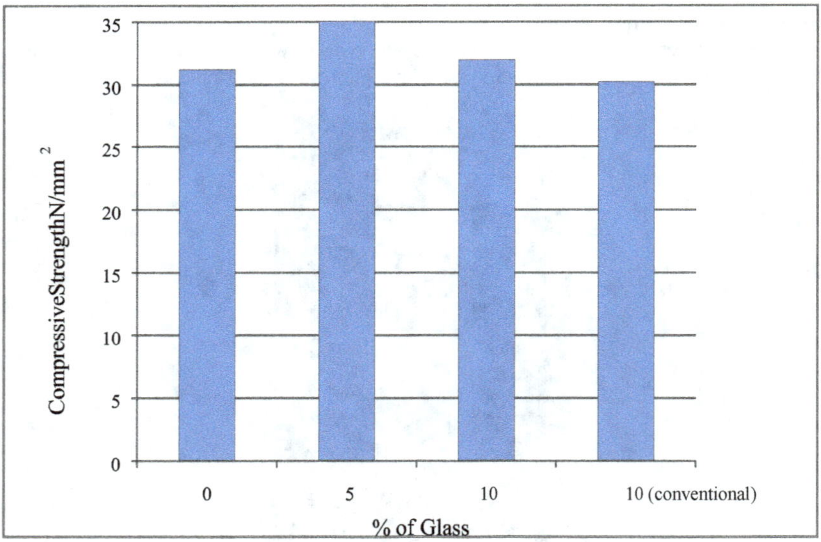

Figure 5 Variation in compressive strength

Water Absorption Test Results
Water Absorption (%) of the specimens were performed and the results are tabulated in Table 3.

Materials Research Forum LLC
https://doi.org/10.21741/9781644901953-15

Table 3 Water Absorption Test Results

Specimen	Water Absorption (%)
Conventional concrete	1%
Translucent Concrete (Block 1)	2%
Translucent Concrete (Block 2)	2%
Translucent Concrete (Block 3)	1.2%

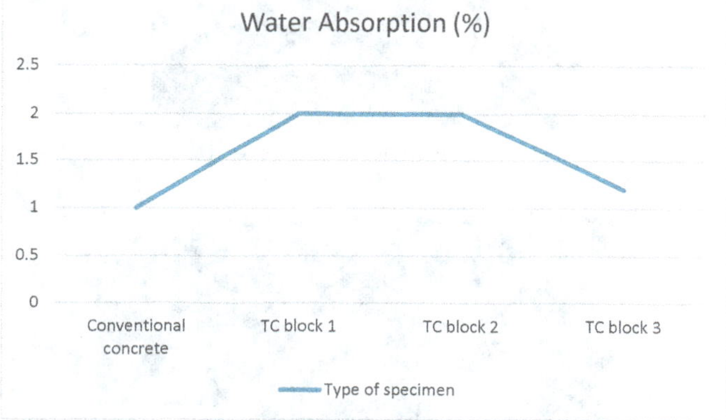

Figure 6 Variation in Water Absorption Capacity of Blocks

From the test results, it was observed that addition of glass particles contributes to the compressive strength of concrete. Glass aggregates when replaced up to 5% gives good compressive strength. Water absorption capacity of the blocks are also within allowable limit.

Light Emitting Property

The light transmission of the block is examined. When light is supplied to one end of the block, the light is transferred to the opposite end without major loss in transmission. As a result, light transmission is a successful one in this type of blocks. Light emitting property of block is shown in Figure 7.

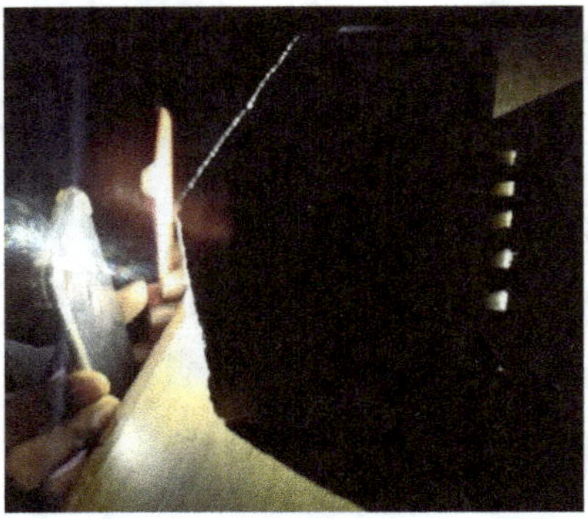

Figure 7 (a) Light emitting property of translucent block

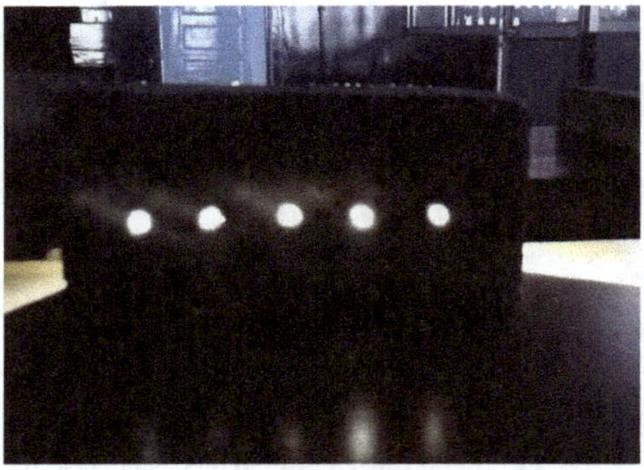

Figure 7 (b) Translucent block

Conclusion

An attempt has been made to prepare a concrete using waste glass pieces which also contribute to the translucent properties. Optical fibers are incorporated inside concrete without affecting the conventional properties. Concrete blocks were prepared of dimension 200*100*100 mm and properties like compression strength and water absorption were studied. When coarse aggregate are replaced with 5% glass aggregates, the compressive strength is effective. Light transmission is also provided with the help of optical fibres. The work can be continued with more layers of optical fiber.

References

[1] Ahmed M Tahwia, Ahmed Abdel Raheem, Nirmen Abdel Aziz, Mohamed Amin, "Light transmittance performance of sustainable translucent self-compacting concrete",Journal of Building Engineering (2021), volume 38, June 2021, 102178. https://doi.org/10.1016/j.jobe.2021.102178

[2] Arul Arulrajah, Alireza Mohammadinia, Sahan Perera, Zipeng Zhang, Yat Choy Wong, "Field study on concrete footpath with recycled plastic and crushed glass as filler materials", Construction and Building materials (2020), Volume 243, May 2020, 118277. https://doi.org/10.1016/j.conbuildmat.2020.118277

[3] Hani Nassif, Hongye Gou, Yi Ba, Pengwei Guo, Weina Meng, "New perspectives on recycling waste glass in manufacturing concrete for sustainable civil infrastructure", Construction and Building materials (2020), Volume 257, October 2020, 119579. https://doi.org/10.1016/j.conbuildmat.2020.119579

[4] Jinbu Lian, Ling Zhang, Ping Liang, Zhong bing, "Daylighting performance and simulation and analysis of Translucent Concrete building envelopes", Renewable Energy (2020), volume 154, July 2020, Pages 754-766. https://doi.org/10.1016/j.renene.2020.03.041

[5] Gregory F. Lane-Serff, Lee S. Cunningham, Nicholas Omoding, "Effect of using recycled waste glass coarse aggregates on the hydrodynamic abrasion resistance of concrete", Construction and Building materials (2020), Volume 268, January 2021, 121177. https://doi.org/10.1016/j.conbuildmat.2020.121177

[6] Zhuan Shen and Zhi Zhou, "Performance and Energy savings of Resin Translucent Concrete", Energy Engineering (2020), Volume 146, Issue 3, June 2020. https://doi.org/10.1061/(ASCE)EY.1943-7897.0000652

[7] Han Seung Lee, Mohd Azrun, Mohd Ariffn, Shing Mei Chiew, "Development and Properties of light transmitting concrete", Journal of Cleaner Production (2020), Volume 284, February 2021, 124780. https://doi.org/10.1016/j.jclepro.2020.124780

[8] Amir Mosavi, Danial Navabi, Mahyar Javiduzi, Mohamad Reza Hafezi, "The high performance light transmitting concrete and experimental analysis using Polymethyl methacrylate optical fibers in it", Journal of Building Engineering (2021), Volume 38, June 2021, 102076. https://doi.org/10.1016/j.jobe.2020.102076

[9] Han Seung Lee, Mohd Azrun, Mohd Ariffn, Shing Mei Chiew, "Development and Properties of light transmitting concrete", Journal of Cleaner Production (2020), Volume 284, February 2021, 124780. https://doi.org/10.1016/j.jclepro.2020.124780

[10] Manzoor Ahmad Tantary, Nadeem Gulzar Shahmir , "A Review on Green Buildings Using Translucent Concrete as Energy Efficient Source", Iconic Research and Engineering (2019), Volume 2 Issue 7, ISSN: 2456-8880.

Experimental Study on Strength Characteristics of Fly Ash and Rice Husk Ash added Clay Soil

J. Jayashree[1,a *], V. Jeevanantham[2,b]

[1]Department of Civil Engineering, Sri Ramakrishna Institute of Technology, Coimbatore, INDIA

[2] Department of Civil Engineering, Bannari Amman Institute of Technology, Sathyamangalam, INDIA

*[a]jjayashree16@gmail.com, [b]jeeva.civil.tpgit@gmail.com

Keywords: Clay Soil, Flyash, Rice Husk Ash, Dry Density, Moisture Content, Unconfined Moisture Content

Abstract. Clay soil is usually stiff in dry state, but in wet state, it loses its stiffness and becomes softer. Various types of clay minerals have unique characteristics of low bearing capacity and certain compressibility. Especially Montmorillonite clay mineral exhibits more degree of swelling and shrinkage and loss of strength when moisture added. According to recent research, rice husk ash is a potential improving material based on its pozzolanic activity. In this study, to improve the engineering properties of expansive clay the admixture such as fly ash and rice husk ash will be added in different percentages as 5, 10, 15 and 20. For the laboratory test that includes atterberg's limits test, standard proctor compaction test and unconfined compression test were conducted on locally collected expansive soil with various percentages of these two stabilizers in order to examine their influences. The maximum values of Unconfined Compressive Strength were obtained corresponding to 15% Rice Husk Ash and 20% Fly ash.

Introduction

Building foundations bears the entire super structure load. The soils beneath the foundations hold play a crucial part in holding the foundation. But Expansive soils, such as montmorillonite deforms more provides lesser resistance against the foundation load. One-fifth of the Indian land is covered by these expansive soils, produces difficulty to building & Road construction activities.

A combination of swelling and shrinkage occurs under moisture variations, causing structural failures. They are not recommended due to low strength, high compressibility and volumetric changes. Improvement of soils may be done through modification of soils or stabilization technique. When cement, fly ash, etc. is added to a soil, it changes its properties called soil modification.

Soil stabilization

Soil stabilization alters the soil properties by physical or chemical means to improve the engineering quality of the soil. We shall increase the bearing capacity of the soil, reduces the compressibility & permeability. The structure long-term performance will be boosted. We shall ensure the good stability of the superstructure particularly in case expansive soil, the chemical stabilization of clay soil using lime is becomes common method and requires more investment, when used for road constructions.

Materials used

Black cotton soil

Black cotton soil, which is rich in montmorillonite clayey mineral and exhibits higher degree of shrinkage, swelling and compressibility nature under loading due to its chemical composition. Higher the montmorillonite presence, it becomes black or blackish grey in colour. Montmorillonite has the structural composition of 1 octahedral alumina sheet and 2 tetrahedral silica sheets in the structural term defined as 2:1 composition. Due to the Isomorphous substitution they are highly adsorbs the water and forms a diffusible double layer around the surface. It becomes more expansive in nature with increase in moisture content. This water imbibing phenomena is occurring in higher degree comparatively with kaolinite and bentonite clay mineral. Soil with higher montmorillonite mineral quickly deforms when moisture added and results in poor stability.

The sample used for laboratory investigation was collected at a depth of 2 meter from ground surface of the site located in Vadavalli and Veerakeralam of Coimbatore district. Location: 11.006729755508342 N, 76.91312369798602 E

Rice husk ash

Figure.1 shows the rice husk ash collected for our study. From the rice mill industry we shall collect the rice husk, which comes out of the industry as a waste product, after separating the rice millets. In statistical point of view, yearly hundred tonnes of rice husk is produced worldwide. The burn product of rice husk taken ash for this study. The ash holds more than 60% of silica, 49 % of alumina with pozzolanic property. Pozzolanic property of this material aids to use this as a partial complementary constituent of cement. Replacing the cement by considerable amount shall result in economic construction with added advantage of strength improvement of soil sub layer.

We have collected rice husk from Coimbatore for conducting the laboratory investigation on the black cotton soil.

Fig.1. Rice husk ash

Fly ash

Fly ash is a soft and powder like materials results from the coal burning at the power stations. Figure.2 shows that collected locally available Flyash poured on the sample prior to mix. Fly ash also exhibits pozzolanic nature. When it reacts with lime, it becomes cementitious materials. Nowadays flyash are directly used for brick production as well, since it is cheap and exhibiting engineering property.

Fig.2.Fly ash

Literature review

Harichane et al. (2011) compared the utilization of natural pozzolana, lime and combination of both, with 2 types of cohesive soils. Consistency, undrained triaxial test, UCC strength test and compaction test was carried out on samples mixed with additives and cured for 1,7,28 and 90 days. The combined action of lime and pozzolana improved the properties of soils.

Phanikumar (2009) compared the effect of lime and flyash in counteracting the problems associated with expansive soil. Base Exchange took place while adding lime and fly ash to clay. This reduces the Plasticity Index values, reduction in degree of swell, increase of dry density, decrease of OMC. Consolidation characteristics such as compression index Cc and coefficient of consolidation, Cv increased and then decreased with increase of % addition. The optimum dosage could be fixed as 5% for rice husk ask and 20% for flyash.

Canakcia et al. (2015). A study on stabilization of highly expansive soil prepared by mixing red clay (70%) and bentonite (30%) was carried out. The additives used were lignin, rice husk ash and rice husk powder. The percentages of 5, 10, 15 and 20 of lignin and rice husk powder were used. Rice husk ash in increasing order of 2.5%, 5%, 7.5% and 10% was also used. A moisture content of 25% was adopted for all mixtures. Unconfined Compression strength tests, Swelling test, Atterberg limit test were conducted on the samples cured for 3 and 7 days. The additives improved the strength significantly. Liquid limit decreased and plastic limit increased with increase in percentage of additives. Swelling also reduced upon addition of additives.

Kolay et al. (2011) described the results of laboratory investigation of peat soil stabilized using quick lime (QL), Ordinary Portland cement (OPC) and Class F flyash. UCS of stabilized peat soil increased with increasing percentage of OPC whereas it decreased beyond 6% of QL and 15% of FA. Curing also increases the UCS value.

Methodology

Earlier research works concludes that 20% fly ash will be optimum for stabilizing the clay soil. In our study, we kept fly ash additive as 20% as constant and we increased the rice hush ash in different quantities i.e (5%,10%,15% and 20%)

 Materials Research Forum LLC
https://doi.org/10.21741/9781644901953-16

Laboratory Investigation

Preliminary tests are conducted for representative clay soil in order to determine the Index and Engineering properties. The laboratory investigation includes Atterbergs limits, particle size distribution, Standard Proctor Compaction test, UCS test as per IS Codes

Properties of Representative Clay

S.No.	Property	Clay	Fly ash	Rice husk ash
1	Liquid Limit	40%	-	-
2	Plastic Limit	29.4%	-	-
3	Plasticity Index	12.6%	-	-
4	Shrinkage Limit	10.8%	-	-
5	Specific gravity	2.76	2.02	0.86
6	Maximum dry density	1.642 g/cc		
7	Optimum Moisture content	25%		
8	Unconfined compressive strength	103.9 kN/m^2		

Influence of Fly ash & Rice husk ash in Properties of Soil

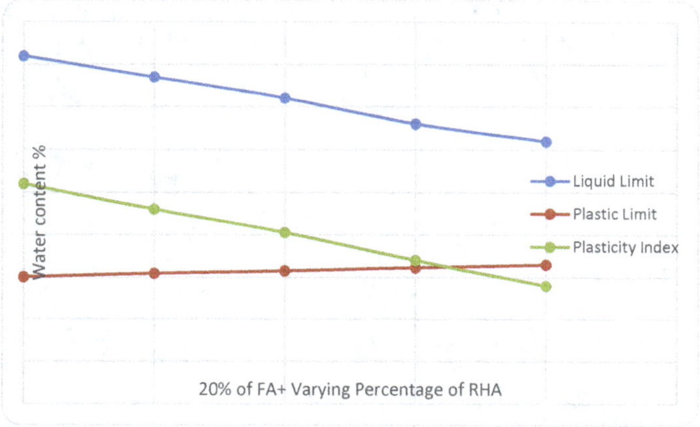

Fig. 3 Influence of Fly ash and Rice husk ash on LL,PL and PI

Figure 3 shows the variation of fly ash and rice husk ash composition influenced the atterberg's limits and plasticity index significantly. For example, Liquid Limit decreased from 82 to 62% with

Materials Research Proceedings **23** (2022) 122-127 https://doi.org/10.21741/9781644901953-16

the content of 20% fly ash and rice husk ash from 5% to 20%. Plastic Limit increased slightly from 30 to 33%. The Plasticity Index decreased from 52 to 28%, indicating negligible plasticity.

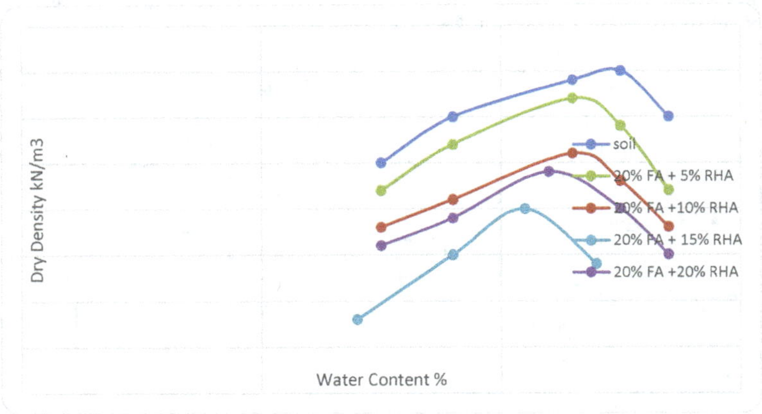

Fig. 4 Influence of Fly ash and Rice husk ash on Compaction characteristics

Figure 4 shows the Comparative results of Proctor compaction for the representative soil and the soil mixed with 20% Fly ash and different rice hush ash contents (5%,10%,15% and 20%). From the graph we observe that the Proctor compaction curves shifting downwards, which clears the increase in additives reduces optimum moisture content and dry density.

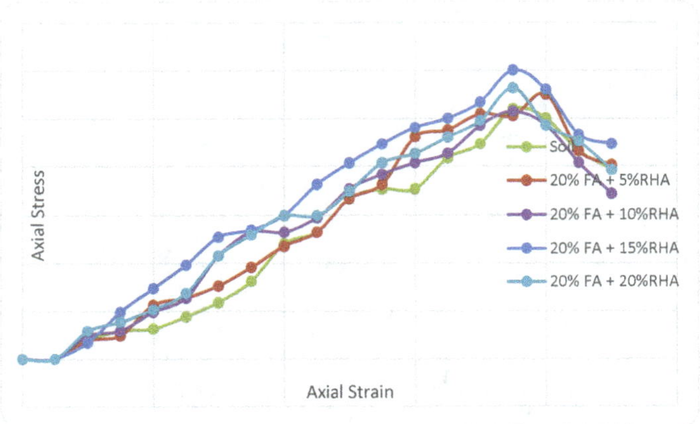

Fig. 5 Influence of Fly ash and Rice husk ash on compressive strength

Materials Research Proceedings **23** (2022) 122-127
https://doi.org/10.21741/9781644901953-16

From the Figure 5, we shall observe that with the increase in % addition of fly ash and rice husk ash content, Compression strength value keep on increasing upto 20 % of fly ash and 15% of rice husk ash after that peak point further addition, decreases the unconfined compressive strength. As a result, combining effect of fly ash and rice husk ash into clay soil improves the soil's strength.

Results & Discussion

It has been found through various laboratory experiments that by keeping the constant fly ash constant and changing the addition percentage of Rice Husk Ash at 5%, 10%, 15%, and 20%, expected engineering properties of soil improved.

The following conclusions were drawn:

1. With increasing percentages of Fly ash and Rice husk ash in black cotton soil, Atterberg's limits are decreases.
2. The shrinkage values were low. Having a low shrinkage rate indicates a lower risk of defect, such as warping and cracking.
3. The maximum dry density value of the soil decreases from 15.5 kN/m^3 to 14kN/m^3 with increase in Rice Husk Ash content and the maximum dry density was obtained in 20% of fly ash and 15% of rice husk ash.
4. The maximum Unconfined Compressive Strength values were obtained corresponding to 15% Rice Husk Ash and 20% Fly ash.

Based on the above said results, Soil properties were improved by using 20% of fly ash and 15% rice husk ash. Hence the optimum value of rice husk ash and fly ash which can be adopted to obtain improved soil properties is the proportion of 20% fly ash and 15% rice husk ash.

References

[1] B. R. Phanikumar . T. V. Nagaraju, " Effect of Fly Ash and Rice Husk Ash on Index and Engineering Properties of Expansive Clays," Journal of Geotechnical and Geological Engineering, pp. 3425–3436 (2018). https://doi.org/10.1007/s10706-018-0544-5

[2] Erdal Cokca, "Use of Class C Fly Ashes for the Stabilization – of an Expansive Soil" Journal of Geotechnical and Geo-environmental Engineering Vol. 127, July, pp. 568-573. https://doi.org/10.1061/(ASCE)1090-0241(2001)127:7(568)

[4] E.A Basha, R.Hashim, H.B Mahmud, A.S Muntohar, "Stabilization of residual soil with Rice Husk Ash and Cement", Construction and Building Materials, July 2005. https://doi.org/10.1016/j.conbuildmat.2004.08.001

[5] Basha AM, Hashim R, Muntohar AS, "Effect of the cement rice husk ash on the plasticity and compaction of soil". Electronic Journal of Geotechnical Engineering 8:1–8.

Experimental Investigation on Properties of Herbocrete by using Terminalia Chebula (Kaddukai)

M.P. Indhu[1,a,*], S. Krishnamoorthi[2,b], M.P. Iniya[3,c], T. Shanmugapriya[3,d], A. Sanjeevi[3]

[1]Kongunadu College of Engineering and Technology (Autonomous), Trichy

[2]Kongu Engineering College, Erode

[3]Kongunadu College of Engineering and Technology (Autonomous), Trichy

[a]indhuperiasamy98@gmail.com, [b]skmoor.civil@kongu.ac.in, [c]iniyaviji@gmail.com, [d] shanmu555shanmu@gmail.com

Keywords: Terminalia Chebula (Kadukkai), Polypropylene Fiber, Compressive Strength, Flexural Strength, Tensile Strength

Abstract. Additives are ingredients and chemicals that are mixed into concrete to improve its fresh state and change its properties. The majority of admixtures are sold as chemical compositions on the market. These chemical admixtures are exceedingly costly and harm the environment significantly. Adding admixtures to concrete, on the other hand, is not a new notion; it has been used and developed in numerous countries in the past. In this project, we are investigating the possibilities of using natural materials like "Terminaliachebula" (Kadukkai). Terminalia chebula extract is added in different concentrations of 2.5%, 5%, 7.5% to the cement concrete, and its effects on the workability and strength of concrete are studied. Terminalia chebula when added to cement mortar does not affect however, it improves workability while also increasing compressive strength by 1.832 times of reference concrete. But the initial setting time of concrete is increased by 30 minutes of nominal OPC concrete.

Introduction

India's building industry is the country's second largest. Chemicals have been used in construction in the past to improve the fresh condition and hardened qualities (maximum compressive strength). This has resulted in various types and quantities of contamination. For more than a century, our forefathers have employed numerous plants as an admixture in construction to increase the overall efficiency of the structure. Researchers want to look into the feasibility of employing They're put to the test on kadukkai-infused concrete specimens, which are then assessed for concrete mix, serviceability, and moisture. It had the best results when compared to the comparable coarse aggregate. Terminaliachebula was used as a concrete additive to create a greener, more environmentally friendly construction with enough quality and durability at a fair cost. Terminaliachebula does not affect the workability of cement mortar, however it does improve the compressive strength of reference mortar. To make the plaster, this mixture was combined with lime. Natural admixtures were used in both mortar and concrete as a result of this research review. I intend to test the Concrete's crushing, bending, and splitting tensile strengths were measured the 10 combinations of kadukkai and eggshell powder.

One of the approaches for improving the structural qualities of concrete is fibre dispersion. Polypropylene fibres are synthetic fibres that come from the textile industry as a waste product. They are inexpensive and come in a variety of aspect ratios. Inexpensive specific gravity and low cost are two properties of polypropylene fibres. Its application enables the material's natural tensile and flexural strength to be reliably and effectively utilised, as well as a significant reduction in plastic shrinkage cracking and heat cracking. It protects the concrete structure from damage and avoids spalling in the event of a fire by providing reinforcement and protection. Pulling wires with a circular cross section or extruding plastic films with a rectangular cross section are used to make the fibres. Fibrillated bundles or monofilament are the two forms they take. The fibrillated polypropylene fibres are created by slitting a plastic film that has been divided into strips. The fibre bundles are fibrillated and sliced into predetermined lengths. When buttons are added to the ends of monofilament fibres, the pull-out load is increased.

Methodology

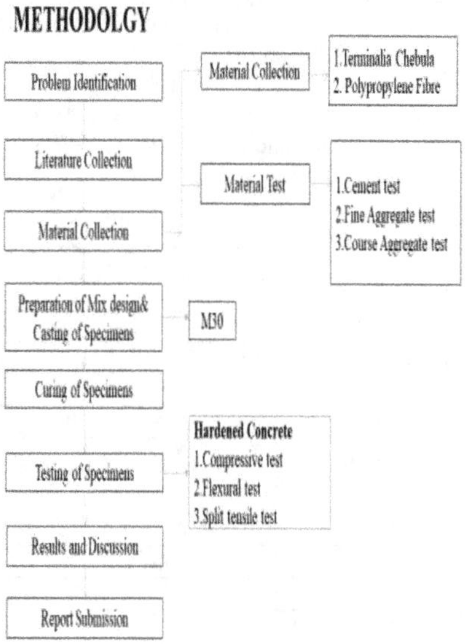

Material Used

Terminalia Chebula (Kaddukai):
Kadukkai is a fruit that can be height. It is used in the product of many herbal remedies for its good medicinal properties. It has extraordinary binding and hardened properties. Kadukkai can be used in construction only after it is made into a dried powder form for its effective usage. The dried kadukkai powder is mixed with sufficient water and enclosed for three days for making kadukkai paste. This paste is used for mortar preparation.

Polypropylene Fiber:
Polypropylene fibres have a tendency to keep the concrete mixture together. This decreases bleeding by slowing the settlement of coarse material. Slower bleeding equals slower drying, which means less plastic shrinkage breaking. Polypropylene fibres serve as crack stoppers in hardened concrete.

Experimental

Mixing Proportions:

The mix proportion for this study is 1:3 and it is the rich mortar ratio. The various proportions of natural admixtures like polypropylene fiber and kadukkai vary from 5% and 7.5%. Each proportion of cement mortar cast three specimens for testing.

Mixing, Casting, and Curing:

The required amounts of materials are taken as per the mix proportion. The natural admixtures like kadukkai paste were mixed with the water content for their effective usage. The percentage of natural admixtures added to the water content for the weight of fiber taken for the mortar. For the conventional mortar the desired quantity of cement is mixed with aggregates and water (Without Natural Admixtures), but for cement mortar, the quantity of 50% cement was replaced by the polypropylene fiber and they are mixed with fine aggregates(sand) and water. Then the fresh mortars were cast into 150mm cube molds for compression testing and were removed after 48 hours. The specimens were separated from the mold and were kept in water for 7-28 days for water curing. These steps are repeated for each proportion(5%, 7.5%) of cement mortar.

Result

Compressive Strength Test:

	No., of Days	**Compressive Strength Test**
5% of Kadukkai & 1% of PP	7	28.97 (N/mm^2)
7.5% of Kadukkai & 1.5% PP	7	27.95 (N/mm^2)
5% of Kadukkai & 1% of PP	14	36.59 (N/mm^2)
7.5% of Kadukkai & 1.5% PP	14	26.784 (N/mm^2)
5% of Kadukkai & 1% of PP	28	36.59 (N/mm^2)
7.5% of Kadukkai & 1.5% PP	28	26.784 (N/mm^2)

Split Tensile Test:

	No., of Days	**Split Strength Test**
5% of Kadukkai & 1% of PP	7	3.394 (N/mm^2)
7.5% of Kadukkai & 1.5% PP	7	4.38 (N/mm^2)
5% of Kadukkai & 1% of PP	14	3.368 (N/mm^2)
7.5% of Kadukkai & 1.5% PP	14	3.825 (N/mm^2)
5% of Kadukkai & 1% of PP	28	3.358 (N/mm^2)
7.5% of Kadukkai & 1.5% PP	28	3.938 (N/mm^2)

Flexural Strength Test:

	No., of Days	**Flexural Strength Test**
5% of Kadukkai & 1% of PP	7	3.358 (N/mm^2)
7.5% of Kadukkai & 1.5% PP	7	3.839 (N/mm^2)
5% of Kadukkai & 1% of PP	14	3.452 (N/mm^2)
7.5% of Kadukkai & 1.5% PP	14	3.512 (N/mm^2)
5% of Kadukkai & 1% of PP	28	3.363 (N/mm^2)
7.5% of Kadukkai & 1.5% PP	28	2.997 (N/mm^2)

Materials Research Proceedings **23** (2022) 128-132 https://doi.org/10.21741/9781644901953-17

Conclusion

- The addition of an additive to concrete changes the hardness and freshness of the material. The use of natural additive reduces CO2 emissions from concrete and alters its mechanical performance. According to the findings, concrete with 5% Kadukkai and 1% PP had the highest compressive strength of 28.97 N/mm2 after 7 days. After 28 days, the highest concrete with 5% Kadukkai and PP of fibre strength properties is 36.59 N/mm2, according to the findings.
- After 7 and 28 days of fermentation with 5% Terminaliachebula extract and Polypropylene extract, the compressive strength increased by about 1.23 times that of reference concrete. When compared to a compressive strength of 5%, the compressive strength was lowered by 7.5 percent.
- The retarder is to blame for the 7.5 percent reduction in compressive strength. Natural admixtures are safe for the environment. When compared to chemical admixture, they were treated as emission admixture. It acts as a retarder in concrete, allowing it to adapt to hot climatic circumstances while maintaining the workability of the material.

Reference

[1] Thirumalini et all (2011) "Study on the performance enhancement of lime mortar used in ancient temples and monuments in India" Indian journal of science and technology. https://doi.org/10.17485/ijst/2011/v4i11.23

[2] Sabine ruckstuhl et all (2001) "Environmental Exposure assessment of Sulfonated Naphthalene formaldehyde condensates and Sulfonated Naphthalene applied as concrete Super plasticizers" Gordon research conference "Environmental science water" in Plymouth, N.I.L., USA,2000.

[3] C. Gunasekara, D.W. Law, S. Setunge, J.G. Sanjayan, Zeta potential, gel formation and compressive strength of low calcium fly ash geopolymers, Constr. Build. Mater. 95 (2015) 592–599. https://doi.org/10.1016/j.conbuildmat.2015.07.175

[4] M. Schneider, M. Romer, M. Tschudin, H. Bolio, Sustainable cement production present and future, Cem. Concr. Res. 41 (7) (2011) 642–650. https://doi.org/10.1016/j.cemconres.2011.03.019

[5] L.K. Turner, F.G. Collins, Carbon dioxide equivalent (CO2-e) emissions: a comparison between geopolymer and OPC cement concrete, Constr. Build. Mater. 43 (2013) 125–130. https://doi.org/10.1016/j.conbuildmat.2013.01.023

[6] P.W. Ken, M. Ramli, C.C. Ban, An overview on the influence of various factors on the properties of geopolymer concrete derived from industrial by-products, Constr. Build. Mater. 77 (2015) 370–395. https://doi.org/10.1016/j.conbuildmat.2014.12.065

[7] D. Ravikumar, S. Peethamparan, N. Neithalath, Structure and strength of NaOH activated concretes containing fly ash or GGBFS as the sole binder, Cem. Concr. Compos. 32 (2010) 399–410. https://doi.org/10.1016/j.cemconcomp.2010.03.007

[8] S. Thirumalini, R. Ravi, M. Rajesh, Experimental investigation on physical and mechanical properties of lime mortar: Effect of organic addition, Journal of Cultural Heritage, 31 (2018) 97–104. https://doi.org/10.1016/j.culher.2017.10.009

[9] C. Nunes, Z. Slížková, Freezing and thawing resistance of aerial lime mortar with metakaolin and a traditional water-repellent admixture, Construction and Building Materials, 114 (2016) 896–905. https://doi.org/10.1016/j.conbuildmat.2016.04.029

[10] P. Pahlavan, S. Manzi, A. Sansonetti, M. C. Bignozzi, Valorization of organic additions in restorative lime mortars: Spent cooking oil and albumen, Construction and Building Materials, 181 (2018) 650–658. https://doi.org/10.1016/j.conbuildmat.2018.06.089

Sustainable Materials and Smart Practices - NCSMSP-2021 Materials Research Forum LLC
Materials Research Proceedings **23** (2022) 133-147 https://doi.org/10.21741/9781644901953-18

Investigation on Strength of Concrete with Addition of Fly Ash and Natural Fiber

N. Chandramohan[1], S. Loganayagan[1], S. Sridhar[1], M. Poornima[1]

[1]Department of Civil Engineering, Bannari Amman institute of Technology, Sathyamangalam, India

Abstract. Green buildings are becoming rising trends in the construction industries. Now-a-days the construction projects focus on the use of recycled and efficient materials in order to reduce the emission of carbon-di-oxide. Our objective is to make a product using fly ash and naturally available material like coconut fiber and rice husk ash which can be used as admixtures with cement. These substitute materials are quickly available in many areas at low cost. They are eco-friendly and also can reduce environmental pollution. They are renewable and less weight. Fibers have been used in the construction field for a long time. The usage of fibers in addition with construction materials like cement can promote eco-friendly and sustainable solutions. Our definitive target is to explore the effectiveness of fiber, rice husk ash and fly ash in concrete for structures. This project primarily centers the investigation of impacts of replacement of the concrete with various rates of fly ash and naturally available materials like rice husk ash and coconut fiber.

Keywords: Green Buildings, Coconut Fiber, Rice Husk Ash, Eco-Friendly and Sustainable Solution

Introduction

Research significance

As natural fibers are replacing a certain level of concrete, the outflow of ozone harming substances (CO2) can be diminished. To develop the normal fiber built up materials to composite dependent on items to substitute their customary designing materials [1]. Fly-ash is an eco-accommodating and green valuable material, which is utilized for different applications in the development field in light of its initial strength acquiring properties. The coconut fiber has the best strength among all the normal fiber. It can attain particular mechanical and actual properties at minimal price. The materials like coconut fiber, rice husk ash are promptly accessible in many places [2]. So it very well may be acquired for minimal price. By development of structures with these materials, it will upgrade the eco-accommodating climate in future.

Methodology

The materials have been finalized and collected. Initially tests for materials that are being used are done in order to check its properties. The mix design has been made as per IS 10262: 2009 with suitable water-cement ratio. Mixing has been done as per standard procedure from IS code. As soon as the mix is prepared, a slump test is done to check the workability of concrete. Concrete is poured into the molds of size 150 × 150 × 150 mm. The cubes were tested for compression strength after 7 days of curing.

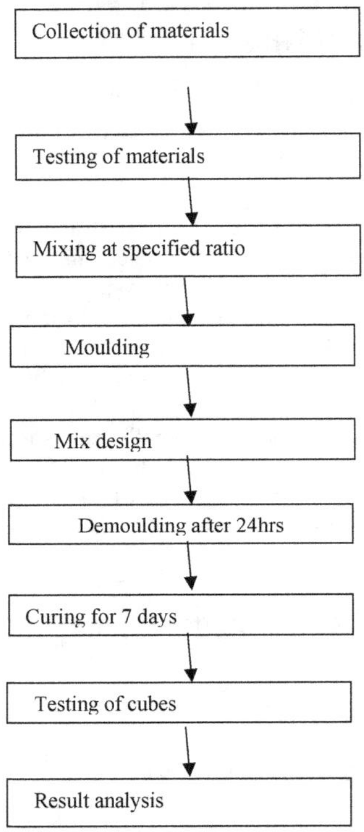

Fig 1: schematic representation of methodology

Materials

Cement

"Ultra-Tech" Ordinary Portland concrete of 43 grade was utilized. Concrete is predominantly utilized as a limiting material in concrete. It is the fundamental material for the wide range of development [3]. Cement in concrete provides strength, can harden early, possess good plasticity, have well moisture-resistant and acts as a good building material. The cement was confirmed to be IS 269-1976.

Properties of cement

Various mixes of concrete utilized in development are portrayed by their actual properties. Some boundaries control the nature of concrete[4]. The actual properties of good concrete depend on:

- Fineness of concrete
- Adequacy
- Consistency

- Strength
- Setting time
- Hotness of hydration

These actual properties are talked about in subtleties in the accompanying section. The detailed test procedures and results are discussed below.

Advantages

- Has great resistance to cracking and shrinkage.
- The initial setting is faster.

Disadvantages

- It can't be utilized for enormous scope cementing on account of its hotness of hydration property.
- The durability is lower.
- Has higher permeability and lower durability.
- Costlier.

Fine aggregate

Fine aggregates are originally known as sand. Sand is a granular surface made from finely divided stone and mineral particles. The squashed stones less than 4.75mm in size are known as fine totals[5]. The base sand size is 0.5mm to 0.25mm. It is of two types. One is naturally available river sand and the other is Manufactured sand (M sand).Sand has assorted structures anyway is depicted by its grain length. Sand grains are more modest than rock and coarser than residue [6]. Locally available river sand affirming to IS 393-1970 has been used.

Fig 2: River sand

Coarse aggregate

Coarse aggregates are bigger in size than fine aggregate materials in construction. Brick chips (broken blocks), stone chips (broken stones), rock, stones, clinkers, soot and so forth are utilized as coarse total in concrete [7]. Dolomite totals, squashed rock or stone, regular crumbling of rock are the significant wellsprings of coarse aggregate. Coarse total goes about as inactive filler material for concrete [8]. Coarse totals are essentially utilized in concrete, rail route track stabilizer, and so on.

Fig 3: 20mm coarse aggregate

Water

As per IS 456 : 2000, water used for mixing the concrete will be spotless and liberated from damaging the measures of sugar, oils, acids, soluble bases, salts, natural materials or different substances that might be pernicious to the steel. Portable water with pH value at least 6 is most suitable for blending the concrete.

Fly ash

Fly ash is a fine powder which is gray in color composed of round, shiny particles that are created as a result in thermal power stations. Fly ash has pozzolanic properties, implying that it responds with lime to react as shape cementitious mixtures [9]. It is usually known as an advantageous cementitious material. Fly ash is very helpful for various uses, including: Dams, Mines concrete, Landfills.

Fig 4: Fly ash

Advantages of fly ash

- It can add more strength to the structure.
- The substantial combination creates an extremely low heat of hydration which prevents warm breaking.
- Fly Ash concrete is impervious to corrosive and sulfate assaults.
- The shrinkage of fly debris concrete is extremely less.

Rice husk ash

Silica exists richly in rice husks. RHA was produced by calcining rice husk at temperature of 500 °C and fused in substantial combinations to supplant diverse proportions of silica seethe (SF) by weight. Flow ability and air content of the new blend and the compressive and flexural strength at various restoring ages were estimated [10]. Porousness prior and then afterward stacking were evaluated by water assimilation and chloride particle entrance. In this experiment rice husk ash is used in different percentages as 15%, 10%, 7%, 3%.

Materials Research Proceedings **23** (2022) 133-147 https://doi.org/10.21741/9781644901953-18

Fig 5: Rice Husk Ash (RHA)

Properties

The fuse of rice husk ash in concrete makes it into an eco-friendly valuable cementitious material.

Benefits

- The hotness of hydration is decreased. This itself helps in drying shrinkage and work with sturdiness of the substantial blend.
- The decrease in the penetrability of substantial design. This will help in infiltration of chloride particles, along these lines staying away from the deterioration of the substantial design.
- There is a higher expansion in the chloride and sulfate assault opposition.
- The rice husk remains in substantial response with the calcium hydroxide to bring more hydration items.

Coconut fiber

Coconut fiber is exceptionally modest and locally accessible in many nations. The utilization of coconut fiber brings about minimal expense and all the more harmless to the ecosystem concrete [11]. The advantages of normal strands may likewise be considered according to a natural point of view.

Fibers are collected from a farm, cleaned, dried in sun and removed dust to analyze the properties. Coconut has a high water absorbing property. Because of this property, the coconut fibers were pre-soaked in water for one day [12]. After this fibers are grained into fine powders. In this experiment coconut fiber is used in different percentages as 10%, 5%, 3%, 2%.

Fig 6: raw coconut fiber

Fig 7: powdered coconut fiber

Properties & advantages

- The benefit of utilizing such filaments gives common minimal expense development and the disposal of the requirement for garbage removal in landfills.
- Use of these filaments in substantial quantities prompts a successful strong waste administration Technique.
- As coconut fibers have good water absorbing capacity, it also can act as an insulating material for buildings.

Material testing

Table 1: basic tests of cement

S. S.NO	PROPERTY	VALUE OBTAINED	STANDARD VALUE
1.	Fineness	6.5	do not exceed 10
2.	Consistency	30mm	35mm
3.	Specific Gravity	3.15	3.15
4.	Initial Setting Time	30mins	30mins
5.	Final Setting Time	800 mins	600 mins

Tests for fine aggregate

Specific gravity test

Weight of the empty pycnometer (W_1) = 681g
Weight of the pycnometer + sand (W_2) = 1393g
Weight of the pycnometer + aggregate in water (W_3) = 1968g
Weight of the pycnometer + water (W_4) = 1534g

Calculations

$$\text{Specific gravity (G)} = \frac{W_2 - W_1}{(W_4 - W_1) - (W_3 - W_2)}$$

$$= \frac{(1393 - 681)}{(1534 - 681) - (1968 - 1393)}$$

Specific gravity = 2.56

Fig 8: pycnometer for finding specific gravity of fine aggregate

Tests for coarse aggregate

Specific gravity test

Relative thickness, or specific gravity, is the proportion of the thickness (mass of unit degree) of a substance to the thickness of a given material. Explicit gravity for drinks is almost continually estimated as for water at its densest (at 4 °C or 39.3 °F); for gasses, the reference is air at room temperature (20 °C or 68.5 °F).

Weight of the empty pycnometer (W_1) = 420g
Weight of the pycnometer + CA (W_2) = 918g
Weight of the pycnometer + CA + water (W_3) = 1782g
Weight of the pycnometer + water (W_4) = 1466g

Calculation

$$\text{Specific gravity (G)} = \frac{W_2 - W_1}{(W_4 - W_1) - (W_3 - W_2)}$$

$$= \frac{(918 - 420)}{(1466 - 420) - (1782 - 918)}$$

Specific gravity = 2.7

Mix design

STIPULATIONS FOR MIX DESIGN CONCRETE

Grade of concrete	:	M25
Type of the cement	:	OPC 53 grade confirming to code IS 8112
Nominal size of aggregate	:	20 mm
Water-cement ratio	:	0.50

Workability : 75 mm (true slump)

TEST DATA FOR MATERIALS
Specific gravity test of the cement = 3.15
Initial setting time of the cement = 30 minutes
Final setting time of the cement = 800 minutes
Specific gravity of coarse aggregate = 2.7
Specific gravity of Fine aggregate = 2.60

MIX CALCULATIONS
Step 1 : Determination Of The Target Strength
target = f_{ck} + 1.65 x S
target= 31.6 N/mm2
Where, S = standard deviation in N/mm^2 = 4 (as per IS 10262- 2009)
Step 2 : Selection of water / cement ratio:-
From code IS 456-2000,
Maximum water-cement ratio = 0.55
w/c = 0.5
Step 3 : Water Content
From IS-10262,
Water content = 190 liters
Step 4 : Calculation of Cement Content
From water-cement ratio, w/c = 0.5
186/c = 0.5
C = 372 kg.
Step 5: Estimation of The Aggregate proportion
As per IS10262,
Coarse Aggregate ratio = 0.62
Fine Aggregate ratio = 1-0.62 = 0.38

MIX DESIGN
1. Volume of concrete = 1m^3
2. Total volume of the cement = cement / (S.G ×1000)
= 372/(3.14 ×1000)
= 0.118 cu.m
3. Volume of water = Water /(S.G×1000)
= 168/(1×1000) = 0.186 cu.m.
4. Total volume of aggregates = 1 – (B+C)
= 1 - (0.118+0.186)
= 0.696 cu.m.
5. Coarse Aggregate (CA) = TA × CA ratio × S.G × 1000
= 0.696 ×0.62 ×2.7 ×1000
= 1165 kg.
6. Fine Aggregate (FA) = TA × FA ratio × S.G × 1000
= 0.696×0.38×2.7×1000
= 687.64 kg.

5) MIX PROPORTION

Cement = 372 kg/m^3
Fine Aggregate = 687.64 kg/ m^3
Coarse Aggregate = 1165 kg/m^3
Water = 186kg/ m^3
Water Cement Ratio = 0.5

Table 2: quantity of materials needed for each mix

MATERIALS	MIX 1 PLAIN CONCRETE	MIX 2 FLY ASH – 25% CF – 10% RHA –15%	MIX 3 FLY ASH – 30% CF – 5% RHA – 10%	MIX 4 FLY ASH– 10% CF – 3% RHA – 7%	MIX 5 FLY ASH– 10% CF – 2% RHA – 3%
Cement	3.75kg	3.84kg	4.42kg	3.102kg	3.29kg
Water	1.88	3.72	3.72	1.88	1.88
Fine Aggregate	6.96kg	13.92kg	13.92kg	6.96kg	6.96kg
Coarse Aggregate	11.79kg	23.58kg	23.58kg	11.79kg	11.79kg
RHA	-	1.14kg	0.724kg	0.27kg	0.114kg
CF	-	0.72kg	0.367kg	0.114kg	0.075kg
Fly Ash	-	1.93kg	2.322kg	0.38kg	0.38kg

Casting and testing

Mixing procedure

Firstly the materials are weighed with a weighing balance. Then cement, fine aggregate are mixed up. The fibers and fly ash are mixed together with cement along with coarse aggregate. All the dry ingredients are mixed together. After mixing thoroughly water is added with the correct water cement ratio to form a concrete mix.

Fig 9: mixing of materials

Slump cone test

As soon as the concrete mix is ready it is tested for slump. It is done to decide the workability of the concrete mix prepared. The slump test is the most basic workability test for concrete mix which involves minimal expense and provides immediate outcome. The concrete mix is poured in a slump cone as 3 layers and it is damped using a damping rod of 25 blows for each layer. After that the value for slump is noted and the cone is removed.

Fig 10: slump cone test

Curing

After checking the slump cone, if it is a true slump the mix can be taken for casting cubes. The standard molds sizes 150×150×150mm which are completely fitted and oiled are utilized. As like slump cones the cube also has to be casted by placing 3 layers of concrete mix with 25 blows using a damping rod. The cubes are de-molded after 24 hours and placed in water curing for 28 days. By universal testing machine the cured cubes were tested after 28days.

Materials Research Proceedings **23** (2022) 133-147 https://doi.org/10.21741/9781644901953-18

Fig 11: Demoulding of cubes

Compression test

Compressive strength is known as load pressure of the concrete to axial loading. After curing 7 days, 14 days and 28 days the compressive test is done. Cubes were placed in the Universal Testing Machine (U.T.M), and load was applied. The readings were noted up to when the final crack of the cube were stopped and then compressive strength was calculated.

Fig 12: compression testing using UTM

Results and discussion

Slump test result

Table 3: slump cone test result

S.NO	MIX DESIGNATION	SLUMP IN mm
1.	Plain concrete	75mm
2.	Cement (50%) + fly ash (25%) + coconut fiber (10%) + rice husk ash (15%)	35mm
3.	Cement (55%) + fly ash (30%) + coconut fiber (5%) + rice husk ash (10%)	40mm
4.	Cement (80%) + fly ash (10%) + coconut fiber (5%) + rice husk ash (7%)	55mm
5.	Cement (85%) + fly ash (10%) + coconut fiber (2%) + rice husk ash (3%)	80mm

Table 3 shows the slump test result. It is observed that the slump value decreases with increasing the percentage of fly ash, rice husk ash and coconut fiber. This shows that the concrete becomes less workable as the fly ash, rice husk ash and coconut fibers are increased.

Cube compression test

Compressive Strength = Maximum load/Cross Sectional Area = P/A

Table 4: compression test results

S.NO	MIX DESIGNATION	COMPRESSIVE STRENGTH AFTER 7 DAYS (N/mm^2)
1.	Plain concrete (OPC)	16.28 N/mm^2
2.	Cement (50%) + fly ash (25%) + coconut fiber (10%) + rice husk ash (15%)	Failure while de-moulding
3.	Cement (55%) + fly ash (30%) + coconut fiber (5%) + rice husk ash (10%)	Failure while de-moulding
4.	Cement (80%) + fly ash (10%) + coconut fiber (5%) + rice husk ash (7%)	10.7 N/mm^2
5.	Cement (85%) + fly ash (10%) + coconut fiber (2%) + rice husk ash (3%)	16.45 N/mm^2

From the above table 4, it is observed that at 7 days, the concrete with 10% fly ash, 3% rice husk ash% and 2% coconut fiber gives the maximum result as 16.45 N/mm^2 compared to OPC 16.28 N/mm^2. The variation mixed with fly ash, rice husk ash and coconut fiber gives different results as the maximum strength reached with replacement of 15% cement and continued to decrease on further replacement of cement.

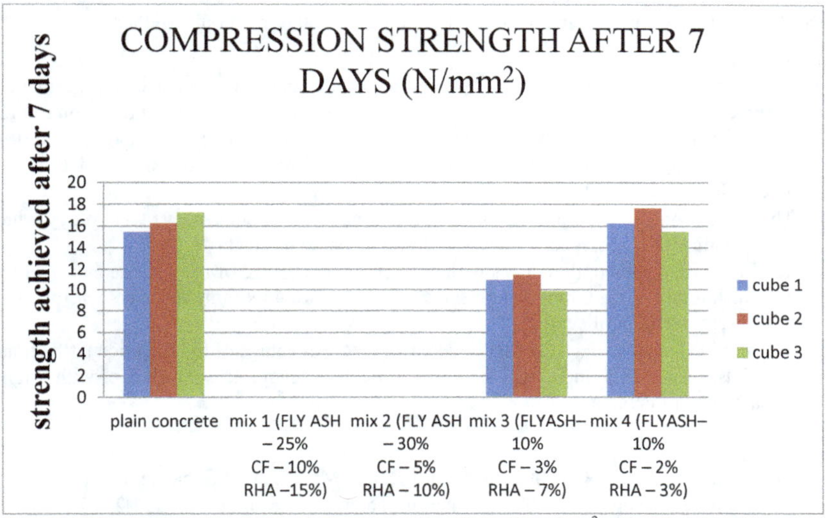

Fig 13: graph representing the compression strength (N/mm^2) after 7 days of curing

The result is graphically presented in figure 13 to analyze the relationship between compressive strength of concrete with various percentages of cement replacement of FA, RHA and CF. The maximum compressive strength of concrete for 7 days was at 15% replacement is 16.45 N/mm^2 while the lowest value of compression strength was at 20% reduction of cement with 10.7N/mm^2. The failure occurred with 50% and 55% reduction of cement resulting in failure while de-moulding the cubes.

Fig 14: Failure while de-moulding

The figure 14 shows the failure while de-moulding. This failure occurs with 50% and 55% reduction of cement with fly ash and fibers. This happened because the percentage of cement is reduced to a higher range. To reduce this failure the cement content has been increased as the cement acts as binding material in concrete which holds the other material together.

Conclusion

The need to create environmentally friendly materials which can serve as a partial replacement for cement in concrete mixtures has gained a lot of attention due to cement's contribution to the CO_2 emissions. Fly ash, coconut fiber and rice husk ash were added in percentage reduction of cement with 50%, 45%, 20% and 15% as shown in table 6.2. From the results obtained, the following conclusion is drawn:

- The use of coconut fiber and rice husk ash in concrete is an innovative technology which is economical, low-cost, eco-friendly and a sustainable product.
- The compressive strength reduces with increase in percentage of fly ash, coconut fiber and rice husk ash. The optimum compressive strength of 16.45N/mm^2 at 7 days was obtained at 15% replacement of cement.
- The slump value also decreases with increasing the percentage of fly ash, coconut fiber and rice husk ash. This indicates that the concrete became less workable with higher replacement of cement.

References

[1] Krishna N K, Prasanth M, Gowtham R, Karthic S and Mini K 2018 Enhancement of properties of concrete using natural fibers Materials Today: Proceedings 5 23816-23 https://doi.org/10.1016/j.matpr.2018.10.173

[2] Vajje S 2013 Study on addition of the natural fibers into concrete International Journal of Scientific & Technology Research 2 213-8.

[3] Pandi S S A, Roja S Y, Jenitha G and Alagusankareswari K 2017 Experimental Study on Behavior of Coir Fiber Reinforced Concrete International Journal of Civil Engineering and Technology 8

[4] SavastanoJr H, Warden PG, Coutts RSP. Microstructure and mechanical properties of waste fiber-cement composites, Cem. Concr. Compos. 2005; 27(5):583-592. https://doi.org/10.1016/j.cemconcomp.2004.09.009

[5] Siddique R. properties of concrete incorporating high volumes of class F fly ash and san fibers, Cem. Concr.Res. 2004; 34(1):37-42. https://doi.org/10.1016/S0008-8846(03)00192-3

[6] Filho RD, Toledo, Sanjuan MA. Effect of low modulus sisal and polypropylene fiber on the free and restrained shrinkage of mortars at early age, Cem. Concr. Res. 1999; 29(10):1597-1604. https://doi.org/10.1016/S0008-8846(99)00136-2

[7] Bhowmick B.B. and Debnath C. R. (2007), "Properties of coir", Indian coconut journal, 15, (5), 2007. p.12.

[8] Yalley, P. P. and Kwan, A.S K (2008), "Use of coconut fiber as an enhancement &. ofconcrete",InternationalJournalofSustainableConstructionEngineering Technology, 27(5), 2008, pp. 575-582

[9] MajidAli,AnthonyLiu,HouSou,NawawiChouw(2011),"Mechanicaland dynamic Properties of coconut fiber reinforced concrete" Construction and Building Materials 30 (2011) 814-825. https://doi.org/10.1016/j.conbuildmat.2011.12.068

[10] Wongsa A, Kunthavai Wong R, Naenudon S, Sata V, Chindaprasirt P. Natural fiberreinforced high calcium fly ash geopolymer mortar. Construction and Building Materials. 2020 Apr 30;241:118143. https://doi.org/10.1016/j.conbuildmat.2020.118143

[11] Atiş CD, Karahan O. Properties of steel fiber reinforced fly ash concrete. Construction and Building Materials. 2009 Jan 1;23(1):392-9. https://doi.org/10.1016/j.conbuildmat.2007.11.002

[12] Barbuta M, Bucur R, Serbanoiu AA, Scutarasu S, Burlacu A. Combined effect of fly ash and fibers on properties of cement concrete. Procedia Engineering. 2017 Jan 1;181:280-4. https://doi.org/10.1016/j.proeng.2017.02.390

Behaviour of Polypropylene and Steel Fibre Reinforced High Strength Concrete Exterior Beam Column Joint

L. Andal[1,a,*], M. Sathish Kumar[1,b], T. Nithiya Sree[2,c], V. Preetha[3,d]

[1]Department of Civil Engineering, Velammal College of Engineering and Technology, Madurai - 625009, Tamil Nadu, India

[2]Thiagarajar College of Engineering, Madurai -625015, Tamil Nadu, India

[3]Department of Civil Engineering, Bannari Amman Institute of Technology, Erode, Tamil Nadu, India

[a] la@vcet.ac.in, [b] msr@vcet.ac.in, [c] nithiyasree1989@gmail.com, [d] preethav@bitsathy.ac.in

Keywords: H gh Strength Concrete, Hybrid Fibre Reinforcement, Beam Column Joint, Joint Shear Force

Abstract. The present work aims at studying the performance of polypropylene and steel fibre reinforced high strength concrete in exterior joint of beam-column. Individually and in combination, steel fibre with 0.25 %, 0.50 %, 0.75 %, and 1 % by volume of concrete and polypropylene fibre with 0.16 % and 0.32 % by volume of cement were induced into the concrete. The characteristics investigated are energy absorption capacity, displacement ductility and shear carrying capacity. From the obtained test results, it is concluded that 0.16% polypropylene fibre with 0.75% steel fibre in hybrid fibre composites show desired strength in all aspects of investigation.

Introduction

A critical component in framed structure is a joint in beam-column, which is subjected to compressive force, tensile force and shear force. In general, under loading condition, these joints are expected to withstand high load without any significant loss in deformation. But in nature under cyclic loading, these joints will fail in shear with spalling of concrete in beam column junction and loss of bond between the reinforcement and concrete matrix. To overcome this problem closely spaced stirrups near the joint is the conventional way to avoid the shear failure in the junction, but it leads to congestion of steel reinforcement and it affects the compaction of concrete. [1] concluded that the possible solution to avoid the congestion of steel reinforcement in joint of beam-column is the use of steel fibres in concrete. Addition of metallic and nonmetallic fibres in concrete intense the tensile capacity of concrete by bridging the micro and macro cracks [2]. Similarly [3] reported increase in load carrying capacity and energy dissipation with increased stiffness retention in exterior joint of beam-column for hybrid cementitious composites compared to non fibre reinforced specimen. [4] concluded that random dispersion of fibres inside the concrete provides three-dimensional reinforcing network which improves better diffusion of stresses both internally and externally. Fibres in joint of beam-column in hybrid form increases the first crack load, ultimate load and ductility factor of the composite [5]. However higher percentage of steel fibre more than 1% volume of concrete combined with 0.2% of polypropylene fibre leads concrete to become harsh due to the balling effect of fibres [6].

[7-9] reported that available codes for seismic and non-seismic design inexact the joint shear strength of the non-seismically designed detailed joints. The main objective of this study is to

determine the energy absorption capacity, displacement ductility and shear carrying capacity in exterior joint of beam-column.

Materials and mix proportions

Materials Used

Ordinary Portland Cement, 53 grade, with a specific gravity of 2.97, was employed in this study. According to IS: 4031-Part 4 [10] and IS: 4031- Part 5 [11], the standard consistency of the cement was 34.5% and the initial and final setting time of the cement were 54 and 280 minutes respectively. 5% weight of cement was partially replaced with microsilica to improve the homogenous property of concrete. River sand having specific gravity of 2.66 and fineness modulus of 3.21, and coarse aggregate with a specific gravity of 2.84 and a fineness modulus of 6.68 was graded in a blend of 70% - 12.5mm passing and 10mm retaining and 30% - 10mm passing and retaining on 4.75mm. Polycarboxylic based ether superplasticizer with maximum dosage of 1.5% by weight of cement was used in this study. 20mm length with 0.04mm lateral dimension polypropylene fibre having a tensile strength of 500 N/mm^2 and 35mm length with 0.45mm diameter steel fibre having a tensile strength of 1000 N/mm^2 were used in this study.

Mix Proportion

According to ACI 211.4R-93 [12], a very high strength concrete matrix of M80 grade was achieved. Table 1 presents the concrete mix proportion. Details of mix designation are given in Table 2. In the mix designation, CM stands for control mix, P1 for concrete with 0.16% and P2 for concrete with 0.32% polypropylene fibre; S1 for concrete with 0.25%, S2 for 0.50%, S3 for 0.75% and S4 for 1.00% steel fibre. Initially, powder materials like dry cement were mixed with microsilica and fine aggregate for one minute in a laboratory pan mixer machine; following that, fibres were unevenly scattered to the dry mix and the process was repeated for another two minutes, and then coarse aggregate with 75% volume of water was added and mixed continuously for three minutes. Superplasticizer was dissolved in the remaining 25% volume of water and this mixer was poured into the wet concrete. Once again, it was mixed thoroughly for another 4 to 5 minutes and finally it was poured into the mould. Removal of mould was done after 24 hours and specimens were cured for 28 days testing. Concrete specimens were prepared for all the fibre combinations like P1, P2, S1, S2, S3, S4, P1S1, P1S2, P1S3, P1S4, P2S1, P2S2, P2S3, P2S4 including CM and all the mechanical characteristics like compressive strength, splitting tensile strength and flexural strength were determined. Based on this, only P1, P2, S1, S2, S3, S4 and best performed hybrid combinations P1S3 and P2S3 alone were used in this study.

Table 1 *Mix proportion*

Material	Weight (kg/m^3)
Cement	691.15
Microsilica	36.38
Fine aggregate	657.64
Coarse aggregate	630 (12.5mm passing and 10mm retaining) 270 (10mm passing and 4.75mm retaining)
Water	189
Superplasticizer	10.9

Materials Research Proceedings **23** (2022) 148-158 https://doi.org/10.21741/9781644901953-19

***Table 2** Mix designation*

Mix Designation	% of Polypropylene Fibre (V_f)	% of Steel Fibre (V_f)
CM	0	0
P1	0.16	0
P2	0.32	0
S1	0	0.25
S2	0	0.50
S3	0	0.75
S4	0	1
P1S3	0.16	0.75
P2S3	0.32	0.75

V_f - Volume Fraction

Experimental Program

Totally, eighteen exterior joint of beam-column were investigated in this experimental investigation. Each specimen consisted of a beam of 100mm wide and 150mm deep framing in to the column of 100mm x 100mm cross section. All beams were reinforced with a same amount of main reinforcing bar of 2 Nos of 8mm diameter bar at top and bottom sides of the cross section, and column with 4 Nos of 8mm diameter. 6mm diameter bar with 90mm centre to centre spacing were used as stirrups and lateral ties in beam and column respectively. According to IS 13920-1993 [13], the top and bottom reinforcement of beam were bent to 90^0 and anchored to the column with the anchorage length of 50 times of reinforcing bar diameter from the face of the column. Fig. 1 shows joint of beam-column detailing. To investigate the structural behaviour of exterior joint of beam-column, the specimen was placed in a 500 kN loading frame with both ends of the column in fixed condition. The load was applied gradually at a distance of 100mm from the end of beam with an increase of 1 kN and the resulting deflection was observed under the load point. Fig. 2 shows joint of beam-column test set up.

Materials Research Forum LLC
https://doi.org/10.21741/9781644901953-19

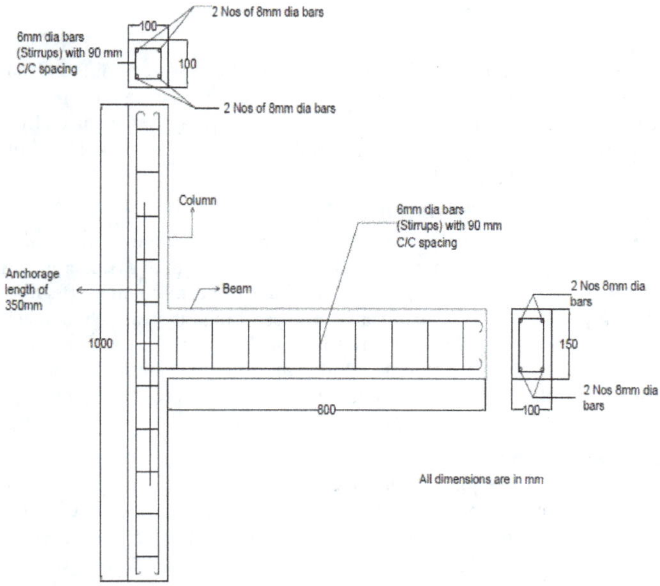

***Fig. 1.** Reinforcement detailing of exterior joint of beam-column*

***Fig. 2.** Test set up of exterior joint of beam-column*

Materials Research Proceedings **23** (2022) 148-158 https://doi.org/10.21741/9781644901953-19

Formulae for calculating joint shear force

Joint shear force

The joint shear force in exterior joint of beam-column is calculated according to Kuang. J.S [7], is expressed as Shear force in the joint core $V_j = T - V_{col}$. where T - tensile force in steel of the beam; and V_{col} - shear force of the column. The values of T and V_{col} may be calculated by $T = (P\, l_b / 0.9\, d_b)$ and $V_{col} = [P\,(l_b + 0.5\, h_c) / l_c]$. where P - applied load in the beam in kN; d_b - effective depth of beam in mm; h_c - effective depth of column in mm; l_b & l_c - effective length of beam and column.

Prediction of shear force by ACI 318R-05:2002

As per ACI 318R-05:2002 [14], the joint shear force in the exterior joint of beam-column calculated by $V_n = 0.85\, \gamma \sqrt{(f_c')} A_j$. where f_c' is compressive strength of cylinder in N/mm^2 ($f_c' = 0.8\, f_{cu}$) while designing the joint of beam-column; A_j - joint of beam-column effective area in mm^2 (100mm x 100mm); and γ is a coefficient based on type of joint in beam-column. The values of γ are $\gamma = 1.0$ for corner joint, $\gamma = 1.25$ for exterior joint and $\gamma = 1.67$ for interior joint. In this study, the specimen is classified as exterior joint. While calculating the ultimate shear capacity in the joint of beam-column, the factor of safety is not taken in to account and the equation can be rewritten as $V_n = 1.25 \sqrt{(f_c')} A_j$.

Prediction of shear force by NZS 3101: 1995

From NZS 3101:1995 [15], the joint shear force in the exterior joint of beam-column for seismic design code was calculated using the formula $V_n = 0.2\, f_c' A_j$. where f_c' is cylindrical specimen compressive strength in N/mm^2 ($f_c' = 0.8\, f_{cu}$); A_j - joint of beam-column effective area in mm^2.

Experimental results and discussions

Failure mode of test specimens

Fig. 3 to Fig. 6 show the failure pattern of exterior joint of beam-column concrete specimens. All the specimen developed tensile crack at the interface between column and beam. Similar to Ganesan. N et al. [5], a clear vertical cleavage was formed at the junction of all the specimens. From the observation, it was clearly noted that polypropylene fibre arrested the formation of micro cracks under load in initial stage and steel fibre arrested both micro and macro cracks under load in initial and ultimate stage. The first crack load increased by 6.48% for P1 and 4.16% for P2 specimens. Similarly for steel fibre based mix, except S2 mix all specimens showed increase in first crack load by 11.11% for S1, 27.31% for S3 and 38.88% for S4 specimens compared to control mix. It was discovered that while adding hybrid fibres to a hybrid fibre mix does not raise the first crack load, it does resist a larger peak load in the ultimate stage in comparison to the control specimen. For specimen P1S3, the crack width is less compared to all other specimen and it withstands higher peak load with large deflection the joint stayed unscathed during the test with minor crack and it is shown in Fig. 6.

Materials Research Proceedings **23** (2022) 148-158 https://doi.org/10.21741/9781644901953-19

Fig. 3. *Cracks in CM Specimen*

Fig. 4. *Cracks in P1 Specimen*

Fig. 5. *Cracks in S3Specimen*

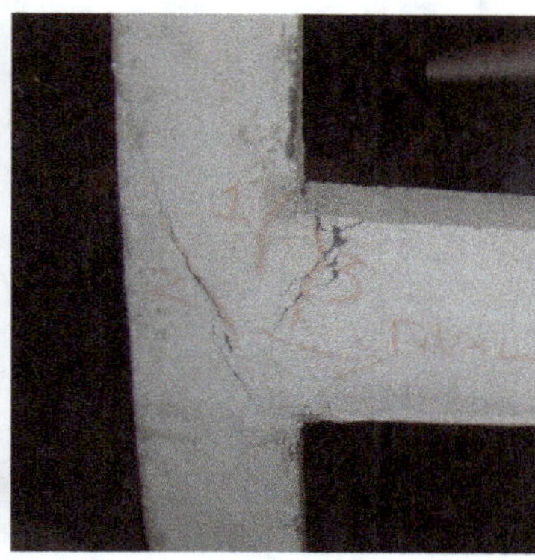

Fig. 6. *Cracks in P1S3 Specimen*

Load deflection behavior

The yield point load and the ultimate point load of all the concrete specimens are given in Table 3. The curve of load deflection of each specimen with and without fibres is shown in Fig. 7 to Fig. 9. It is perceived from the results that there is considerable improvement in load carrying capability after the induction of fibres Siva Chidambaram. R et al. [3]. P1 and P2 specimen showed 3.84% and 5.76% increase in load carrying capacity and ultimate deflection is more than 1.5 times compared to the control mix. Steel fibre based mix shows increase in load carrying capacity in the range of 34.61%, 42.30%, 68.26% and 80.76% respectively for S1, S2, S3 and S4 compared to the control mix. In hybrid fibre mix, P1S3 shows higher ultimate crack load of 25.4 kN with larger deflection of 43.05mm compared to all other mix, this was attributed to polypropylene fiber's capacity to resist micro fractures from forming and steel fibre to resist macro fractures Vahid Afroughsabet et al. and Ganesan. N et al. [2&5].

Table 3
Experimental results of joint in beam-column

Mix Designation	Compressive strength of companion cubes N/mm² (f_{cu})	Yield load and corresponding deflection		Ultimate load and corresponding deflection		Energy absorption capacity kNm	Displacement ductility $\mu = \delta_u / \delta_y$
		Load kN	Deflection mm (δ_y)	Load kN (P_u)	Deflection mm (δ_u)		
CM	80.21	4.32	3.29	10.4	16.22	0.055	4.93
P1	80.49	4.6	4.3	10.8	30.6	0.056	7.12
P2	80.44	4.5	4.53	11	24.75	0.053	5.46
S1	81.04	4.8	4.22	14	29.72	0.083	7.04
S2	83.44	4	3.6	14.8	30.22	0.106	8.39
S3	85.63	5.5	4.81	17.5	33.64	0.133	6.99
S4	85.78	6	3.36	18.8	36.83	0.182	10.96
P1S3	85.16	4	3.23	25.4	43.05	0.342	13.33
P2S3	85.97	3.5	2.5	16	30.26	0.176	12.10

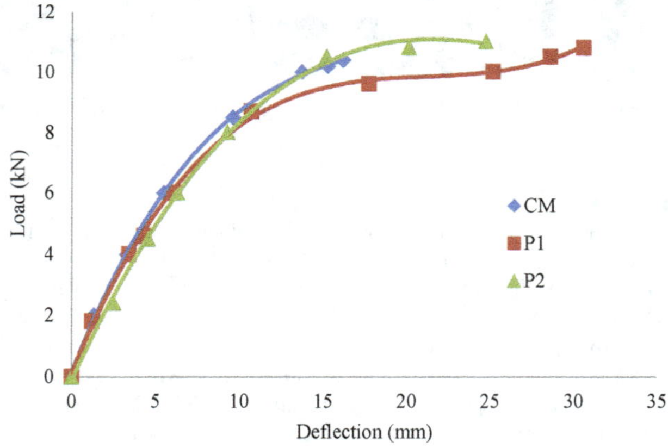

Fig. 7. *Load deflection curves of beams with control mix and polypropylene fibre mix*

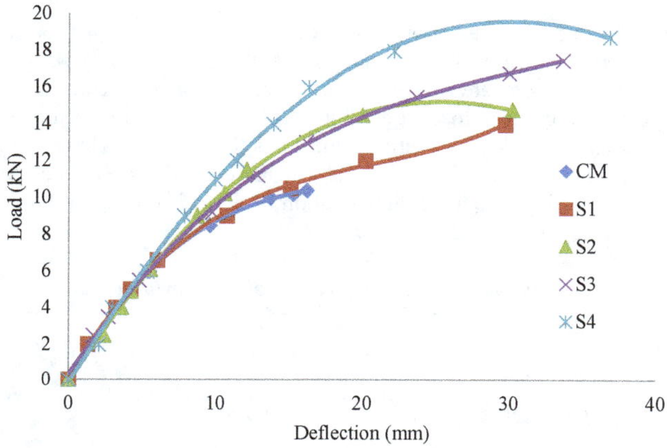

Fig. 8. *Load deflection curves of beams with control mix and steel fibre mix*

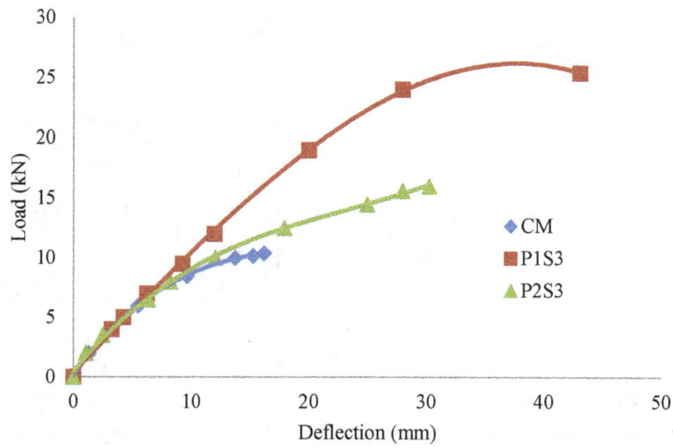

Fig. 9. *Load deflection curves of beams with control mix and hybrid fibre mix*

Energy absorption capacity and displacement ductility

To measure energy absorption capacity, the area beneath the load deflection curve was computed, and the findings are shown in Table 3. From the test results, it is clearly noticed that addition of fibres increases the energy absorption capacity consistently and it reaches a maximum for P1S3 mix, which is around 6.2 times greater than control mix. Displacement ductility refers to the ductility of a structure or an element and it is defined as the ratio of ultimate deflection to yield deflection. Table 3 shows the displacement ductility for all mix. It is observed that P1S3 mix showed increase in displacement ductility value of 170.38% compared to all other mix.

Joint shear force

The joint shear force values obtained through the experiments are shown in Table 4. It is clear that adding fibres to concrete increases the joint shear force. P1 and P2 mix shows 3.83% and 5.75% increase in joint shear force compared to control mix. For steel fibre based mix, S4 showed 80.75% increase in strength. In high peak load, it exhibits steel fibres suppressing the growth of micro cracks and bridging larger cracks. This directly increases the ductility behavior at joint of beam-column. In hybrid fibre mix, P1S3 shows 144.22% increase in joint shear force compared to the control mix. The increase in shear force with addition of fibres is due to the formation of strong interface zone between the fibres and the concrete matrix.

Table 4

Experimental and theoretical joint shear force

Mix designation	Joint shear force kN	Joint shear force kN		V_{exp}/V_{ACI}	V_{exp}/V_{NZS}
		ACI 318R-05:2002	NZS 3101:1995		
CM	62.55	100.13	128.34	0.62	0.49
P1	64.95	100.31	128.78	0.65	0.50
P2	66.15	100.27	128.70	0.66	0.51
S1	84.20	100.65	129.66	0.84	0.65
S2	89.01	102.13	133.50	0.87	0.67
S3	105.25	103.46	137.01	1.02	0.77
S4	113.06	103.55	137.25	1.09	0.82
P1S3	152.76	103.17	136.26	1.48	1.12
P2S3	96.22	103.66	137.55	0.93	0.70

Comparison with design codes

The joint shear force calculated from the experimental data in the exterior joint of the beam-column was compared to existing seismic codes of design ACI 318R-05:2002 and NZS 3101:1995 (Table 4). As illustrated in Fig. 10, both the ACI and NZS codes overestimate the joint shear force. Experimental results obtained were lesser by 50% than the predicted values by these two codes for the control mix, P1 and P2 mix and thus the obtained results are coinciding with the results obtained by Kuang. J.S et al. [7]. According to ACI 318R-05:2002 code, for steel fibre based mix, except S3 and S4 mix the specimens showed increase in predicted values of 19.54% for S1 and 14.71% for S2 mix respectively. Similarly, for NZS 3101:1995 code, S1 and S2 mix showed increase in predicted values by 54% and 49.99%. In hybrid fibre mix, the predicted values from both codes showed decrease in value by 32.45% and 10.80% for P1S3 mix. Therefore, it is found that ACI and NZS codes for predicting joint shear in joint of beam-column the concrete without fibres was overestimated, while the hybrid fibre reinforced concrete was underestimated.

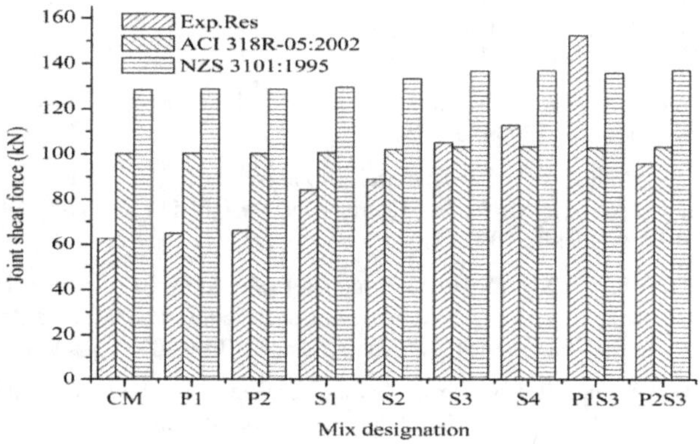

Fig. 10. *Theoretical and experimental comparisons of joint shear force with codes*

Conclusions

The impacts of polypropylene and steel fibre reinforced concrete in the joint of an exterior beam column were analyzed experimentally and numerically in this paper. The following conclusions are drawn from this study.

(1) The test specimen P1S3 (0.16 % polypropylene fibre with 0.75 % steel fibre) has a relatively small crack width than the other specimens, and it can withstand a higher peak load of 25.4 kN and a larger deflection of 43.05mm than the control mix, which can only withstand a peak load of 10.4 kN and a deflection of 16.22mm.

(2) The addition of fibres enhances the energy absorption capacity in the exterior joint of beam-column, reaching a maximum of 6.2 times that of the control mix for P1S3 (0.16 % polypropylene fibre with 0.75 % steel fibre). Similarly, P1S3 (0.16% polypropylene fibre with 0.75% steel fibre) mix shows increase in displacement ductility value of 170.38% compared to control mix.

(3) The increase in joint shear force with addition of fibres is due to the formation of strong interface zone between the fibres and the concrete matrix. In this P1S3 (0.16% polypropylene fibre with 0.75% steel fibre) mix shows 144.22% increase in joint shear force compared to the control mix.

(4) Seismic design codes ACI 318R-05:2002 and NZS 3101:1995 for predicting joint shear in joint of beam-column over estimates for the concrete without fibres and under estimates for the hybrid fibre reinforced concrete.

References

[1] C. Röhm, B. Novák, S. Sasmal, R. Karusala, V. Srinivas, Behaviour of fibre reinforced beam- column sub-assemblages under reversed cyclic loading, Construction and Building Materials. 36 (2012) 319-29. https://doi.org/10.1016/j.conbuildmat.2012.04.114

[2] Vahid Afroughsabet, Togay Ozbakkaloglu, Mechanical and durability properties of high-strength concrete containing steel and polypropylene fibers, Construction and Building Materials. 94 (2015) 73-82. https://doi.org/10.1016/j.conbuildmat.2015.06.051

[3] R. Siva Chidambaram, Pankaj Agarwal, Seismic behavior of hybrid fiber reinforced cementitious composite beam column joints, Materials and Design. 86 (2015) 771-781. https://doi.org/10.1016/j.matdes.2015.07.164

[4] V. Ramakrishnan, Structural synthetic fibres for three-dimensional reinforcement of concrete, ACI Special Publication. 261 (2009) 55-72.

[5] N. Ganesan, P.V. Indira, M.V. Sabeena, Behaviour of hybrid fibre reinforced concrete beam column joints under reverse cyclic loads, Materials and Design. 54 (2014) 686-693. https://doi.org/10.1016/j.matdes.2013.08.076

[6] N. Ganesan, P.V. Indira, M.V. Sabeena, Bond stress slip response of bars embedded in hybrid fibre reinforced high performance concrete, Construction and Building Materials. 50(2014) 108-115. https://doi.org/10.1016/j.conbuildmat.2013.09.032

[7] J.S. Kuang, H.F. Wong, Effects of beam bar anchorage on beam column joint behavior, Structures & Buildings. 159(2006) 115-124. https://doi.org/10.1680/stbu.2006.159.2.115

[8] J.S. Kuang, H.F. Wong, Effects of beam column depth ratio on joint seismic behavior, Structures & Buildings. 161(2008) 91-101. https://doi.org/10.1680/stbu.2008.161.2.91

[9] S.R. Uma, Sudhir K. Jain, Seismic design of beam column joints in RC moment resisting frames - review of codes, Structural Engineering and Mechanics. 23(2006) 579-597. https://doi.org/10.12989/sem.2006.23.5.579

[10] Indian Standard Method, Physical tests for hydraulic cement, determination of consistency of standard cement paste. IS 4031; Part 4: 1988.

[11] Indian Standard Method, Physical tests for hydraulic cement, determination of initial and final setting times. IS 4031; Part 5: 1988.

[12] American Concrete Institute, Guide for selecting proportions for high-strength concrete with portland cement and fly ash. ACI committee 211: Reapproved 1998.

[13] Indian Standard, Ductile detailing of reinforced concrete structures subjected to seismic forces - code of practice. IS 13920:1993.

[14] American Concrete Institute. Building code requirements for reinforced concrete and commentary. ACI 318 R-05:2002.

[15] Standards Association of New Zealand. The Design of Concrete Structures. NZS 3101:1995.

A Research on Partial Replacement of Fine Aggregate by Waste Foundry Sand

S. Ramkumar[1,a*], A. Sridhar[1], M. Vignesh[1]

[1]Department of Civil Engineering, M.Kumarasamy College of Engineering, Karur, India

[a*]ramkumars.civil@mkce.ac.in

Keywords: Foundry Sand, Chemical Admixture, Metal Casting, Environmental Problems, Tensile Strength, Conventional Concrete, M-Sand

Abstract. The by-products from metal casting industries, waste foundry sand is generated in huge amount which causes many pollution to the environment like infertility of sand, unsightliness, awful odour, etc., because of improper disposal. Such kind of environmental problems can be reduced when those wastes are used as building material during concrete production. So, a research work was carried out in concrete containing waste foundry sand in the range of 15% to 25% with 5% increase, as a partial replacement for fine aggregate(M-sand) for M-20 grade concrete. The concrete made of foundry sand in the proposed mix design was tested and compared with ordinary concrete for workability, compressive strength, Flexural strength and Tensile strength. The cubes were tested on 14th and 28th day for mix of 1:1.54:2.97 at a water-cement ratio of 0.45 and the results were carried out with comparison.

Introduction

The scope of the project is to reduce the environmental pollution caused by waste foundry sand and use them as partial replacement of fine aggregate in concrete. This concrete is subjected to compression, tensile, and flexural test with some percentage of partial replacement of fine aggregate by waste foundry sand[1]. The objective of the live foundry sand project is to foster the green procurement, waste prevention and recycling by studying. It is to improve acceptance of the generally valuable cleaned and recycling soil material to use geo engineering in future. It is having good durability and workability. Keep away contaminated materials from landfills. It is improved fertility, and to substitute synthetic fertilizers.

Materials

Admixtures

An unique pink colored liquid which is used for integral water-resistant component for mortar, plaster and concrete of surface-active agents, additives, and special selective polymers. The admixture has many characteristics as Compatibility enables easy to disperse and compatible with any type of Portland cement, including Slag Cement, Cohesiveness will minimize the mortar loss caused by rebound in plaster work, which makes it very easy for application and saving[2,8]. Economically necessary for a very low dose, low permeability will improve Water-tightness in large measure, the water reducer helps to reduce the water content, will improve workability without increasing the water content, Reduced Shrinkage cracks in concrete and mortar, a better finish, works to disturb the causing leakage capillary Concrete-formed structures that make it waterproof and durable, help improve strength, make concrete more cohesive and therefore waterproof better protects the steel against corrosion, less time is needed for plaster, which saves on the cost of labor per square foot, supplied under conditions favourable to the consumer[3,9].

The admixture can be used in various fields such as

- Slabs and roof supports
- Lower ground floor
- Water reservoir and water containment structures
- Washrooms and terraces
- External plastering
- Repairing and renovating.

Waste Foundry Sand

For metal castings, usually recycled sands will be used in order to reduce the usage of the natural materials. After several times of usage of the sand, it will lose its characteristics and it will become unsuitable for the manufacturing process of metal castings[4,6]. These unsuitable sands will become waste and will be discarded away. For compaction, molding sand will be used because it leads to the desired shape required during the casting process.

Table No. 1: Properties of Waste Foundry Sand

S. NO.	PROPERTY	RESULT
1	Water Absorption	0.42 %
2	Moisture Content	1.6 %
3	Specific Gravity	2.4
4	Fineness Modulus	3.92

Casting and Curing

Proper casting and curing are very important factor which controls the strength and durability of concrete. We decided to make three types of specimens, partially replaced natural fine aggregate by waste foundry sand with 15 % , 20 %, and 25%[5,7]. These specimens are made with water cement ratio 0.45 and are cured for 28 days, hardened concrete tests were carried out at an interval of 14 days and 28 days.

Concreting

All the materials are properly weighed and taken in separate bonds. The mixing tray is placed in position and is ensured to be dry and clean of dust and other particles. First fine aggregate and cement are mixed thoroughly, and then coarse aggregate is added and mixed. Finally, water of required quantity is added and the concrete paste is obtained.

Casting of Specimen

Concrete of M20 grade (1:1.54:2.97) with water cement ratio 0.45 is used. Cubes, cylinders, and beams (14 days and 28 days) were prepared by partial replacement of fine aggregate by foundry sand waste. After mixing fresh concrete tests are carried out and then the mixture ready to be casted if the test results are within the allowed standards. The moulds are taken cleaned and tightened completely after which oil or grease is applied on the inner face of the mould which ensures smooth de-moulding of specimens. The mould is filled in three layers of good annual compaction or mechanical compaction and the top of the mould is well finished.

Materials Research Proceedings **23** (2022) 159-166 https://doi.org/10.21741/9781644901953-20

Table No. 2: Size of Mould

S. No.	SPECIMEN	SIZE (mm)
1	Cube	150*150*150
2	Cylinder	150*300
3	Beam	100*100*500

Table No. 3: No. of Specimens Casted

S.NO.	PERCENTAGE OF REPLACEMENT	NAME OF SPECIMEN	NO. OF SPECIMENS		
			14 DAYS	28 DAYS	TOTAL
1	0	Cube	3	3	6
		Cylinder	3	3	6
		Beam	3	3	6
2	15% Foundry Sand	Cube	3	3	6
		Cylinder	3	3	6
		Beam	3	3	6
3	20% Foundry Sand	Cube	3	3	6
		Cylinder	3	3	6
		Beam	3	3	6
4	25% Foundry Sand	Cube	3	3	6
		Cylinder	3	3	6
		Beam	3	3	6

Fresh Concrete Test

Slump Cone Test

Table 4 Slump cone test

S. NO.	% OF WASTE FOUNDRY SAND	SLUMP VALUE (mm)	DEGREE OF WORKABILITY
1	0	110	True
2	15	62	Shear
3	20	80	Shear
4	25	73	Shear

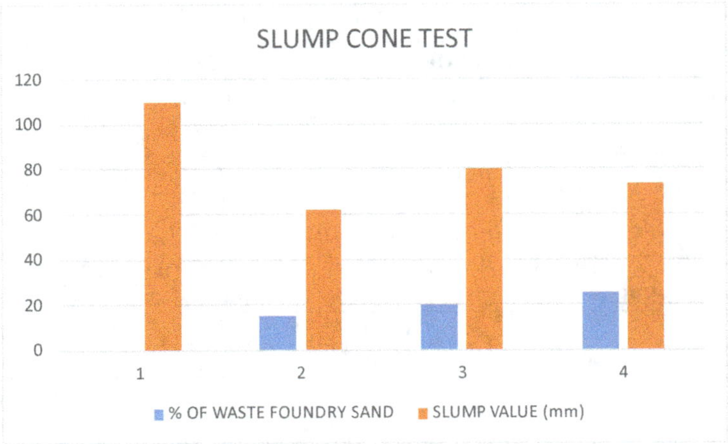

Fig. 1 Slump Cone Test Result

Hardened Concrete Test

Compressive Strength Test

Fig. 2: Compressive Test of Concrete Cube

Table No. 4 Compressive Strength Test Results

No. of Days	Sample (%)	Load (kN)				Cross-sectional Area (mm^2)	Compressive Strength (N/mm^2)
		Trial-1	Trial-2	Trial-3	Mean		
14	0	480	471	475	474.5	150*150	18.0
	15	295	296	294	295	150*150	13.1
	20	306	306	307	306	150*150	13.6
	25	299	301	302	300	150*150	13.33
28	0	486	468	477	476.6	150*150	21.3
	15	301	302	303	302	150*150	13.42
	20	308	307	307	306	150*150	13.6
	25	304	305	304	304	150*150	13.511

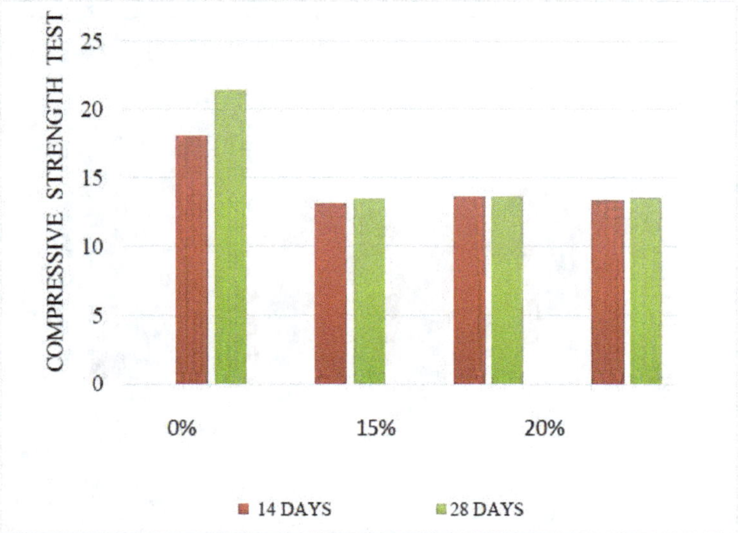

Fig. 3 Compressive Strength Test Result

Tensile Strength Test

Table No. 5 Tensile Strength Test Results

No. of Days	Sample (%)	Load (kN)				Cross-sectional Area (mm^2)	Tensile Strength (N/mm^2)
		Trial-1	Trial-2	Trial-3	Mean		
14	0	215	208	214	211.5	150*300	2.99
	15	183	176	181	180	150*300	1.79
	20	191	188	197	192	150*300	1.84
	25	180	171	184	178.3	150*300	1.71
28	0	233	239	236	236	150*300	3.34
	15	187	183	185	185	150*300	1.81
	20	193	192	196	194.33	150*300	1.86
	25	176	169	181	175.33	150*300	1.68

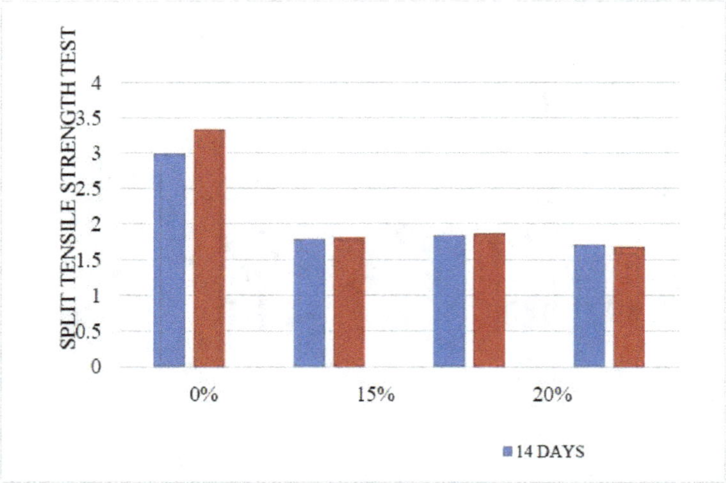

Fig. 4 Tensile Strength Test Result

Materials Research Forum LLC
https://doi.org/10.21741/9781644901953-20

Flexural Strength Test

Table No. 6 Flexural Strength Test Results

No. of Days	Sample (%)	Load (kN)				Cross-sectional Area (mm^2)	Flexural Strength (N/mm^2)
		Trial-1	Trial-2	Trial-3	Mean		
14	0	18	17	16	17.5	100*100*500	3.75
	15	10	11	11	10.6	100*100*500	1.95
	20	11	12	12	11.6	100*100*500	2.13
	25	10.5	10.5	10	10.3	100*100*500	2.2
28	0	24	23	22	23.5	100*100*500	4.15
	15	12	12.5	12	12.1	100*100*500	2.35
	20	13	13.5	13	13.1	100*100*500	3.2
	25	11.5	11	11	11.1	100*100*500	2.8

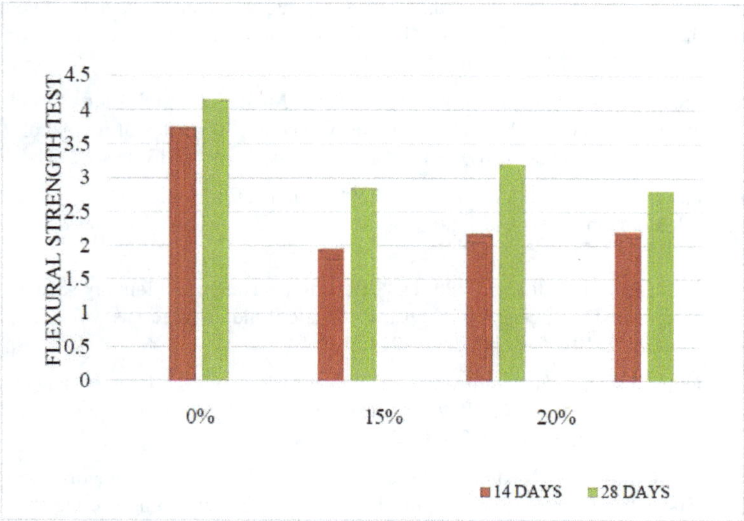

Fig. 5 Flexural Strength Test Result

Conclusion

- Comparing to the controlled mix of nominal concrete, there will be a slight decrease in the strength of the concrete when foundry sand is used as a partial replacement at 20%.
- Usage of waste foundry sand will also lead to eco-friendly environment since the wastes are not buried or thrown away after metal castings are done.
- By this point, the concrete can also be termed as a Green Concrete.

References

[1] Eknath P Salokhe, D. B .Desai "Application of Foundry Waste Sand In Manufacture of Concrete," IOSR Journal of Mechanical and Civil Engineering (IOSR-JMCE) ISSN: 2278-1684, PP: 43-48.

[2] Gurpreet Singh and Rafat Siddique, Effect of waste foundry sand (WFS) as partial replacement of sand on the strength, ultrasonic pulse velocity and permeability of concrete, Journal of Construction and Building Materials 26 (2012), 416-422. https://doi.org/10.1016/j.conbuildmat.2011.06.041

[3] Han-Young Moon, Yun-Wang Choi, Yong-Kyu Song and Jung-Kyu Jeon, Fundamental properties of Mortar and Concrete using Waste Foundry Sand, Journal of the Korea Concrete Institute, Vol.17 No.1, February,2005, 141-147. https://doi.org/10.4334/JKCI.2005.17.1.141

[4] Jayachandra , Shashi kumar.A , Sanjith.J , DG.Narayana, " Strength Behaviour of Foundry Sand on Modified High Strength Concrete," International Journal of Research in Engineering and Technology eISSN: 2319-1163 | pISSN: 2321-7308.

[5] Pathariya Saraswati C, Rana Jaykrushna, Shaha Palas, Mehta Jay, Patel Ankit , "Application Of Waste Foundry Sand For Evolution Of Low Cost Concrete," International Journal of Engineering Trends and Technology(IJETT)-Volume 4 Issue 10 Oct 2013, ISSN 2231-5381.

[6] Pranita Bhandari, Dr.K. M. Tajne "Use of Foundry Sand in Conventional Concrete," Volume 2, No 1, 2011 International Refereed Journal of Engineering and Science Volume 2, Issue 2 (February 2013), PP.45-53.

[7] Rafat Siddique and El-Hadj Kadri, (2011), "Effect of metakaolin and foundry sand on the near surface characteristics of concrete", Construction and Building Materials, vol. 25, pp 3257–3266. https://doi.org/10.1016/j.conbuildmat.2011.03.012

[8] Rafat Siddique, Geert de Schutter, Albert Noumowe, "Effect of used-foundry sand on the mechanical properties of concrete", Elsevier, Construction and Building Materials 23 (2009). https://doi.org/10.1016/j.conbuildmat.2008.05.005

[9] Smit M. Kacha, Abhay V. Nakum, Ankur C. Bhogayata, "Use Of Used Foundry Sand in Concrete: A State Of Art Review," International Journal of Research in Engineering and Technology eISSN: 2319-1163 | pISSN: 2321-7308.

Materials Research Proceedings **23** (2022) 167-174 https://doi.org/10.21741/9781644901953-21

Shear Behaviour of Fiber Reinforced Concrete Beams Using Steel and Polypropylene Fiber

S. Ramkumar[1,a*], A. Dinesh[2,b], A. Karthick[3,c]

[1]Department of Civil Engineering, M.Kumarasamy College of Engineering, Karur, India

[2]Department of Civil Engineering, Sri Ramakrishna Engineering College, Coimbatore, India

[3]Department of Civil Engineering, K.Ramakrishna College of Technology, Trichy, India

[a*]ramkumars.civil@mkce.ac.in, [b]dinesh.a@srec.ac.in; dinesh.a.civil@gmail.com, [c]karthicka.civil@krct.ac.in

Keywords: Dramix, Fiber, Polypropylene, Steel, Concrete

Abstract. The experimental study provides a series of tests for characterizing the structural behavior of fibre reinforced concrete beams subjected to shear loads. The paper involves usage of 2 types of fibers - polypropylene and steel fiber. The work suggests that the shear cracking resistance of the materials used are significantly improved by the fibers. The fibers reduced the crack width to about one quarter of the width in the shear-reinforced girders. Reliance on steel fibres increases the ductility of concrete. Adding steel fibres to concrete improves its post-tensile cracking behaviour. Shear strength is increased with the increase in fiber aspect ratio and fiber volume fraction. The concrete beams are casted for the size of 150 mm x 250 mm x 2100 mm. The behavior of fiber reinforced concrete beams for the addition of 0.4 percentage of fibers in both PFRC and SFRC under loading condition were observed and the load carrying capacity was increased compared to reinforced concrete.

Diagonal Cracking Capacity

The diagonal tension failure occurs immediately after the diagonal crack is formed. Therefore, the shear stress at the diagonal cracking can be assumed to be the ultimate shear strength in the case of the diagonal tension failure.

From numerous experimental data about the shear resistance of the RC beams without shear support, the empirical equation was proposed by Okamura and Hagias in 1980[2,3]. Based on this equation, the modification has been made to incorporate the size effect directly in 1986. This modification was proposed by Niwa and Okamura. This revised equation has been adopted into JSCE Shear Design Specification[1,4].

Shear stress at the formation of diagonal crack

In an RC beam with no shear strengthening under shear force, once diagonal crack is formed, the beam will fail very suddenly[5,7]. However, the the design shear stress when a diagonal crack is created cannot be obtained by the elastic theory, because it involves many factors such as concrete strength, shear span-effective depth ratio (a/d), longitudinal reinforcement ratio, effective depth, etc[12].

Materials

Metal fibers are made from steel or stainless steel. The polymer fibres used are acrylic, aramide, carbon, nylon, polyester, polyethylene and polypropylene fibres[8,10]. Glass fibre is the most widely used mineral fibre[6]. Different kinds of organic and inorganic natural fibers such as cellulose are also used to strengthen the cement matrix.

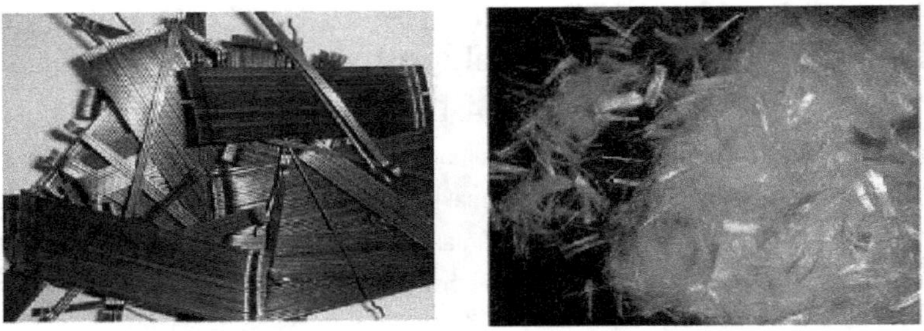

Fig. 1 Glued DRAMIX Steel Fiber and Polypropylene Fiber

Mixture Proportion

The Concrete mix proportions of fiber reinforced concrete with and without ferrous and non-ferrous fibers[9,11] with the design compressive strength of 25Mpa is adopted for this test. The following table shows the proportion range.

Table 1: Design Mixture proportions

Design Mix	Cement (Kg/m³)	Fine Aggregate (Kg/m³)	Coarse Aggregate (Kg/m³)	60% 0f 20mm C.A (Kg/m³)	40% of 12.5mm C.A (Kg/m³)	Water (Kg/m³)
M25	414	552	1173	704	469	186

Experimental set up

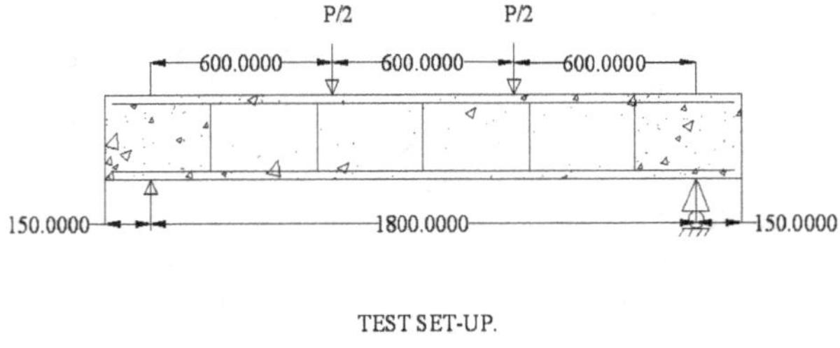

Fig. 2 Experimental set-up of beam (2-point loading)

Fig. 3 Experimental set-up of beam with loading frame in laboratory

Experimental Results

Load and Deflection for RCC Beam

Table 2 Load and Displacement values for Reinforced Concrete Beam

S. No.	Load (KN)	Point 1 Displacement (mm)	Central Displacement (mm)	Point 2 Displacement (mm)
1.	8.24	0.27	0.29	0.27
2.	16.45	0.64	0.65	0.60
3.	24.67	0.97	0.98	0.90
4.	32.94	1.40	1.46	1.31
5.	41.00	2.24	2.40	2.12
6.	49.30	3.0	3.24	2.86
7.	57.40	3.69	4.01	3.56
8.	65.15	4.72	5.10	4.55
9.	72.73	6.98	8.00	6.85
10.	80.301	7.50	8.62	7.20
11.	78.50	7.90	9.00	7.60

Materials Research Proceedings **23** (2022) 167-174 https://doi.org/10.21741/9781644901953-21

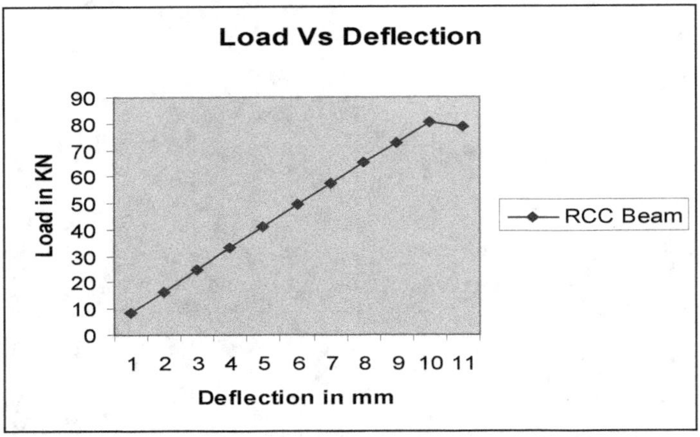

Fig. 4 Load Deflection curve for RCC beam 1

Load and Deflection for RCC Beam with minimum stirrups

Table 3 Load and Displacement values for Reinforced Concrete Beam with minimum stirrups

S. No.	Load (KN)	Point 1 Displacement (mm)	Central Displacement (mm)	Point 2 Displacement (mm)
1.	8.24	0.46	0.46	0.38
2.	16.45	0.79	0.80	0.71
3.	24.67	1.26	1.30	1.15
4.	32.94	1.82	1.96	1.72
5.	41.00	2.83	3.12	2.75
6.	49.30	3.58	4.05	3.65
7.	57.40	4.25	4.75	4.10
8.	65.15	5.34	6.00	5.25
9.	72.73	5.90	6.71	6.08

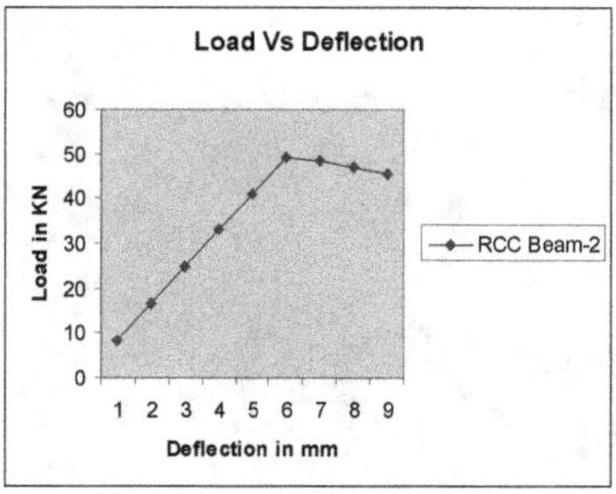

Fig. 5 Load Deflection diagram for RCC Beam 2

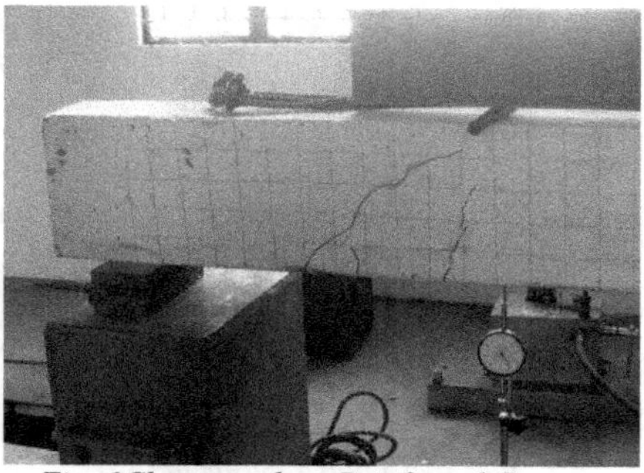

Fig. 6 Shear crack on Reinforced Concrete with minimum stirrups beam

Load and Deflection for SFRC Beam

Table 4 Load and Displacement values for SFRC beam

S. No.	Load (KN)	Point 1 Displacement (mm)	Central Displacement (mm)	Point 2 Displacement (mm)
1.	8.36	0.42	0.45	0.41
2.	17.21	0.92	0.94	0.88
3.	25.26	1.17	1.22	1.13
4.	33.25	1.85	2.02	1.82
5.	41.56	2.39	2.60	2.32
6.	49.62	3.05	3.25	2.95
7.	58.21	3.96	4.32	3.75
8.	65.35	4.63	5.21	4.52
9.	63.25	5.85	6.32	5.85
10.	61.75	6.45	7.25	6.42

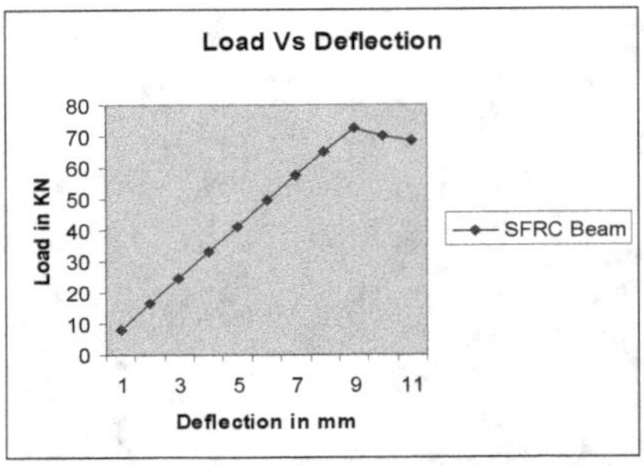

Fig. 7 Load Deflection diagram for SFRC Beam

Fig. 8 Shear Crack on SFRC Beam

Load and Deflection for PFRC Beam

Table 5 Load vs Displacement value for PFRC beam

S. No.	Load (KN)	Point 1 Displacement (mm)	Central Displacement (mm)	Point 2 Displacement (mm)
1	8.36	0.54	0.62	0.43
2	17.21	0.91	0.86	0.74
3	25.26	1.13	1.16	1.08
4	33.25	1.94	2.1	1.86
5	41.56	2.31	2.48	2.20
6	49.62	2.71	2.95	2.61
7	58.21	3.71	4.25	3.64
8	65.35	4.51	4.94	4.36
9	63.25	5.30	6.80	5.70
10	61.75	8.80	9.95	8.85

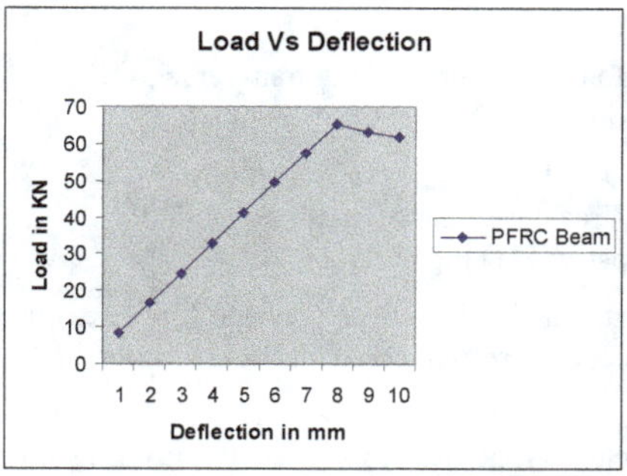

Fig. 9 Load Deflection diagram for PFRC Beam

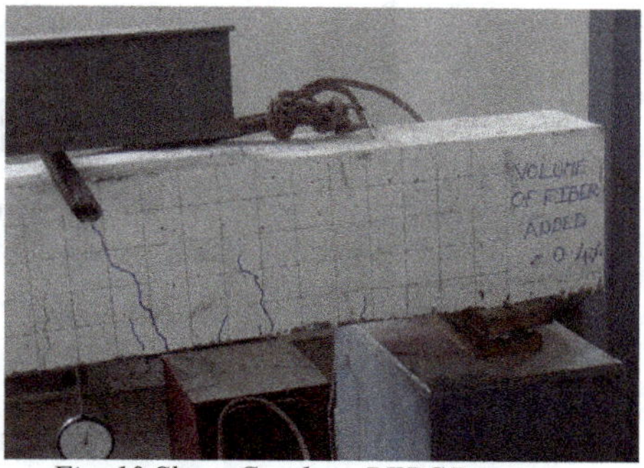

Fig. 10 Shear Crack on PFRC Beam

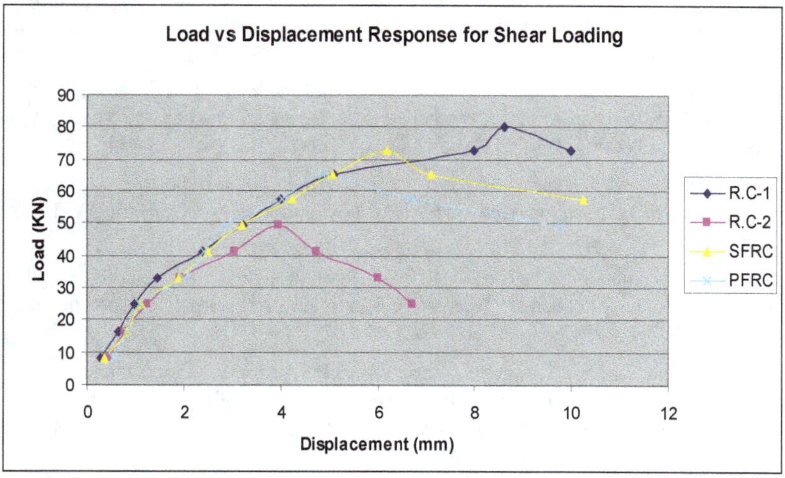

Fig. 11 Load-Displacement Response under Shear Loading

Conclusion

- The cracking mechanism occurs early in PFRC.
- The compressive strength increases in comparison to the tensile strength marginally due to addition of fiber content.
- The toughness in both tension and compression increases because of the incorporation of fibers in concrete which results in greater resistance to concrete shear.

Materials Research Proceedings **23** (2022) 167-174 https://doi.org/10.21741/9781644901953-21

References

[1] Adebar, P., Collins, M.P. (1996). 'Shear strength of members without transverse reinforcement', Canadian Journal of Civil Engeenering, Vol. 23, pp. 30-41. https://doi.org/10.1139/l96-004

[2] Al-Ahmed,A.H.,Al -Gasham, S.T.,Alternative Cracking Shear Strength Equation for Reinforced Concrete Normal Beams without Stirrups‖ V. 8(1) PP 44-49.

[3] Anbuvelan, M.M. Khadar. M.H, M. Lakshmipathy and K.S. Sathyanarayanan ,"Studies on properties of concretes containing polypropylene, steel and reengineered plastic shred fibre", Indian Concrete Journal, Vol.81, issue 4, pp 1724.

[4] Chunxiang Qian and Piet Stroeven, 2000, 'Fracture Properties of Concrete Reinforced with steel –polypropylene hybrid fibers, Cement and Concrete Composites, pp. 343-351. https://doi.org/10.1016/S0958-9465(00)00033-0

[5] Gracia, S.A., Ma, R.J., Romero, M., Talero, R., 2005. Characterisationof a polypropylene fibred cement compositeusing ESEM, FESEM and mechanical testing. Constructionand Building Materials, pp396-403. https://doi.org/10.1016/j.conbuildmat.2004.07.023

[6] Krishna Rao, N.R. Dakhshina Murthy And V. Santhosh Kumar (2011)'Behaviour of Polypropylene Fibre Reinforced Fly ash Concrete Deep Beams In Flexure And Shear', Asian Journal of Civil Engineering (Building And Housing) Vol. 12, No. 2 pp 143-154.

[7] Mehul J. Patel, S. M. Kulkarni 'Effect of Polypropylene Fibre on The High Strength Concrete', Journal of Information, Knowledge And Research in Civil Engineering Volume 2, Issue 2, pp 127.

[8] Murahari, Rama mohan Rao p (2013) 'Effects of Polypropylene fibres on the strength properties Of fly ash based concrete', International Journal of Engineering Science Invention, Volume 2 Issue 5 , May. 2013, pp.13-19.

[9] Peng Zhang and Qingfu Li, 'Fracture Properties of Polypropylene Fiber Reinforced Concrete Containing Fly Ash and Silica Fume', Research Journal of Applied Sciences, Engineering and Technology 5(2)pp 665-670, 2013. https://doi.org/10.5281/zenodo.345692. https://doi.org/10.19026/rjaset.5.5006

[10] Priti A. Patel., Dr. Atul K. Desai., and Dr. Jatin A. Desai., "Evaluation Of Engineering Properties for Polypropylene Fibre Reinforced Concrete", International Journal of Advanced Engineering Technology, Vol. 3, Issue 1, pp. 42-45.

[11] Reineck K. H., Kuchma D., Kim K.S. and Marx S.. —Shear Database for Reinforced Concrete Members without Shear Reinforcement‖ ACI Structural Journal. Pp.240-249.

[12] Thirumurugan.S, Siva Kumar.A (2013), 'Compressive Strength Index of Crimped Polypropylene Fibers in High Strength Cementitious Matrix' World Applied Sciences Journal,pp 698-702.

An Experimental Investigation of Lightweight Self Compacting Concrete with Replacement of Coarse Aggregate as Pumice Stone- A Review

M. Arun Kumar[1,a,*], M. Preethi[1,b,*], S. Pavithran[1,c] and M. Praveen[1,d]

[1]Department of Civil Engineering, Kongu Engineering College, Perundurai, Erode, Tamil Nadu, India

[a]rsmanoharun@gmail.com, [b]preethimuthukumar31@gmail.com, [c]pavithran12121@gmail.com, [d]praveenrajj272@gmail.com

Keywords: Pumice Stone, Coarse Aggregate, Lightweight SCC, Fresh Properties, Hardened Properties

Abstract. Due to its unique properties as compared to ordinary concrete, lightweight concrete play a major role in construction sector. Here, this research explain the development of lightweight self-compact concrete by replacing the coarse aggregate together the pumice stone, which is used as a lightweight material in various proportions. An investigation on the effect of coarse aggregate on the partially replaced with pumice stone in lightweight self-compact concrete is carried out. The fresh and hard property of this lightweight self-compacting concrete have been studied and compared with the results of normal concrete. Pumice stone is used due to its special property such as unit weight, heat insulation property, resistance against fire when we combined with the coating substance the properties of this concrete has been improved. Several properties of lightweight self-compacting concretes such as unit weight, flow diameter, flow diameter after an hour, V-funnel and L-box tests, 28 days split- tensile strength, dry unit test, water absorption 7- and 28-day compressive strength, and ultrasonic pulse velocity test was investigated. According to the study, lightweight self-compacting concrete properties include flow strength, segregation resistance, and filling capability of fresh concrete. Pumice stone is used by replacing with natural coarse aggregate, at the levels of 20%, 40%, 60%, 80% by volume with fly ash and blast furnace slag minerals at the constant rate of 40%. 28 days compressive strength, dry unit weights, thermal conductivity in addition to ultra-sonic velocity of self-compacting concrete were obtained. The compression, flexural, and split tensile strengths of cubes, cylinders, along with prisms are tested for 7, 14, and 28 days. Results shows that pumice stone met the requirements for structural applications.

Introduction

General

Concrete is a heterogeneous substance that is made by mixing cement, sand and aggregate with water. In most cases, cement is used as a binder and sand is used to fill the gaps between the aggregates. Aggregate is the most common material in concrete, accounting for more than 70% to 80% of the total volume and providing the concrete with strength. The high demand for concrete in construction is raised due to population growth, resulting in increased demand for different construction materials. So, it is essential to develop or materials that can be used substitute for sand and aggregate in concrete. As aggregate is the most common material in concrete, it is obtained by grinding stones in stone quarries, which requires a large amount of manpower and mechanical equipment. When stones are crushed, very small dust particles with low density are produced, which easily mix with air and have a negative impact on the environment as well as the workers' health. In order to overcome this issue, various lightweight aggregates are used. Pumice stone is one of the many lightweight aggregates used in concrete to minimise self-weight and dead load, as well as to develop the mechanical and durability property of concrete, resulting in environmental and economic benefits.

Properties of concrete
Concrete has three main characteristics. They are workability, strength, and durability. Hardened concrete is supposed to have higher strength and durability, whereas fresh concrete is thought to have higher workability. Mix design and fresh properties can both be attributed to hardened properties. [3]. It is critical to evaluate the characteristics of hardened concrete at earlier. The issue is due to consistency and mechanical property does not improve after the hardening process. Concrete's structural behaviour is determined by the mixed proportions and material properties of composite system, which does not alter after the process of hardening. Workability, and fundamental rheological properties, reversible and non-reversible evolution, thixotropy, slump loss, setting time, bleeding, segregation and practical issues related to formwork filling is the fresh properties [4]. Compressive strength, tensile strength, elastic properties, shrinkage, cracking resistance, electrical, thermal, creep, transport and few properties comes under the properties of hard concrete [4].

Properties of pumice
Pumice has numerous potential features for lightweight aggregate, including limited density, strong thermal insulation, excellent resistance of fire and non-combustibility, adequate compressive strength, appropriate elasticity, a structure has pores with low permeability. Pumice stone as coarse aggregate has a specific gravity of 0.82. Pumice exists in a range of colours, ranging from brown to black and white to gold. This is largely determined by the pumice's chemical composition. Pumice ore primarily contains low amounts of TiO2 and SO3, as well as CaO (1–2%), Fe2O3 (1–3%), Na2O-K2O (7–8%), Al2O3 (13–17%), and SiO2 (60–75%). Heat insulation, reduced alkali-silica reaction (ASR) expansion, heat resistance, freeze/thaw resistance, chemical resistance, and lightweight matrix are all benefits of pumice in concrete and mortar. The addition of pumice stone to the matrix, on other hand, decreased the mechanical strength, drying shrinkage, water absorption and workability.

Self Compacting Concrete (SCC)
Self-compacting concrete (SCC) is a distinctive category of concrete, which is also called self-consolidating concrete that could be placed and compacted without the need of vibrations. Due to their excellent deformability and cohesiveness which allowed it to be treated without segregation or bleeding [1]. The properties of Self-compacting concrete are used to fill in the gaps, passing ability and immune to separate [1]. For the last few years, usage in concrete that compacts itself has increased. When processing SCC, the use of chemical admixtures is often needed to improve workability and minimise segregation [9]. Two types of admixtures are used. They are mineral and chemical admixture. water-reducing chemical admixture such as superplasticizer, used in SCC [3].

Lightweight concrete (LWC)

Lightweight concrete which utilizes the lightweight aggregates that can help to reduce the density of concrete and to reduce dead load acting over the structure [3]. The significant issue is concrete's density when controlling the dead load of concrete structure, in reality it's troublesome throughout the construction of tall structures [2]. Consequently, lightweight concrete have replaced the standard concrete in several contemporary buildings. Generally, a concrete is often regarded to as lightweight if the density in range of 1440–1840 kg/m3 [2]. In the preparation of lightweight concrete, lightweight aggregates for instance enlarged clay shale, paper waste, vermiculite, rice husk, perlite, pumice stone, stuff, scoria, and cinders were utilised [2]. Lightweight aggregate concrete with outstanding durability with superior advantages such as the use of natural materials, lightness, and ease of workability [11].

Lightweight Self Compacting Concrete (LWSCC)

Several investigations in development of Lightweight self-compacting concrete, LWSCC have been conducted in recent decades with the aim of combining the advantages of both self-compacting concrete (SCC) and Lightweight concrete (LWC) [3]. Lightweight self-compacting concrete (LWSCC), a type of high-performance concrete which get established through self-compacting concrete (SCC) [3]. According to studies, usage of various mineral admixtures enhances the concretes workability , which additionally enhances the qualities of fresh and hardened concrete [4]. Also, it gives better mechanical efficiency and advantage such as thermal properties, economic viability is also improved. Another benefit is that LWSCC effectively decreases the structures self-weight and acoustic rate also its possible to use it for repairs and upkeep [1]. According to research, the compressive strength of Lightweight Self Compacting Concrete is affected because of aggregate type, w/b ratio proportion. Packing density theory which is a concrete mixture design method which has been utilised victoriously in lightweight self compacting concrete to assess ideal mortar to aggregates packing voids ratio [3]. Key procedures in this process for achieving the LWSCC mix design are: (1) reducing void volumes associated with coarse aggregate, (2) lowering the water/cement ratio, (3) increasing the density of cementitious materials, (4) improving the discharge ability and necessities of fresh concrete [3]. Self-compacting lightweight concrete research and assessment processes are familiar to which used by SCC. Extensive research on the practicability of SCC has been conducted in North America and Europe. The L-box, U-box, and J-ring tests have been stated to be useful in evaluating SCC passing capacity and, to a lesser extent, deformability and resistance to segregation [1]. The L-box test is ideal for on-site SCC quality control when compared with slump flow test. For estimating resistance of SCC to segregation, the visual stability index, wet sieve segregation test, and penetration test are commonly used [1]. Parhizkar et al. (2011) made research regarding few characteristics of pumice as lightweight aggregates in concretes and concluded that Pumice in lightweight concrete meets the requirements structural lightweight concrete specifications [5]. On the basis of the above-mentioned information, an analysis was studied and reported.

Literature Assessment

This literature study reveals that pumice stone is replaced with natural coarse aggregate. It is a new style of concrete which have the advantages of both lightweight and self-compacting concrete. The lightweight concrete is to decrease the dead load acting over the structure and also to remove the construction issues. If a concrete density is lies between 1440 and 1840 kg/m^3 then it is considered as a lightweight aggregate [2]. The main aim is to arrive mix design to test the workability of the concrete. Shear flow velocity increases as the binder content increases [1]. While adding different mineral admixtures to the concrete it increases the workability and also improving its properties [3]. When we replaced as coarse aggregate it does not affect the properties of concrete and thereby it increases the mechanical properties. From this study, lightweight and mineral aggregate cause a reduction in 14% of the concrete specific weight. While replacing 30 to 40% of the coarse aggregate with pumice, it increases its density, compressive strength, split tensile strength and also flexural strength in LWSCC that is lightweight self-compacting concrete [6]. It is studied that pumice stone absorbs more water than the natural course aggregate, in order to solve this issue various superplasticizers or coating materials or polymers are used. When we replace normal aggregate completely the compressive strength of the concrete is reduced. So, we can replace the pumice stone with the percentage of 50 to gain the compressive strength [9]. Usually, 60 to 100% replacement of the pumice stone can be done only for the -structural purposes. Pumice Stone in concrete mix, the workability of the self-compacting concrete is get improved slightly. Polycarboxylic ether-based superplasticizer used in this study [9]. Lightweight self-compacting concrete containing pumice stone had shown the better flow property [9]. Lightweight aggregate has promising future in self-compacting concrete is observed from the study because they have good passing ability, segregation, compressive strength and filling capacity [9]. Fly ash is a complementary material used in this concrete to increase the compressive strength. LWSCC, lightweight self-compacting concrete is less labour-intensive solution that can be implemented quickly. When Mixture containing 30% pumice stone has the high compressive strength is also assessed during the study. When determining the effectiveness of internal curing the volume of water observed by pumice is an important factor to consider. When the mix containing pumice shows specific increasing in durability of the concrete after 90 days. Usage of pumice indicates that decrease in global warming potential and it seems to be an environmentally sustainable concrete. So, it is aimed to replace the pumice as partial replacement with coarse aggregate with coating of cement paste in that aggregate to reduce water absorption into SCC, to enhance the fresh and hard property of SCC using lightweight aggregate.

Materials

In this study, Portland pozzolana cement (PPC) is used. M-Sand is used as fine aggregate. It has to pass through sieve of size 4.75 millimetre and has a specific gravity of 2.6. The modulus of fineness is 2.94 [9]. The coarse aggregate with size of maximum 12.5 mm is preferred also it is replaced by pumice stone with varying percentages. It has the specific gravity range of 0.69-1.89 [11]. Pumice is a volcanic rock, is widely used to make lightweight concrete and insulate low-density cinder blocks. The usage of pumice stone in lightweight self-compacting concrete are reducing. The major reason for this its increasing water absorption rate. So, we came up with the idea of coating the pumice stone with any one of the polymers or cement paste in order to decrease the rate. The coarse aggregate's modulus of elasticity is one of its properties. A concrete with a higher aggregate modulus would have a higher modulus. Fly ash has a specific gravity of 2.20. Particles vary in size from 10 to 100 microns. It is a thin grey powder with spherical glassy fragments that float in the flue gases. When water and cement are mixed, a paste is formed that binds the concrete together. It causes concrete to harden and this mechanism is called as hydration. As a result, provide workability during mixing and placing. It is often assumed that water suitable for drinking is often necessary for the development of concrete. PCE (polycarboxylate super plasticizer), also known as polycarboxylate ether super plasticizer, is a new type of advanced world-level environment friendly concrete admixture. PCE water reducer is a liquid super plasticizer that is ready to use. It is light yellow colour in appearance. It may even be fully submerged in water. PCE concrete admixture has a water removal potential of up to 25%.

Mix Proportion of LWSCC

The LWSCC mix proportion is critical in the application since they may affect the necessary characteristics in fresh and hard states. ,Hence for satisfying the self-compacting requirement, LWSCC, as SCC, must achieve the necessary new characteristics such as filling ability, passing ability, and segregation resistance. [11]. The proportion mix were obtained using the EFNARC (European Federation of National Associations Representing Concrete) guidelines [9]. Thus the consistency was perhaps critical aspect in ensuring the created mixtures easy handling for practical deployment. The goal is to attain superior workability which would result in self-flowing concrete that did not require compaction. Another crucial component was to keep the mixture simple by employing only a few common materials and ordinary equipment [28].

Behnam Vakhshouri et al. Method

When it comes to the mechanical properties of hard concrete, mixed design and new concrete characteristics are the very important factors for consideration. Because LWSCC is extremely affective to change the mix component characteristics and proportions, it demands more quality control. The optimum mortar to aggregates packing voids ratio was determined using packing density theory, a technique of concrete mix design which have been victoriously used in LWSCC. In this approach, the key phases to achieve the LWSCC mix design are: (a) limiting void volumes linked to coarse aggregate (b) minimising water to cement ratio, (c) maximising the density of cementitious materials, and (d) optimising the discharge ability and necessities of fresh concrete. Limit the fine aggregate content and water to powder ratio, as well as using super plasticizer, viscosity modifying, and air entraining agents in the mix design, prepares the required fluidity and viscosity of mortar in LWSCC, similar to mix design of SCC [3].

T.Z.H. Ting et al. Method

When creating LWSCC mix design, there are a few common challenges to consider. Furthermore, no set of guidelines or guideline for developing LWSCC mix design has been provided. As a result, identifying the appropriate aggregate size distribution is crucial for designing a LWSCC mix. The physical characteristics of the LWA included have a huge effect on mix proportions of LWSCC and also with the performance in both fresh and hard states. Specific gravity, size gradation, form, texture, and water absorbing ability are all properties of LWA that can dramatically affect the quantity of material needed in mix design. The most of the existing LWSCC mix design approaches in the literature is based on close aggregate packing. Number of researcher utilise the trial-and-error technique to build the LWSCC mix design because most of the suggested methods aren't suitable for usage after the application's demands change. The mix is suited for areas that require a great deal of early strength [11].

S.P. Sangeetha et al. Method

M30 concrete was employed, with a nominal mix according to IS 456-2000 and IS 10262:2009. The specimens are casted by replacing coarse aggregates with pumice aggregates in various percentages, such as 0%, 10%, 20%, 30%, 40%, 50%, 60%, 80%, and 100%, respectively. The higher the replacement of pumice stone, the lighter the mix and the greater the water absorption [5].

Materials Research Proceedings **23** (2022) 175-190 https://doi.org/10.21741/9781644901953-22

Remarks

To minimise segregation and retain strength, Mazaheripour et al. proposed using a high-performing concrete mix design technique for LWSCC. However, the method cannot generate the optimal LWSCC mixture portion in the views of new and hard characteristics. The majority of proposed methods for portioning LWSCC mix design is dependent on the close aggregate packing principle, as shown by the studies discussed above. The aggregate packing concept is used to select the aggregates with the least void to reduce the void created by LWA and to decide the right coarse to fine aggregates ratio for making the lowest density LWSCC. For simplifying and developing performance based LWSCC mix design technique, further statistical data analysis is required. The coarse to fine aggregate ratio employed is normally in the range of 0.5 to 0.6, and most researchers recommend the ratio of 0.6 because it is majorly cost effective. [11]. LWA is commonly found in an angular and flaky form, most studies have set the maximum coarse aggregate size at 12.5 or 16 mm [6].

Fresh Properties of LWSCC

Fresh concrete is a form of concrete that may be moulded and is still in a plastic state.

Workability criteria

This chapter discusses the fresh concrete test. In a fresh concrete test, the workability is a crucial consideration.For better evaluation the work ability of SCLC, the two dynamic and static stability tests were used.Static stability is concerned with characteristics of SCLC throughout time from casting to initial set, whereas dynamic stability is very much related to the characteristics of SCLC on the occasion of the activity of mixing, transportation, and casting [1]. LWSCC should be examined on filling capacity, passing capacity, and segregation resistance. These factors are utilised for determining practicability of LWSCC [11]. The main property of fresh concrete are consistency, workability, settlement and bleeding, plastic shrinkage and loss of consistency. Slump flow, V-funnel test are the methods used for determining filling ability [10]. It is the most standard test for SCC to determine its consistency. The capacity of SCC to pass through boundaries and into narrow places is also shown by its passing ability. Its stability against segregation allows concrete to retain the consistency of aggregates and mortar [10]. L- box, U-box and J-ring testsare made for evaluating the passing capacity. These tests also called as dynamic stability tests. Segregation resistance is also assessed through various tests.

Review of dynamic tests

LWSCC characteristics such as both fresh and hardened properties were studied in both review and research papers and listed below.

Slump flow test-Zhimin Wu et al. [1] studied that the slump test is unsuitable for determining the fluidity of SCLC, the slump flow test is used instead. Flowability or deformability of concrete mix was tested using a slump flow test. A normal slump cone and a steel plate with dimensions of 900 mm x 900 mm composes the testing apparatus. The mean of greatest diameter (D1) of spread concrete and the diameter (D2) of spread concrete at a right angle to the largest diameter were used to calculate the slump flow. During slump flow test, the segregation and bleeding of SCLC mixtures were visually examined. SCLC blends with slump flows in the range of 550–850 mm was found to be adequate. T.Z.H. Ting et al. [11] studied the impact of steel fibres over the properties of LWSCC in both the fresh and hardened states. In their study, they used expanded clay as aggregates. The addition of steel fibres to LWSCC increased the slump flow spread slightly.

The viscosity of fresh concrete and the quantity of superplasticizer necessitated for producing appropriate slump flow are both enlarged by enhancing the replace level of additives from 10% to 15% [18].Fariborz Hedayatinia et al. [10] studied according to Eurocode, slump flow range is 650 to 800 mm, and the T_{50} cm range is 2–5 seconds. Fig.1 shows the slump flow tests,

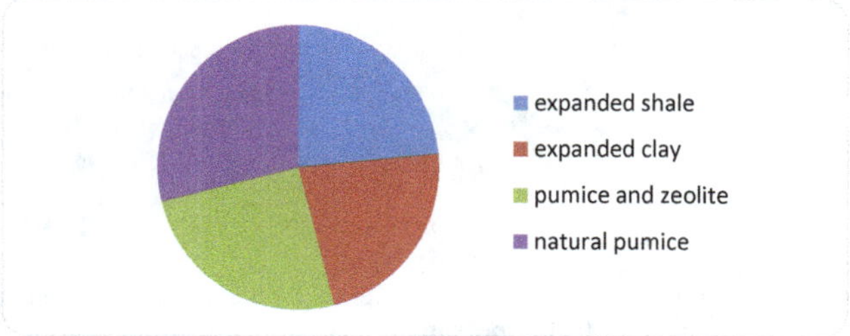

Figure.1. Slump flow tests

V-funnel test-Murat Kurt et al. [5] observed SCC has high passing ability, based on EFNARC, when the ratio of h2 to h1 is more than 0.8. it is used to assess the self-levelling and the capacity to pass the fresh SCLC. Zhimin Wu et al. [1] evaluated SCC T_V, in class 1 is less than 8 seconds, whereas SCC T_V in class 2 is 9–25 seconds. When polypropylene fibres are used, the V-funnel test time is increased [51]. In the control sample and the additives-containing specimens, the V-funnel flow times were 13 s and 8.8–12.8 s, respectively. The addition of pumice reduced the flow time, indicating increased workability [8]. Because of obstruction of steel fibres inside the V-funnel limited region, the V-funnel flow time reduced because of the increment in steel fibres [11]. Fig.2 shows the V-funnel tests,

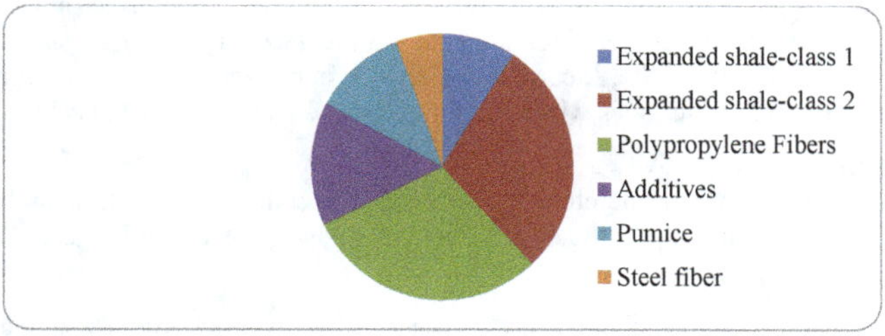

Figure.2. V-funnel tests

L-box test-L-box tests were carried out based on the EFNARC Committee's recommendations. Zhimin Wu et al. [1] studied the height of concrete in the chimney, h1, the height of concrete in the channel section, h2 and the time it takes for SCLC to reach 400 mm from three steel bars, T_{400}.For normal SCC, the L-box ratio is suggested within 0.7 - 0.9 [8].Obstructions are more significant in determining the L-box test. For the control mix and additive-containing specimens, the L-box (h2=h1) test ratios were 0.87 and 0.86–0.94, respectively [8]. L-box flow times improved more when Portland cement (PC) was replaced with fine aggregate (FA) than when ground granulated blast furnace slag(S) was used, which was better than binary FA + S [38].In an extremely fluid substance, the L-box (h2/h1) ratio equals 1. According to the EFNARC Committee report, if this ratio is less than 0.8, the aggregate may be blocked [6]. Fig.3 shows the L-box tests,

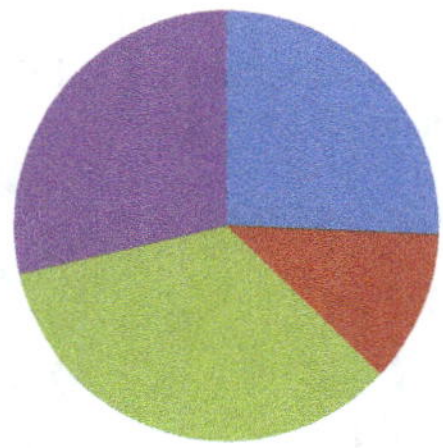

Figure.3. L-box tests

U-box test- Zhimin Wu et al. [1] observed the difference of height in concrete between the two boxes Δh, is the main parameter to be measured. In U-box test, increasing the volume % of polypropylene fibres in LWSCC decreases the filling height [51]. The H2-H1 value in the U box test must be less than 30 cm [13].

5.3 Remarks

The investigations described so far in this indicates the work ability of LWSCC is very much reliant on packing density, void volume of the aggregates. As SCC, the water - binder ratio, super plasticizer dose, and total binder content all of these has a significant impact on LWSCC workability in terms of filling capacity, passing capacity, and segregation resistance. Application of a thin cementations crust to pumice aggregates appears to improve pumice aggregate self-compacting concrete (PASCC) workability retention. Different types of additional resources are included, has a variety of effects on the LWSCC's workability Additionally, while including fibres such as steel and synthetic fibres might increase filling ability, it has a detrimental impact on passing ability. As the binder content increased, so did the necessity for super plasticizer in order to maintain the same filling ability. Incorporating fly ash in replace to binary or ternary blend can increase entire 3 new characteristics while it also lowers the quantity of SP necessitated. [11].

Hardened properties of LWSCC

Hardened concrete is a type of concrete that must be able to endure the structural and service loads that will be applied to it, as well as the environmental exposure for which it was intended.

Tests on hardened concrete

This chapter discusses the hardened concrete test which includes compressive strength test, flexural strength test, split tensile test and ultrasonic pulse velocity test. Generally the properties of hardened concrete are strength of concrete ,concrete creep, shrinkage, modulus of elasticity, water tightness, rate.

Tests criteria

Compressive strength test

The compressive strength of any fresh material is the most crucial component. Concrete's compressive strength has a significant impact on its structural performance [11]. Zhimin Wu et al. [1] used expanded shale as the replacement material and observed SCLC1 had a compressive strength of 42.6 MPa and a density of 1879 kg/m3 after 28 days, whereas SCLC2 had a compressive strength of 50.1 MPa and a density of 1920 kg/m3. Revathy.S [50] observed the 7-day and 28-day compressive strengths of concrete are seen to rise initially as the replacement percentage of pumice powder increases, reaching at around 15% and then decreasing. It is clear that after 15% replacement, there is a decrease in strength. As the amount of aluminium powder is increased, the compressive strength decreases [50]. Despite their high pozzolanicity, zeolite-containing SCCs have a poor compressive strength when compared to control SCCs. Pumice-containing mixes have a higher compressive strength than control mixtures. The compressive strength of concrete containing 10% metakaolin (MK) was higher than that of plain concrete [8].The process of inclusion of synthetic fibres in LWSCC with expanded clay as the aggregates and recycled concrete aggregate as the partial replacement enhances the strength by addition of silica fume [11]. Fig.4 shows the compressive strength tests,

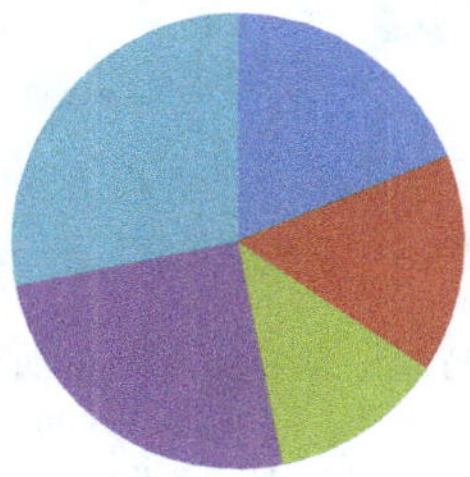

Figure.4. Compressive strength tests

Flexural strength test

An essential criteria for determining tensile strength of concrete is flexural strength [11]. With the inclusion of synthetic fibres, there was no substantial increase in the flexural strength of LWSCC [11]. For maximum volume percentage of the fibres, flexural strength is increased by 8.7% and 10.7%, respectively (0.3 %) [51]. When the volume percentage of polypropylene fibres in LWSCC is raised, flexural strength improves [51]. Among those 3 categories of aggregates, LWSCC along with furnace slag as aggregates had highest flexural strength, but LWSCC with enlarged clay as aggregates had low strength [11]. Fig.5 shows the flexural strength tests,

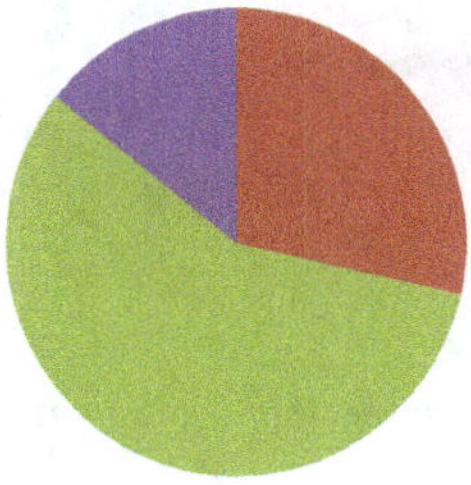

Figure.5. Flexural strength tests

Split tensile test

Concrete is usually weak under tensile action. Under flexural loading, the tensile strength of concrete is usually used for estimating the load which causes cracking in member [11]. As the proportion of polypropylene fibres in LWSCC increases, the splitting tensile strength increases [51].The inclusion of synthetic fibre did not improve the tensile strength of LWSCC.But, additional research into LWSCC's tensile strength is required before it can be used to completely replace conventional concrete in any structure. [11]. Fig.6 shows the split tensile tests,

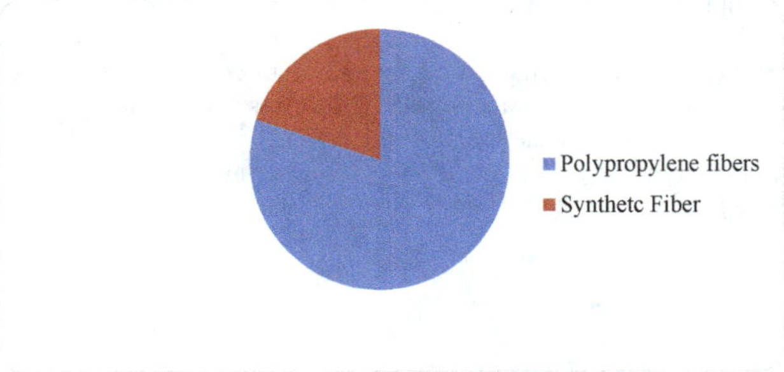

Figure.6. Split tensile tests

Ultrasonic pulse velocity test

Non-destructive concrete testing uses ultrasonic pulse velocity to assess the homogeneity, quality, void structure, damage mechanism, and compressive strength of concrete mixes [20]. Nominal frequency of the transducer pair was 54 kHz. Send a wave pulse into concrete then monitor time it takes for the pulse to propagate through the samples is how ultrasonic pulse velocities are measured. Murat Kurt et al. [6], Pumice stone is used to assess the properties and evaluated that in the 1st and 2nd groups of tests, the ultrasonic pulse velocities of SCLC were determined to be between 2611 and 2884 m/s and 3022–4770 m/s, accordingly. Ultrasonic pulse velocity was affected by the variables that affected compressive strength. As a result, the ultrasonic pulse velocities increases linearly with compressive strength [6]. Fig.7 shows the ultrasonic pulse velocity tests,

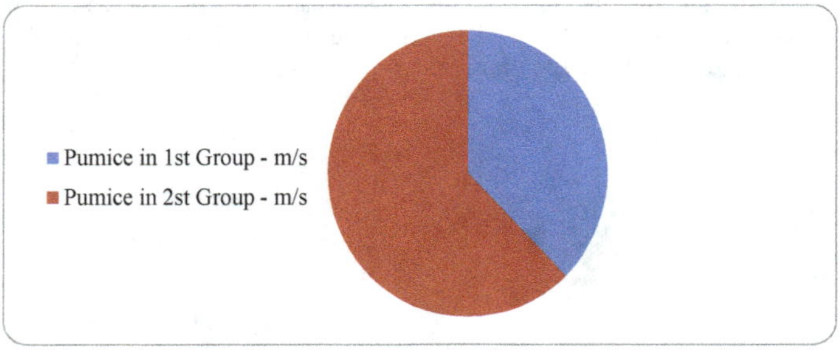

Figure.7. Ultrasonic pulse velocity tests

Remarks

The homogeneity of the batch concrete determines compressive strength of LWSCC. The mixing period, methodology determine the uniformity and homogeneity of LWSCC.The strength variability of LWSCC can be linked to the distribution of their aggregates, and it is a function of segregation resistance. Changes in mix component properties and proportions, such as water to binder ratio, binder content, and the incorporation to supplemental cementitious materials, affect the compressive strength of LWSCC. The strength of LWA have a significant effect on compressive strength of LWSCC. [11].The unit weight, thermal conductivity, and ultrasonic pulse

Materials Research Proceedings **23** (2022) 175-190 https://doi.org/10.21741/9781644901953-22

velocity values of concrete samples reduced as a result of using pumice and natural aggregate combined in SCLC, but compressive strengths and water absorption ratios increases[6].

Advantages, Limitations and Applications of LWSCC

The benefits are decrease in structure's self-weight, lowering the dead weight of the parts and making building easier, thermal properties have been improved, fire resistance has been improved, on-site transportation and storage of precast units was reduced, formwork and propping are being reduced [52]. Also, reduction in dead load saves money on foundations and reinforcement. It also has high durability, cost effectiveness, and site productivity. Few applications are listed below. LWSCC are used to construct the high-rise structures, partition walls, Residential and non – residential buildings in seismic areas. As thermal insulation, cellular lightweight aggregate is used in the form of safety laws for non-loading flat roofs. For old sewage pipes, wells, unused cellars and basements, storage tanks, tunnels, and subways, bulk feeling is achieved by using very low strength material. Development of light-weight heat-insulating wall panels. Maintain concrete's acoustic balance. Produce light plates made of cement and plaster. Use in the bridge to prevent it from freezing. Lightweight concrete is used in the building of roof slabs and small houses with load bearing walls because of its low strength. It's also used in the design of staircases, windows, and garden walls. This is also used in the design of partition walls in large buildings. Inside the structure, these are formed into slabs and used as thermal insulators. Thus, LWSCC are employed in these areas [6]. In developing LWSCC mix design, there are numerous common problems [11]. Because of the porosity and angularity of the aggregate, it is more difficult to place and finish. The dry grains absorb water quickly and can disturbs the mixture's workability [11]. It also slows the drying time. The main disadvantage is water absorption. Flooring contractors would have to delay their work for a prolonged period of time. The mixing period is longer than conventional concrete to ensure optimal mixing. It shows that they are resistant to lightweight concrete or porous concrete.

Conclusion

The focus of the research is to evaluate fresh and hard characteristics of self-compacting concrete (SCC) made with pumice stone as a coarse aggregate replacement with cement paste coating. The following hypotheses can be taken from the study:

We can understand the characteristics of the LWSCC. And we know about the importance of fly ash and polycarboxylate ether-based super plasticisers added in the mix and also the coating of aggregate. The literature study provided experimental result in performance of pumice aggregate in SCC in the fresh and hardened properties. Pumice concrete has a lower thermal conductivity, which indicates it provides less heat. Pumice aggregate coating using a thin cementations crust (though difficult in practice) seems to add the workability retention of (LWSCC). Due of its low density and high porosity, adding pumice the parts of aggregate reduced workability, unit weight, and mechanical strength. Replacement of 30% pumice yields the best performance of hardened and fresh test results. The practicability of entire LWSCC mixes are studied and within the specified standard limits, indicating that the addition of mineral admixtures to the superplasticizer improves LWSCC workability. As a result, noise emissions from vibrators used for normal weight concrete compaction can be reduced by this concrete technology. It is clear that pumice stone, while being slightly weak in compression, can be used in combination with other pozzolanic materials such as GGBS and RHA for construction purposes. According to the results, when using GGBS, RHA, less than 30-40% substitution of coarse aggregate by pumice stone is preferred for better density and strength. To get the expected results of the desired flowability, fresh and hardened properties, various types and range of mineral and chemical ad-mixtures (super plasticizer, air entraining agent, and viscosity modifying agent) are applied in mix.

Reference

[1] Muhammet Gokhan Altun, Suleyman Ozen, Ali Mardani-Aghabaglou, Effect of side chain length change of polycarboxylate-ether based high range water reducing admixture on properties of self-compacting concrete, Construction and Building Materials, 246, (2020), 118427, https://doi.org/10.1016/j.conbuildmat.2020.118427

[2] Hatice Oznur Oz, Properties of pervious concretes partially incorporating acidic pumice as coarse aggregate, Construction and Building Materials, 166, (2018), 601–609, https://doi.org/10.1016/j.conbuildmat.2018.02.010

[3] Kaizhi Liu, Rui Yu, Zhonghe Shui a, c, Xiaosheng Li, Cheng Guo a, b, Bailian Yu, Shuo Wu, Optimization of autogenous shrinkage and microstructure for Ultra-High-Performance Concrete (UHPC) based on appropriate application of porous pumice, Construction and Building Materials, 214, (2019), 369–381, https://doi.org/10.1016/j.conbuildmat.2019.04.089

[4] Ahmed T. Omar, Assem A.A. Hassan, Use of polymeric fibres to improve the mechanical properties and impact resistance of lightweight SCC, Construction and Building Materials, 229, (2019), 116944, https://doi.org/10.1016/j.conbuildmat.2019.116944

[5] O. R. Khaleel1a, S. A. Al-mishhadani, and H. Abdul Razak1, The Effect of Coarse Aggregate on Fresh and Hardened Properties of Self-Compacting Concrete (SCC), Procedia Engineering,14, (2011),805-813, http://doi.org/10.1016/j.proeng.2011.07.102

[6] Farhad Aslani, Shami Nejadi, Mechanical properties of conventional and self-compacting concrete: An analytical study, Construction and Building Materials, 36, (2012), 330–347, http://dx.doi.org/10.1016/j.conbuildmat.2012.04.034

[7] Hesam Madani, Mohammad Naser Norouzifar, Jamshid Rostami, The synergistic effect of pumice and silica fume on the durability and mechanical characteristics of eco-friendly concrete, Construction and Building Materials, 174, (2018),356–368, https://doi.org/10.1016/j.conbuildmat.2018.04.070

[8] Mehmet Karatas, Ahmet Benli, Abdurrahman Ergin, Influence of ground pumice powder on the mechanical properties and durability of self-compacting mortars, Construction and Building Materials, 150, (2017), 467–479, http://dx.doi.org/10.1016/j.conbuildmat.2017.05.220

[9] Katrin Schumacher, Nils Sabmannshausen, Christian Pritzel, Reinhard Trettin, Lightweight aggregate concrete with an open structure and a porous matrix with an improved ratio of compressive strength to dry density, Construction and Building Materials, 264, (2020), 120167, https://doi.org/10.1016/j.conbuildmat.2020.120167

[10] Christin Remayanti Nainggolan, Indriids Wijatmiko, and Ari Wibowo1, Flexural Behaviour of Reinforced Concrete Beam with Polymer Coated Pumice, https://doi.org/10.1063/1.5003509

[11] Behnam Vakhshouri and Shami Nejadi, Compressive strength and mixture proportions of self-compacting lightweight concrete, Computers and Concrete,19, (2017), 555-566, https://doi.org/10.12989/cac.2017.19.5.555

[12] Ramkumar K.B. a, Kannan Rajkumar P.R, Noor Ahmmad Shaik a, Jegan M., A Review on Performance of Self-Compacting Concrete – Use of Mineral Admixtures and Steel Fibres with Artificial Neural Network Application, Construction and Building Materials, 261, (2020), 120215, https://doi.org/10.1016/j.conbuildmat.2020.120215

[13] Roberta Occhipinti a, Antonio Stroscio a, Claudio Finocchiaro a, Maura Fugazzotto a,b, Cristina Leonelli c, Maria José Lo Faro d, Bartolomeo Megna e, Germana Barone a, Paolo Mazzoleni, Alkali activated materials using pumice from the Aeolian Islands (Sicily, Italy) and their potentiality for cultural heritage applications: Preliminary study, Construction and Building Materials, 259, (2020), 120391, https://doi.org/10.1016/j.conbuildmat.2020.120391

[14] Xiaoxiao Wang, Yao Wu, Xiangdong Shen, Hailong Wang, Shuguang Liu, Changwang Yan, An experimental study of a freeze-thaw damage model of natural pumice concrete, Powder Technology, S0032-5910, (18),30595-3, http://doi.org/10.1016/j.powtec.2018.07.096

[15] Chaabane Lynda Amel a, El-Hadj Kadri b, Yahia Sebaibi a, Hamza Soualhi b, Dune sand and pumice impact on mechanical and thermal lightweight concrete properties, Construction and Building Materials, 133, (2017), 209–218, http://dx.doi.org/10.1016/j.conbuildmat.2016.12.043

[16] Hanifi Binici, Effect of crushed ceramic and basaltic pumice as fine aggregates on concrete mortars properties, Construction and Building Materials, 21, (2007), 1191–1197, http://doi.org/10.1016/j.conbuildmat.2006.06.002

[17] Mucip Tapan a, Zeynel Yalcın b, Orhan Içelli b, Husnu Kara b, Salim Orak c, Ali Ozvan a, Tolga Depci, Effect of physical, chemical and electro-kinetic properties of pumice samples on radiation shielding properties of pumice material, Annals of Nuclear Energy, 65, (2014), 290–298, http://dx.doi.org/10.1016/j.anucene.2013.11.021

[18] Armin Azad a, Amir Saeedian a, Sayed-Farhad Mousavi a, Hojat Karami a, Saeed Farzin a, Vijay P. Singh b, Effect of zeolite and pumice powders on the environmental and physical characteristics of green concrete filters, Construction and Building Materials, 240, (2020), 117931, https://doi.org/10.1016/j.conbuildmat.2019.117931

[19] Aseel Madallah Mohammed a, Diler Sabah Asaad b, Abdulkader I. Al-Hadithi c, Experimental and statistical evaluation of rheological properties of self-compacting concrete containing fly ash and ground granulated blast furnace slag, Engineering Sciences, https://doi.org/10.1016/j.jksues.2020.12.005

[20] Sandra Juradin, Goran Baloevic, and Alen Harapin, Experimental Testing of the Effects of Fine Particles on the Properties of the Self-Compacting Lightweight Concrete, Advances in Materials Science and Engineering,2012,8, http://doi.org/10.1155/2012/398567

[21] Kianoosh Samimi a, Siham Kamali-Bernard, Ali Akbar Maghsoudi, Mohammad Maghsoudi, Hocine Siad, Influence of pumice and zeolite on compressive strength, transport properties and resistance to chloride penetration of high strength self-compacting concretes, Construction and Building Materials, 151, (2017), 292–311, http://dx.doi.org/10.1016/j.conbuildmat.2017.06.071

[22] Abdullah Bilal, Basharat Jamil, Nadeem Ul Haque, Md Azeem Ansari, Investigating the effect of pumice stones sensible heat storage on the performance of a solar still, Groundwater for Sustainable Development, 9, (2019), 100228, https://doi.org/10.1016/j.gsd.2019.100228

[23] H. Binici a, ∗, M.Y. Durguna, T. Rızaoglub, M. Koluçolak, Investigation of durability properties of concrete pipes incorporating blast furnace slag and ground basaltic pumice as fine aggregates, Scientia Iranica A, (2012) ,19 (3), 366–372, https://doi.org/10.1016/j.scient.2012.04.007

[24] Hasan Oktay a, Recep Yumrutas b, Abdullah Akpolat, Mechanical and thermophysical properties of lightweight aggregate concretes, Construction and Building Materials, 96, (2015), 217–225, http://dx.doi.org/10.1016/j.conbuildmat.2015.08.015

[25] Saif I. Mohammed, Khalid B. Najim, Mechanical strength, flexural behavior and fracture energy of Recycled Concrete Aggregate self-compacting concrete, Structures, 23, (2020), 34–43, https://doi.org/10.1016/j.istruc.2019.09.010

[26] Ozlem Sally Bideci, Alper Bideci, Ali Haydar Gultekin, Sabit Oymael, Hasan Yildirim, Polymer Coated Pumice Aggregates and Their Properties, Composites, S1359-8368, (13),00584-2, http://dx.doi.org/10.1016/j.compositesb.2013.10.009

[27] L. Cavaleri, N. Miraglia, M. Papia, Pumice concrete for structural wall panels, Engineering Structures, 25, (2003), 115–125, https://doi.org/10.1016/S0141-0296(02)00123-2

[28] Kozo Onoue, Hiroki Tamai, Hendro Suseno, Shock-absorbing capability of lightweight concrete utilizing volcanic pumice aggregate, Construction and Building Materials, 83, (2015), 261–274, http://dx.doi.org/10.1016/j.conbuildmat.2015.03.019

[29] Abhishek Jain a, Rajesh Gupta a, Sandeep Chaudhary, Sustainable development of self-compacting concrete by using granite waste and fly ash, Construction and Building Materials, 262, (2020), 120516, https://doi.org/10.1016/j.conbuildmat.2020.120516

[30] Ozlem Sallı Bideci, The effect of high temperature on lightweight concretes produced with colemanite coated pumice aggregates, Construction and Building Materials, 113, (2016), 631–640, http://dx.doi.org/10.1016/j.conbuildmat.2016.03.113

[31] Revathy.S1, Josina Thomas2, Lightweight characteristics of self-compacting concrete using aluminium powder and fine pumice powder, structural Engineering and Construction Management ,03,2016,2395-0072,

[32] H. Mazaheripour a, S. Ghanbarpour b, S.H. Mirmoradi a, I. Hosseinpour a, the effect of polypropylene fibers on the properties of fresh and hardened lightweight self-compacting concrete, Construction and Building Materials, 25, (2011) ,351–358, https://doi.org/10.1016/j.conbuildmat.2010.06.018

[33] Xiaoxiao Wang, Yao Wu, Xiangdong Shen, Hailong Wang, Shuguang Liu, Changwang Yan, An experimental study of a freeze-thaw damage model of natural pumice concrete, Powder Technology, S0032-5910, (18),30595-3, http://doi.org/10.1016/j.powtec.2018.07.096

[34] Chaabane Lynda Amel a, El-Hadj Kadri b, Yahia Sebaibi a, Hamza Soualhi b, Dune sand and pumice impact on mechanical and thermal lightweight concrete properties, Construction and Building Materials, 133, (2017), 209–218, http://dx.doi.org/10.1016/j.conbuildmat.2016.12.043

[35] Hanifi Binici, Effect of crushed ceramic and basaltic pumice as fine aggregates on concrete mortars properties, Construction and Building Materials, 21, (2007), 1191–1197, http://doi.org/10.1016/j.conbuildmat.2006.06.002

[36] Mucip Tapan a, Zeynel Yalçın b, Orhan Içelli b, Husnu Kara b, Salim Orak c, Ali Ozvan a, Tolga Depci, Effect of physical, chemical and electro-kinetic properties of pumice samples on radiation shielding properties of pumice material, Annals of Nuclear Energy, 65, (2014), 290–298, http://dx.doi.org/10.1016/j.anucene.2013.11.021

[37] Armin Azad a, Amir Saeedian a, Sayed-Farhad Mousavi a, Hojat Karami a,, Saeed Farzin a, Vijay P. Singh b, Effect of zeolite and pumice powders on the environmental and physical characteristics of green concrete filters, Construction and Building Materials, 240, (2020), 117931, https://doi.org/10.1016/j.conbuildmat.2019.117931

[38] Aseel Madallah Mohammed a, Diler Sabah Asaad b, Abdulkader I. Al-Hadithi c, Experimental and statistical evaluation of rheological properties of self-compacting concrete containing fly ash and ground granulated blast furnace slag, Engineering Sciences, https://doi.org/10.1016/j.jksues.2020.12.005

[39] Sandra Juradin, Goran Baloevic, and Alen Harapin, Experimental Testing of the Effects of Fine Particles on the Properties of the Self-Compacting Lightweight Concrete, Advances in Materials Science and Engineering,2012,8, http://doi.org/10.1155/2012/398567

[40] Kianoosh Samimi a, Siham Kamali-Bernard, Ali Akbar Maghsoudi, Mohammad Maghsoudi, Hocine Siad, Influence of pumice and zeolite on compressive strength, transport properties and resistance to chloride penetration of high strength self-compacting concretes, Construction and Building Materials, 151, (2017), 292–311, http://dx.doi.org/10.1016/j.conbuildmat.2017.06.071

[41] Abdullah Bilal, Basharat Jamil∗, Nadeem Ul Haque, Md Azeem Ansari, Investigating the effect of pumice stones sensible heat storage on the performance of a solar still, Groundwater for Sustainable Development, 9, (2019), 100228, https://doi.org/10.1016/j.gsd.2019.100228

[42] H. Binici a, ∗, M.Y. Durguna, T. Rızaoglub, M. Koluçolak, Investigation of durability properties of concrete pipes incorporating blast furnace slag and ground basaltic pumice as fine aggregates, Scientia Iranica A, (2012) ,19 (3), 366–372, https://doi.org/10.1016/j.scient.2012.04.007

[43] Hasan Oktay a, Recep Yumrutas b, Abdullah Akpolat, Mechanical and thermophysical properties of lightweight aggregate concretes, Construction and Building Materials, 96, (2015), 217–225, http://dx.doi.org/10.1016/j.conbuildmat.2015.08.015

[44] Saif I. Mohammed, Khalid B. Najim, Mechanical strength, flexural behavior and fracture energy of Recycled Concrete Aggregate self-compacting concrete, Structures, 23, (2020), 34–43, https://doi.org/10.1016/j.istruc.2019.09.010

[45] Ozlem Sallı Bideci, Alper Bideci, Ali Haydar Gultekin, Sabit Oymael, Hasan Yildirim, Polymer Coated Pumice Aggregates and Their Properties, Composites, S1359-8368, (13),00584-2, http://dx.doi.org/10.1016/j.compositesb.2013.10.009

[46] L. Cavaleri ∗, N. Miraglia, M. Papia, Pumice concrete for structural wall panels, Engineering Structures, 25, (2003), 115–125, https://doi.org/10.1016/S0141-0296(02)00123-2

[47] Kozo Onoue a, ⇑, Hiroki Tamai b, Hendro Suseno, Shock-absorbing capability of lightweight concrete utilizing volcanic pumice aggregate, Construction and Building Materials, 83, (2015), 261–274, http://dx.doi.org/10.1016/j.conbuildmat.2015.03.019

[48] Abhishek Jain a, Rajesh Gupta a, Sandeep Chaudhary, Sustainable development of self-compacting concrete by using granite waste and fly ash, Construction and Building Materials, 262, (2020), 120516, https://doi.org/10.1016/j.conbuildmat.2020.120516

[49] Özlem Sallı Bideci, The effect of high temperature on lightweight concretes produced with colemanite coated pumice aggregates, Construction and Building Materials, 113, (2016), 631–640, http://dx.doi.org/10.1016/j.conbuildmat.2016.03.113

[50] Revathy.S1, Josina Thomas2, Lightweight characteristics of self-compacting concrete using aluminium powder and fine pumice powder, structural Engineering and Construction Management ,03,2016,2395-0072,

[51] H. Mazaheripour a, S. Ghanbarpour b, S.H. Mirmoradi a, I. Hosseinpour a, The effect of polypropylene fibers on the properties of fresh and hardened lightweight self-compacting concrete, Construction and Building Materials, 25, (2011) ,351–358, https://doi.org/10.1016/j.conbuildmat.2010.06.018

Sustainable Materials and Smart Practices - NCSMSP-2021 | Materials Research Forum LLC
Materials Research Proceedings **23** (2022) 191-204 | https://doi.org/10.21741/9781644901953-23

Development of Self-Sensing Cement Composite using Nanomaterials for Structural Health Monitoring of Concrete Columns – A Comprehensive Review

A. Dinesh[1,a,*], A. Mohanraj[2,b], S. Veeraraghavan[3,c], K.G. Naveena[1,d]

[1]Department of Civil Engineering, Sri Ramakrishna Engineering College, Coimbatore - 641022, India

[2]Department of Civil Engineering, Bannari Amman Institute of Technology, Sathyamangalam – 638401, India

[3]Independent Researcher (Cement composite and Materials), Coimbatore - 641022, India

[a]dinesh.a@srec.ac.in; dinesh.a.civil@gmail.com, [b]mohanraja@bitsathy.ac.in, [c]veeraraghavan.civil@gmail.com, [d]naveena.1909025@srec.ac.in

Keywords: Self-Sensing, Nanomaterials, Carbon, Health Monitoring, Strength and Conductivity

Abstract. Due to age, structural deterioration, and other factors, concrete constructions such as beams and columns will inevitably deteriorate. The growth of nanomaterials and recent advances in multidisciplinary research has broadened cement composites' applicability in various fields. A self-sensing cement composite can detect its own deformation, strain, and stress by changing its electrical characteristics, which may be measured with electrical resistivity. Carbon-based nanomaterials, such as carbon fiber, carbon black, and carbon nanotube, have a strong potential to increase cement composite's mechanical (strength) and electrical (resistivity, sensitivity) potentials due to their remarkable strength and conductivity. Due to the artificial integration of conductive carbon-based components will generate piezoresistive properties in typical cement composites, transforming them into self-sensing cement composites. As a result, the review focuses primarily on the development of nanoparticle-based self-sensing cement composites and their use in the health monitoring of structural columns. This research critically examines the materials used, fabrication techniques, strength, and sensing methodologies used to develop the self-sensing cement composite. The difficulties of commercializing self-sensing cement composites, as well as potential solutions, are also highlighted. According to the review, the difference in Poisson ratio and youngs modulus between the self-sensing cement composite and columns leads the self-sensing cement composite to have different strength and conductivity before and after embedding in columns. According to the study, the addition of conductive material diminishes the composite's workability due to its large specific surface area. Because of the well-distributed conductive network, the composite's resistivity is significantly lowered. The study also shows that the inclusion of a self-sensing cement composite has no bearing capacity influence on the column. Finally, according to the review, the self-sensing cement composite has the ability to monitor the health of structural columns.

Introduction

Environmental risks and concrete aging cause catastrophic structural collapses in concrete structures. Many breakthroughs have been made in concrete construction, both structurally [1] and in terms of material [2–4], akin to those made in structural steel [5,6]; however, concrete deterioration is unavoidable. In the domains of bricks [7,8], soil [9–11], and concrete [12–14], material modification for property enhancement is also increasing. The building's safety must be thoroughly and continuously evaluated through frequent inspection and maintenance to avoid catastrophic structural failures. Structural health monitoring systems are applied in critical infrastructures and buildings to analyze the structure's performance and detect aberrant responses [15]. To monitor structural performance, piezoceramic sensors [16], ultrasonic sensors [16], and other conventional sensors [17] are offered.

On the other hand, structural material evolution has gone through several stages, from high-strength materials to materials with high strength and low weight, and most recently, materials with both strength and self-sensing [3,18]. Self-sensing materials may detect their own stress, strain, and damage [19,20]. Compared to conventional sensors that are included or attached, self-sensing cement composites are a sort of smart material that is innately smart. Those self-sensing cement composites are extremely robust, more compatible, have a huge sensing volume, and have no mechanical properties loss [18,20]. Damage detection is important for structural health monitoring and predicting service life.

Steel fiber (SF) [21,22], carbon fiber (CF) [23,24], carbon nanofiber (CNF) [20,25,26], carbon black (CB) [27,28], and nickel powder (NP) [29,30] are some of the conductive elements that can be used to make self-sensing cement composites. As seen in Fig. 1, these materials in the composite create a conductive network [31]. If the conductive particle concentration is high enough, a conductive network will form before applying the load, as shown in Fig. 1 (a). However, if only a low concentration of conductive materials is utilized, an external load is required to produce total continuous conduction, as shown in Fig. 1 (b). Furthermore, the generated hydration products will act as a barrier to the creation of the composite's conductive network. Because aggregates operate as a significant barrier to the continuous network channel in the composite, coarse aggregates are rarely used in self-sensing cement composites. Consequently, conductive and non-conductive phases exist in self-sensing cement composites [18,20,32]. Conductive materials like CF and CB produce the conductive phase [20]. Conventional materials like cement and hydration products like calcium-silicate hydrate (CSH) and portlandite, on the other hand, create the non-conductive phase. In some composites, a hybrid material combination improves the composite's self-sensing capacity [33,34].

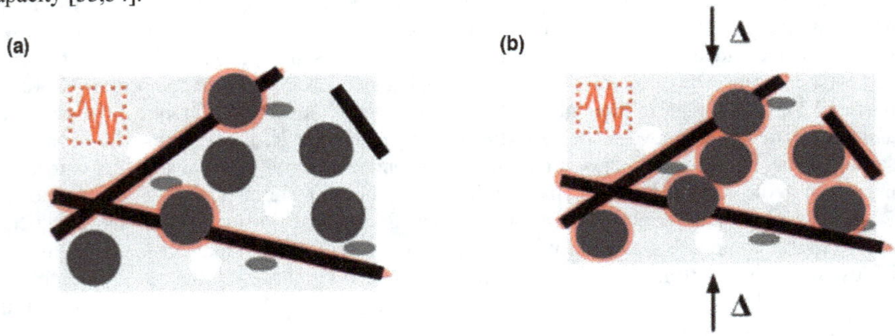

Fig. 1 (a) Formation of the conductive pathway before loading and (b) formation of the conductive pathway under strain [31]

Thus, the main objective of this paper is to investigate the creation of self-sensing cement composites utilizing an interdisciplinary approach and a range of conductive materials. The next phase focuses on using the self-sensing cement composites that have been produced to monitor the health of columns.

Materials and Methods

Cement, sand, and water are the most common raw materials used to make self-sensing cement composites [35–37]. Fly-ash/silica fume has been employed as a binder material in fabricating sensors on occasion [23,35]. As a water lowering agent, a polycarboxylate superplasticizer was utilized, which can reduce water content by 30%. Conductive fillers such as CF, SF, and carbon nanotube (CNT) are often utilized [18]. On the other hand, dispersion of conductive components in the composite is time-consuming. Hybrid CNT/NCB composite fillers are occasionally used because nano carbon black (NCB) self-assembles onto the surface of CNT by electrostatic adsorption, resulting in the formation of a grape-bunch structure as seen in Fig. 2. Mechanical stirring was also employed to achieve dispersion of CNTs and NCBs embedded in the cement matrix without using ultrasonic treatment. Finally, the use of fly ash and water reducers had evident effects on the dispersion of conductive materials in the cement matrix.

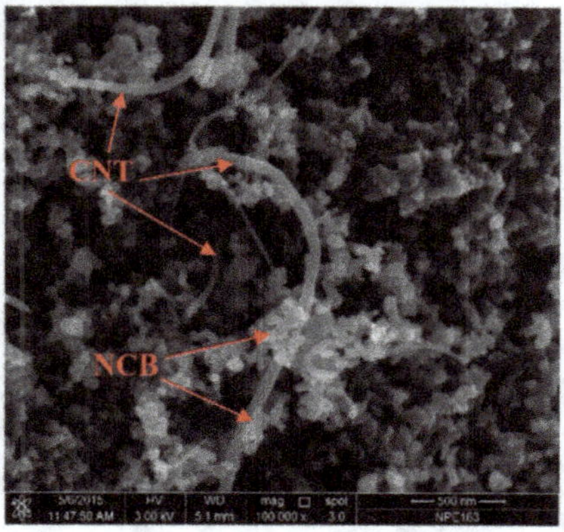

Fig. 2 SEM image of cement composite with grape brunch shape of CNT and NCB [27]

The self-sensing cement composite is typically made using two probe or four-probe technologies [18,20,38–40]. The cementitious composite, including conductive elements, was cast and cured according to the design calculations. Two stainless steel gauge electrodes separated by a distance of 10 mm were introduced into the specimens using the two-probe method, as illustrated in Fig. 3. Four copper or steel electrodes are utilized in the four-probe approach. In the two probe techniques, both current supply and voltage measurement are performed on the same electrodes [27].

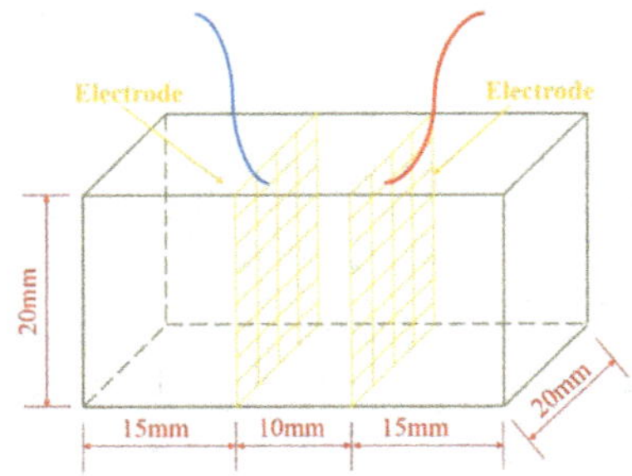

Fig. 3 Schematic diagram of cementitious composite with electrode position in the composite (Two probe method) [27]

Self-sensing cement composite columns were cast in bulk form according to the design calculations. The concrete column is subsequently subjected to mechanical and electrical testing simultaneously. Fig. 4 shows the experimental setup for compressive strength, elastic modulus, and resistivity measurements. In compliance with IS code requirements, the concrete columns were tested using hydraulic testing equipment with a capacity of 2000 kN. Through a PC interface, the measurements were automatically recorded [27].

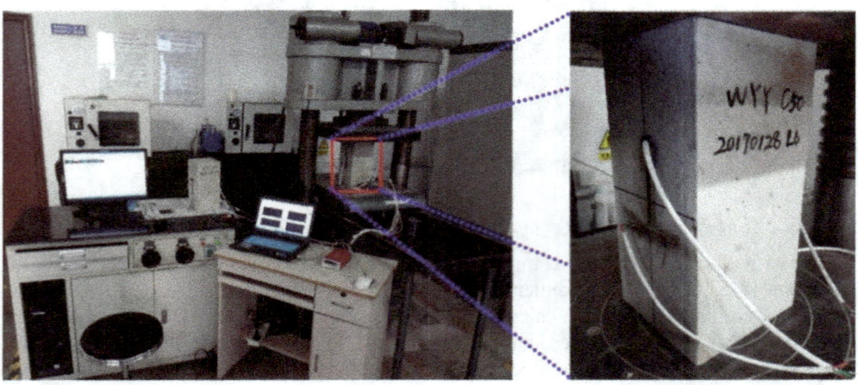

Fig. 4 Setting up an experiment to determine the strength and elastic modulus of concrete columns [27]

Results and Discussions

Workability

The conductive materials can reduce the relative slump of the composite. The use of hybrid shape memory alloys (SMA) and SF in cement composites can reduce relative slump [2]. Because of the interplay between CFs and the restriction in aggregate movement imposed by CFs, adding 0.1% CF resulted in a 43% reduction in slump diameter [41]. Even though CF-based specimens may have a more controllable disruption of the electrically conductive network, uniform dispersibility, initial manufacturing costs, and mechanical property impacts must be considered when choosing a carbon-based material.

Compressive strength

The CNFs/epoxy composites' compressive strength, poisons ratio, and elastic modulus was measured using a uniaxial compression test. As illustrated in Fig. 5 (a), increasing the CNF volume content to 0.58% resulted in a considerable increase in compressive strength. However, as the volume content of the CNF increased by 1.16%, there was a steady decrease in compressive strength. Because of the inhomogeneous dispersion of conductive materials in the composite, particle agglomeration occurs, lowering the composite's compressive strength at greater concentrations [42,43]. The composites filled with 0.58% CNF had a compressive strength of 77.6 MPa, which was determined to be 16% higher than the control specimen. This is due to an interfacial connection between the CNF and the epoxy matrix, which enables the load to be transmitted to the CNF via the interface. It was eventually observed that when the concentration of CNFs grew, so did the elastic modulus, as seen in Fig. 5 (b). When the elastic modulus of the specimen containing 1.16% CNFs was compared to the control specimen, the elastic modulus increased by 26%. This is because the CNF has a stronger strength and elastic modulus than the epoxy matrix [44]. The CNF pullout limits the composite's transverse distortion, which improves the composite's mechanical properties.

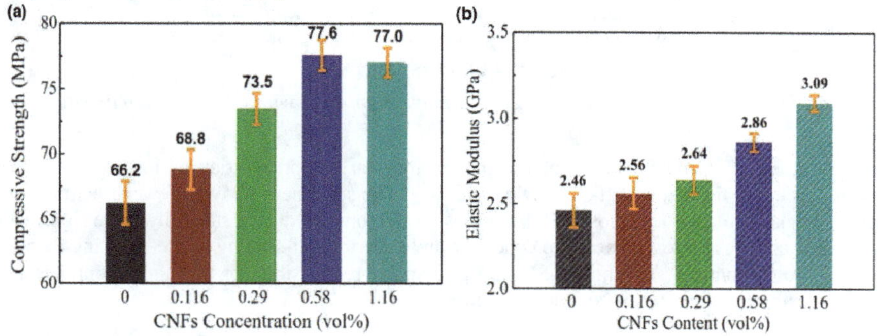

Fig. 5 Epoxy composite showing (a) variation of compressive strength and (b) variation of the elastic modulus with different vol% of CNF [44] (modified form)

When MWCNT was embedded, the composites' compressive and flexure strengths increased by 36.07% and 18.11%, respectively. This was most likely due to microfracture bridging, nanopore filling, and other processes caused by the distributed MWCNT. CFs are simpler to distribute in dense ECC systems uniformly and are reported to increase the mechanical characteristics of the mixtures.

Flexural strength

Fig. 6 depicts the rise in flexural (and compressive) strength of cement mortars, including CNT/NCB composite fillers. As the CNT concentration in the composites was 0.77 vol.% and 1.52 vol.%, there was a significant improvement in flexural strength of 6.7% and 18.3%, respectively, when compared to ordinary cement mortar. The crack-bridging impact of CNT, the draw-out effect of CNT, and the capillary filling effect of NCB all contribute to the increase in flexural strength

[45]. Compared to normal cement mortar, when the CNT content in composites was increased by 2.40 vol.% and 3.12 vol.%, the flexural strength was lowered by 15% and 45%, respectively. The composite's flexural strength is reduced due to the inhomogeneous distribution of CNT in the composite and an excess water binder ratio.

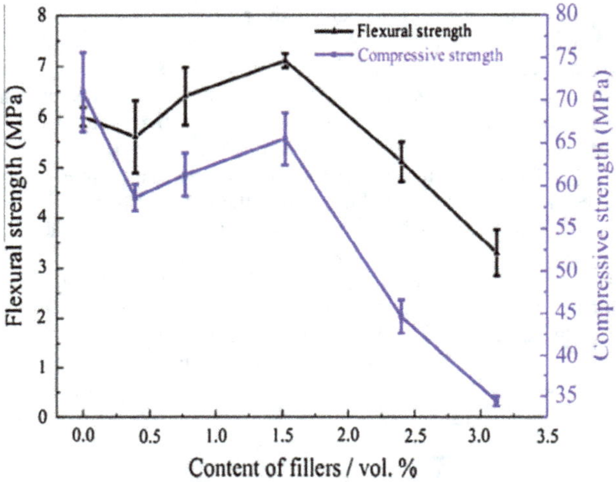

Fig. 6 Flexural and compressive strength of cement composite with CNT/NCB composite fillers [45]

The use of CB and SF in a diphasic pattern may be an optimization option for improving concrete's crack formation self-sensing capabilities. The specimens showed fewer fractures, and larger cracks as the CF level rose. CNT composites demonstrated a substantially more significant fractional change at a lower moisture condition and superior self-sensing to pre-cracking than CF and control specimens. The macro-SFs are principally responsible for the high endurance and variety of fracture features of conductive concrete.

Conductivity

The electrical resistivity of smart concrete containing MWCNT and SF changes when the compressive load is raised from 20 to 100 MPa, as shown in Fig. 7 [31]. When the compressive strength was increased to 20 MPa, no changes in electrical resistivity were found. Electrical resistivity fell by 3.04%, 5.09%, 7.88%, and 9.30% as compressive strength was increased by 40 MPa, 60 MPa, 80 MPa, and 100 MPa, respectively. When the compressive stress and compressive strain increase proportionately, as shown in Fig. 1 (a) and Fig. 1 (b), the distance between the conductive fillers decreases, increasing the connection between the conductive fillers.

The addition of SF to the composite improves the connection between the conductive components, resulting in a significant reduction in the composite's resistivity. Similarly, the quantum tunneling effect improves the piezoelectric response when fine steel slag aggregates (FSSA) are used. The composite will exhibit combination conductivity and piezoresistivity behavior if hybrid SF and FSSA are utilized [29]. A lower piezoelectric response shows that the composite's piezoelectric response has been reduced due to the production of a few microfractures, which could potentially harm the conducting network. The composite's conductivity is enhanced by a consistent distribution of conductive components, fewer defects, and fewer cracks.

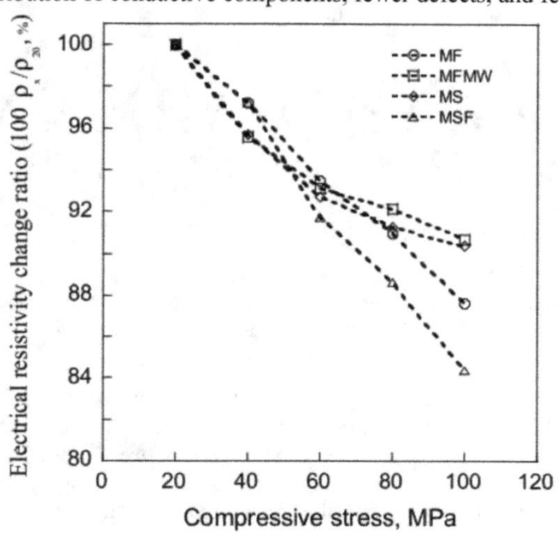

Fig. 7 Electrical Resistivity variation of the steel fibers (MF), fine steel slag aggregates (MS), steel fibers and fine steel slag aggregates (MSF), carbon nanotube and steel fibers (MFMW) embedded composite concerning the change in stress [31]

The ECC system's flexural strength and conductivity have been proven to be improved by SFs or CB particles. Because of the bridging phenomena generated by the long bar shape, SFs may provide superior mechanical and electrical performance. On the other hand, the total conductive admixture had the opposite effect; higher gauge factors were seen for the lowest CF additions tested, demonstrating that CFRCC self-sensing capabilities are affected by percolation conditions. Increase the length of the fiber, and the conductivity will increase if the CF concentration remains constant. Percolation has been achieved for all PAN CF. Percolation was seen at lower CF values as the aspect ratio of the fibers increased. As the size and duration of the loading increased, the fractional change in resistivity (FCR) likewise increased.

Piezoresistivity

Piezoresistivity refers to the change in electrical characteristics of a composite as a result of a change in electrical properties. Fig. 8 (a) depicts the change in FCR concerning stress, while Fig. 8 (b) depicts the change in FCR concerning strain (b). The change in FCR shows good agreement with the change in stress and strain of the composite, as shown in Fig. 8 (a) and 8 (b). It has been noticed that as the composite's stress/strain increases, the FCR value decreases [27,46]. This is because as stress/strain increases, the conductive components in the composite converge and form a continuous conductive channel, increasing conductivity and lowering the composite's FCR. When compressive stress/strain reaches its maximum value, the accompanying FCR value reaches the polyline's bottom.

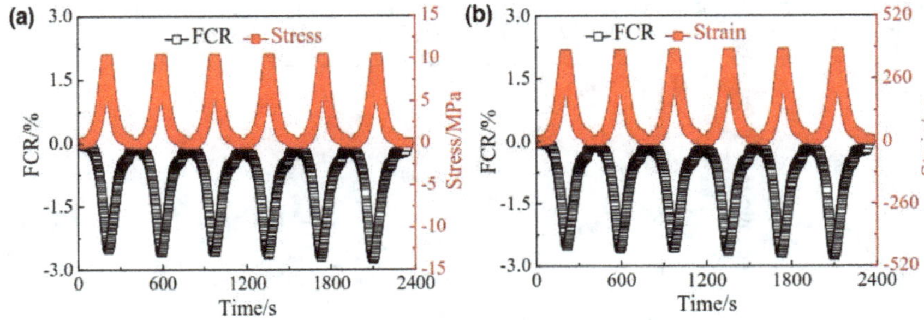

Fig. 8 Concrete column self-sensing signals under cyclic loading with a stress amplitude of 10 MPA with the representation of (a) FCR (%) vs. Stress (MPa) concerning time and (b) FCR (%) vs. Strain (με) concerning time [27]

Fig. 9 (a) shows the experimental setup for measuring piezoresistivity and conductivity in the self-sensing column. Fig.9 (b) and 9 (c) depict a concrete column's real and schematic failure. When the strain is 0.002, the cement composite exhibits peak stress of 9.1 MPa, as shown in Fig. 9 (d). It has been discovered that when concrete deforms, the space between conductive components shortens, lowering the composite's resistance [47]. However, when the load on the composite increases, additional fractures occur, increasing the composite's resistance at greater stress levels. The specimen in Fig. 9 (c) has a crack between the two inner probes of the structure, and the purpose of providing mesh in this study is to prevent the crack from propagating through the structure indefinitely, as seen in Fig. 9 (b).

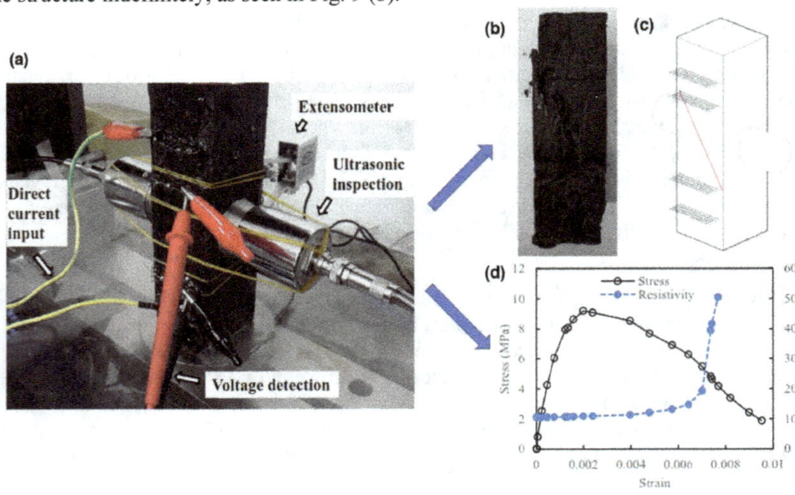

Fig. 9 (a) Experimental setup of concrete columns (b) failure of concrete columns – actual (c) failure of concrete columns – schematic and (d) Stress-strain and resistivity curve of concrete columns (modified form) [47]

The specimen had low electrical conductivity under no loading conditions because there were only a few CNFs installed and no contacts developed between them. In contrast, the epoxy matrix showed a much more significant deformation under loading conditions, causing the distance between the CNFs to decrease, as shown in Fig. 10. (a). Consequently, as shown in Fig. 10 (c), the electrical conductivity of the composites rises as the CNFs get much more extensive. As a consequence, the electrically conductive pathways generated by a few CNFs touching each other (blue lines in Fig. 10 (a) and 10 (c)) were assumed to be a significant cause of the reduction in electrical resistivity of composites containing a large number of CNFs.

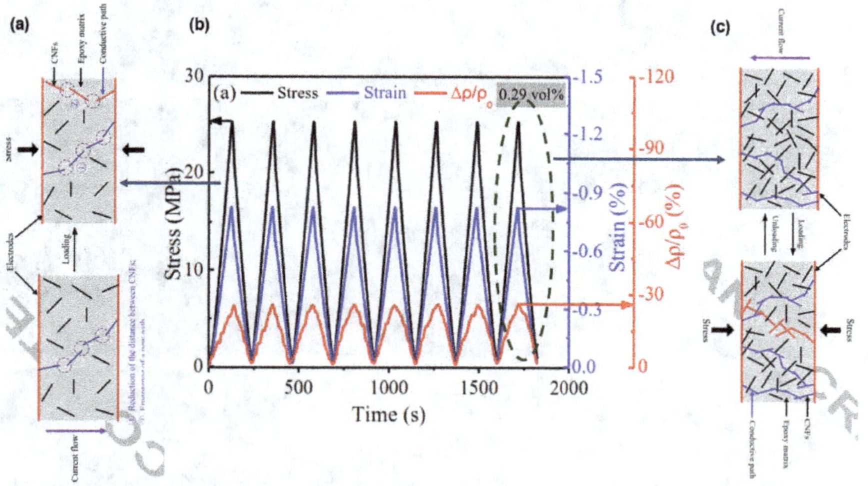

(a) Schematic depiction of CNF orientation change before and during loading (b) Time history association between FCR and cyclic compressive stress/strain with 0.29 vol. % CNF (c) Schematic of CNF orientation between electrodes during loading and unloading (modified form - [44])

Scanning Electron Microscopy (SEM) Analysis
The SEM images of nanocomposites filled with varying amounts of CNFs (0 vol.%, 0.29 vol.%, 0.58 vol.%, and 1.16 vol.%) employed in their investigation are shown in Fig. 11 (a) - (i). It was discovered in their investigation that specimens containing the same amount of CNFs had microscopic features on the fracture surface [44]. According to fig. 11 (b) – (e), the conductive materials were equally disseminated in the composite at 0.29% and 0.58% CNFs, respectively. Fewer CNFs scrambled together at higher CNF concentrations (=1.16%), resulting in CNF agglomeration, as illustrated in Fig. 11. (f) – (i). The dispersion of CNFs was impressive, increasing the nanocomposites' characteristics. As a result, the SEM picture proved to be a suitable tool for demonstrating material dispersion and composite property enhancement.

SEM micrographs of the surface layer of repaired cracks revealed that CF and CNT composites had better physical self-healing than control composites, resulting in more nucleation zones. SEM investigation of the core layer revealed bridged CFs across the healed microcracks, which is most likely the cause of low pre-cracking sensitivities with CF content [48]. Thus SEM investigation proves to be the better solution to identify the modified conductive network inside the composite.

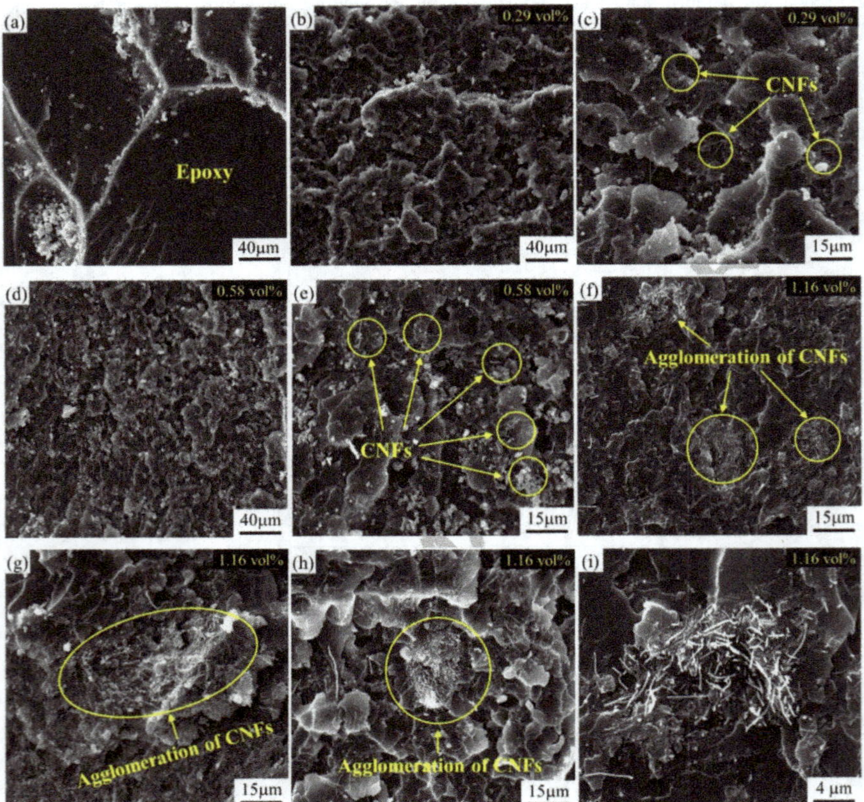

Fig. 11 SEM images of CNF embedded nanocomposites of volume (a) 0%, (b), (c) 0.29 vol.% (d), (e) 0.58 vol.% (f) to (i) 1.16 vol.% [44]

Conclusions

The review concentrates on developing nanoparticle-based self-sensing cement composites for use in column health monitoring. The self-sensing cement composite components, production techniques, strength, and sensing approach are all investigated. The challenges of commercializing self-sensing cement composites, as well as potential solutions, are also highlighted. From the review, the following conclusions are summarized:

1. Due to the high specific surface area of the composite, the addition of conductive material reduces its workability.
2. Due to crack bridging and pore filling, the composite's compressive strength rises to a certain level (depending on the composite's specific potential). Compressive strength

decreases beyond a certain point due to poor dispersion of conductive components and the production of air bubbles in the composite.
3. Due to the addition of conductive material, the composite's flexural strength tends to rise because the conductive substance improves crack resistance, which positively impacts fatigue life and energy absorption capacity.
4. The composite's resistivity is significantly reduced due to the well-distributed conductive network.
5. The conductivity of the composite tends to improve as the length of the conductive material increases. The creation of a conductive network and the formation of hydration products in the composite are clearly visible in SEM analysis.

As a result, the review paper investigates the numerous potentials required by the self-sensing cement composite and the potential of particular fillers in improving the composite's strength and conductive properties.

Acknowledgments
The authors would like to thank the Management, Sri Ramakrishna Engineering College, Coimbatore, for their support. The authors would like to express their gratitude to the Management, PSG College of Technology, Coimbatore, for their assistance in carrying out this study.

References
[1] A. Dinesh, R. Prasanth Kumar, S.R. Abijith, Experimental investigation on bubble deck concrete using plastic waste, 2020. https://doi.org/10.1007/978-981-15-0990-2_14.
[2] J.M. Rius, A. Cladera, C. Ribas, B. Mas, Shear strengthening of reinforced concrete beams using shape memory alloys, Constr. Build. Mater. 200 (2019) 420–435. https://doi.org/10.1016/j.conbuildmat.2018.12.104.
[3] A. Professor, Study on Strength properties of High performance concrete, n.d. www.ripublication.com/ijaer.htm.
[4] A.S. S.D., D. A., S.B. V., Investigation of waste marble powder in the development of sustainable concrete, Mater. Today Proc. (2020). https://doi.org/10.1016/j.matpr.2020.10.536.
[5] P. Eswaramoorthi, P. Magudeaswaran, A. Dinesh, PUSHOVER ANALYSIS OF STEEL FRAME, n.d.
[6] A. Professor, Analytical study on the behaviour of cold formed steel double channel beam sections, n.d. www.ripublication.com/ijaer.htm.
[7] M. Sutcu, H. Alptekin, E. Erdogmus, Y. Er, O. Gencel, Characteristics of fired clay bricks with waste marble powder addition as building materials, Constr. Build. Mater. 82 (2015) 1–8. https://doi.org/10.1016/j.conbuildmat.2015.02.055.
[8] D. Sellakutty, A. Professor, UTILISATION OF WASTE PLASTIC IN MANUFACTURING OF BRICKS AND PAVER BLOCKS, 2016. https://www.researchgate.net/publication/303273892.
[9] K. Sathish Raja, Dinesh. A, Study on Self Compacting Concrete – A Review, Int. J. Eng. Res. V5 (2016). https://doi.org/10.17577/ijertv5is020349.
[10] A. Professor, Stabilization of soil by using solid waste-A Review, © 2017 IJEDR |. 5 (2017). www.ijedr.org.
[11] Individual_chapter.pdf, (n.d.).
[12] A. Dinesh, S. Harini, J.P. Jasmine, J. Jincy, J. Shagufta, International journal of engineering sciences & research technology experimental study of blast furnace slag concrete, Int. J. Eng. Sci. Res. Technol. 6 (2017) 42–50.

https://doi.org/10.5281/zenodo.345692.
[13] A. Dehghani, F. Aslani, The effect of shape memory alloy, steel, and carbon fibres on fresh, mechanical, and electrical properties of self-compacting cementitious composites, Cem. Concr. Compos. 112 (2020) 103659. https://doi.org/10.1016/j.cemconcomp.2020.103659.
[14] X. Xi, D.D.L. Chung, Piezoelectret-based and piezoresistivity-based stress self-sensing in steel beams under flexure, Sensors Actuators, A Phys. 301 (2020) 111780. https://doi.org/10.1016/j.sna.2019.111780.
[15] Y. Zhang, L. Yuan, Z. Zhou, A. Kaur, Sensors for Structural Health Monitoring in Civil Infrastructure, J. Sensors. 2018 (2018) 1–2. https://doi.org/10.1155/2018/6023059.
[16] Y. Chen, Y. Wen, P. Li, Characterization of PZT ceramic transducer embedded in concrete, Sensors Actuators, A Phys. 128 (2006) 116–124. https://doi.org/10.1016/j.sna.2006.01.001.
[17] Y.H. Huang, L. Liu, F.C. Sham, Y.S. Chan, S.P. Ng, Optical strain gauge vs. traditional strain gauges for concrete elasticity modulus determination, Optik (Stuttg). 121 (2010) 1635–1641. https://doi.org/10.1016/j.ijleo.2009.03.002.
[18] A. Dinesh, S.T. Sudharsan, S. Haribala, Self-sensing cement-based sensor with carbon nanotube: Fabrication and properties – A review, Mater. Today Proc. (2021). https://doi.org/10.1016/j.matpr.2021.02.722.
[19] T. Shi, Z. Li, J. Guo, H. Gong, C. Gu, Research progress on CNTs/CNFs-modified cement-based composites – A review, Constr. Build. Mater. 202 (2019) 290–307. https://doi.org/10.1016/j.conbuildmat.2019.01.024.
[20] A. Dinesh, B. Abirami, G. Moulica, Materials Today : Proceedings Carbon nanofiber embedded cement composites : Properties and promises as sensor – A review, Mater. Today Proc. (2020). https://doi.org/10.1016/j.matpr.2020.10.526.
[21] R.K. Biswas, F. Bin Ahmed, M.E. Haque, A.A. Provasha, Z. Hasan, F. Hayat, D. Sen, Effects of Steel Fiber Percentage and Aspect Ratios on Fresh and Harden Properties of Ultra-High Performance Fiber Reinforced Concrete, Appl. Mech. 2 (2021) 501–515. https://doi.org/10.3390/applmech2030028.
[22] G. Batson, Steel fiber reinforced concrete, Mater. Sci. Eng. 25 (1976) 53–58. https://doi.org/10.1016/0025-5416(76)90051-3.
[23] S. Wen, D.D.L. Chung, Carbon fiber-reinforced cement as a strain-sensing coating, 31 (2001) 665–667.
[24] S. Wen, D.D.L. Chung, Strain-Sensing Characteristics of Carbon Fiber-Reinforced Cement, (2006) 2006.
[25] E. Hammel, X. Tang, M. Trampert, T. Schmitt, K. Mauthner, A. Eder, P. Pötschke, Carbon nanofibers for composite applications, Carbon N. Y. 42 (2004) 1153–1158. https://doi.org/10.1016/j.carbon.2003.12.043.
[26] I. Kang, Y.Y. Heung, J.H. Kim, J.W. Lee, R. Gollapudi, S. Subramaniam, S. Narasimhadevara, D. Hurd, G.R. Kirikera, V. Shanov, M.J. Schulz, D. Shi, J. Boerio, S. Mall, M. Ruggles-Wren, Introduction to carbon nanotube and nanofiber smart materials, Compos. Part B Eng. 37 (2006) 382–394. https://doi.org/10.1016/j.compositesb.2006.02.011.
[27] S. Ding, Y. Ruan, X. Yu, B. Han, Y.Q. Ni, Self-monitoring of smart concrete column incorporating CNT/NCB composite fillers modified cementitious sensors, Constr. Build. Mater. 201 (2019) 127–137. https://doi.org/10.1016/j.conbuildmat.2018.12.203.

[28] W. Li, W. Dong, L. Shen, A. Castel, S.P. Shah, Conductivity and piezoresistivity of nano-carbon black (NCB) enhanced functional cement-based sensors using polypropylene fibres, Mater. Lett. 270 (2020) 127736. https://doi.org/10.1016/j.matlet.2020.127736.
[29] B. Han, K. Zhang, X. Yu, E. Kwon, J. Ou, Nickel particle-based self-sensing pavement for vehicle detection, Meas. J. Int. Meas. Confed. 44 (2011) 1645–1650. https://doi.org/10.1016/j.measurement.2011.06.014.
[30] K. Zhang, B. Han, X. Yu, Nickel particle based electrical resistance heating cementitious composites, Cold Reg. Sci. Technol. 69 (2011) 64–69. https://doi.org/10.1016/j.coldregions.2011.07.002.
[31] S.Y. Lee, H.V. Le, D.J. Kim, Self-stress sensing smart concrete containing fine steel slag aggregates and steel fibers under high compressive stress, Constr. Build. Mater. 220 (2019) 149–160. https://doi.org/10.1016/j.conbuildmat.2019.05.197.
[32] B. Han, S. Ding, X. Yu, Intrinsic self-sensing concrete and structures: A review, Meas. J. Int. Meas. Confed. 59 (2015) 110–128. https://doi.org/10.1016/j.measurement.2014.09.048.
[33] L. Shi, Y. Lu, Y. Bai, Mechanical and Electrical Characterisation of Steel Fiber and Carbon Black Engineered Cementitious Composites, Procedia Eng. 188 (2017) 325–332. https://doi.org/10.1016/j.proeng.2017.04.491.
[34] L. Zhang, B. Han, J. Ouyang, X. Yu, S. Sun, J. Ou, Multifunctionality of cement based composite with electrostatic self-assembled CNT/NCB composite filler, Arch. Civ. Mech. Eng. 17 (2017) 354–364. https://doi.org/10.1016/j.acme.2016.11.001.
[35] S.J. Lee, I. You, G. Zi, D.Y. Yoo, Experimental investigation of the piezoresistive properties of cement composites with hybrid carbon fibers and nanotubes, Sensors (Switzerland). 17 (2017). https://doi.org/10.3390/s17112516.
[36] H.B. Birgin, S. Laflamme, A. D'alessandro, E. Garcia-Macias, F. Ubertini, A weigh-in-motion characterization algorithm for smart pavements based on conductive cementitious materials, Sensors (Switzerland). 20 (2020). https://doi.org/10.3390/s20030659.
[37] P. Park, Y. Rew, A. Baranikumar, Controlling conductivity of asphalt concrete with graphite, 7 (2014) 106.
[38] S. Wen, D.D.L. Chung, Piezoresistivity-based strain sensing in carbon fiber-reinforced cement, ACI Mater. J. 104 (2007) 171–179. https://doi.org/10.14359/18580.
[39] H. Deng, H. Li, Assessment of self-sensing capability of Carbon Black Engineered Cementitious Composites, Constr. Build. Mater. 173 (2018) 1–9. https://doi.org/10.1016/j.conbuildmat.2018.04.031.
[40] O. Galao, F.J. Baeza, E. Zornoza, P. Garcés, Strain and damage sensing properties on multifunctional cement composites with CNF admixture, Cem. Concr. Compos. 46 (2014) 90–98. https://doi.org/10.1016/j.cemconcomp.2013.11.009.
[41] G. Yıldırım, M.H. Sarwary, A. Al-Dahawi, O. Öztürk, Ö. Anıl, M. Şahmaran, Piezoresistive behavior of CF- and CNT-based reinforced concrete beams subjected to static flexural loading: Shear failure investigation, Constr. Build. Mater. 168 (2018) 266–279. https://doi.org/10.1016/j.conbuildmat.2018.02.124.
[42] X. Xin, M. Liang, Z. Yao, L. Su, J. Zhang, P. Li, C. Sun, H. Jiang, Self-sensing behavior and mechanical properties of carbon nanotubes/epoxy resin composite for asphalt pavement strain monitoring, Constr. Build. Mater. 257 (2020) 119404. https://doi.org/10.1016/j.conbuildmat.2020.119404.
[43] A. Yazdanbakhsh, Z. Grasley, B. Tyson, R.K. Abu Al-Rub, Distribution of carbon

nanofibers and nanotubes in cementitious composites, Transp. Res. Rec. (2010) 89–95. https://doi.org/10.3141/2142-13.

[44] Y. Wang, Y. Wang, B. Wan, B. Han, G. Cai, Z. Li, Properties and mechanisms of self-sensing carbon nanofibers/epoxy composites for structural health monitoring, Compos. Struct. 200 (2018) 669–678. https://doi.org/10.1016/j.compstruct.2018.05.151.

[45] B. Han, L. Zhang, S. Sun, X. Yu, X. Dong, T. Wu, J. Ou, Electrostatic self-assembled carbon nanotube/nano carbon black composite fillers reinforced cement-based materials with multifunctionality, Compos. Part A Appl. Sci. Manuf. 79 (2015) 103–115. https://doi.org/10.1016/j.compositesa.2015.09.016.

[46] Y. Ding, Z. Chen, Z. Han, Y. Zhang, F. Pacheco-Torgal, Nano-carbon black and carbon fiber as conductive materials for the diagnosing of the damage of concrete beam, Constr. Build. Mater. 43 (2013) 233–241. https://doi.org/10.1016/j.conbuildmat.2013.02.010.

[47] Q. Liu, W. Wu, J. Xiao, Y. Tian, J. Chen, A. Singh, Correlation between damage evolution and resistivity reaction of concrete in-filled with graphene nanoplatelets, Constr. Build. Mater. 208 (2019) 482–491. https://doi.org/10.1016/j.conbuildmat.2019.03.036.

[48] H. Siad, M. Lachemi, M. Sahmaran, H.A. Mesbah, K.A. Hossain, Advanced engineered cementitious composites with combined self-sensing and self-healing functionalities, Constr. Build. Mater. 176 (2018) 313–322. https://doi.org/10.1016/j.conbuildmat.2018.05.026.

Structural Health Monitoring Using Sensors with Application of Wavelet Analysis

S. Saranya[1,a,*], M. Ranjitham[2,b], A. Dinesh[3,c], S. Divyapriya[4,c]

[1]Department of Electronics and Communication Engineering, Sri Ramakrishna Engineering College, Coimbatore - 641022, India.

[2]Department of Civil Engineering, Bannari Amman Institute of Technology, Sathyamangalam – 638401, India.

[3,4]Department of Civil Engineering, Sri Ramakrishna Engineering College, Coimbatore - 641022, India.

[a*]Saranya.s@srec.ac.in , [b]@bitsathy.ac.in, [c]dinesh.a@srec.ac.in; dinesh.a.civil@gmail.com, [d]divyapriya.1909008@srec.ac.in

Keywords: Structural Health Monitoring, Fourier Transform, Wavelet Transforms, Signal Processing, Damage, Corrosion

Abstract. Structural Health Monitoring (SHM) establishes a damage detection system to maintain safety in the current structure. Owners and maintenance managers are looking for cost-effective and dependable inspection and monitoring solutions to ensure the safety and reliability of capital-intensive assets. A longer overhaul time is envisaged with today's software technology and design codes. The visualization aims to put advanced technology into practice to provide consumers and the government with value-added services. Meanwhile, the wavelet transforms, a signal processing technique based on a windowing approach using enlarged 'scaled' and shifted wavelets, is being applied in various industries. By bypassing many of the restrictions of the Fourier transform (FT), the wavelet transform has proven to be effective in SHM systems. Structures tend to get damaged in situations such as accidental fire, aggregate contraction, salinity exposure, corrosion due to bacterial influence, physical and material damage. Also, structures tend to lose their tensile strength when exposed to long-term factors such as moisture, heat, rains, storms, etc. Structural Health Management plays a vital role here to monitor the health conditions of structures to prevent any loss. To stand up for this need, it is imminent to provide a safe structure for people to ply through. The proposed methodology shows a clear picture of how to assess any structure condition at any time and gives a clear view of its current stature on whether it is damaged. Hence in this article, the behavior of the structure is assessed using wavelet transformation. The hardware configurations, including the MSP430FR6989 microcontroller with TDC1000-TDC7200EVM, are embedded with the aggregate, making it smart enough to detect the defects through the software interpretation with a signal processing toolbox of MATLAB coding.

Introduction

In recent past decades, concrete has been used as the foremost construction material. In contrast, the worsening or weakening of these concrete structures is said to be unavoidable due to the atmospheric and external climatic features [1]. Cementitious composite [2–4] similar to soil [5–7] can change its properties when there is a modification in constituents of materials. Many innovations are brought in the infrastructure in materials [8–10] as well as structures [11–13]. But the damage to concrete structures is inevitable. The methodology described here monitors the health of concrete structures and paves the way for detecting the imperfections in the tunnels, wind energy plants, pavements, aircraft structures, etc. The initial work has experimented with the concrete structure, which can be implemented in the above-mentioned areas based on feasibility.

The structures, including bridges, energy plants, pipelined architectures like water, gas, and oil pipelines, tunnels, rig wells, pavements, or others, are subjected to various factors that malfunction as the days pass on. The funds spent on the maintenance and repair is a vast sum that can be substituted by an innovative SHM approach with the help of the sensors and the signal processing techniques [14,15]. The flaws can be detected in the early stages, making it cost-effective and avoiding making it functionally obsolete. The idea of replacing routine-based maintenance with circumstance-based maintenance is the future of civil infrastructure, providing the following benefits: Uninterrupted observation, Automation, Facilitating the proactive response by identifying dent in an early stage, economic and feasible.

Structures tend to get damaged in climatical situations such as fire, aggregate expansion, sea water effects, bacterial corrosion, physical damage, and chemical damage. Also, structures tend to lose their tensile strength when exposed to long-term factors such as moisture, heat, rains, storms., etc. Structural Health Management plays a vital role here to monitor the health conditions of structures to prevent any loss [16]. To stand up for this need, it is imminent to provide a safe structure for people to ply through. This is done as the product assesses any structure condition at any time and gives a clear view of its current stature on whether it is damaged.

The research article describes the various conventionally used non-destructive techniques for the health monitoring of structures. In this work, the specimen of concrete was baked in the laboratory. It was embedded with a few electronic modules like MSP430FR6989 with inbuilt ultrasonic sensor to check its strength using the continuous wavelet transform resulting from signal processing. The irregularities in the output waveforms characterize the internal flaws in the concrete material.

Traditional Non-destructive methodologies of Health monitoring:

The non-destructive testing of concrete has been in use for a few decades past. It is said to be the influential tool for monitoring or assessing the concrete in terms of strength, durability, longevity, and structure quality. This method is used to know the cracks in terms of depth and severity, deformations, and even the micro defects that can be identified. The traditional non-destructive health monitoring methodologies include surface hardness detection, rebound test, penetration, pull-out techniques, combined test, dynamic test or vibration testing, radioactive or nuclear methodology, Magnetic and electric method, acoustic emission technique, etc.

The material characteristics can be better understood by performing the non-destructive tests on the various scale for assessing the parameters like electrical characteristics, material absorbency, scattering, the transmission of waves like Gamma-rays and X-rays, acoustic waves, ability to transmit and receive ultrasonic signals inform of pulse velocity thereby determining the moisture density and thickness. As shown in Fig. 1, many non-destructive techniques are available for the health monitoring of concrete structures. To elaborate on a few of these tests before entering into the details of the proposed methodology is listed as follows:

Surface hardness test: The concrete strength is a replication of its age; hence it can be tested by traditional methodologies like William's pistol test and the various hammer tests. This helps us calculate the strengths of concrete, but it requires a variety of equipment and manpower to derive at the estimation.

Materials Research Proceedings **23** (2022) 205-216 https://doi.org/10.21741/9781644901953-24

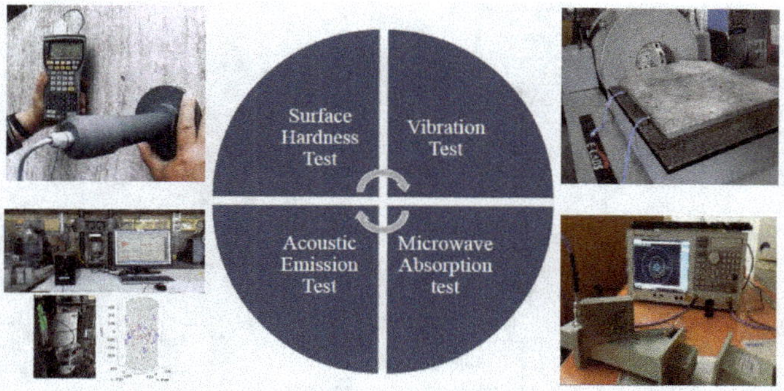

Fig. 1. Traditional Non-destructive methodologies of Health monitoring

Dynamic test or vibration testing: Under this category, a list of parameters could be analyzed like resonant frequency, ultrasonic signals inform of pulse velocity, and mechanical nature of sound waves. This helps in determining the lifespan, homogenous nature, and elasticity property of the concrete material.

Acoustic Emission test: Acoustic sensors play a vital role in performing this test. These sensors are embedded onto the concrete and checked for the reception of the same by applying force. If no deformation occurs, the original and the retrieved signals will be a replica.

Microwave absorption test: The electrical and magnetic tests are performed to know the concrete material's moisture and thickness. Microwave absorption is one such credential for performing the above characteristic.

Proposed methodology

The methodology proposed for the SHM can be termed as an innovative means to assess the health of any structure in concrete materials [17–19]. The work involved in the composite development and health monitoring is shown in Fig. 2. The methodology has an overall workflow where the work progressed from casting a specimen as per the standard norms of IS10262:2009 with fine aggregate, coarse aggregate, and cement. The concrete structures are baked and then tested with the hardware comprising of MSP430FR6989 microcontroller and TDC1000-TDC7200EVM, which has a built-in ultrasonic sensor module for the PWM generation and analysis. The signal interpretation was followed by the above-mentioned process, thereby detecting the flaws in the concrete.

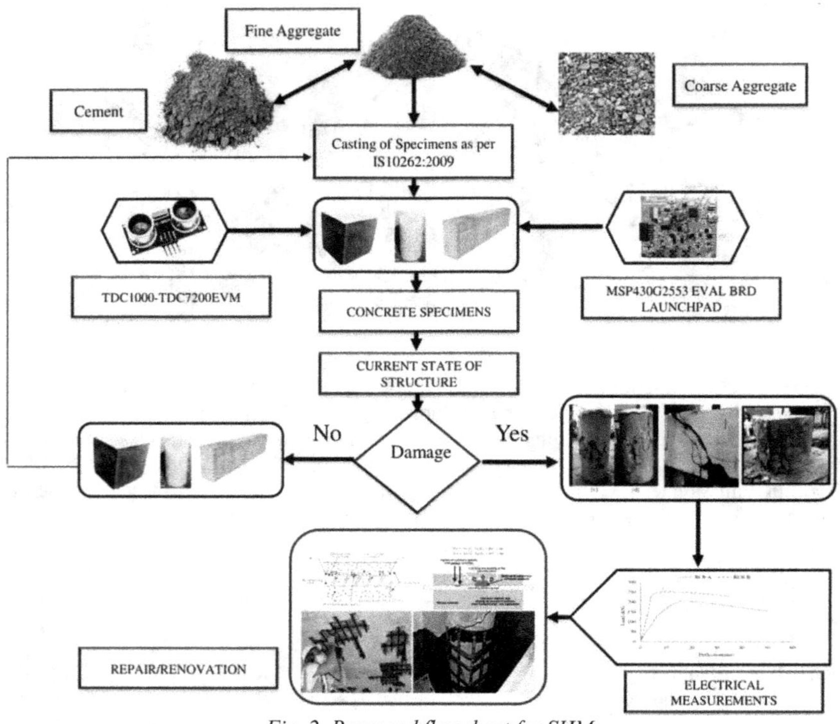

Fig. 2. Proposed flowchart for SHM

Sensor components for SHM

MSP430FR6989: It is a launchpad kit that is under evaluation mode by the Texas instruments. The benefit of the toolkit is it consumes ultra-low power comprising of FRAM, which can act as a real-time emulator for varying dynamic scenarios of atmospheric and climatic conditions. It also features a 320 segment LCD display. The FRAM and non-volatile memory can ensure high access speed and good endurance, so the replacement of hardware requirements will be minimal [20].

TDC1000-TDC7200EVM: TDC1000 is an ultrasonic wave generator with Time-of-Flight applications acting as an analog ultrasonic device that is functional up to 4MHz frequency. TDC7200 is a digital converter of time used to measure the time elapse and can function as a stopwatch device [21]. The TDC 1000 provides a complete wireless ultrasonic sensing solution.

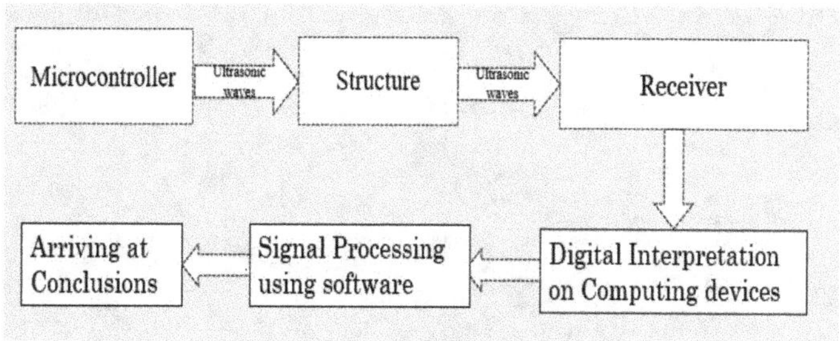

Fig. 3. The layout of innovative Structure

As shown in Fig. 3, the concept behind the ultrasonic wave is the piezoelectricity which is produced by the vibration of quartz crystal. The piezoelectric strain sensors help generate strong voltage signals without requiring intermediate devices like gauges, bridges, amplification modules, and conditioning units [22]. This can provide a significant result in terms of vibration and dynamic means. The piezoelectric equations are expressed as the strain and electric displacement in terms of applied stress, electric field, and temperature using the constitutive tensile factor as shown in equation (1) and equation (2).

$$S_{ij} = s^{D}_{ijkl} T_{kl} + g_{ikl} D_k + \delta_{ij}\, \alpha_i^{D\theta} \quad (1)$$

$$E_i = - g_{ikl} T_{kl} + \beta^{T}_{ik} D_k + \breve{E}_{i\theta} \quad (2)$$

Where g_{ikl} is the piezoelectric voltage coefficient

β^{T}_{ik} is the impermittivity coefficient.

$\breve{E}_i$ is the pyroelectric voltage coefficient

The above equation is called the sensor equation, which predicts how much electric field, i.e., voltage per unit thickness, is generated by "squeezing" the piezoelectric material. Fig. 4 and Fig. 5 depict the coding for ultrasonic wave generation using pulse width modulation and reception of the same, respectively.

```
const byte ANTENNA = 9;

void setup()
  {
  // set up Timer 1
  TCCR1A = bit (COM1A0);  // toggle OC1A on Compare Match
  TCCR1B = bit (WGM12) | bit (CS10);   // CTC, no prescaler
  OCR1A =  9;         // compare A register value to 10 (zero relative)
  }  // end of setup

void loop()
  {
  pinMode (ANTENNA, OUTPUT);
  delay (500);
  pinMode (ANTENNA, INPUT);
  delay (300);
  }  // end of loop
```

Fig. 4. Ultrasonic Waves production using pulse-width modulation

```
const int pingPin = 7; // Trigger Pin of Ultrasonic Sensor
const int echoPin = 6; // Echo Pin of Ultrasonic Sensor

void setup() {
   Serial.begin(9600); // Starting Serial Terminal
}

void loop() {
   long duration, inches, cm;
   pinMode(pingPin, OUTPUT);
   digitalWrite(pingPin, LOW);
   delayMicroseconds(2);
   digitalWrite(pingPin, HIGH);
   delayMicroseconds(10);
   digitalWrite(pingPin, LOW);
   pinMode(echoPin, INPUT);
   duration = pulseIn(echoPin, HIGH);
   inches = microsecondsToInches(duration);
   cm = microsecondsToCentimeters(duration);
   Serial.print(inches);
   Serial.print("in, ");
   Serial.print(cm);
   Serial.print("cm");
   Serial.println();
   delay(100);
}

long microsecondsToInches(long microseconds) {
   return microseconds / 74 / 2;
}

long microsecondsToCentimeters(long microseconds) {
   return microseconds / 29 / 2;
}
```

Fig. 5. Receiving Module Reception of Waves in Ultrasonic sensor

Structure: The work initially commenced with a laboratory-baked concrete cube of dimensions 150mm x 150mm x 150 mm (length x breadth x width). The cube is fully baked with mortar and cement. This is considered as a test structure, and the initial testing was run on it.
Receiver module: The Values obtained from the MSP430FR6989 controller were interpreted using MATLAB, and later the plot was done and analyzed. The digital interpretation and the signal processing modules go hand in hand under the receiver module, resulting in a conclusion of measured results.

The codes for producing the ultrasonic waves and their corresponding reception are shown below. In the transmitter end, the output and the input pin are given a delay of 500µsec, thereby avoiding interference. The PINs 6 and 7 are activated in the reception module to felicitate the trigger mode and echo signal.

Signal Processing: This is carried out by the MATLAB and the signal processing toolbox. The Wavelet transform function is used to plot the data received from the structures used for testing.

Wavelet Transform for SHM

The issue with Fourier transforms, specifically, the FFT is that it can change a signal of the entire signal length. The signal acquired and decomposed cannot give the information related to the time of their occurrence of the transient state. The DFT lags in time resolution but has a good frequency resolution. If DFT is used, the damage detection will be based only on the frequency domain analysis, and the time analysis will be lost. If this is the case, then the vibration signal will not be considered, leading to the loss of all-time related information. Hence a wavelet-based analysis was taken for experimentation purposes. Fig. 6 (a) – (d) shows the various wavelet types based on time frequency response.

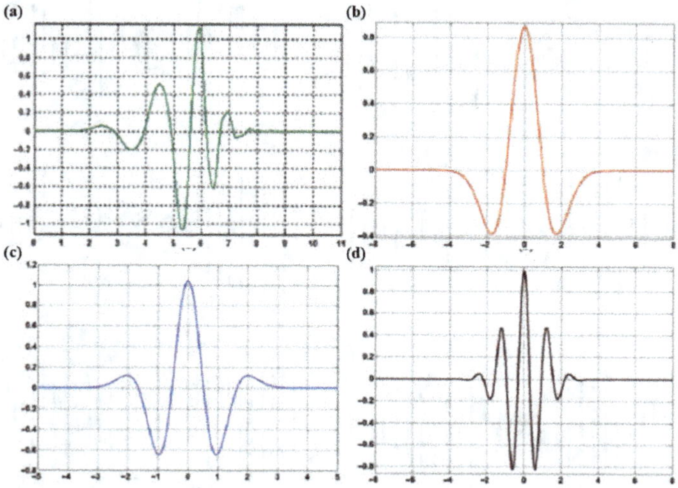

Fig. 6. Example representation of some mother wavelets (a) Daubechies wavelet (b) Mexican hat wavelet (c) Gaussian wavelet (d) Morlet wavelet [23]

The main difference between them is CWT uses an infinite number of scales and locations while the DWT uses only a finite set of scales and locations. Equation (3) represents the continuous wavelet transform (CWT), and equation (4) represents the discrete wavelet transform (DWT) is given below.

$$T(a,b) = \frac{1}{\sqrt{a}}\int_{-\infty}^{\infty} x(t)\psi\frac{(t-b)}{a}\,dt \qquad (3)$$

$$T_{m,n} = \int_{-\infty}^{\infty} x(t)\psi_{m,n}(t)\,dt \qquad (4)$$

Besides the hardware module and data reception from the microcontroller, data analysis plays a vital role in this proposed work. The main objective is to analyze the wavelets bypassing the ultrasonic wave in a proper and damaged structure and compare them and conclude their deformations with the help of wavelet transform. The below diagram represents the comparison of waveforms when passed through an appropriate and defective structure. The waveform with inadequate structure has undergone an extreme level of stress and strain. The model represents that the waveform passing through a proper structure can be reproduced harmonically. In contrast, Fig.7 (b) on the right side depicts that the waveforms are lagging in harmony and the original waveform cannot be replicated, indicating a level of deformation.

The structure was subjected to heavy mechanical stress under the influence of specific mechanical instruments held in laboratory use. For every particular unit of stress applied to the structure, there was a certain amount of deformity. Each step was monitored digitally and on a continuous time pattern by using the sensor and the controller. The analysis of undamaged and damaged signals to be obtained is shown in graphical means as shown in Fig 7 (a) and (b).

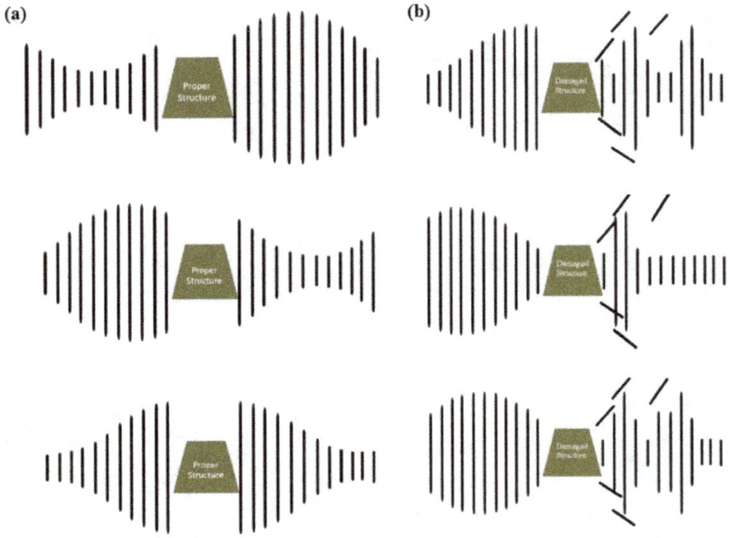

Fig. 7. Signal passing through (a) undamaged structure and (b) deformed structure

The values obtained from the MSP430FR6989 controller after passing the ultrasonic wave through a cube baked in the laboratory as shown in the below Fig. 8 (a) and (b) (damaged and undamaged respectively) is retrieved in the excel sheet is then plotted with the wavelet transformation function (DCT) to obtain the plot as shown in Fig. 9 (a) and (b). The test structure was baked with the dimension of 150mm x 150mm x 150 mm and then applied pressure under the mechanical conditions to undergo deformation; hence the test results can be compared and analyzed.

Fig. 8. Concrete specimens (a) before damage and (b) after damage

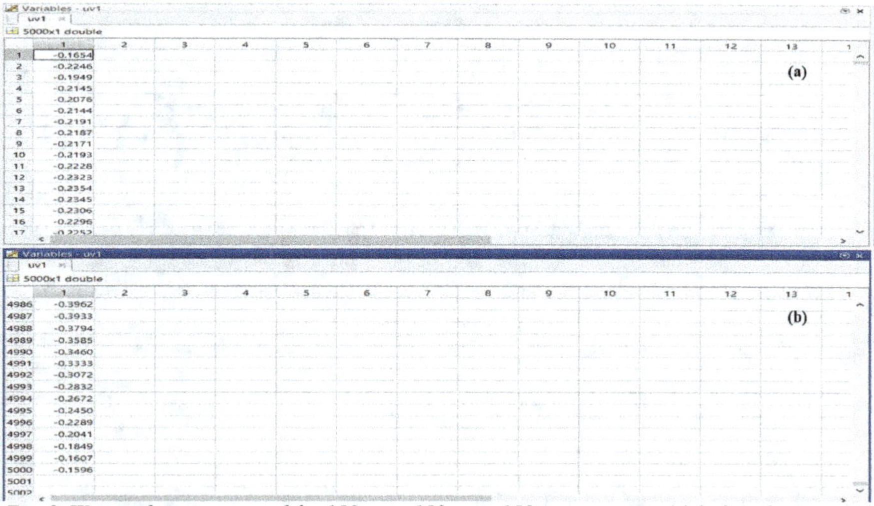

Fig. 9. Wave values interpreted for 150mm x 150mm x 150 mm structure (a) before damage and (b) after damage

The wave readings interpret time and voltage along the x and y-axis, as shown in Fig.10. Fig.10 (a) has a constant spike value of wavelets ranging from 0.8V to -0.2V depicting the outcome from the ultrasonic reception where the cube was in a proper shape (as shown in the test structure a), and Fig.10 (b) depicts the wavelets after passing through an irregular cube as shown in the test structure b in Fig. 8 (b). The latter influences noise signals which degrade the wavelets with irregular spikes so as to mention the deformation. These resultant wavelets are scaled and shifted versions of the mother wavelets

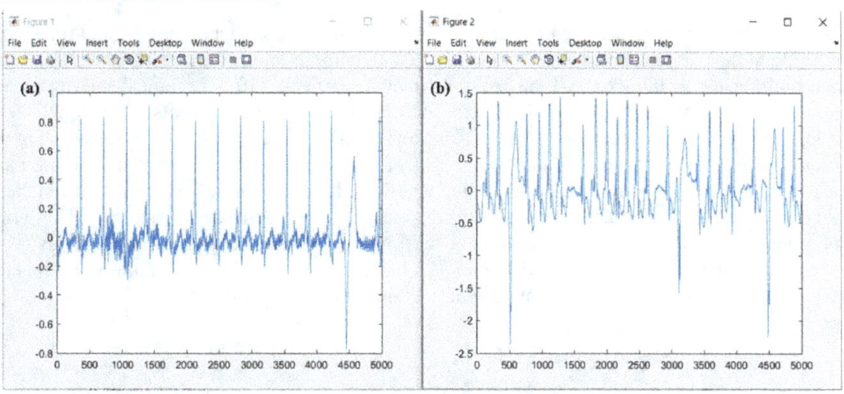

Fig. 10. The wave readings of the cube (a) before deformation and (b) after the deformation

Materials Research Proceedings **23** (2022) 205-216
https://doi.org/10.21741/9781644901953-24

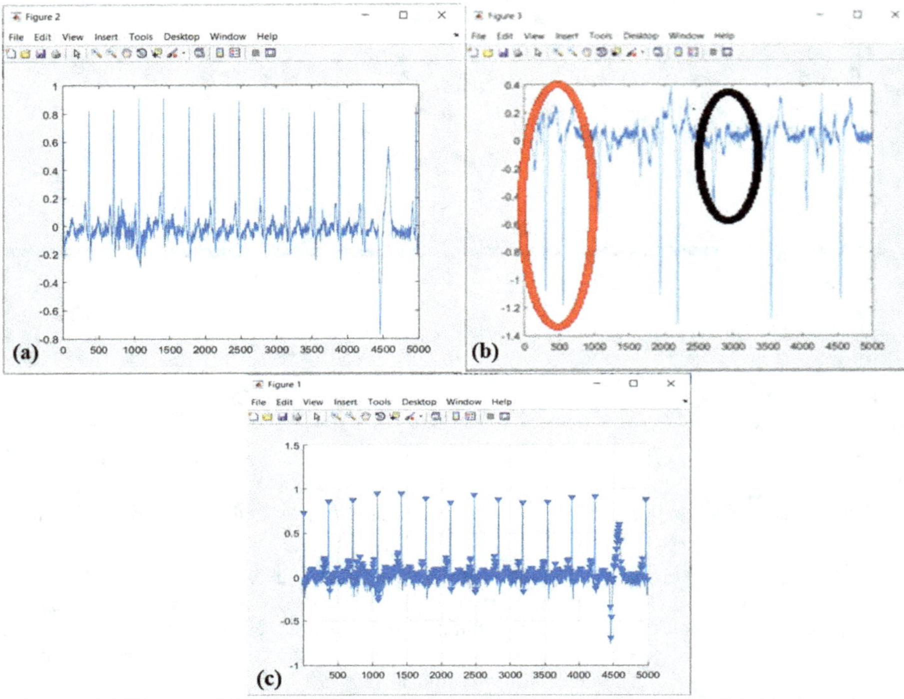

Fig. 11. (a) Wavelet of the normal cube, (b) Wavelet of the wrecked cube, and (c) After wavelet analysis

The wavelet functions are used to detect events like anomalies, change points, and transients. They extend their application with de-noising and data compression. Along with Wavelets, many other multi-scale techniques can be used to analyze data at different times and frequency resolutions, thereby decomposing them into various functional components. The dimensional features can be reduced, and extracting the discrimination features can also be performed. Fig. 11 (a) shows the wavelet of the normal cube. In Fig.11 (b), the wavelets represented in the red and black markings cannot be reproduced by using a definite function: the influence of noise in them, which indirectly shows the deformation or cracks in the concrete structures. The same experimentation was done in a 200 mm thick solid wall in our college premises which has given the result as shown in Fig.12 (a) and (b). Fig.12 (a) has a proper wavelet resulting in a healthy structure, whereas Fig 12(b) represents a wall with minor cracks and flaws.

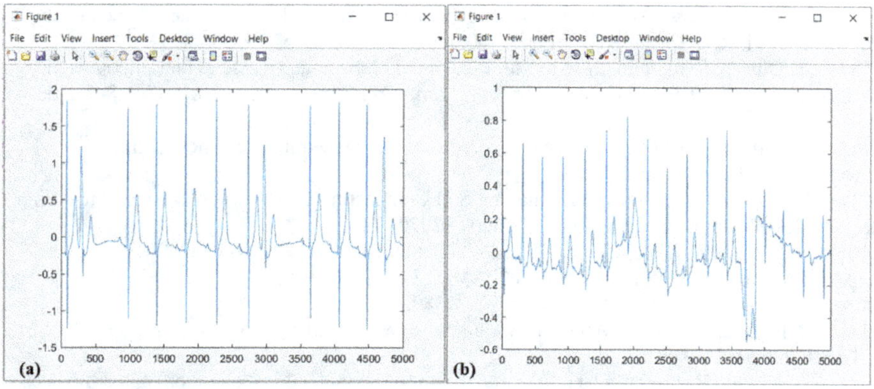

Fig. 12. Waveform of a 200 mm thick solid wall (a) without visible cracks and (b) with visible cracks

Conclusion and future scope

The waveforms of the test structure (the cube) taken before and after the impact of applying a force under laboratory conditions were tested. The wall with the defect was also tested, and the results were obtained. The waveforms have significant differences between them. From this, it can easily be proven that the structure has undergone severe skeptical damages that may not be shown to the human eye but can be easily portrayed. The Piezo-electric concept, instead of the traditional vibration method, is cost-effective. Signal processing gives more accurate and reliable results and consumes less energy. This work can also be further developed to monitor the signal in continuous means for a more extended period, and the toolkit protection has also to be ensured.

Acknowledgements

The authors would like to express their gratitude to the management, Sri Ramakrishna Engineering College, Coimbatore. The authors would like to thank the management, PSG College of Technology, Coimbatore for their support in conducting this work.

References

[1] K. Kovler, V. Chernov, Types of damage in concrete structures, Woodhead Publishing Limited, 2009. https://doi.org/10.1533/9781845697037.1.32

[2] A. Dinesh, S.T. Sudharsan, S. Haribala, Self-sensing cement-based sensor with carbon nanotube: Fabrication and properties – A review, Mater. Today Proc. (2021). https://doi.org/10.1016/j.matpr.2021.02.722

[3] A. Dinesh, B. Abirami, G. Moulica, Materials Today : Proceedings Carbon nanofiber embedded cement composites : Properties and promises as sensor – A review, Mater. Today Proc. (2020). https://doi.org/10.1016/j.matpr.2020.10.526

[4] A.S. S.D., D. A., S.B. V., Investigation of waste marble powder in the development of sustainable concrete, Mater. Today Proc. (2020). https://doi.org/10.1016/j.matpr.2020.10.536

[5] A. Professor, Stabilization of soil by using solid waste-A Review, © 2017 IJEDR |. 5 (2017). www.ijedr.org

[6] A. Dinesh, S. Harini, J.P. Jasmine, J. Jincy, J. Shagufta, International journal of

engineering sciences & research technology experimental study of blast furnace slag concrete, Int. J. Eng. Sci. Res. Technol. 6 (2017) 42–50. https://doi.org/10.5281/zenodo.345692
[7] A. Dinesh, S. Indhumathi, Moorthi Pichumani, Performanc e assessemnt of copper slag and sawdust ash in stabilization of black cotton soil, 2021.https://doi.org/10.1007/978-981-33-4590-4_21
[8] A. Professor, Study on Strength properties of High performance concrete, n.d. www.ripublication.com/ijaer.htm.
[9] K. Sathish Raja, Dinesh. A, Study on Self Compacting Concrete – A Review, Int. J. Eng. Res. V5 (2016). https://doi.org/10.17577/ijertv5is020349
[10] D. Sellakutty, A. Professor, UTILISATION OF WASTE PLASTIC IN MANUFACTURING OF BRICKS AND PAVER BLOCKS, 2016. https://www.researchgate.net/publication/303273892
[11] P. Eswaramoorthi, P. Magudeaswaran, A. Dinesh, PUSHOVER ANALYSIS OF STEEL FRAME, n.d.
[12] A. Dinesh, R. Prasanth Kumar, S.R. Abijith, Experimental investigation on bubble deck concrete using plastic waste, 2020. https://doi.org/10.1007/978-981-15-0990-2_14
[13] A. Professor, Analytical study on the behaviour of cold formed steel double channel beam sections, n.d. www.ripublication.com/ijaer.htm
[14] K. Chilamkuri, V. Kone, Monitoring of varadhi road bridge using accelerometer sensor, Mater. Today Proc. 33 (2020) 367–371. https://doi.org/10.1016/j.matpr.2020.04.159
[15] R.B. Randall, W.A. Smith, Detection of faulty accelerometer mounting from response measurements, J. Sound Vib. 477 (2020) 115318. https://doi.org/10.1016/j.jsv.2020.115318
[16] Z.H. Warsi, S.M. Irshad, F. Khan, M.A. Shahbaz, M. Junaid, S.U. Amin, Sensors for structural health monitoring: A review, 2019 2nd Int. Conf. Latest Trends Electr. Eng. Comput. Technol. INTELLECT 2019. (2019) 1–6. https://doi.org/10.1109/INTELLECT47034.2019.8955453
[17] R. Min, Z. Liu, L. Pereira, C. Yang, Q. Sui, C. Marques, Optical fiber sensing for marine environment and marine structural health monitoring: A review, Opt. Laser Technol. 140 (2021) 107082. https://doi.org/10.1016/j.optlastec.2021.107082
[18] A. Di Graziano, V. Marchetta, S. Cafiso, Structural health monitoring of asphalt pavements using smart sensor networks: A comprehensive review, J. Traffic Transp. Eng. (English Ed. 7 (2020) 639–651. https://doi.org/10.1016/j.jtte.2020.08.001
[19] G.N. Devi, M.M. Vijayalakshmi, Smart structural health monitoring in civil engineering: A survey, Mater. Today Proc. 45 (2020) 7143–7146. https://doi.org/10.1016/j.matpr.2021.02.095
[20] F.M. Launchpad, D. Kit, User's Guide MSP430FR6989 LaunchPad™ Development Kit (MSP- -EXP430FR6989), (2015). www.ti.com
[21] E.V.M. User, HDC1080EVM User ' s Guide User ' s Guide, (2015).
[22] D. Zou, T. Liu, C. Liang, Y. Huang, F. Zhang, C. Du, An experimental investigation on the health monitoring of concrete structures using piezoelectric transducers at various environmental temperatures, J. Intell. Mater. Syst. Struct. 26 (2015) 1028–1034. https://doi.org/10.1177/1045389X14566525
[23] M.M. Reda Taha, A. Noureldin, J.L. Lucero, T.J. Baca, Wavelet transform for structural health monitoring: A compendium of uses and features, Struct. Heal. Monit. 5 (2006) 267–295. https://doi.org/10.1177/1475921706067741

Carbon-Based Nanomaterial Embedded Self-Sensing Cement Composite for Structural Health Monitoring of Concrete Beams - A Extensive Review

A. Dinesh[1,a,*], S. Durgadevi[2,b], S. Veeraraghavan[3,c], S. Janani Praveena[1,d]

[1]Department of Civil Engineering, Sri Ramakrishna Engineering College, Coimbatore - 641022, India

[2]Department of Civil Engineering, Bannari Amman Institute of Technology, Sathyamangalam – 638401, India

[3]Independent Researcher (Cement composite and Materials), Coimbatore - 641022, India

[a]dinesh.a@srec.ac.in; dinesh.a.civil@gmail.com , [b]durgadevi@bitsathy.ac.in, [c]veeraraghavan.civil@gmail.com, [d]jananipraveena.1909014@srec.ac.in

Keywords: Structural Health Monitoring, Self-Sensing Cement Composites, Carbon-Based Nanomaterials, Strength, Conductivity, Beams

Abstract. Structural health monitoring has proven to be a dependable source for ensuring the integrity of the structure. It also aids in detecting and estimating the progression of cracks and the loss of structural performance. The most compelling components in the structural health monitoring system are sensing material and sensor technology. In health monitoring systems, fiber optic sensors, strain gauges, temperature sensors, shape memory alloys, and other types of sensors are commonly used. Even though the sensors bring monetary value to the system, they have some apparent drawbacks. As a result, self-sensing cement composite was established as a sensor alternative with better endurance and compatibility than sensors. Carbon nanotubes, nanofibers, graphene nanoplates, and graphene oxide are carbon-based nanomaterials with unique mechanical and electrical properties. As a result, this review comprises a complete assessment of the fresh, mechanical, and electrical properties of self-sensing cement composite developed using carbon-based nanoparticles. The research also focuses on the self-monitoring performance of cement composite in concrete beams, both bulk and embedded, by graphing the deviation of fractional change in resistivity with strain. The network channel development of carbon-based nanomaterials in cement composites and their characterization acquired using scanning electron microscopy (SEM), and X-Ray diffraction spectroscopy (XRD) research are also comprehensively discussed. According to the study, increasing carbon-based embedment decreased the relative slump and flowability while increasing the composite's compressive, split tensile, flexural, and post-peak performance. Also, the amount of carbon in the carbon-based nanomaterial directly relates to the composite's conductivity. As a result, the development of piezoresistive and sensing capabilities in carbon-based self-sensing cement composites not only improves mechanical and conductive properties but also serves as a sensor in structural health monitoring of flexural members.

Introduction

Concrete is the most widely utilized material for numerous infrastructures, including residential, government, and commercial buildings, all of which contribute to the nation's economic development. However, the concrete structures were damaged due to weather conditions, a lack of improved design and condition assessment methods, and non-periodic maintenance [1]. As a result, enormous efforts are required to restore the failing infrastructure to a safe and serviceable state. Although numerous concrete repair procedures, such as epoxy injection and grouting, are available, they are not continuous and are only used once structural deterioration is evident. As a result, a continuous monitoring system is required to check the concrete structure's performance [2,3] which is called as Structural health monitoring. Metallic sensors are commonly employed in the structural health monitoring procedure [4]. Sensors such as accelerometers [5], corrosion sensors [6], ultrasonic sensors [7], fiber optic sensors [8], electrochemical sensors [9], strain gauges [10], and wireless strain sensors [11] are used in structural health monitoring. However, due to incompatibility and low durability, the use of specific sensors is limited. As a result, an alternative solution for sensors is necessary, leading to the invention of self-sensing cement composite.

The application of self-sensing cement composite in developing intelligent infrastructure with conductive and sensing capabilities is excellent [12,13]. The creation of these composites increases the infrastructure's serviceability, safety, and durability [14]. Similar to structural steel [15,16], concrete construction has experienced significant advancements, both structurally [17] and in terms of material [18–20], yet concrete degradation is unavoidable. In the areas of bricks [21,22], soil [23–25], and concrete [26–28], material modification for property improvement is also on the growth. Nanomaterials have the power to change the mechanical and electrical properties of any composite due to their unique features. As a result, the nanomaterial's advantages are utilized to develop the self-sensing cement composite. To improve the conductivity properties of the cement composite, nano conductive materials such as carbon fiber [27], carbon nanofiber [13], carbon nanotube [12], steel fiber [29], graphite powder [30], and nickel powder [31] are utilized. Such a composite can sense stress, strain, cracks, and damage [32,33]. Those composites can also improve the mechanical properties of the composite, such as compressive strength, tensile strength, and flexural strength [12,13,34]. The integrated conductive components in the composite form a conductive network, which boosts the composite's conductivity. Although nanoparticles build the conductive network, aggregates and hydration products act as a barrier to the conductive network's creation in the composite.

As a result, this review paper focuses on the conductive substance that aids in the construction of self-sensing cement composites used in beams. The barriers that become a hurdle for creating self-sensing cement composite are also discussed in the article. Workability, compressive strength, flexural strength, conductivity, and piezoresistivity of the self-sensing cement composite are also reviewed, along with their morphological characterization.

Materials and Methods

Cement, fine aggregate, and water are the conventional components used to make cement composites [35,36]. The water-to-cement ratio is kept at 0.4. To increase the composite's workability, polycarboxylate ether polymers (=1% mass of cement) were added as a superplasticizer. Since dispersion of carbon nanomaterials in aqueous solutions is difficult, a specific fabrication technology, as illustrated in Fig. 1, is used to produce good dispersion [12,13]. The nanoparticles are dispersed with preliminary mechanical mixing (sonication for 30 mins) in deionized water with a superplasticizer as the dispersion. The cement powder was then mixed with water and plasticizer before being poured into oiled moulds. The cement composite is allowed to cure for a total of 28 days. The cement composite with electrodes was utilized for mechanical and electrical tests once solidified.

Materials Research Proceedings **23** (2022) 217-230 https://doi.org/10.21741/9781644901953-25

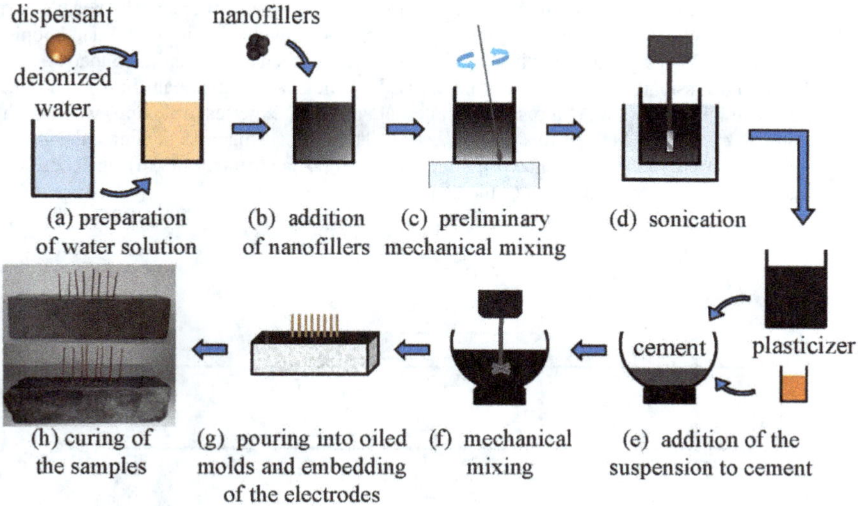

Fig. 1. Fabrication Methodology for developing self-sensing cement composite [37]

The conductivity of the cement composite is typically determined using a two or four-probe approach. The most frequent method is the four-probe method, which uses four electrodes, as illustrated in fig. 2 (a) and (b) [27]. Current is supplied by the outer two electrodes, while the inner two electrodes measure voltage. To test the electrical resistance, electrodes are inserted in the fresh mix [38,39].

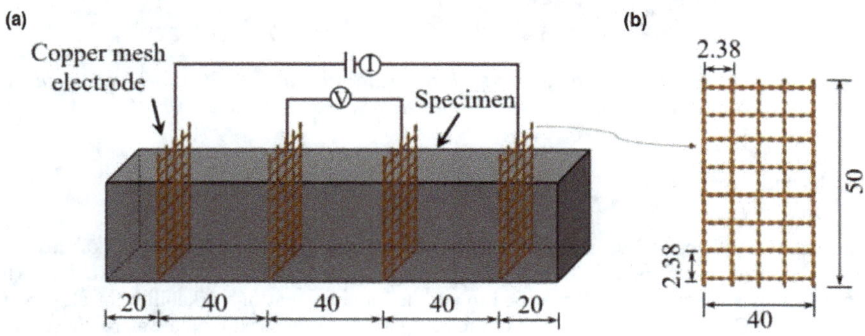

Fig. 2. (a) Schematic representation of electrical resistivity specimens using the four-probe method and (b) Configuration of electrode embedded in the composite [27]

Portland cement, fine aggregate (fineness modulus = 2.94), and coarse aggregate (maximum radius = 20mm) are used to make reinforced concrete beams of grade C40. A water reducing agent and silica fume (radius = 100nm) are commonly employed to prepare concrete. The beam's clear span, width, and depth are 2400mm, 150mm, and 300mm, respectively. The upper bar and hooping bar are made of 235 MPa yield strength steel. During the experiment, a concentrated load is given to the beam and dissipated through two points, dividing the span of the beam into three equal spans, as illustrated in Fig. 3. As a result, the internal span will experience flexure, whereas the outer span will experience both flexure and shear [40]. The self-sensing cement composite is used in the uniaxial compression zone (center top), uniaxial tension zone (center bottom), and combined compression and shear zone (side top) of the beam.

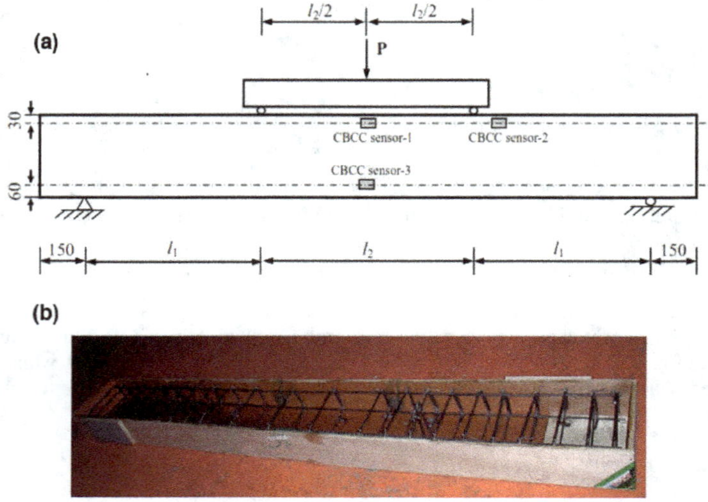

Fig. 3. (a) Schematic of the loading arrangement and location of CBCC sensors location in the beam and (b) CBCC sensor location in the concrete beam model [40]

Results and Discussions

Workability

The presence of conductive materials substantially impacts the composite's workability. As illustrated in Fig. 4, including CF in the composite diminishes the composite's workability. The superplasticizer dosage must be modified to get the same amount of workability for all percent inclusions of CF [41]. The workability of the composite is influenced by the water-cement ratio and the high concentration of fine particles [35,42]. The slump values for different aspect ratios of CF are shown in the graph with regard to CF concentration. It has been identified that when the length of CF and the concentration of conductive materials increases, the composite's workability decreases. When the length of CF is extended for the same concentration of CF, the workability is lowered. Similarly, if the CF dosage is doubled for the same period of time, the workability is lowered.

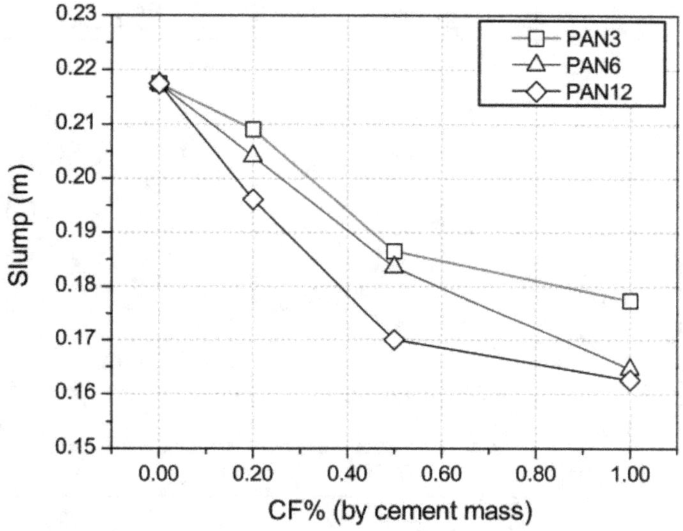

Fig. 4. Variation of the slump concerning increase in the concentration of CF (%) [41]

Strength

The flexural strength of the composite, as shown in Fig. 5., varies according to the addition of CF (PAN3 = 3mm, PAN6 = 6mm, PAN12 = 12mm). The flexural strength of the composite increases as the concentration of CF in the composite increases, as shown in the diagram. The higher the failure load levels, the shorter the CF [41]. Because shorter fibers have more fibers to support the crack area. As a result, the short fibers have a more significant impact on the composite's flexural strength.

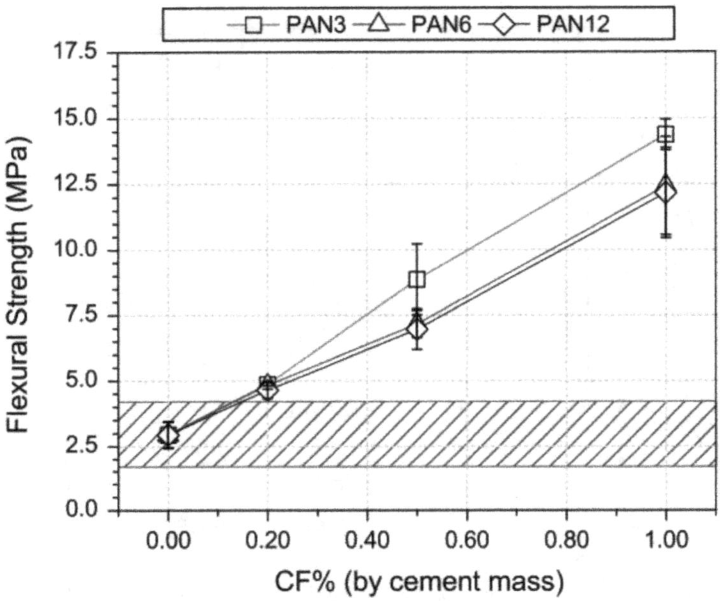

Fig. 5. Variation of flexural strength concerning increase in the concentration of CF (%) [41]

Because of the inclusion of CF of various lengths, the compressive strength of the CF embedded cement composite varies (3mm, 6mm, 12mm), as shown in Fig. 6. According to the data, the strength increments range from 2.6% to 25%. At 1% PAN3 cement paste, the maximum compressive strength values are observed. The higher concentration of PAN3 CF in the composite improves the strength of the cement composite in PAN3 embedded cement paste. Surprisingly, when the concentration of CF in the composite increases for PAN6, the strength of the cement composite decreases. The compressive strength of the PAN12 embedded cement composite is raised first, followed by drop in compressive strength at 1%. The highest strength of the cement composite is attained at 0.63% PAN12 CF, assuming a parabolic curve. Beyond that, the compressive strength of the composite was lowered because the CF embedded composite tended to form clusters due to the longer embedded CF.

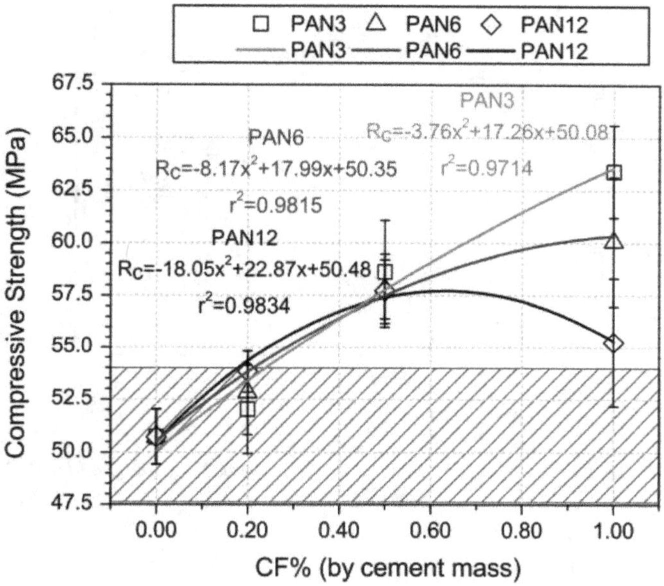

Fig. 6. Variation of compressive strength concerning increase in the concentration of CF (%) [41]

Conductivity

The conductivity of traditional cement composites is extremely high, necessitating some lowering to convert them to self-sensing cement composites. The researchers shown that adding up to 1.5% SMA and SF in the composite did not improve its self-sensing characteristics considerably. The conductivity of the composite may be enhanced to 5.79 X 10^{-6} S/cm by including 1.5% SMA, which is comparable to the conventional composite. This is because the lower concentrations of SMA and SF could not improve the composite's conductivity. The addition of 0.1% CF, on the other hand, considerably increases the composite's conductivity by forming conductive channels. The percolation transition zone of the CF embedded cement composites is S-shaped, as shown in Fig. 7. Fig. 7 shows the conductivity of the SMA and SF embedded cement composites. The random orientation in the number and quality of interconnections of the CF utilized in the composite causes the S-Shape curve. According to the research, the fibers are consistently disseminated in the composite at lower concentrations of conductive elements. However, at more significant concentrations, the fibers agglomerate and form clusters, which improves the composite's conductivity. The clusters are generated individually until a certain concentration of CF is reached, after which they are cross-connected and form a conductive channel in the composite, as illustrated schematically in fig. 7.

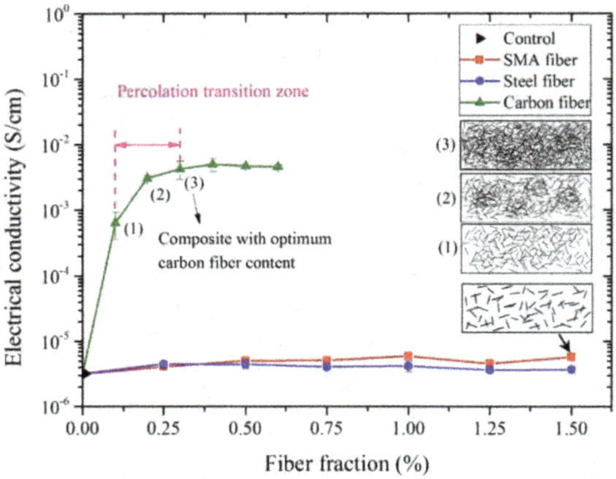

Fig. 7. Variation of electrical conductivity of the cement composite without and with SMA, SF, and CF [27]

Piezoresistivity

Under external load with frequency ranging from 0.1 Hz to 10 Hz, the piezoresistive behavior of cement composites with embedded MWCNT, CNF, CB, and GNP is shown in Fig. 8. The strain is measured with two strain gauges, and the electrical resistance is monitored with electrodes. The resistance varies with strain for all the cement composites embedded with MWCNT, CNF, CB, and GNP, as shown in fig. 8. This indicates that the conductive materials utilized are more strain-sensitive. MWCNT embedded composites have a higher sensitivity to strain than composites implanted with CNF, CB, or GNP.

In comparison to other samples, the MWCNT embedded composite has a more excellent gauge factor (=4139) and strain sensitivity (=3.763e10), according to the research. MWCNT embedded composite has excellent mechanical properties in addition to electrical ones. The elastic modulus of the MWCNT cement composite (=23410 MPa) is about four times that of the ordinary cement composite (=8821 MPa) [37]. The study also found that the elastic modulus of CNF, CB, and GNP embedded composites is comparable to that of the conventional cement composites. This demonstrates that the form and aspect ratio of conductive materials significantly impact the composite's strength increase.

The use of CF and CB yields similar results. They showed that combining CF and CB in the composite leads to good piezoresistivity repeatability. The sensitivity of such cement composites with CF and CB is 0.0138%, and the repeatability is 4.36% [43]. In the case of beams, it has been discovered that when the strain value exceeds 0.2%, the elastic modulus decreases, resulting in a significant increase in resistance owing to CF fracture [44]. This demonstrates that CF embedded composites are inappropriate for sensing composite strain and stress. The piezoresistive behavior of the hybrid conductive materials implanted in cement composites is also good. Because of their high dispersion in the composite, the cement composite with CNT/NCB has a good piezoresistive behavior (=55.28). [36].

Materials Research Forum LLC
https://doi.org/10.21741/9781644901953-25

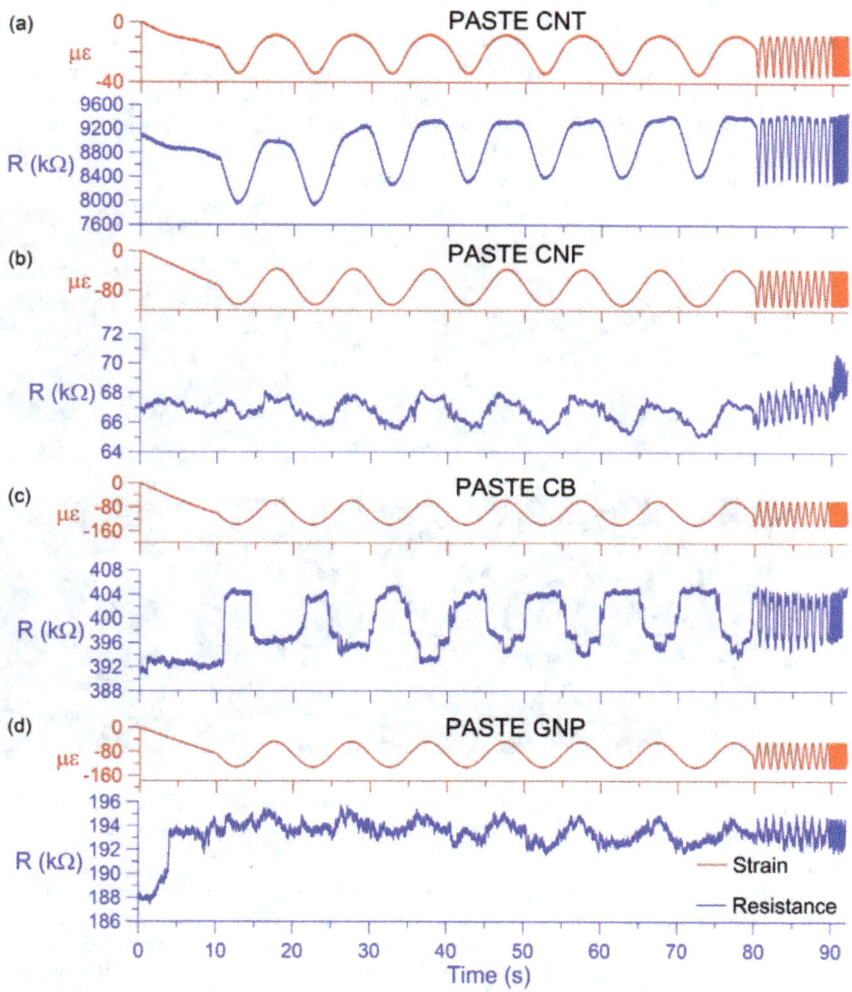

Fig. 8. Sample with time records of measured strain and electrical resistance were recorded. with (a) MWCNT, (b) CNF, (c) CB, and (d) GNP from strain sensing tests [37]

Morphology Characterization

The cement composite's effective dispersion of conductive components is depicted in the SEM image [37]. As seen in the picture, the conductive particles (MWCNT, CNF, CB, and GNP) used

Materials Research Proceedings **23** (2022) 217-230
https://doi.org/10.21741/9781644901953-25

in the composite are not fractured throughout the mixing process. Due to their geometric characteristics, CB particles, on the other hand, are less well-known than other materials. Furthermore, the fig. 9. shows that the particles are adequately dispersed in the matrix and that no particle agglomeration in the form of bundles has occurred in the composite. A SEM picture of the cement paste, for example, might reveal a non-uniform distribution of CNF [45]. According to SEM, the microcracks in the designed cement composite have also been amended, according to SEM [46]. They may also identify the presence of CFs bridged across healed microcracks. The development of hydration products can also be seen in the SEM images.

Fig. 9. SEM images of nano modified cement composite with (a) MWCNT, (b) CNFs, (c) CB and (d) GNPs [37]

Conclusions

The fresh, mechanical, and electrical properties of a self-sensing cement composite produced with carbon-based nanoparticles are investigated in detail. By graphing the deviation of fractional change in resistivity with strain, the study also focuses on the self-monitoring performance of cement composite in concrete beams, both bulk and embedded. SEM and XRD are used to characterize network channels in carbon-based nanomaterials embedded cement composites. As a consequence of the investigation, the following conclusions were made.

Mechanical Characteristics: Self-sensing cementitious composites can be made using both individual and hybrid fillers. The embedment of self-sensing cement composite does not affect the load-bearing capacity of structural components. The composite's self-sensing qualities alter before and after embedding due to the shift in poisson ratio and youngs modulus. The implanted self-sensing cement composite can reflect the stress/strain state of beams/columns, implying that it could be used as a sensor in SHM. The ultrasonic velocity through concrete might depict the damage development. When cracks in the ultrasonic transmission channel are blocked, the velocity might soon drop to zero. This shows that ultrasonic velocity can detect cracks formation in the composite.

Electrical Characteristics: The damage evolution process can be depicted using resistivity, including an elastic phase with little damage propagation, a moderate damage development stage, an expeditisve following the force peak, and a relaxed stage with macro fractures. The self-sensing cementitious composite displays a robust and repeatable self-sensing feature under cyclic and monotonic compression. The innovative concrete with steel fibres and tiny steel slag particles demonstrated a clear reverse electrical resistivity response when compressed. The use of FSSAs instead of silica sand enhanced the piezoelectric sensitivity due to the quantum tunneling effect, while the addition of short steel fibers considerably increased the conductive network.

Morphology: According to SEM data, most CNFs can be evenly disseminated in the epoxy matrix to form conductive networks. However, when the CNF concentration was high, there were more agglomerations of CNFs, which had a negative impact on the mechanical and piezoresistive properties of the nanocomposites.

Thus, the review concludes that by creating a high-sensitivity, well-compatible, and long-serve life self-sensing cement composite, infrastructure monitoring can be done on a wide scale and at a cheap cost.

Acknowledgments

The authors would like to express their gratitude to the management, Sri Ramakrishna Engineering College, Coimbatore. The authors would like to thank the management, PSG College of Technology, Coimbatore, for their support in conducting this work.

References

[1] R. Capozucca, Damage to reinforced concrete due to reinforcement corrosion, Constr. Build. Mater. 9 (1995) 295–303. https://doi.org/10.1016/0950-0618(95)00033-C

[2] P.C. Chang, A. Flatau, S.C. Liu, Review paper: Health monitoring of civil infrastructure, Struct. Heal. Monit. 2 (2003) 257–267. https://doi.org/10.1177/1475921703036169

[3] J.M.W. Brownjohn, Structural health monitoring of civil infrastructure, Philos. Trans. R. Soc. A Math. Phys. Eng. Sci. 365 (2007) 589–622. https://doi.org/10.1098/rsta.2006.1925

[4] V. Agarwal, A. Shelke, B.S. Ahluwalia, F. Melandsø, T. Kundu, A. Habib, Damage localization in piezo-ceramic using ultrasonic waves excited by dual point contact excitation and detection scheme, Ultrasonics. 108 (2020) 106113. https://doi.org/10.1016/j.ultras.2020.106113

[5] R.B. Randall, W.A. Smith, Detection of faulty accelerometer mounting from response measurements, J. Sound Vib. 477 (2020) 115318. https://doi.org/10.1016/j.jsv.2020.115318

[6] Z. Pei, X. Cheng, X. Yang, Q. Li, C. Xia, D. Zhang, X. Li, Understanding environmental impacts on initial atmospheric corrosion based on corrosion monitoring sensors, J. Mater. Sci. Technol. (2020) 1–8. https://doi.org/10.1016/j.jmst.2020.01.023

[7] J. Chakraborty, A. Katunin, Detection of structural changes in concrete using embedded ultrasonic sensors based on autoregressive model, Diagnostyka. 20 (2019) 103–110. https://doi.org/10.29354/diag/100448

[8] A. Barrias, J.R. Casas, S. Villalba, Fatigue performance of distributed optical fiber sensors in reinforced concrete elements, Constr. Build. Mater. 218 (2019) 214–223. https://doi.org/10.1016/j.conbuildmat.2019.05.072

[9] F. Guan, X. Zhai, J. Duan, M. Zhang, B. Hou, Influence of sulfate-reducing bacteria on the

corrosion behavior of high strength steel eq70 under cathodic polarization, PLoS One. 11 (2016) 1–22. https://doi.org/10.1371/journal.pone.0162315

[10] S.Z. Chen, G. Wu, D.C. Feng, Damage detection of highway bridges based on long-gauge strain response under stochastic traffic flow, Mech. Syst. Signal Process. 127 (2019) 551–572. https://doi.org/10.1016/j.ymssp.2019.03.022

[11] Z. Herrasti, I. Gabilondo, J. Berganzo, I. Val, F. Martínez, Wireless Sensor Nodes for Acceleration, Strain and Temperature Measurements, Procedia Eng. 168 (2016) 1659–1662. https://doi.org/10.1016/j.proeng.2016.11.484

[12] A. Dinesh, S.T. Sudharsan, S. Haribala, Self-sensing cement-based sensor with carbon nanotube: Fabrication and properties – A review, Mater. Today Proc. (2021). https://doi.org/10.1016/j.matpr.2021.02.722

[13] A. Dinesh, B. Abirami, G. Moulica, Materials Today : Proceedings Carbon nanofiber embedded cement composites : Properties and promises as sensor – A review, Mater. Today Proc. (2020). https://doi.org/10.1016/j.matpr.2020.10.526

[14] B. Han, S. Ding, X. Yu, Intrinsic self-sensing concrete and structures: A review, Meas. J. Int. Meas. Confed. 59 (2015) 110–128. https://doi.org/10.1016/j.measurement.2014.09.048

[15] P. Eswaramoorthi, P. Magudeaswaran, A. Dinesh, PUSHOVER ANALYSIS OF STEEL FRAME, n.d.

[16] A. Professor, Analytical study on the behaviour of cold formed steel double channel beam sections, n.d. www.ripublication.com/ijaer.htm

[17] A. Dinesh, R. Prasanth Kumar, S.R. Abijith, Experimental investigation on bubble deck concrete using plastic waste, 2020. https://doi.org/10.1007/978-981-15-0990-2_14

[18] J.M. Rius, A. Cladera, C. Ribas, B. Mas, Shear strengthening of reinforced concrete beams using shape memory alloys, Constr. Build. Mater. 200 (2019) 420–435. https://doi.org/10.1016/j.conbuildmat.2018.12.104

[19] A. Professor, Study on Strength properties of High performance concrete, n.d. www.ripublication.com/ijaer.htm

[20] A.S. S.D., D. A., S.B. V., Investigation of waste marble powder in the development of sustainable concrete, Mater. Today Proc. (2020). https://doi.org/10.1016/j.matpr.2020.10.536

[21] M. Sutcu, H. Alptekin, E. Erdogmus, Y. Er, O. Gencel, Characteristics of fired clay bricks with waste marble powder addition as building materials, Constr. Build. Mater. 82 (2015) 1–8. https://doi.org/10.1016/j.conbuildmat.2015.02.055

[22] D. Sellakutty, A. Professor, UTILISATION OF WASTE PLASTIC IN MANUFACTURING OF BRICKS AND PAVER BLOCKS, 2016. https://www.researchgate.net/publication/303273892

[23] K. Sathish Raja, Dinesh. A, Study on Self Compacting Concrete – A Review, Int. J. Eng. Res. V5 (2016). https://doi.org/10.17577/ijertv5is020349

[24] A. Professor, Stabilization of soil by using solid waste-A Review, © 2017 IJEDR |. 5 (2017). www.ijedr.org

[25] A. Dinesh, S. Indhumathi, Moorthi Pichumani, Performance assessemnt of copper slag and sawdust ash in stabilization of black cotton soil, 2021.https://doi.org/10.1007/978-981-33-4590-4_21.

[26] A. Dinesh, S. Harini, J.P. Jasmine, J. Jincy, J. Shagufta, International journal of engineering sciences & research technology experimental study of blast furnace slag concrete, Int. J. Eng.

Sci. Res. Technol. 6 (2017) 42–50. https://doi.org/10.5281/zenodo.345692

[27] A. Dehghani, F. Aslani, The effect of shape memory alloy, steel, and carbon fibres on fresh, mechanical, and electrical properties of self-compacting cementitious composites, Cem. Concr. Compos. 112 (2020) 103659. https://doi.org/10.1016/j.cemconcomp.2020.103659

[28] X. Xi, D.D.L. Chung, Piezoelectret-based and piezoresistivity-based stress self-sensing in steel beams under flexure, Sensors Actuators, A Phys. 301 (2020) 111780. https://doi.org/10.1016/j.sna.2019.111780

[29] R.K. Biswas, F. Bin Ahmed, M.E. Haque, A.A. Provasha, Z. Hasan, F. Hayat, D. Sen, Effects of Steel Fiber Percentage and Aspect Ratios on Fresh and Harden Properties of Ultra-High Performance Fiber Reinforced Concrete, Appl. Mech. 2 (2021) 501–515. https://doi.org/10.3390/applmech2030028

[30] I. Papanikolaou, C. Litina, A. Zomorodian, A. Al-Tabbaa, Effect of natural graphite fineness on the performance and electrical conductivity of cement paste mixes for self-sensing structures, Materials (Basel). 13 (2020) 1–19. https://doi.org/10.3390/ma13245833

[31] B.G. Han, B.Z. Han, J.P. Ou, Experimental study on use of nickel powder-filled Portland cement-based composite for fabrication of piezoresistive sensors with high sensitivity, Sensors Actuators, A Phys. 149 (2009) 51–55. https://doi.org/10.1016/j.sna.2008.10.001

[32] E. Teomete, O.I. Kocyigit, Tensile strain sensitivity of steel fiber reinforced cement matrix composites tested by split tensile test, Constr. Build. Mater. 47 (2013) 962–968. https://doi.org/10.1016/j.conbuildmat.2013.05.095

[33] A.O.S. Solgaard, M. Geiker, C. Edvardsen, A. Küter, Observations on the electrical resistivity of steel fibre reinforced concrete, Mater. Struct. Constr. 47 (2014) 335–350. https://doi.org/10.1617/s11527-013-0064-y

[34] I.L. Larsen, R.T. Thorstensen, The influence of steel fibres on compressive and tensile strength of ultra high performance concrete: A review, Constr. Build. Mater. 256 (2020) 119459. https://doi.org/10.1016/j.conbuildmat.2020.119459

[35] J. Song, D.L. Nguyen, C. Manathamsombat, D.J. Kim, Effect of fiber volume content on electromechanical behavior of strain-hardening steel-fiber-reinforced cementitious composites, J. Compos. Mater. 49 (2015) 3621–3634. https://doi.org/10.1177/0021998314568169

[36] B. Han, L. Zhang, S. Sun, X. Yu, X. Dong, T. Wu, J. Ou, Electrostatic self-assembled carbon nanotube/nano carbon black composite fillers reinforced cement-based materials with multifunctionality, Compos. Part A Appl. Sci. Manuf. 79 (2015) 103–115. https://doi.org/10.1016/j.compositesa.2015.09.016

[37] A.L. Pisello, A. D'Alessandro, S. Sambuco, M. Rallini, F. Ubertini, F. Asdrubali, A.L. Materazzi, F. Cotana, Multipurpose experimental characterization of smart nanocomposite cement-based materials for thermal-energy efficiency and strain-sensing capability, Sol. Energy Mater. Sol. Cells. 161 (2017) 77–88. https://doi.org/10.1016/j.solmat.2016.11.030

[38] X. Fan, D. Fang, M. Sun, Z. Li, Piezoresistivity of carbon fiber graphite cement-based composites with CCCW, J. Wuhan Univ. Technol. Mater. Sci. Ed. 26 (2011) 339–343. https://doi.org/10.1007/s11595-011-0226-0

[39] F. Azhari, N. Banthia, Cement-based sensors with carbon fibers and carbon nanotubes for piezoresistive sensing, Cem. Concr. Compos. 34 (2012) 866–873. https://doi.org/10.1016/j.cemconcomp.2012.04.007

[40] H. Xiao, H. Li, J. Ou, Strain sensing properties of cement-based sensors embedded at

various stress zones in a bending concrete beam, Sensors Actuators, A Phys. 167 (2011) 581–587. https://doi.org/10.1016/j.sna.2011.03.012

[41] F.J. Baeza, O. Galao, E. Zornoza, P. Garcés, Effect of aspect ratio on strain sensing capacity of carbon fiber reinforced cement composites, Mater. Des. 51 (2013) 1085–1094. https://doi.org/10.1016/j.matdes.2013.05.010

[42] A. Al-Dahawi, G. Yıldırım, O. Öztürk, M. Şahmaran, Assessment of self-sensing capability of Engineered Cementitious Composites within the elastic and plastic ranges of cyclic flexural loading, Constr. Build. Mater. 145 (2017) 1–10. https://doi.org/10.1016/j.conbuildmat.2017.03.236

[43] J. Ou, B. Han, Piezoresistive cement-based strain sensors and self-sensing concrete components, J. Intell. Mater. Syst. Struct. 20 (2009) 329–336. https://doi.org/10.1177/1045389X08094190

[44] G. Yıldırım, M.H. Sarwary, A. Al-Dahawi, O. Öztürk, Ö. Anıl, M. Şahmaran, Piezoresistive behavior of CF- and CNT-based reinforced concrete beams subjected to static flexural loading: Shear failure investigation, Constr. Build. Mater. 168 (2018) 266–279. https://doi.org/10.1016/j.conbuildmat.2018.02.124

[45] A. Yazdanbakhsh, Z. Grasley, B. Tyson, R.K. Abu Al-Rub, Distribution of carbon nanofibers and nanotubes in cementitious composites, Transp. Res. Rec. (2010) 89–95. https://doi.org/10.3141/2142-13

[46] H. Siad, M. Lachemi, M. Sahmaran, H.A. Mesbah, K.A. Hossain, Advanced engineered cementitious composites with combined self-sensing and self-healing functionalities, Constr. Build. Mater. 176 (2018) 313–322. https://doi.org/10.1016/j.conbuildmat.2018.05.026

Sustainable Materials and Smart Practices - NCSMSP-2021 Materials Research Forum LLC
Materials Research Proceedings **23** (2022) 231-237 https://doi.org/10.21741/9781644901953-26

Nano Silica Particles as A Structural Build Up Agent for 3D Printing Cement Paste with Additional Accelerators

Varun Sabu Sam, G.Hemalatha, Vincent Sam Jebadurai S, Arunraj E, J. Jensika Rani

Department of Civil Engineering, Karunya Institute of Technology, Coimbatore, Tamil Nadu

Keywords: 3D Printing, Additive Manufacturing, Rheological Proprieties. Concrete Printing. Nano Particle

Abstract. 3D printing also commonly known Additive manufacturing (AM) is the layering of materials by a computer to build 3D shapes of desired materials. It's very handy for prototyping and working with geometrically complicated materials. It was founded and developed in the 1980s, but it was a costly and demanding procedure at the time, with few uses. It became quiet relevant and inexpensive in the 2000 and a variety of applications which includes product development, component manufacturing, and tool manufacturing, plastic, aerospace engineering metalworking, Foot wear, dental and medical applications, and electronics. Even the household uses of 3D printers were happening. The construction sector also started using 3D technology and the systems used were called 3D printers for construction Computer-aided design, or CAD design, or the usage of a 3D scanner are both used to build a 3D model.

Introduction

The 3D printing printer reads and takes the design and lays down consecutive layers of a printing medium, which can be a liquid or a sheet material, which are eventually united to make the finished result. Although the whole procedure is slow, it can be utilized to make any shape. Based on the technique we select we can make multiple components continuously and we can use more than on material and color. Accuracy of the process can be increased by various methods like using Removes material from an enormous printed item using a high-resolution subtractive method. Use of dissolved materials that support overhanging features during fabrication is effective in some process. 3D printing can be used to make building components or to make the whole building. It gained popularity in the construction field since it's already experienced in the computer aided design and manufacturing. The building information modelling or BIM can make use of 3D printing in to a new level.

By the use of 3D printing we can achieve faster and accurate construction of components by lowering man power and waste generated and thus decreasing the overall expense. It can also decrease the risk factor in harsh and dangerous places which is not ideal for human forces. Recent researches towards automated construction have paved the way to the advancement of many strategies and ideas in additive manufacturing. The various techniques now available are contour crafting, D shape and concrete printing. These are considered very nature friendly and effective which makes the construction faster . The use of Nano silica in the development of cement-based matrices for use in 3D printing is found to better the cement's rheological qualities and thus increase the efficiency of the 3D printing process. Further studies have to be done to see the whether there is any change in cement rheological characteristics with the addition of chemical accelerator with Nano silica particle which is added to the cement.

Objectives

To see if combining a chemical accelerator with Nano silica particles in a cement-based matrix formulation for 3D printing provides access to the Nano particle's full potential.

Literature

3D printing is a computer based production of 3 dimensional shapes and sizes.3D printing is considered as the next big thing in construction and has a huge scope for research. Recent discovery and advancement of artificial intelligence has paved Techniques like contour shaping, D-shape, and concrete printing have paved the path for additive manufacturing to advance. These methods are quite effective. can attain eco-friendly cheaper and reliable ways of construction. Different researches considered many set of parameters. This chapter gives salient features of experimental investigations.

Underwater 3D printing of cement based mortar –Ibrahim mazhoud 30th July 2019

The development and advancement of renewable marine energies is needed to face the global warming situations going continuously with the development of automated and digital based construction methods such as 3D printing. Advancement in this area could lead to the change in offshore construction methods. Through studies researchers have reported, the stability and strength of the structure through printing can be seen as a competition between the rarity of cement-based material structural build-up and the rate of increase of load acting on the entire structure.

Extrusion-based 3D printing with micro wire reinforced geopolymer composite January 2019

To enhance and increase cementitious materials' toughness and post-crack moment capacity for extrusion-based 3D printing, additive manufacturing is used in the methodology process. The material used is a fly ash based geopolymer. This research proposes a method for improving cementitious material toughness and post-crack moment capacity.

Nano silica particles as structural fillers in Portland cement pastes for 3D printing January of the next year Oscar A. Mendoza

This paper gives a 3D printing-based comparative investigation of the influence of nano silica particles on the characteristics of Portland cement. Different solid substitutions of cement by pozzolanic particles were used to make cement pastes. After varying amounts of resting time, their yield stress was measured. The rate of thixotropic accumulation of each paste was calculated using the rheological results. The parameters of a perfect 3D printing process were calculated using these data.

Jian Gong uses cellulose nanocrystals as a support material for 3D printing complexly formed structures using multiple materials. August of this year

To produce very high complicated structures sacrificial support materials is required Traditional petroleum-based support materials cannot be taken as sustainable and it cannot by recycled. In place of them cellulose nanocrystals gel can be used as a 3D printing due to its sustainable and renewable properties. CNC is produced from forestry and natural products which is water based and with the use of this material the entire 3D printing process can be made environmentally safe. This paper sees the potential of CNC material as a 3D printing material and use of it for advance applications

3D printing for building with a polymer-foam and concrete complicated wall August 2019 Yulin Huang
This paper introduces a new additive manufacturing approach for concrete structure construction.. The materials used for this proposed idea consist of a two polymer foam printer walls which is used to contain a 3rd wall. This method uses a mobile polyarticulated robot. The foam used is for internal as well as external insulation to the structure. This paper also talks about the potential of doing 3D printing using clay or other earthly materials.

Methodology

Materials

Portland cement

This study looked at class g cement pastes that met API specification 10A/IS010426-1.is used. Portland cement can be used. It is the most common type of cement used around the world as a basic element in concrete, mortar, stucco, and non-specialty grout. It was created by Joseph Aspdin in the late 1800s in England from several types of hydraulic lime, primarily limestone. It is made by forming clinker from limestone and clay minerals in a kiln, grinding it, and adding a specified percentage of gypsum. The most prevalent type is ordinary Portland cement, or OPC. Cement should be free of impurities, lumps, and other foreign materials.

Nano silica

Although nano silica is not a typical material used in concrete, silica fume, which is hazardous, has been employed in micro silica for decades, and high-performance concrete has been produced. It helps refine the pore structure and enhance characteristics even at a low level of replacement because to its fine particle size, which speeds the hydration process early. Using micro silica in concrete is getting popular even though its high cost due to tis pozzolanic behavior. Due to its finer size micro silica fills the remaining voids and thus decreases porosity. The replacement of Nano silica reduces CO2 emissions.

Accelerator

A cement accelerator is a sort of additive used in concrete and mortar that reduces the time it takes for the concrete to set, allowing it to cure sooner. This is beneficial in the winter since it reduces the possibility of frost damage to concrete. Common type of chemicals used in accelerators are calcium nitrate, calcium chloride, calcium nitrite, and calcium formate.

Viscosity modifying agent

VMAs (viscosity modifying agents) are admixtures that change qualities like viscosity workability and cohesion. It has no effect on the yield point of concrete. Some of the type of concrete which uses VMA are

- self-compacting concrete
- pumped concrete
- under water concrete
- light weight concrete
- semi dry concrete
- porous concrete
- concrete with poorly graded aggregates

Cement pastes were made with a 0.45 water-to-cementitious-materials ratio, 0.60 percent VMA addition by mass of cement, and solid cement replacement with Nano silica for this experiment.

Table 1 Formulation of paste used for study

Particle type	Cement	Nano particle	VMA	accelerator	w/c ratio
Nano silica	99%	1.00%	0.60%	1.00%	0.45%

Using the paste shear test was conducted to find shear stress.

With this experiment we have to find the shearing angle and shearing stress. Using equations we have to find rheological properties like rate of thixotropic build up, maximum layer height, and maximum horizontal velocity.

Fig 1. VMA or viscosity modifying agent

Fig 2. Accelerator

Results

Nano silica is substituted to the cement particles and with the addition of vma and accelerator pastes are made. The quantity of substances added to the cement is specific and are recorded. Shear test and shearing angle have to be found out of this paste at different resting times and the data's should be analyzed.

Table 1 formulations

Particle type	Cement	Nano particle	Viscosity modifying agent	Accelerator	Water cement ratio
Nano silica	99%	1%	0.60%	1%	0.45%

The specific mass and theoretical density of Nano silica particle is noted down.

Table 2 specific mass

Particle	Specific mass(g/cm^3)
Nano silica	2.50

Sustainable Materials and Smart Practices - NCSMSP-2021 Materials Research Forum LLC
Materials Research Proceedings **23** (2022) 231-237 https://doi.org/10.21741/9781644901953-26

Table 3 theoretical density

Sample	ρ (g/cm³)
Nano silica	1.885

$$\rho = \frac{\varepsilon mi}{\varepsilon \frac{mi}{pi}} \qquad (1)$$

Where,

VMA is the mass of the ith component, including water. Mi is the mass of the ith component, including water.

The specific mass of the ith paste component is Pi.

Density is calculated and compared with theoretical density

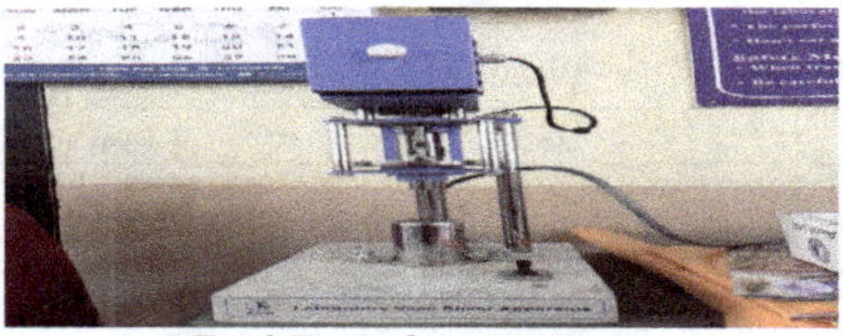

Fig 3 Vane shear apparatus

Vane shear test is done for the paste to find the shear stress and shearing angle with different resting times of the paste after mixture. The resting time is taken from other journal papers which have conducted similar tests. Initial reading and final readings are taken of the pastes with different resting times. Using formulas from the given data torque T and shear stress are calculated. Obtained data's are recorded in table 4

Vane shear test results for finding shear stress and shearing angle

D=12mm H=24mm

Table 4 vane shear test to find shear stress and shearing angle

Sl no	Resting Time in minutes	Initial reading ϴ1	Final reading ϴ2	ϴ	Torque T	Shear stress
1	0	165	190	25	1.11	0.1943
2	23	172	219	47	2.08	0.3652
3	45	227	276	49	2.177	0.3808
4	68	178	237	56	2.48	0.4352

$$T = \frac{\theta \times k}{180} \qquad (2)$$

Materials Research Forum LLC
https://doi.org/10.21741/9781644901953-26

$$s = \frac{T}{\pi\{\left[D^2\frac{H}{2}\right] + \left[\frac{D^3}{6}\right]\}} \qquad 3$$

From earlier studies it was found that With just Nano silica, the initial yield stress and rate of thixotropic paste building are increasing. 1% of Nano silica was used for such studies as Due to the rheometer's restrictions on the greatest shear stress that could be applied, bigger levels of Nano silica could not be evaluated. 4

$$\tau_{0,t} = A_{thix} \times t + \tau_{0,fit}$$

Where Athix is the linear equation's slope, t is the resting period of zero, and fit is the static yield stress obtained from the fitted equation at t=0.

Using this formula Athix is calculated of pastes with different resting time and is recorded in table 5.

Table 5 rheological properties

SL no	Resting time	A thix
1	23	$7.43x10^3$
2	45	$4.14x10^3$
3	68	$3.542x10^3$

From this data we can calculate h max using formula and check the change is rheological properties

$$h_{max} = \frac{\tau_{0,fit}\sqrt{3}}{\rho g} \qquad 5$$

H max=0.0714
$\tau_{0,fit}$=0.1943
ρ=2.50
g=1.885

It has been found that cement paste with resting time 23 minutes is showing better rheological properties as it's A thix and hma values are higher than other resting time values. Athix for cement paste without accelerator at resting time 23 minutes was 6.45x103. The resting time is the time between densification and rheological testing.

Conclusion

Nano silica is found as an effective thickener for paste of cement and is proved in order to increase all the parameters and properties related to a material's structural build-up agent. By the experiment done its been found that using Nano silica with the addition of accelerator rheological properties of cement paste is increasing for one of the resting times.

Scope of future study

Experiments should be done with a higher concentration of NS It was unable to be tested due to the rheometer's limitations on highest shear stress. The type of change in characteristics that will occur on cement paste with increasing concentrations of NS is unknown, but the linear trend is expected to continue.

Reference

[1] A. Kazemian "Cementitious materials for construction scale 3D printing: Laboratory testing of fresh printing mixture" 2017, pp.639-647 https://doi.org/10.1016/j.conbuildmat.2017.04.015

[2] B. Zareiyan "Effects of interlocking on interlayer adhesion and strength of structures in 3D printing concrete"2017, pp.212-221 https://doi.org/10.1016/j.autcon.2017.08.019

[3] D.G soltan "A self-reinforced cementitious composite for building scale 3D printing "2018, pp.1-13 https://doi.org/10.1016/j.cemconcomp.2018.03.017

[4] D. Marchon "Hydration and rheology control of concrete for digital fabrication: potential admixtures and cement chemistry" 2018, pp.96-110 https://doi.org/10.1016/j.cemconres.2018.05.014

[5] J.G sanjayan" Effect of surface moisture on inter layer strength of 3D printed concrete" 2018, pp.468-475 https://doi.org/10.1016/j.conbuildmat.2018.03.232

[6] L.Reiter "The role of early age structural build up in digital fabrication with concrete"2018, pp.89-95 https://doi.org/10.1016/j.cemconres.2018.05.011

[7] N. labonnote "Additive construction state of the art challenges and opportunities" 2016, pp.347-366 https://doi.org/10.1016/j.autcon.2016.08.026

[8] N. Roussel "Rheological requirements for printable concretes" 2018, pp.76-85 https://doi.org/10.1016/j.cemconres.2018.04.005

[9] Oscar A. Mendoza Reales" Nano silica particle as structural buildup agents for 3D printing with portland cement pastes" volume219, 20 September 2019 https://doi.org/10.1016/j.conbuildmat.2019.05.174

[10] R.Duballet "Classification of building systems for concrete 3D printing"2017, pp.247-258 https://doi.org/10.1016/j.autcon.2017.08.018

[11] S.C.Paul "Fresh and hardened properties of 3D printable cementitious materials for building and construction "2018, pp.311-319 https://doi.org/10.1016/j.acme.2017.02.008

[12] S.H.Ghaffar "Additive manufacturing technology and its implementation in construction as an eco-innovative solution" 2018, pp.1-11 https://doi.org/10.1016/j.autcon.2018.05.005

[13] S.lim "Developments in construction scale additive manufacturing process" 2012, pp.262-268 https://doi.org/10.1016/j.autcon.2011.06.010

[14] V.Vaitkevicius "Effect of ultra-sonic activation on early hydration process in 3d concrete printing technology" 2018, pp.35 https://doi.org/10.1016/j.conbuildmat.2018.03.007

Sustainable Materials and Smart Practices - NCSMSP-2021 Materials Research Forum LLC
Materials Research Proceedings **23** (2022) 238-245 https://doi.org/10.21741/9781644901953-27

A Constructive Analysis on Topology Optimization of Structural Beam Element Using Different Support Condition

J. Jensika Rani[1], G. Hemalatha[1], B. Praveen[1], S. Vincent Sam Jebadurai[1]

[1]Dept. of Civil Engineering, Karunya Institute of Technology and Sciences, Coimbatore, Tamil Nadu 641114, India

Keywords: Topology Optimization, Structural Optimization, ESO, MATLAB

Abstract. Topology Optimization (TO) is the method used for maximizing the performance of the system by optimizing material arrangement inside an appropriate topology design operation, for an appropriate set of loads, boundary restraints, and constraints. This helps in the reduction of the materials which in turn reduces the cost of materials. In the current scenario, the price of the ingredients of the traditional concrete is very high. The implementation of TO gives a good impact on the economy of construction, as the materials are optimally placed in the plane of load distribution as per various loading condition. The Evolutionary Structural optimization (ESO) is used for topology optimization. The main merits of implementing topology optimization are that the cost, as well as the weight of the structural member, becomes more effective. Customarily MATLAB software is used for analysis and designing the member. This method will highly useful for the society and the environment.

Introduction

Topology optimization is a design process, where a mathematical concept in which the size and shape of the member are optimized. In 1994, the concept of topology optimization was introduced and implemented successfully at the University of Michigan. At first, the software Opti-struct was launched and officially used for topology optimization in the same year. The conventional design of the structural members leads to huge usage of materials which is expensive. On implementing the concept, the shape and size of the member are optimized which results in the reduction of the materials. Hence the total cost of the material will be reduced. This method uses the mechanism of finite element analysis. The technology of Topology Optimization (TO) is to increase the performance of the system by arranging the materials inside a topology optimization design operation, for an appropriate set of loads, boundary restraints, and constraints. This helps to reduction of materials which in turn reduces the cost of materials. In the current scenario, the price of the raw materials of the traditional concrete is very high in number.

The implementation of TO gives a good impact on the economy of construction, as the materials are optimally placed in the plane of load distribution as per various loading condition. The MATLab software is used for topology optimization. The main merits of implementing topology optimization are the cost, as well as the weight of the structural member, becomes more effective. This method will be highly useful for society and the environment. In TO, the overall structure are designed by removing excess material from the structure such that the final design is lighter without compromising with the performance. The design is optimized using either gradient-based mathematical program techniques or the method of moving asymptotes or non-gradient-based algorithms such as genetic algorithms.

In 3-D TO Perle Geoffroy-Donders, et. al., researched about the homogenization process modified and orientated periodic microstructures. using the compliance minimization. Zi-Long

Zhao, et. al., deliberated the direct approach to control the topology in structural optimization using the control of complexity in the structure. Hao Li, et. al., had studied the multiphase composite frequency response structures with level sets using a new multiscale TO method. In a novel study of topology optimization, Khader M. Hamdia et al. looked into the flexoelectric nanostructures are designed using a computational material design process.. Willem Roux, et. al., studied the spatial kernel approach in topology optimization using the simply supported beam with a concentrated load. In concurrent optimization Yunfeng Luo et al. explored the discrete-continuous parameterization (DCP) in continuous material orientations and structural topologies. The method of consecutive iteration of analysis and design for large-scale TO using Eigen frequencies was explored by Zhan Kang et al. Shouyu Cai studied the adaptive bubble method in structural shape and topology optimization by introducing a hole in the specimen. Yingjun Wang et al. used a multistory mesh, MGCG, and a local-update technique to examine efficient isogeometric topology optimization.

In Bi-material topology optimization Ryan Alberdi, et. al., explored the detrition criteria with inertia and material rate effects using finite deformations. Hyeong Seok Koh, et. al., deliberated the study of TO in multicomponent structure using sub structuring-based model order reduction in an efficient way. Andrew T. Gaynor et al. debated the need of a projection-based TO approach to minimize occluded voids in additive manufacturing design. Jackson L. Jewett et. al., used hybrid bi-linear topology optimization to study the behavior in deep RC beams of strut-and-tie configurations. Weisheng Zhang, et.al, deliberated the study of explicit TO using the MMV (moving morphable void) method based on IGA. In inverse topology optimization Tsuyoshi Nomura et al. looked into the structure's design and fibre orientation using tensor field variables. In a reinforced cement concrete model, Hayoung Chung et al. explored the investigation of level-set topology optimization utilizing nonlinear thermoelectricity. Yan Zhang, et. al., deliberated how to maximize the natural frequencies of non-homogeneous cellular structures using Kriging-assisted multiscale TO.

Zeng Meng et al. researched a To discuss epistemic and aleatory uncertainty, a new hybrid reliability-based TO technique combines fuzzy and probabilistic models. Using topology optimization Quoc Khanh Nguyen et al. investigated the zero lower bound on the physical density in material distribution. In Stiffness-based optimization Vasileios S. Papapetrou, et. al., studied the Continuous fibre composites provide a foundation for topology and fibre routes. Siraj-ul-Islam, et. al., studied the parameterized level set based on structural optimization for localised radial basis functions. Jackson L. Jewett and colleagues investigated how to determine TO design, construction, and experimental evaluation of concrete beams. Lei Zhao et. al., examined the structure topological optimization accompanied by dynamic fatigue restrictions under dynamic random loads. In topology optimization Jong Woo Lee, et. al., calculated the stress constraint using the layer wise theory of composite laminates. Using regional stress and displacement solidity constraints, Haijun Xi et al. investigated uncertainty-oriented TO in parametric structures with intervals. In additive manufacturing Baoshou Liu, et. al., studied the TO of structures using local material uncertainties.

The TO of stress-constrained structural parts was addressed the risk-factor technique by Tommaso Pastore et al. Carl-Johan Thore used the inertia relief approach to study the TO of freely floating elastic continua. Through snap-through behaviour, Hao Deng et al. investigated TO for energy dissipation design of lattice structures. In multiscale optimization Yan Zhang, et. al., explored the topological sketch in sandwich structures using graded cellular cores. Using movable morphable components, Xianda Xie et al. checked hierarchical spline-based is geometric TO.

Oded Amir studied the procedure in TO for reinforced concrete structures. From the above survey, it is understood that many types of research have been done in the field of TO in various types of beams in various aspects.

Mechanism in topology optimization

The mechanism involved in the topology optimization is the compliant mechanism that is often applied in the field of mechanical engineering. This signifies that the compliant mechanism is a flexible method that utilizes elastic deformation that transforms force and motion.

Procedure for topology optimization

In software, the step-by-step procedure is as follows:

- Modelling
- Analysis
- Optimizing
- Remodelling
- Producing manufacturing design

Modelling:

A model, in the broadest sense, is a three-dimensional representation of a structure, system, or technique. Model-based design is a method of developing digital representations to aid in the development of designs and decision-making. The commonly used software for modelling is:

- ANSYS
- AUTOCAD
- ABAQUS
- SOLIDWORKS, etc.

Analysis:

The activity of evaluating the effect of loads on structural systems and its constituents is characterized as analysis. The output of the study were used to confirm the structure's appropriateness for use, usually without the need for physical inspections.

Optimizing:

The act of optimization is defined as the process of finding the ideal response from a set of eligible selections. In many engineering challenges, the best answer is the problem's objective function's minimum or maximum value.

Remodeling:

Remodeling is described as the process of creating an optimized model with the same modelling software as before. The software used for modelling of the model and remodeling the optimized model is the same.

Producing manufacturing design:

Once the optimized model is remodeled in the software, it is ready to be fabricated for the construction purpose. The fabrication method is typically performed via 3D printing technology. In the field of construction, Construction components or complete buildings can be 'printed' using 3D printing. The development of 3D printing could result in more precise and faster manufacture of complicated or bespoke objects, as well as lower labor costs and waste output.

Analytical study

The Evolutionary structural optimization method is used for optimization. The program done in MATLAB software was taken from Liang Xia, et. al. in the year 2016. In this program, the function of six values is given as input, the material property of the specimen such as young's modulus and poison's ration is fixed value from the experiment. According to our simply support condition, the case falls into third condition. After giving the input the software starts, topology optimization of the structural beam element. Here a point to be noted is that general ratio of 1:8 is adopted for fixing the dimension of the specimen. Finally, the output obtained from the MatLab software will be done experimentally.

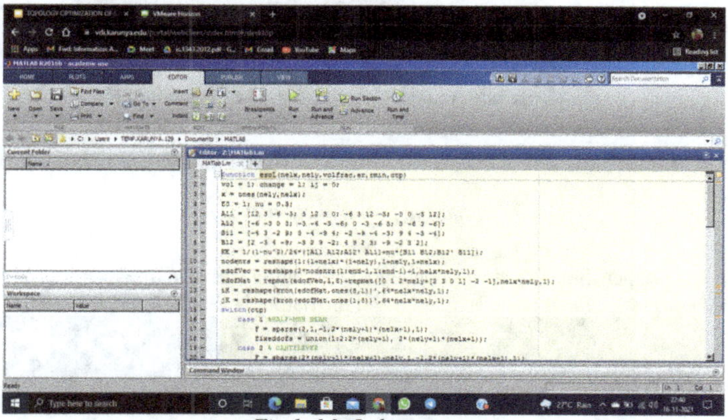

Fig 1: MatLab program

Output from the software:

The program done in MATLab software consist of 3 different cases, each has a different support condition and point loading conditions.

Case 1:

The benchmark test conducted for this case, a propped cantilever beam condition, in which PL (point load) was put in at the left side of the support. The corresponding result obtained by calling esoL (nelx, nely, volfrac, er, rmin, ctp) was found as shown in the fig 2. Here nelx and nely are the number of elements to be present in vertical and horizontal direction respectively; volfrac represents the volume fraction prescribed; er indicates the evolutionary ratio; rmin denotes the filter radius; and ctp denotes the benchmark design type.

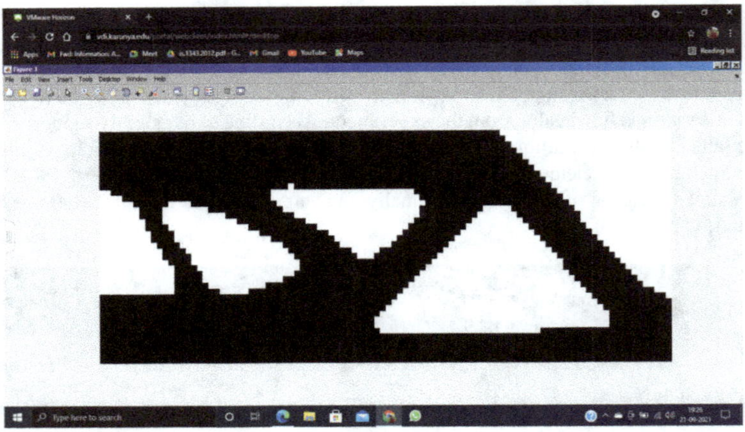

Fig 2: Optimized propped cantilever Beam

Case 2:
In the second case, benchmark test was conducted for cantilever beam, the load was applied at the free end. The same six functions were called and the output is shown fig 3.

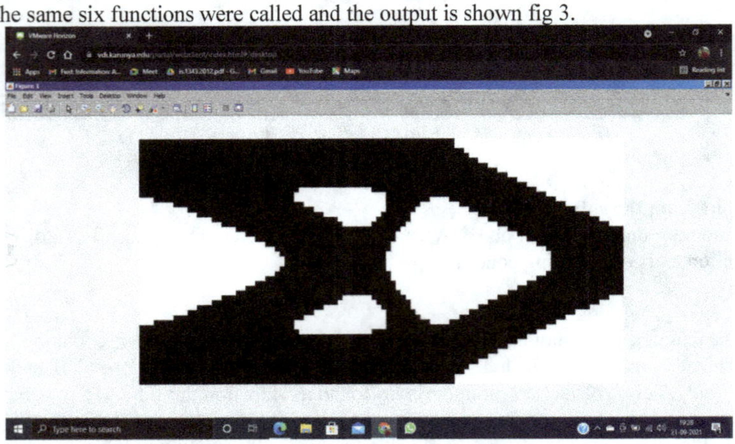

Fig 3: Optimized Cantilever Beam

Case 3:
Simply support condition was considered with a roller-support at one end. The type of loading condition is to be point load at the middle of the span, according to the conditions the output was found as shown in fig 4.

 Materials Research Forum LLC
https://doi.org/10.21741/9781644901953-27

Fig 4: Optimized simply-supported beam

Conclution:

The topology optimization of the beam varies based on the protocols used based on the method selection. Here, the principle of mass topology optimization is used in the evolutionary structural optimization (ESO) method. The boon in using TO is cost-effective as the materials are provided only in the necessary areas. The raw materials get reduced. The design process is very rapid and gives a good sustainability. The fabrication process of the optimized beam is typical which the only bane in TO is. In this research the TO of beam is done using point load under different support conditions. The support conditions used were propped cantilever, cantilever and simply-supported. The application of the ESO method is facile. The methodology or the phenomena used is to inflate the stiffness and shrink the mass of the element. The load organization is in the form of truss action. The materials required for the truss action is retained and the remaining part in the specimen is abolished. From the outcomes, it is understood that ESO method optimizes the structural beam element based on the principle of truss action. Hence, truss-like topologies are obtained in the ESO method.

Reference

[1] Perle Geoffroy-Donders, Grégoire Allaire, Olivier Pantz, 3-d topology optimization of modulated and oriented periodic microstructures by the homogenization method, (2020). https://doi.org/10.1016/j.jcp.2019.108994

[2] Zi-Long Zhao, Shiwei Zhou, Kun Cai, Yi Min Xie, A direct approach to controlling the topology in structural optimization, (2020).

[3] Hao Lia, Zhen Luo, Mi Xiao, Liang Gao, Jie Gao, A new multiscale topology optimization method for multiphase composite structures of frequency response with level sets, (2019), 116 - 144. https://doi.org/10.1016/j.cma.2019.07.020

[4] Khader M. Hamdia, Hamid Ghasemi, Yakoub Bazi, Haikel AlHichri, Naif Alajlan, Timon Rabczuk, A novel deep learning based method for the computational material design of flexoelectric nanostructures with topology optimization, (2019), 21 - 30. https://doi.org/10.1016/j.finel.2019.07.001

Materials Research Forum LLC
https://doi.org/10.21741/9781644901953-27

[5] Willem Roux, Guilian Yi, Imtiaz Gandikota, A spatial kernel approach for topology optimization, (2020).

[6] Yunfeng Luo, Wenjiong Chen, Shutian Liu, Quhao Li, Yaohui Ma, A discrete-continuous parameterization (DCP) for concurrent optimization of structural topologies and continuous material orientations, (2020).

[7] Zhan Kang, Jingjie He, Lin Shi, Zhaohui Miao, A method using successive iteration of analysis and design for large-scale topology optimization considering eigenfrequencies, (2020). https://doi.org/10.1016/j.cma.2020.112847

[8] Shouyu Cai, Weihong Zhang, An adaptive bubble method for structural shape and topology optimization, (2020).

[9] Yingjun Wang, Zhongyuan Liao, Ming Ye, Yu Zhang, Weihua Li, Zhaohui Xia, An efficient isogeometric topology optimization using multilevel mesh, MGCG and local-update strategy, (2020).

[10] Ryan Alberdi, Kapil Khandelwal, Bi-material topology optimization for energy dissipation with inertia and material rate effects under finite deformations, (2019). https://doi.org/10.1016/j.finel.2019.06.003

[11] Hyeong Seok Koh, Jun Hwan Kim, Gil Ho Yoon, Efficient topology optimization of multicomponent structure using substructuring-based model order reduction method, (2020).

[12] Andrew T. Gaynor, Terrence E. Johnson, Eliminating occluded voids in additive manufacturing design via a projection-based topology optimization scheme, (2019).

[13] Jackson L. Jewett, Josephine V. Carstensen, Experimental investigation of strut-and-tie layouts in deep RC beams designed with hybrid bi-linear topology optimization, (2019).

[14] Weisheng Zhang, Dingding Li, Pilseong Kang, Xu Guo, Sung-Kie Youn, Explicit topology optimization using IGA-based moving morphable void (MMV) approach, (2020).

[15] Tsuyoshi Nomura, Atsushi Kawamoto, Tsuguo Kondoh, Ercan M. Dede, Jaewook Lee, Yuyang Song, Noboru Kikuchi, Inverse design of structure and fiber orientation by means of topology optimization with tensor field variables, (2019). https://doi.org/10.1016/j.compositesb.2019.107187

[16] Hayoung Chung, Oded Amir, H. Alicia Kim, Level-set topology optimization considering nonlinear thermoelasticity, (2020).

[17] Yan Zhang, Liang Gao, Mi Xiao, Maximizing natural frequencies of inhomogeneous cellular structures by Kriging-assisted multiscale topology optimization, (2020). https://doi.org/10.1016/j.compstruc.2019.106197

[18] Zeng Meng, Yongsheng Pang, Yuxue Pu, Xuan Wang, New hybrid reliability-based topology optimization method combining fuzzy and probabilistic models for handling epistemic and aleatory uncertainties, (2020). https://doi.org/10.1016/j.cma.2020.112886

[19] Quoc Khanh Nguyena, Stefano Serra-Capizzano, Eddie Wadbroa, On using a zero lower bound on the physical density in material distribution topology optimization, (2019).

[20] Vasileios S. Papapetrou, Chitrang Patel, Ali Y. Tamijani, Stiffness-based optimization framework for the topology and fiber paths of continuous fiber composites (2020). https://doi.org/10.1016/j.compositesb.2019.107681

[21] Siraj-ul-Islam, Wajid Khan, Baseer Ullah, Zahur Ullah, The localized radial basis functions for parameterized level set based structural optimization, (2020). https://doi.org/10.1016/j.enganabound.2020.01.008

[22] Jackson L. Jewett, Josephine V. Carstensen, Topology-optimized design, construction and experimental evaluation of concrete beams, (2019).

[23] Lei Zhao, Bin Xu, Yongsheng Han, Jingdan Xue, Jianhua Rong, Structural topological optimization with dynamic fatigue constraints subject to dynamic random loads, (2019). https://doi.org/10.1016/j.engstruct.2019.110089

[24] Jong Wook Lee, Jong Jin Kim, Gil Ho Yoon, Stress constraint topology optimization using layerwise theory for composite laminates, (2019).

[25] Haijun Xi, Lei Wang, Yaru Liu, Uncertainty-oriented topology optimization of interval parametric structures with local stress and displacement reliability constraints, (2020).

[26] Baoshou Liu, Chao Jiang, Guangyao Li, Xiaodong Huang, Topology optimization of structures considering local material uncertainties in additive manufacturing, (2020).

[27] Tommaso Pastore, Valentina Mercuri, Costantino Menna, Domenico Asprone, Paola Festa, Ferdinando Auricchio, Topology optimization of stress-constrained structural elements using risk-factor approach, (2019). https://doi.org/10.1016/j.compstruc.2019.106104

[28] Carl-Johan Thore studied the topology optimization of freely floating elastic continua using the inertia relief method, (2020).

[29] Hao Deng, Lin Cheng, Xuan Liang, Devlin Hayduke, Albert C. To, Topology optimization for energy dissipation design of lattice structures through snap-through behavior, (2020). https://doi.org/10.1016/j.cma.2019.112641

[30] Yan Zhang, Mi Xiao, Xiaoyu Zhang, Liang Gao, Topological design of sandwich structures with graded cellular cores by multiscale optimization, (2020). https://doi.org/10.1016/j.cma.2019.112749

[31] Xianda Xie, Shuting Wang, Manman Xu, Ning Jiang, Yingjun Wang, A hierarchical spline-based isogeometric topology optimization using moving morphable components, (2020).

[32] Oded Amir, A topology optimization procedure for reinforced concrete structures, (2019).

[33] Liang Xia, Qi Xia, Xiaodong Huang, Yi Min Xie, Bi-directional Evolutionary Structural Optimization on Advanced Structures and Materials: A Comprehensive Review, (2016).

Experimental Analysis of Glass Fibre in Concrete

V.M. Gnanasundar[1,*], T. Palanisamy[2], G.S. Thirugnanam[3], C. Vishalachi[2]

[1]Department of Civil Engineering, Bannari Amman Institute of Technology, Erode, Tamil Nadu, India

[2]Department of Civil Engineering, National Institute of Technology, Karnataka, India

[3]Department of Civil Engineering, Institute of Road Transport Technology, Erode, Tamil Nadu, India

*vmgnanasundar20@gmail.com

Keywords: Admixture, Cement, Glass Fibre, Polymer, GFRC

Abstract. Compared compared to concrete in a construction, the essential portion of the structure has higher weight, however steel utilised as reinforcement has no weight. To address this problem, the Glass Fibre Reinforced Concrete (GFRC) material was developed. Polymers and glass fibre are impregnated in the cementation framework of GFRC, which is a material. Glass fibre, Fly ash, silica sand, Portland cement and water are all components of concrete. The glass content, mix procedure, and casting process all have an impact on the qualities of GFRC concrete. We present the fibre glass as well as other characteristic synthetic chemicals in GFRC to develop a material that is extremely solid and adaptableto construction.By this research, using 0.5 percent and 0.1 percent glass fibres increases compressive and flexural strength of concrete for 7,14 and 28 days with no admixtures.

Introduction

Fibre reinforced concrete, as well as a variety of cement types, is currently used in a number of tough settings for future Structural Engineering applications. One of the most essential aspects of Fibre Reinforced Concrete (FRC) is its superior resistance to cracking and crack propagation, as well as the presence of fibrous material that promotes structural integrity [1].Fibreglass is a resilient, flexible, and durable material. When compared to metals, its bulk strength and weight qualities are also quite beneficial, and it can be easily manufactured utilizing moulding procedures [2].Fibre-reinforced cementation composites with excellent performance for a variety of Structural Engineering applications, including great durability and impact load resistance [3]. It is now feasible to develop the characteristics of both fresh and hardened concrete primarily to the development of a variety of chemical admixtures [4].Fibre-reinforced concrete has shown better fracture resistance than non-fibre concrete, as well as higher durability, shear strength, and fatigue resistance [5]. Concrete can be strengthened and made more ductile by adding fibre. Several methods for impact testing concrete have been developed in the past [6]. The most common testing devices are drop weight machines, swinging pendulum machines, and split Hopkinson pressure bar equipment.Several admixtures have been utilized to improve the performance of concrete, including fibre reinforcement, super plasticizer, air-entraining agent, accelerators, retarders, waterproofing compounds, and others [7,8]. Compressive, tensile, flexural, and impact strength all improve when admixtures are added to ready-mixed fibre-reinforced concrete. Plastic shrinkage and other types of volume changes in concrete will cause concrete without fibre to fracture [9].

Table 1. Types of fibres and their characteristics

S. NO	FibreType	Characteristics	Remarks
1	PVA (Polyvinyl Alcohol)Fibre	ρ = 900 kg/m^3 σ = 1100 – 1400 N/mm^2 ε = 10,000 – 64,000 N/mm^2	PVA fibres with a high modulus may be made using special manufacturing procedures.
2	Glass Fibre (ECR glass)	ρ = 2720 kg/m^3 σ = 3400 N/mm^2 ε = 80,000 N/mm^2	It outperforms E-glass fibre in terms of mechanical characteristics, heat resistance, water resistance, electrical leakage rate, and surface resistance.
3	Carbon fibre	ρ = ~1,800 kg/m^3 σ = 3,500 N/mm^2 ε = 1.83x10^{5}N/mm^2	It has a high durability.
4	Polyester fibre	ρ = ~1830 kg/m^3 σ = 650 N/mm^2 ε = 920 N/mm^2	Developed for the textile industry but can also be found in the construction materials industry.
5	Basalt fibre	ρ = ~2,700 kg/m^3 σ = 3,700 – 4,840 N/mm^2 ε = 89- 90 GPa	Heat insulators, fire resistance, high tensile strength, high E-modulus, and environmental friendliness are all features of this material.

Fibres used

Fibre

Fibre is a very small strand of reinforcing material that has specific properties. They are flexible with circular cross section. The aspect ratio is a useful metric for defining the fibre. The aspect ratio of a fibre is the length-to-diameter ratio. Generally aspect ratiovaries from 30 to 150. Table 1 lists the fibres that were employed in this study, as well as their properties. Even though all type of fibres can be used in concrete, not all of them can be used as efficiently and economically. Each fibre has its own set of qualities and constraints. Glass fibres, polypropylene, Nylons, asbestos, basalt fibre and polyester fibreare some of the materials that might be employed.

Steel fibre

One of the most often utilizedfibres is steel fibre. Round fibres are commonly utilized. The size ranges from 0.25 to 0.75mm in diameter. When it is employed, rusting of steel fibres is a big issue, and as a result, some strength is lost. The use of polypropylene and nylon fibres to boost impact strength has been discovered to be effective. They have a high tensile strength, but their low modulus of elasticity and increased elongation don't help with flexural strength.

Asbestos
Acid may be used with Portland cement to make asbestos, which is a mineral fibre that has proven to be the most successful of all fibres. Asbestos tensile strength ranges from 560N/mm^2 to 980N/mm^2. Asbestos cement is a composite product with significantly better flexural strength than Portland cement paste.

Admixtures inGFRC
Admixture is becoming an unavoidable part of concrete construction. Admixtures are employed to a higher extent in the GFRC. The use of plasticizer and superplasticizer is necessary since the high dose of glass fibres in the GFRC makes the combination stiff and difficult to deal with.

Plasticizer
Plasticizer is vital substances to minimize the force required to make the flow able mix and also tomake GFRC to flow more smoothly for a particular water-cement ratio.The water-cement ratio is primarily used to determine the workability and hydration of cement. We can lower the water-cement ratio to a certain extent using plasticizer.

Viscosity modifyingadmixtures
Viscosity modifying admixtures keep cement paste from washing out of porous concrete mixes. VMA enhances the binding between cement paste and aggregate, increasing the concrete's strength. It also improves the ability of cement paste to keep its ingredients in suspension by increasing the viscosity of the mixing water. It also used for reducing the possibility of bleeding and segregation.

Polymers
White latex, usually acrylic emulsions with a solid concentration of roughly 50%, is the most common polymer used. It is recommended that a dose of 5-6 percent by weight of cement be used. As flexible polymer bridges help to bind brittle mineral components, the polymers must be resistant to them.

Anti-efflorescence and water-repellentadmixtures
Due to the increased cement content of GFRC, more lime is produced during the setting process. This renders it susceptible to efflorescence, which is caused by lime. Pozzolana injected at a rate of 15-20% in cement usually removes efflorescence and reduce lime production. However, depending on the circumstance, a small amount may be added.

Properties of GlassFibre
Tensile and flexural strength, modulus of elasticity, compressive strength, impact resistance, shear strength, shrinkage and moisture movement, fire endurance, acoustical properties, thermal conductivity, permeability, and moisture absorption are all factors that influence the performance of GFRC materials. Table 2 shows the properties of Glass Fibre.

Materials Research Proceedings **23** (2022) 246-254 https://doi.org/10.21741/9781644901953-28

Table 2. Propertiesof Glass Fibre

Properties	**Value**
Flexural strength	16-32N
Tensile strength	9-19N
Elongation test	1-2.5%
Compressive strength	15-25N
Impact strength	4-12KN
Specific gravity	0.8-1
Aspect ratio	0.67
Co efficient of thermal expansion	12-20

Experimentalwork

General

The total no of 18cubes 150mm×150mm×150mm and 18 beams 100mm×100mm×500mm with the addition of 0.5% and 1% of glass fibre for analyzing compressive strength, flexural strength.

Materials Used

Cement

In the local market, 53 grades of Ordinary Portland cement are available, which have been employed in this study. The cement used here was tested for various amounts according to IS: 4031-1988 [20] and found to meet various IS: 12269-1987 [21] requirements. Cement was discovered to have a specific gravity of 3.15.

Coarseaggregate

Crushed angular granite as source material was used as coarse aggregate.60% of 20mm aggregate (specific gravity-2.78) and 40% of 12mm aggregate (specific gravity-2.80) has been used here for this investigation.

Fineaggregate

River sand was used as fine aggregate. The specific gravity and size for fine aggregate was found to be 2.64 and 4.75mm respectively.

Water

For mixing and curing, only pure, impurity-free water was used. Because the ratio of water to cementitious ingredients determines the strength and durability of concrete, only pure water free of contaminants was used for mixing and curing. Throughout the experiment, the pH of the water was kept between 6.5 and 7.5.

Fibre

Chopped glass fibre is the fibre used here and aspect ratio for glass fibre is 0.67. The specific gravity of this fibre is of 0.8 to 1.

MixProportion

Mix design as per IS: 10262-1982 [22]. All the samples were prepared using designed mix. Design for the M25 (25N\mm^2) grade concrete was done based as I.S code method. Table 3 presents the mix proportions that have been used in this experiment.

Table 3. Mix Proportions

Ingredients	Unit	Quantity
Water	Liter	185.580
Cement	kg/m^3	412.400
Fine aggregate	kg/m^3	446.600
Coarse aggregate	kg/m^3	1320.940

Mixing

All of the components were hand mixed at room temperature in the laboratory. Chopped AR glass fibre – 2–3% by weight for premixed, 4–6% for spray–up. 5 percent acrylic solids by weight of cement in acrylic polymer emulsion At 10 to 25% cement replacement, sand: cement equals roughly 1:1 pozzolana (VCAS).

Admixtures

For face coat and pourable (self–consolidating) back coat, use a superplasticizer (high–range water reducer, such as a polycarboxylate). The mixed concrete is poured into the mould in such a way that no segregation occurs. A tamping rod was used to ensure proper compaction.

CuringProcess

The cement mixture builds strength in glass fibre reinforced concrete by maintaining the correct moisture in temperature, similar to the process in conventional concrete. For the first 24 to 48 hours after moulding, all test specimens must be stored under conditions that are representative of the concrete in the structure or pavement.The specimen can be sent to the laboratory for further curing and testing after the initial cure time. A minimum of 95% humidity and a temperature of 73.4 3°F (230.94°C) are required for laboratory curing.

Discussion of Testresults

Based on the results of the experiment, it was determined that adding glass fibre to concrete reduces bleeding. Reduced bleeding increases the integrity of concrete, improves its homogeneity, and minimizes the probability of cracks forming in areas where settlement is restricted.

Compressivestrength

Compressive strength of cube with addition of 0.5% & 1% glass fibre with results are tabulated below in Table 4 and Table 5.

Table 4. Compressive Strength (Normal concrete)

No of days	Sample A (N/mm²)	Sample A ((N/mm²)	Sample A (N/mm²)	Average (N/mm²)
7	13.130	13.250	13.180	13.240
14	20.190	20.980	20.690	20.960
28	27.640	27.800	27.680	27.770

Table 5. Compressive Strength (Fibre inclusion)

% of fibre added	No of days	Compressive strength (Average)
0.5	3	15.910
	7	21.650
	28	31.010
1	3	18.870
	7	22.990
	28	32.560

Flexuralstrength

Flexural strength of beam with addition of 0.2% & 0.3% glass fibre with results are tabulated below in Table 6, and Table 7 shows the final results of compressive and flexural strength.

Table 6. Flexural strength

%of Fibre Added	No of Days	Sample A (N/Mm²)	Sample B (N/Mm²)	Sample C (N/Mm²)	Average
0.5	7	0.310	0.129	0.125	0.128
	14	0.227	0.217	0.224	0.243
	28	0.586	0.589	0.622	0.604
1	7	0.153	0.158	0.156	0.156
	14	0.269	0.227	0.272	0.282
	28	0.618	0.624	0.630	0.624

Materials Research Proceedings **23** (2022) 246-254 https://doi.org/10.21741/9781644901953-28

Table 7: Final Results

% of Fibre Added	No of Days	Compressive Strength (N/mm²)		Flexural Strength (N/mm²)	
		Without Glass Fibre	With GlassFibre	Without Glass Fibre	With Glass Fibre
0.5	7	13.240	15.910	0.090	0.128
	14	20.960	21.650	0.194	0.243
	28	27.770	31.010	0.512	0.604
1	7	13.240	18.870	0.090	0.156
	14	20.960	22.990	0.194	0.282
	28	27.770	32.560	0.512	0.624

Graph

A graph is drawn for above compressive and flexural strength in such a way that the strength is taken in differently.Fig .1 and Fig .2 shows compressive and flexural strength.

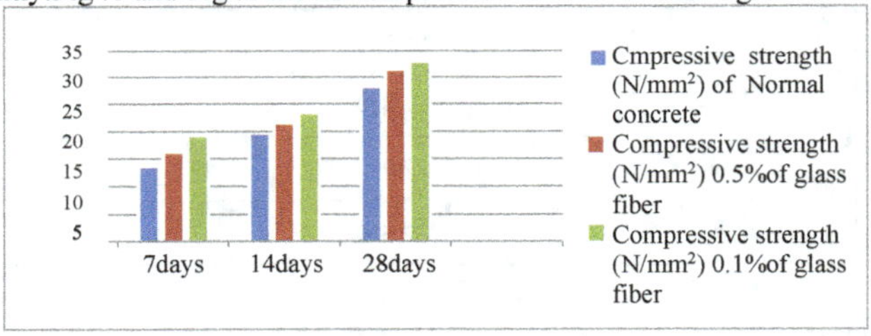

Fig 1. Compressive strength

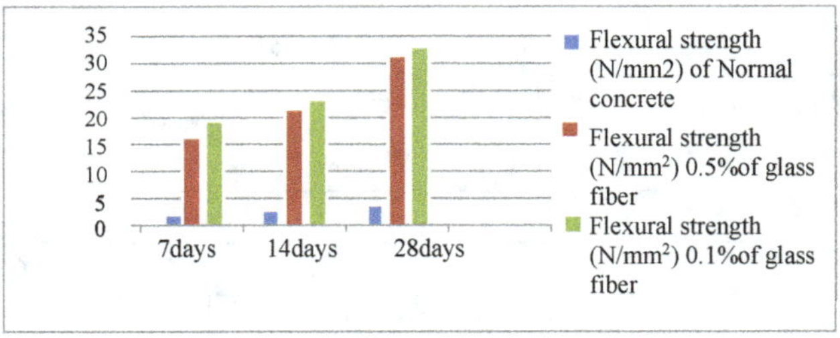

Fig 2. Flexural strength

Conclusion

In this analysis, compressive and tensile strength of concrete with glass fibre are studied and it is concluded as follows

- Adding glass fibre to concrete has made them more workable than conventional concrete.
- The compressivestrength of concrete with 0.5% of glass fibre for 7 days is 15.910N/mm², 14 days is21.650N/mm² and 28 days is 31.010N/mm².
- The compressivestrength of concrete with 0.1% of glass fibre for 7 days is 18.870N/mm²,

14 days is22.990 N/mm^2 and 28 days is32.560N/mm^2.

- The flexural strength of concrete with 0.5% of glass fibre for 7 days is 0.128N/mm^2, 14 days is0.243N/mm^2 and 28 days is 0.604 N/mm^2.
- The flexural strength of concrete with 0.1% of glass fibre for 7 days is 0.156N/mm^2, 14 days is0.282N/mm^2 and 28 days is 0.624N/mm^2.
- In addition of0.5% and 1.0% of glass fibre in concrete efficiently increases the compressive and flexural strength of concrete.
- Compressive and flexural strength increases with increasing the curing days.
- Hence concrete with 0.5% and 1.0% of glass fibre is efficient to resist the loads when compared with conventional concrete without any addition of admixtures.

References

[1] PreranaMurlidharBhagatkar, M.R.Nikhar, V.A.Kalmegh, Review on Experimental Investigation on Glass Fibre Reinforced Concrete, International Journal of Creative Research Thoughts (IJCRT), Volume 8, Issue 11 November 2020, ISSN: 2320-2882.

[2] J.D.Chaitanyakumar, G.V.S. Abhilash, P.Khasim Khan, G.Manikantasai, V.Taraka ram, Experimental Studies on Glass Fibre Concrete, American Journal of Engineering Research (AJER) e-ISSN: 2320-0847 p-ISSN : 2320-0936 Volume-5, Issue-5, pp-100-104, 2016.

[3] Ganesh Babu.K and PavanKumar. D, Behaviorof Glass Fibre Reinforced Cement Composites. ICFRC International Conference on Fibre Composites, High Performance Concretes and Smart materials(2004).8-10 Jan.,Chennai.

[4] GhobarahA, Ghorbel M N, Chidiac S E, Upgrading tensional resistance of reinforcement concrete beam using fibre- reinforcement polymer, Jof composite forconstruction,(2002)257-263. https://doi.org/10.1061/(ASCE)1090-0268(2002)6:4(257)

[5] Griffiths R. An assessment of the properties and degradation behaviour of glass- fibre-reinforcement polyester polymer concrete. Composites Science and Technology .pp(2000.)2747-2753. https://doi.org/10.1016/S0266-3538(00)00147-0

[6] Huang C S, YehY K,Liu G Y, Hu H T , Tsai K C, Weng Y T, Wang S H &Wu M H, Axial load behaviour of stiffened concrete- filled steel c columns, J of structural Engineering,128(9)(2002)1222-1230. https://doi.org/10.1061/(ASCE)0733-9445(2002)128:9(1222)

[7] Maanman A.E. 1985. Fibre Reinforcement for concretes, Concrete International: Design and Construction.7(3):21-25.

[8] Pravin A & Wang W, Behaviour of FRP jacketed concrete columns under eccentric loading, J of composite for construction,4(3) (2000)146-152. https://doi.org/10.1061/(ASCE)1090-0268(2001)5:3(146)

[9] ParvizSoroushian..Fibre-Type Effect on the Performance of Steel Fibre Reinforced Concrete. ACI Materials Journal. March- Apri(l1991).pp.129-134.

[10] Rochettee P &LabossiereP, Axial testing of rectangular column model confined with composites, J of composites for construction, 4(3)(2000)129-136. https://doi.org/10.1061/(ASCE)1090-0268(2000)4:3(129)

[11] SrinivasaRaoand SeshadriSekhar T.2005. Strength and Durability Properties of glass Fibre Reinforcement Concrete. Proceedings of the International Conference on Recent Advances in Concrete and Construction Technology.(December 7-9), SRMIST, Chennai,India.pp.43-50.

[12] Li, V.; Yang, E.; Li, MField Demonstration of Durable Link Slabs for JointlessBridge Decks Based on Strain-HardeningCementitious Composites - Phase 3: Shrinkage Control (PDF), Michigan Department of Transportation. (28 January2008),

[13] ACI 544.3R-93: Guide for Specifying, Proportioning, Mixing, Placing, and Finishing Steel FibreReinforced Concrete, American Concrete Institute,(1998)

[14] Wang, Y.; Wu, HC.;Li, V. "Concrete Reinforcement with Recycled Fibres". Journal of Materials in Civil Engineering. (November 2000). https://doi.org/10.1061/(ASCE)0899-1561(2000)12:4(314)

[15] Ochia, T.; Okubob, S.; Fukuib, K. "Development of recycled PET fibre and its application as concrete-reinforcing fibre". Cement and Concrete Composites. 29(6): (July 2007).448-455. https://doi.org/10.1016/j.cemconcomp.2007.02.002

[16] Toxicological Profile for Synthetic Vitreous Fibres (U.S. Department of Health and Human Services, Public Health Services, Agency for Toxic Substances and Disease Registry),(September) 2004, p.17.

[17] T. W. Hesterberga, G. Chaseb, C. Axtenc, 1, W. C. Millera, R. P. Musselmand, O. Kamstrupe, J. Hadleyf, C. Morscheidtg, D. M. Bernsteinh and P. Thevenaz "Biopersistenceof Synthetic Vitreous Fibres and Amosite Asbestos in the Rat Lung Following Inhalation". Toxicology and Applied Pharmacology. 151(2) (2 August 1998).:262-275. https://doi.org/10.1006/taap.1998.8472

[18] Ilschner, B; et al. Composite Materials. Ullmann'sEncyclopediaof Industrial Chemistry. Weinheim, Germany: Wiley-VCH Verlag GmbH &Co.KGaA(2000).

[19] Erhard, Gunter. Designing with Plastics. Trans. Martin Thompson. Munich: Hanser Publishers,(2006). https://doi.org/10.3139/9783446412828.fm

[20] "Reflective Cracking Treated with GlasGrid" (PDF). CTIP News. 2010. Archived from the original (PDF) on (26 February 2013. Retrieved 1September2013.)

[21] Electric Field-Assisted Orientation of Short Phosphate Glass Fibres on Stainless Steel for Biomedical Applications QiangChen, Jiajia Jing, HongfeiQi, Ifty Ahmed, HaiouYang, Xianhu Liu, T. L. Lu, and Aldo R. Boccaccini ACS Applied Materials &Interfaces, 11529-11538https://doi.org/10.1021/acsami.8b01378

[22] IS 4031-6 (1988): Methods of physical tests for hydraulic cement, Part 6: Determination of compressive strength of hydraulic cement (other than masonry cement) [CED 2: Cement and Concrete],Bureau of Indian Standards, New Delhi.

[23] IS 12269 (1987): 53 grade ordinary Portland cement [CED 2: Cement and Concrete], Bureau of Indian Standards, New Delhi.

[24] IS: 10262 - 2019Indian Standard recommended guidelines for Concrete Mix Design, Bureau of Indian Standards, New Delhi.

Sustainable Materials and Smart Practices - NCSMSP-2021 Materials Research Forum LLC
Materials Research Proceedings **23** (2022) 255-268 https://doi.org/10.21741/9781644901953-29

Structural Health Monitoring of Infrastructures using Sensors as Smart Materials– Review and Perspective

A. Dinesh[1,a,*], A. Karthick[2,b], S. Ramkumar[3,c]

[1] Department of Civil Engineering, Sri Ramakrishna Engineering College, Coimbatore - 641022, India

[2]Department of Civil Engineering, K. Ramakrishnan college of technology, Trichy, India

[3]Department of Civil Engineering, M. Kumarasamy college of engineering, Karur, India

[a]dinesh.a@srec.ac.in; dinesh.a.civil@gmail.com, [b]karthicka.civil@krct.ac.in, [c]ramkumars.civil@mkce.ac.in

Keywords: Structural Health Monitoring, Sensors, Self-Sensing, Cement Composite, Strain, Corrosion

Abstract. Smart materials with sensors can monitor the structure's performance under external loading circumstances. They may also monitor internal deformations or damages caused by environmental factors such as temperature, humidity, etc. As a result, the sensors are linked to structural health monitoring to create automated systems for structural monitoring, inspection, and damage identification. The formulation of this review article was prompted by a growing interest in structural health monitoring and the need to ensure structure safety to detect problems early and avert collapse. The structure, measurement methods, and potential of sensors such as fiber optic, piezoelectric, corrosion, ceramic, and self-sensing cement composite utilized in the health monitoring of concrete structures are discussed in this review paper. This review also includes a brief and comparative analysis of various sensors, as well as the optimal number and location of sensors. The study exposed that choosing a suitable sensor is critical for accurate sensing and long-term structure monitoring. The sensor can detect physical (stress, strain) and chemical (corrosion) variables that affect the structure's endurance. Despite significant advances in damage monitoring approaches utilizing sensors, the study suggests that efficient sensor deployment remains problematic. The review revealed that the type of parameter to be monitored (stress, strain, humidity, etc.) and the structural and climatic conditions in which the sensor will be used determine the sensor's selection. As a result, a self-sensing cement composite based on carbon nanofiber (CNF) has been developed, which has good durability and compatibility with concrete structures. However, increasing the amount of CNF lowers the composite's compressive and flexural strength due to particle agglomeration. As a result, the review covers several sensors used in structural health monitoring with their measurements, applications, benefits, and limitations.

Introduction

Concrete is the most extensively used building material because of its manufacturing, availability, and strength simplicity. Concrete is widely used to construct buildings, pavements, dams, and other structures. Despite its many advantages, concrete lacks significant strength and durability, which leads to damages [1,2]. To withstand external stresses, several advancements have been made in structural design [3,4] and bracings [5]. Despite significant advancements in structures' ability to endure external loads, assuring the structure's safety remains difficult. As a result, numerous advancements in concrete have been made by altering the materials [6–9] in the same way bricks [10] and soil [8,11] have been changed. Conductive materials like carbon nanofiber [12], carbon nanotube [13], carbon fiber [14,15], and carbon black [16,17] are used to enhance the strength and durability of concrete. As a result, various concrete types, such as high-strength concrete [18,19], lightweight concrete, and ultra-high-performance concrete, have been developed. Cracking, deformation, and deterioration are inevitable, despite several material advances to enhance concrete's strength capabilities. As a result, concrete performance is monitored during the structure's maintenance and service life.

Structural health monitoring is a technique for monitoring the state of a structure's safety during its service life [20,21]. They are used to monitor the condition of bridges, buildings, and pipelines that are failing due to earthquakes, aging, and other risks. Accelerometers [22], acoustic emission sensors [23], ceramic sensors [24], corrosion sensors [25], electrochemical sensors [26], fiber optic sensors [27], shape memory alloys [28], and strain gauges [29] are among the sensors used to analyze and monitor the structure. These sensors are used for preventative maintenance, which helps to improve living conditions while preventing financial losses. The two kinds of sensors accessible are wired and wireless sensors. Wired sensors, such as fiber optic, piezoelectric, and ultrasonic sensors, are expensive and have a high chance of intrusion failure. The wireless sensor, which includes an acoustic sensor, can detect flaws across vast distances with the assistance of humans. The requirements for structural health monitoring, which include dependability, robustness, and cost-effectiveness, limit the use of sensors.

The study's main objective is to explore the need for health monitoring of structural components and the method for monitoring structural health using sensors. The following part compares the use of a wireless strain sensor, a ceramic sensor, and a corrosion sensor in structural health monitoring.

Sensors for Structural Health Monitoring

Structural health monitoring is becoming more significant because of the time-consuming and costly manual evaluations performed at predetermined intervals. Sensor and sensing technology has advanced to the point that they can now be integrated into concrete structures, overcoming these constraints. Many studies have been conducted in the field of structural health monitoring utilizing a variety of sensors, including fiber optic sensors [30], accelerometers [31], acoustic emission sensors [23], electrochemical sensors [32], wireless sensors [33], and self-sensing cement composite [34]. The choice of sensor is determined by the structural and environmental conditions in which the sensor will be used and the type of parameter to be monitored (stress, strain, humidity, etc.). [35]. Fig. 1 depicts the numerous sensors, structural specifications, and potential in measuring external parameters.

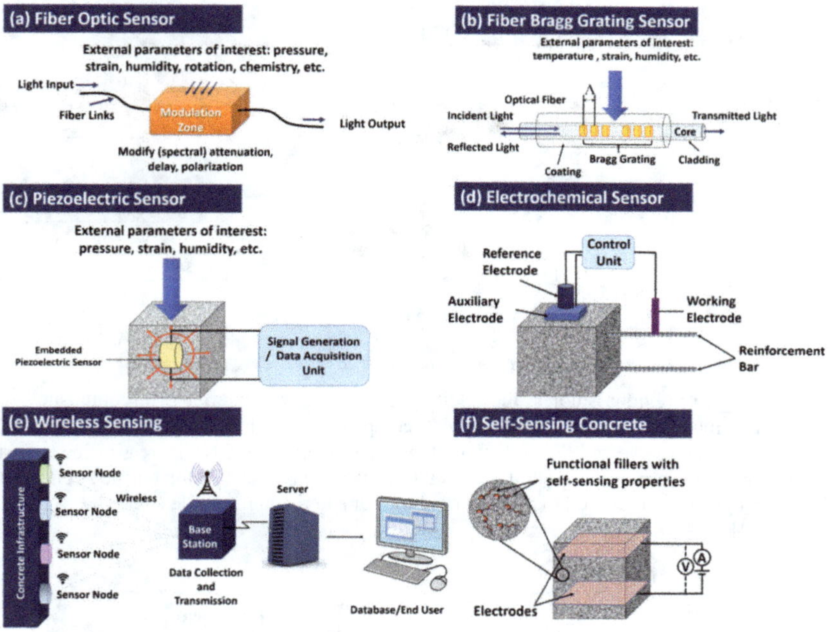

Fig. 1. Demonstration of sensors used in the health monitoring of concrete structures [36]

Self-sensing cement composite

Because of the evolution of interdisciplinary research and the application of nanomaterials in concrete, many new studies have been developed in cement composite. Researchers are quite interested in the field of intelligent concrete with self-sensing capabilities [13]. Nanomaterials such as steel fiber [37], graphite powder [38], carbon black [16], carbon nanofiber [12], carbon nanotube [13], and others are used to create self-sensing capacity in cement composites. The composite will develop piezoresistive properties by incorporating these nanoparticles, transforming them from a traditional cement composite into a self-sensing cement composite. When nanoparticles are included in a composite, they form a conductive network that modifies the composite's resistivity/conductivity as stress/strain varies [13]. Before being placed in the compression testing equipment, the composite is connected to a power supply (LCR meter) and a multimeter to analyze compressive strength and conductivity simultaneously, as shown in Fig. 2 [12]. When the load is applied, the area changes along with the voltage, which may be measured using a multimeter. The fractional change in resistance and sensitivity of the composite is evaluated, as illustrated in Fig. 2. The property that changes resistivity with change in stress is piezoresistivity of the composite, and many researchers have investigated piezoresistivity of the composite with various implanted conductive materials [12,13,39,40]. Scanning electron microscopy (SEM), and X-ray diffraction spectroscopy (XRD) examinations are used to understand better the internal changes in the composite caused by the embedment of conductive components.

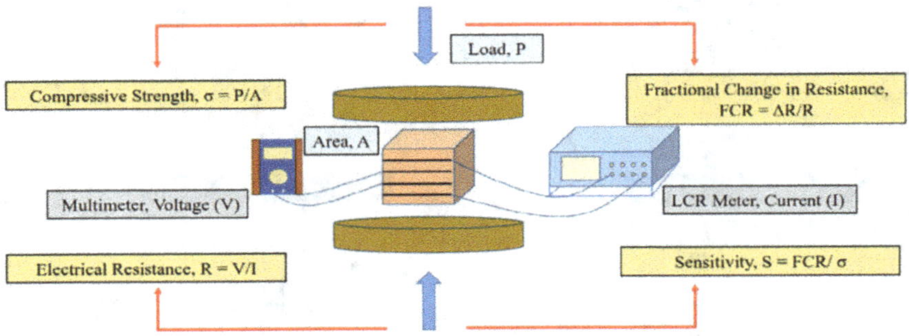

Fig. 2. Schematic diagram depicts methods for analyzing CBS qualities [12]

According to previous studies, the addition of conductive material to a composite greatly impacts the mechanical strength of cement composite. The inclusion of CNF reduces the composite's compressive and flexural strength as the proportion of CNF increases due to particle agglomeration, as seen in fig. 3 [41,42]. When CNF is increased to 1%, the tensile strength of the cement composite is improved to 50 MPa (approx.) due to the high tensile strength (=3800 MPa) of CNF [43].

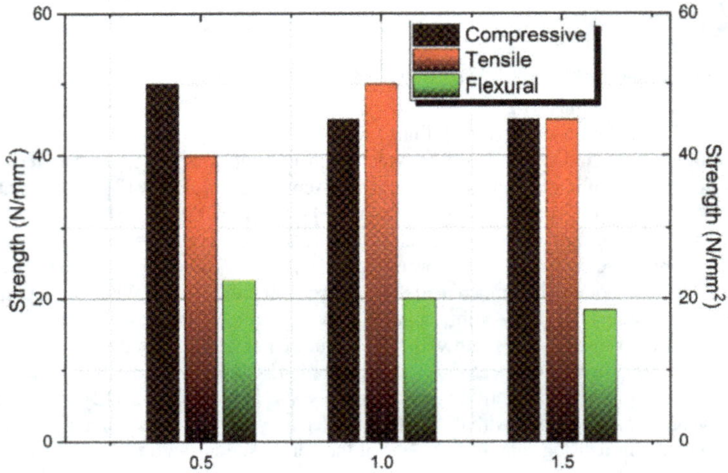

Fig. 3. Compressive, split tensile, and flexural properties of cement composites containing 0.5 %, 1%, and 2% CNF [12]

The electrical conductivity of a composite comprising carbon nanotubes with different electrical characteristics is examined in an analytical study [44]. The effect of aspect ratio (L/D ratio) and its influence on percolation threshold and electrical conductivity is investigated. They identified that the larger the aspect ratio smaller would be the percolation threshold in the composite. Suppose the quantity of conductive materials incorporated in the composite is less than the threshold volume fraction. In such instances, they have the slightest effect on the electrical conductivity of the composite, as indicated by the spherical model's lower bound. CNT is used to develop self-sensing cement composites in similar analytical models [45,46]. Multiwalled carbon nanotubes (MWCNT) of electrical conductivity > 1250 s/cm are used to prepare epoxy matrix composite for strain monitoring in pavements [47]. They identified that epoxy composite developed with CNT is a suitable cement-based sensor with a gauge factor of 26.04, which is far higher than the gauge factor of conventional metal strain sensors with a gauge factor of 2. Those CNT embedded composites can also measure strain ranging within 1000 $\mu\varepsilon$. Similarly, many investigations are done on self-sensing cement composite using CNT [48–51]. Few researches are also focused on the hybrid combination of nanomaterials, including CF/CNT and CF/CB [52,53]. These findings imply that hybrid CNT combinations, rather than individual CNTs, can efficiently improve the flexural strength and electrical conductivity of cement mortars [52]

Ultrasonic sensor

Ultrasonic sensors are used to examine or identify propagating cracks in concrete [6]. The stressed state is used to assess deformation to spot cracks quickly. Fig. 4 (a) shows the DIC values at a tensile stress of 3.2 N/mm^2. Under this load, the correlation factor between the signals sent and received by the embedded ultrasonic sensor is less than 0.8. The dashed lines in Fig 4 (a) show the placement of embedded sensors. The growing surface fracture impacts the signal even if it is not in the wave's direct path. As a result, Fig. 4 (a) depicts the absence of crack formation. The cross-correlation factor fell further in fig. 4 (b) as the tensile stress increased to 3.4 N/mm^2. In the wave path, a crack with a width of 160 μm and a length of 20 mm can be seen. The ultrasonic wave is depicted in Fig. 4 (c) with a tensile strength of 3.8 N/mm^2. The crack development and surface deformation show the spread of the fracture into the direct path of the wave traveling between the implanted sensors. As a result, the study finds that ultrasonic sensors can detect crack propagation in a structure before visible cracks appear on the surface of the concrete.

An ultrasonic sensor based on an autoregressive model (AR) is used to detect structural changes in concrete [7]. This study uses a model to employ an ultrasonic sensor to localize operational changes in massive concrete buildings. When a static force is applied to the top of the sensor, the residual error increases to its maximum value (t=75 minutes). The nonlinearities in the operational change in the structure determine the amplitude of the parameters. The damage-sensitive feature in this system is AR residual error. The level of change in the structure is precisely proportional to the changes in AR parameters. However, the association level appears to vanish if the damage is present with operational and environmental fluctuations. In this study, ultrasonic sensors perform both static and dynamic load tests on reinforced concrete structures. The localized operational changes in the reinforced concrete structures are quantified using an autoregressive model.

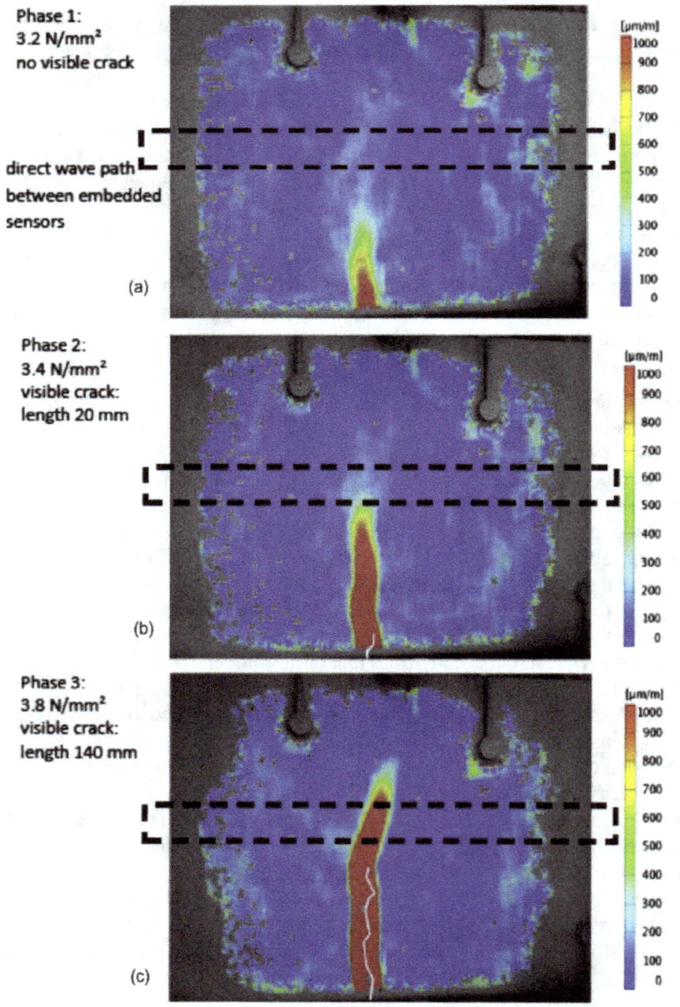

Fig. 4. Tensile stress-causing surface deformation (a) 3.2 N/mm² (b) 3.4 N/mm² and (c) 3.8 N/mm² respectively.

The influence of composite age on ultrasonic wave transmission and reflection coefficients in the concrete-steel-concrete contact is investigated [8]. The structure in question is a multilayer structure with two elastic and isotropic layers with different physical properties. In sand concrete (= 40.65 MPa), conventional concrete (= 45.89 MPa), and high-performance concrete (= 81.02 MPa), the wave propagation velocity is determined. Aging generates a substantial variability in the longitudinal coefficient (TL) in high-performance concrete, as shown in Fig. 5 (a), and sand concrete, as shown in Fig. 5 (c). The evolution of the transverse transmission (TT) coefficient is depicted in Figures 5 (b) and (c). The TT coefficient drops initially, then climbs dramatically before stabilizing at a specific angle. When comparing the effect of concrete age on sand concrete from the fifteenth day to high-performance concrete and regular concrete, the sand concrete exhibits a more substantial influence. Fig. 5 (d) depicts the behavior of ultrasonic waves that characterize the interface's longitudinal reflection coefficient RL. The curve has a falling and then climbing shape. There is a noticeable drop in the coefficient RL during the measuring interval of 0 to 15 days, which is minimal compared to conventional and high-performance concrete. According to both computational and experimental research, the age of concrete and the inner concrete structure directly impact the acoustic properties and angle of incidence.

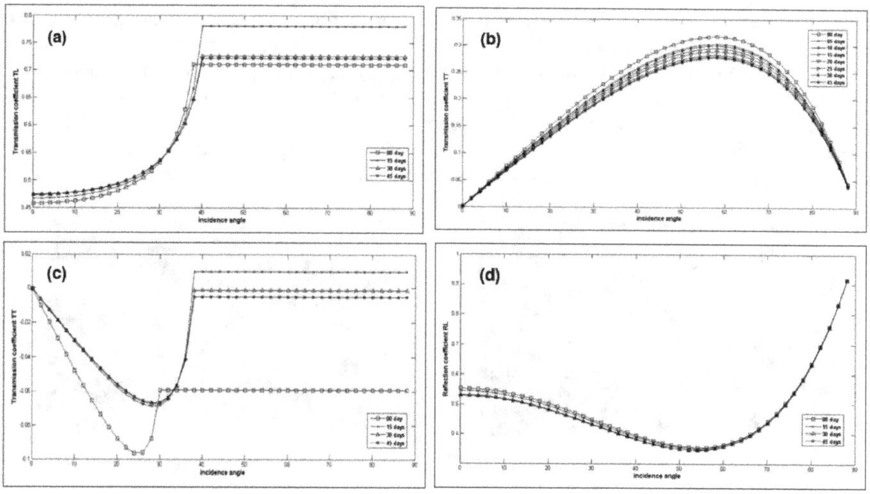

Fig. 5. (a) Variation of the TL coefficient with concrete age (high-performance concrete–steel contact) (b) Variation of the TT coefficient with concrete age (steel–high-performance concrete interface) (c) Variation of the TT coefficient with concrete age (sand concrete–steel contact) and (d) Variation of the RL coefficient with concrete age (steel–sand concrete interface).

Strain gauge

Due to the time-consuming and challenging sample surface preparation and strain gauge adhesion, traditional strain gauges such as electric resistance strain gauges, demec mechanical strain gauges, and compressometers have limited applications in SHM. As a result, the automatic digital image correlation (ADIC) optical strain gauge was developed as practical and cost-effective monitoring of concrete parameters [9]. The ADIC significantly broadens the convergence range and can accurately measure the concrete strain caused by an autonomous searching approach. After obtaining the strain, the concrete stress-strain curve is drawn using the load to determine the elastic modulus of the concrete. According to the research, the produced optic strain gauge has good accuracy, convenience, and reliability when compared to the electronic strain gauge, compressometer, and demec mechanical strain gauge. For C20, C40, and C60 concrete cubes and

Materials Research Forum LLC
https://doi.org/10.21741/9781644901953-29

cylinders, the optic strain gauge has a standard deviation of 0.37, 0.44, and 0.58, which is lower than the electronic strain gauge (0.42, 0.34, and 0.94), compressometer (0.64, 0.43, and 0.71), and Demec mechanical strain gauge (1.96, 1.41, and 1.33).

Strain gauges are also embedded in concrete members to measure the prestressing force in concrete beams and slabs [10]. Vibrating beam strain gauges (VBSG) are similar to vibrating wire strain gauges in that they work on the same principles. The VBSG test frame setup is shown in Fig. 6. The steel bar is extended at both ends and tightened with a circular plate, as shown in Fig. 6 (a), to guarantee that the strain in the VBSG steel bar matches the strain in the concrete at the steel bar level. The strain gauge is positioned in the center of the steel bar, as shown in Fig. 6 (b). Beeswax is applied to the strain gauge to make it water-resistant, as shown in Fig. 6 (c). A PVC pipe with water sealant is installed above the strain gauge, as shown in Fig. 6 (d), to prevent water from entering into the pipes from the cement paste. The current strain in the VBSG can be calculated by comparing the frequency data of the stretched part with static strain measured with electrical strain gauges. The average inaccuracy in calculating the prestressing force is 1.9% from frequency measurements and 1.4% from static strain data. Because both strain and frequency yield precise results, the prestressing force in the structure may be calculated by anyone. The error associated with using a vibrating wire strain gauge is 1.4% to 1.9%, which is slightly higher than the error associated with using a VBSG (around 0.5%). VBSG, on the other hand, can analyze the state of prestressed concrete buildings with pinpoint accuracy.

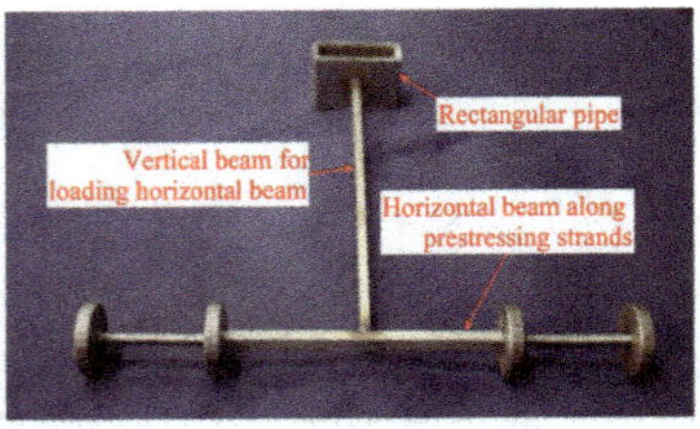

(a) Frame structure of vibrating beam strain gauge.

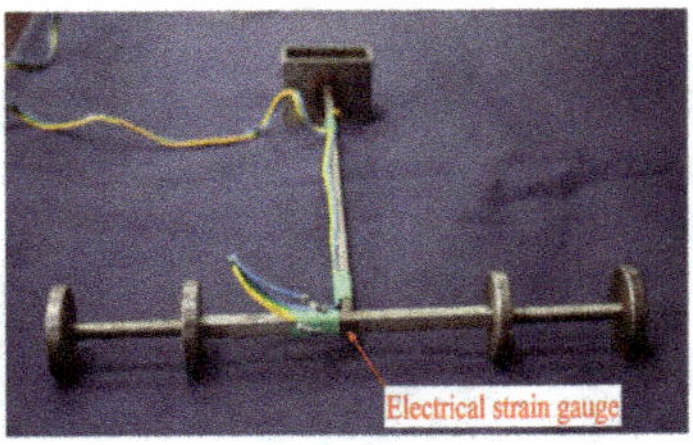

(b) Strain gauge location in VBSG.

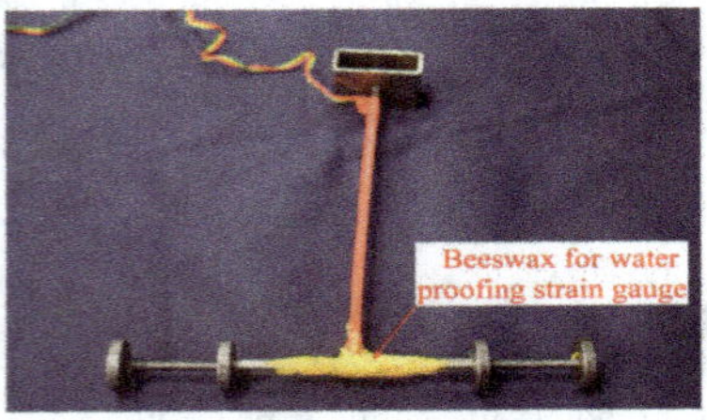

(c) Water proofing the strain gauge.

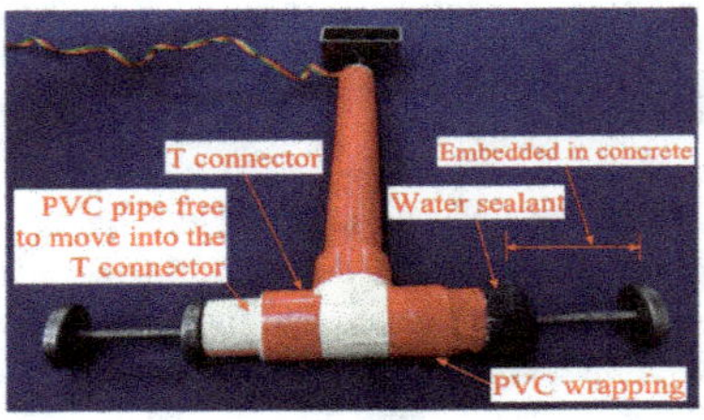

(d) Clear cover for movement of VBSG using PVC pipes.

Fig. 6. Vibrating beam strain gauge [54]

At the start of corrosion, the concrete behavior is studied using strain gauges, SEM, and EDS [11]. They were particularly interested in the penetration of rust agents into pores and the measurement of stresses and strains in the concrete around the rebars when corrosion first began. An analytical model is also proposed to calculate the crack width at the time of cracking. The method is applied to two concretes with different porosity networks and corrosion current densities to analyze the reinforcement corrosion process and discover cracks in the surrounding concrete. SEM and EDS are used to investigate the corrosion advancement and rust layer growth on prismatic samples with a base of 50 X 50mm^2 that includes rebar, as shown in Fig. 7. The researchers discovered that placing the strain gauge close to the rebar provided more information on the circumferential strain in the concrete.

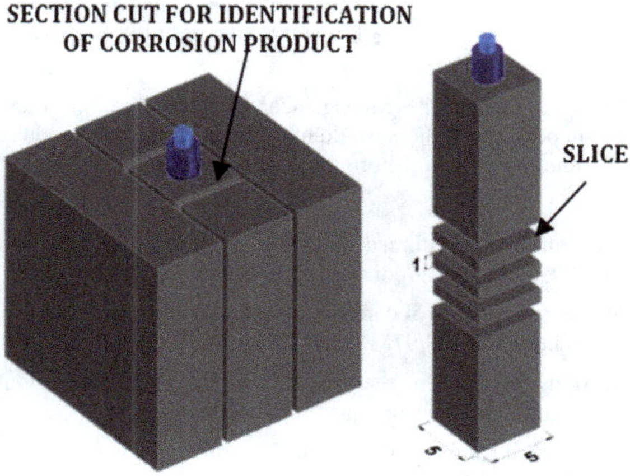

Fig. 7. The cutting technique for the specimens for SEM observations [55]

Conclusions

Thus, the article reviews the different sensors, working, and monitoring vital infrastructure status. Monitoring criteria such as sensor sensing capabilities in detecting mechanical (displacement, strain, vibration, fractures) and durability (corrosion, chloride ion concentration) features of the concrete structure are also carefully examined. Through this review, the following conclusions are revealed

1. The type of parameter to be monitored (stress, strain, humidity, etc.) and the structural and climatic circumstances in which the sensor will be utilized decide the selection of the sensor.
2. In self-sensing cement composite, because of particle agglomeration, increasing the proportion of CNF reduces the compressive and flexural strength of the composite.
3. Researches identified that the higher the aspect ratio, the lower the percolation threshold of the composite. They found that an epoxy composite containing CNT is an excellent cement-based sensor, with a gauge factor of 26.04, which is much greater than the gauge factor of typical metal strain sensors, which is 2.
4. In conventional sensors, the research finds that ultrasonic sensors may detect fracture propagation in a structure before visible cracks appear on the surface of the concrete. The

localized operational changes in the reinforced concrete structures are quantified using an ultrasonic sensor with an autoregressive model.

5. According to both computational and experimental research, the age of concrete and the inner concrete structure directly influence the acoustic properties and angle of incidence.
6. The standard deviation of the optic strain gauge is 0.37, 0.44, and 0.58, which is lower than the electronic strain gauge (0.42, 0.34, and 0.94), compressometer (0.64, 0.43, and 0.71), and Demec mechanical strain gauge (0.42, 0.34, and 0.94). (1.96, 1.41, and 1.33). VBSG, on the other hand, can analyze the state of prestressed concrete buildings with good accuracy.
7. The researchers discovered that placing the strain gauge close to the rebar provided more information on the circumferential strain in the concrete.

Thus, the study demonstrates that self-sensing cement composites, ultrasonic sensors, and strain gauges have promising concrete structure health monitoring applications.

Acknowledgments
The authors would like to thank the Management, Sri Ramakrishna Engineering College, Coimbatore, for their support. The authors would like to express their gratitude to the management of PSG College of Technology in Coimbatore for their assistance in carrying out this study.

References

[1] R. Capozucca, Damage to reinforced concrete due to reinforcement corrosion, Constr. Build. Mater. 9 (1995) 295–303. https://doi.org/10.1016/0950-0618(95)00033-C.

[2] K. Kovler, V. Chernov, Types of damage in concrete structures, Woodhead Publishing Limited, 2009. https://doi.org/10.1533/9781845697037.1.32.

[3] A. Professor, Analytical study on the behaviour of cold formed steel double channel beam sections, n.d. www.ripublication.com/ijaer.htm.

[4] A. Dinesh, R. Prasanth Kumar, S.R. Abijith, Experimental investigation on bubble deck concrete using plastic waste, 2020. https://doi.org/10.1007/978-981-15-0990-2_14.

[5] P. Eswaramoorthi, P. Magudeaswaran, A. Dinesh, PUSHOVER ANALYSIS OF STEEL FRAME, n.d.

[6] A. Professor, Study on Strength properties of High performance concrete, n.d. www.ripublication.com/ijaer.htm.

[7] K. Sathish Raja, Dinesh. A, Study on Self Compacting Concrete – A Review, Int. J. Eng. Res. V5 (2016). https://doi.org/10.17577/ijertv5is020349.

[8] A. Professor, Stabilization of soil by using solid waste-A Review, © 2017 IJEDR |. 5 (2017). www.ijedr.org.

[9] A. Dinesh, S. Harini, J.P. Jasmine, J. Jincy, J. Shagufta, International journal of engineering sciences & research technology experimental study of blast furnace slag concrete, Int. J. Eng. Sci. Res. Technol. 6 (2017) 42–50. https://doi.org/10.5281/zenodo.345692.

[10] D. Sellakutty, A. Professor, UTILISATION OF WASTE PLASTIC IN MANUFACTURING OF BRICKS AND PAVER BLOCKS, 2016. https://www.researchgate.net/publication/303273892.

[11] A. Dinesh, S. Indhumathi, Moorthi Pichumani, Performance assessemnt of copper slag and sawdust ash in stabilization of black cotton soil, 2021.https://doi.org/10.1007/978-981-33-

4590-4_21.

[12] A. Dinesh, B. Abirami, G. Moulica, Materials Today : Proceedings Carbon nanofiber embedded cement composites : Properties and promises as sensor – A review, Mater. Today Proc. (2020). https://doi.org/10.1016/j.matpr.2020.10.526.

[13] A. Dinesh, S.T. Sudharsan, S. Haribala, Self-sensing cement-based sensor with carbon nanotube: Fabrication and properties – A review, Mater. Today Proc. (2021). https://doi.org/10.1016/j.matpr.2021.02.722.

[14] S. Wen, D.D.L. Chung, Strain-Sensing Characteristics of Carbon Fiber-Reinforced Cement, (2006) 2006.

[15] W. Wang, S. Wu, H. Dai, Fatigue behavior and life prediction of carbon fiber reinforced concrete under cyclic flexural loading, Mater. Sci. Eng. A. 434 (2006) 347–351. https://doi.org/10.1016/j.msea.2006.07.080.

[16] A.O. Monteiro, P.B. Cachim, P.M.F.J. Costa, Electrical Properties of Cement-based Composites Containing Carbon Black Particles, Mater. Today Proc. 2 (2015) 193–199. https://doi.org/10.1016/j.matpr.2015.04.021.

[17] Y. Ding, G. Liu, A. Hussain, F. Pacheco-Torgal, Y. Zhang, Effect of steel fiber and carbon black on the self-sensing ability of concrete cracks under bending, Constr. Build. Mater. 207 (2019) 630–639. https://doi.org/10.1016/j.conbuildmat.2019.02.160.

[18] M. Aboubakar, Y. Sato, K. Niitani, Experimental investigation of monotonic behavior and stress-strain models of AE and non-AE high strength concrete with BFS fine aggregates under freezing and thawing, Constr. Build. Mater. 249 (2020) 118679. https://doi.org/10.1016/j.conbuildmat.2020.118679.

[19] J. Zhu, J. Chen, T. Chan, Analytical model for circular high strength concrete filled steel tubes under compression, Eng. Struct. 244 (2021) 112720. https://doi.org/10.1016/j.engstruct.2021.112720.

[20] P.C. Chang, A. Flatau, S.C. Liu, Review paper: Health monitoring of civil infrastructure, Struct. Heal. Monit. 2 (2003) 257–267. https://doi.org/10.1177/1475921703036169.

[21] J. Paek, K. Chintalapudi, R. Govindan, J. Caffrey, S. Masri, A wireless sensor network for structural health monitoring: Performance and experience, Second IEEE Work. Embed. Networked Sensors, EmNetS-II. 2005 (2005) 1–10. https://doi.org/10.1109/EMNETS.2005.1469093.

[22] K. Chilamkuri, V. Kone, Monitoring of varadhi road bridge using accelerometer sensor, Mater. Today Proc. 33 (2020) 367–371. https://doi.org/10.1016/j.matpr.2020.04.159.

[23] N. Beheshtizadeh, A. Mostafapour, S. Davoodi, Three point bending test of glass/epoxy composite health monitoring by acoustic emission, Alexandria Eng. J. 58 (2019) 567–578. https://doi.org/10.1016/j.aej.2019.03.006.

[24] V. Agarwal, A. Shelke, B.S. Ahluwalia, F. Melandsø, T. Kundu, A. Habib, Damage localization in piezo-ceramic using ultrasonic waves excited by dual point contact excitation and detection scheme, Ultrasonics. 108 (2020) 106113. https://doi.org/10.1016/j.ultras.2020.106113.

[25] L. Xie, X. Zhu, Z. Liu, X. Liu, T. Wang, J. Xing, A rebar corrosion sensor embedded in concrete based on surface acoustic wave, Meas. J. Int. Meas. Confed. 165 (2020) 108118. https://doi.org/10.1016/j.measurement.2020.108118.

[26] G.T. Parthiban, T. Parthiban, R. Ravi, V. Saraswathy, N. Palaniswamy, V. Sivan, Cathodic protection of steel in concrete using magnesium alloy anode, Corros. Sci. 50 (2008) 3329–3335. https://doi.org/10.1016/j.corsci.2008.08.040.

[27] A. Barrias, J.R. Casas, S. Villalba, Fatigue performance of distributed optical fiber sensors in reinforced concrete elements, Constr. Build. Mater. 218 (2019) 214–223. https://doi.org/10.1016/j.conbuildmat.2019.05.072.

[28] S. Abouali, M. Shahverdi, M. Ghassemieh, M. Motavalli, Nonlinear simulation of reinforced concrete beams retrofitted by near-surface mounted iron-based shape memory alloys, Eng. Struct. 187 (2019) 133–148. https://doi.org/10.1016/j.engstruct.2019.02.060.

[29] S.Z. Chen, G. Wu, D.C. Feng, Damage detection of highway bridges based on long-gauge strain response under stochastic traffic flow, Mech. Syst. Signal Process. 127 (2019) 551–572. https://doi.org/10.1016/j.ymssp.2019.03.022.

[30] K.T. Lau, C.C. Chan, L.M. Zhou, W. Jin, Strain monitoring in composite-strengthened concrete structures using optical fibre sensors, Compos. Part B Eng. 32 (2001) 33–45. https://doi.org/10.1016/S1359-8368(00)00044-5.

[31] S. Mahjoubi, R. Barhemat, Y. Bao, Optimal placement of triaxial accelerometers using hypotrochoid spiral optimization algorithm for automated monitoring of high-rise buildings, Autom. Constr. 118 (2020) 103273. https://doi.org/10.1016/j.autcon.2020.103273.

[32] A.S.E. Belaidi, B. Benabed, H. Soualhi, Physical and mechanical properties of concrete repair materials in dry and hot-dry environment, J. Adhes. Sci. Technol. 29 (2015) 543–554. https://doi.org/10.1080/01694243.2014.998001.

[33] Z. Herrasti, I. Gabilondo, J. Berganzo, I. Val, F. Martínez, Wireless Sensor Nodes for Acceleration, Strain and Temperature Measurements, Procedia Eng. 168 (2016) 1659–1662. https://doi.org/10.1016/j.proeng.2016.11.484.

[34] M.S. Konsta-Gdoutos, C.A. Aza, Self sensing carbon nanotube (CNT) and nanofiber (CNF) cementitious composites for real time damage assessment in smart structures, Cem. Concr. Compos. 53 (2014) 162–169. https://doi.org/10.1016/j.cemconcomp.2014.07.003.

[35] C. Maraveas, T. Bartzanas, Sensors for structural health monitoring of agricultural structures, Sensors (Switzerland). 21 (2021) 1–32. https://doi.org/10.3390/s21010314.

[36] S. Taheri, A review on five key sensors for monitoring of concrete structures, Constr. Build. Mater. 204 (2019) 492–509. https://doi.org/10.1016/j.conbuildmat.2019.01.172.

[37] F. Köksal, F. Altun, I. Yiğit, Y. Şahin, Combined effect of silica fume and steel fiber on the mechanical properties of high strength concretes, Constr. Build. Mater. 22 (2008) 1874–1880. https://doi.org/10.1016/j.conbuildmat.2007.04.017.

[38] S.H. Liu, M.Q. Wu, M.J. Rao, L.H. Li, H.L. Xiao, Preparation, Properties, and Microstructure of Graphite Powder-Containing Conductive Concrete, Strength Mater. 51 (2019) 76–84. https://doi.org/10.1007/s11223-019-00052-1.

[39] X. Dongyu, C. Xin, H. Shifeng, Investigation of inorganic fillers on properties of 2-2 connectivity cement/polymer based piezoelectric composites, Constr. Build. Mater. 94 (2015) 678–683. https://doi.org/10.1016/j.conbuildmat.2015.07.090.

[40] H. Liu, K. Liu, Z. Lan, D. Zhang, Mechanical and Electrical Characteristics of Graphite Tailing Concrete, Adv. Mater. Sci. Eng. 2018 (2018). https://doi.org/10.1155/2018/9297628.

[41] H. Siad, M. Lachemi, M. Sahmaran, H.A. Mesbah, K.A. Hossain, Advanced engineered cementitious composites with combined self-sensing and self-healing functionalities, Constr. Build. Mater. 176 (2018) 313–322. https://doi.org/10.1016/j.conbuildmat.2018.05.026.

[42] L. Gao, T.W. Chou, E.T. Thostenson, Z. Zhang, A comparative study of damage sensing in fiber composites using uniformly and non-uniformly dispersed carbon nanotubes, Carbon N. Y. 48 (2010) 3788–3794. https://doi.org/10.1016/j.carbon.2010.06.041.

[43] X. Fan, D. Fang, M. Sun, Z. Li, Piezoresistivity of carbon fiber graphite cement-based composites with CCCW, J. Wuhan Univ. Technol. Mater. Sci. Ed. 26 (2011) 339–343. https://doi.org/10.1007/s11595-011-0226-0.

[44] Y. Fang, L.Y. Li, S.H. Jang, Calculation of electrical conductivity of self-sensing carbon nanotube composites, Compos. Part B Eng. 199 (2020) 108314. https://doi.org/10.1016/j.compositesb.2020.108314.

[45] Y. Kuronuma, T. Takeda, Y. Shindo, F. Narita, Z. Wei, Electrical resistance-based strain sensing in carbon nanotube/polymer composites under tension: Analytical modeling and experiments, Compos. Sci. Technol. 72 (2012) 1678–1682. https://doi.org/10.1016/j.compscitech.2012.07.001.

[46] X.F. Sánchez-Romate, J. Artigas, A. Jiménez-Suárez, M. Sánchez, A. Güemes, A. Ureña, Critical parameters of carbon nanotube reinforced composites for structural health monitoring applications: Empirical results versus theoretical predictions, Compos. Sci. Technol. 171 (2019) 44–53. https://doi.org/10.1016/j.compscitech.2018.12.010.

[47] X. Xin, M. Liang, Z. Yao, L. Su, J. Zhang, P. Li, C. Sun, H. Jiang, Self-sensing behavior and mechanical properties of carbon nanotubes/epoxy resin composite for asphalt pavement strain monitoring, Constr. Build. Mater. 257 (2020) 119404. https://doi.org/10.1016/j.conbuildmat.2020.119404.

[48] A. Naqi, N. Abbas, N. Zahra, A. Hussain, S.Q. Shabbir, Effect of multi-walled carbon nanotubes (MWCNTs) on the strength development of cementitious materials, J. Mater. Res. Technol. 8 (2019) 1203–1211. https://doi.org/10.1016/j.jmrt.2018.09.006.

[49] T. Yin, J. Xu, Y. Wang, L. Liu, Increasing self-sensing capability of carbon nanotubes cement-based materials by simultaneous addition of Ni nanofibers with low content, Constr. Build. Mater. 254 (2020) 119306. https://doi.org/10.1016/j.conbuildmat.2020.119306.

[50] X.F. Sánchez-Romate, J. Martín, A. Jiménez-Suárez, S.G. Prolongo, A. Ureña, Mechanical and strain sensing properties of carbon nanotube reinforced epoxy/poly(caprolactone) blends, Polymer (Guildf). 190 (2020). https://doi.org/10.1016/j.polymer.2020.122236.

[51] Alamusi, N. Hu, H. Fukunaga, S. Atobe, Y. Liu, J. Li, Piezoresistive strain sensors made from carbon nanotubes based polymer nanocomposites, Sensors. 11 (2011) 10691–10723. https://doi.org/10.3390/s111110691.

[52] B. Han, L. Zhang, S. Sun, X. Yu, X. Dong, T. Wu, J. Ou, Electrostatic self-assembled carbon nanotube/nano carbon black composite fillers reinforced cement-based materials with multifunctionality, Compos. Part A Appl. Sci. Manuf. 79 (2015) 103–115. https://doi.org/10.1016/j.compositesa.2015.09.016.

[53] J.M. Park, P.G. Kim, J.H. Jang, Z. Wang, J.W. Kim, W. Il Lee, J.G. Park, K. Lawrence

DeVries, Self-sensing and dispersive evaluation of single carbon fiber/carbon nanotube (CNT)-epoxy composites using electro-micromechanical technique and nondestructive acoustic emission, Compos. Part B Eng. 39 (2008) 1170–1182. https://doi.org/10.1016/j.compositesb.2008.03.004.

[54] S. Biswal, A. Ramaswamy, Measurement of existing prestressing force in concrete structures through an embedded vibrating beam strain gauge, Meas. J. Int. Meas. Confed. 83 (2016) 10–19. https://doi.org/10.1016/j.measurement.2016.01.031.

[55] A.M. Bazán, J.C. Gálvez, E. Reyes, D. Galé-Lamuela, Study of the rust penetration and circumferential stresses in reinforced concrete at early stages of an accelerated corrosion test by means of combined SEM, EDS and strain gauges, Constr. Build. Mater. 184 (2018) 655–667. https://doi.org/10.1016/j.conbuildmat.2018.06.195.

Evaluation of Corrosion Inhibition on Rebar using Pomegranate Peel Extract as Organic Inhibitor

B. Jeyanth[1,a*], R. Ashwathi [1,b], V. Prabhakaran [1,c], A. Karthick [1,d]

[1]Department of Civil Engineering, Bannari Amman Institute of Technology, Sathyamangalam, Erode, Tamilnadu, India

[a*]jeyanthbaskaran5890@gmail.com, [b]ashwathir@bitsathy.ac.in, [c]prabhakaran@bitsathy.ac.in, [d]karthickkncetcivil@gmail.com

Keywords: Pomegranate Peel Extract (PPE), Inhibitor, Carbonation, Chlorination, Corrosion

Abstract. Concrete is well known for its binding nature and Compressive strength. However, the tensile strength of the concrete is not so sufficient. Hence to improve the tensile nature and to enhance the flexibility in concrete steel rod is introduced as reinforcement. The rebar provides sufficient flexural strength to the concrete. Thus, the reinforced concrete is used in construction for its ease of access and long life. The major issue and problem faced by the rebar in concrete is corrosion which leads to reduction in the strength and age of the structure. The main objective of this research is to provide the protection for steel rebar against corrosion which improves the life of the structure. Many researchers worked on this to find a solution for corrosion. Corrosion inhibitors is one among the best method which helps us to protect the steel against corrosion. The pomegranate fruit which is highly antioxidant can fight against corrosion. An extract was obtained from pomegranate peel powder which can act as an organic inhibitor is used in this research. The inhibitor opted for this study is cheap, economic and renewable since it is organic. In this project the steel was coated with PPE extract and tested in different environment conditions and the efficiency of corrosion control was obtained with impressed voltage current. The result was compared with conventional uncoated rebar. The result proved that PPE is a good natural inhibitor for the steel, and the efficiency of the inhibitor is good enough to resist the corrosion in the rebar.

Introduction

Corrosion is one of the major distresses to the concrete which leads to the reduction in strength of the concrete. corroded rod will lose its bonding with the concrete, and leads to the structural failure. Thus, the solution for the corrosion in rebar is essential to increase the durability of the concrete and to enhance the life of the structure [1]. The pores in the concrete allows the atmospheric moisture and oxygen to it which is the main cause for corrosion [2]. Oher causes for corrosion are low concrete cover, presence of the salts like calcium, chlorine in the water which form carbonate and chloride. The concrete with pores in it allows the O_2 to pass through it and finally allow it to reach the surface of the steel and forms the ferrous oxide. Further the ferrous oxide compound forms the rust which ends in corrosion. In addition to oxygen the other agents that leads to corrosion are carbon-di-oxide and chloride. The chloride in the concrete when reacts with the steel forms (Fe-Cl) a complex called ferric chloride, this process called chlorination will end in the failure of the rebar. The root cause for chlorination process is thepermeability. It starts with process between the carbonic acid and the hydroxides present in the concrete [3]. The pH level of the concrete will increase rapidly to 12-13 due to the presence slaked lime. Chloride will act as the catalyst for the corrosion if it has sufficient concentration at the rebar surface to interrupt down the passive layer over the rebar. The chloride ions will react with the Fe2+ when it dissolved in the solution and form an iron – chloride Complex. Further hydrolyses of the complex with the metal hydroxide will liberates chloride ions which will attack the steel [4]. Many researchers were working from a decade to find a better solution to resist the corrosion like steel coatings, cathodic protection, Electrical resistivity method, and inhibitors. Corrosion can restrict and the steel can be protected by using inhibitor is a well-known practice which has been used since 1960 [5]. Inhibitors are classified as organic and chemical inhibitors, organic type of inhibitors are preferred for their non-toxic nature, easy availability and low cost. Pomegranate is the fruit which has a rich anti-oxidant property, thus in this research an inhibitor is developed from the pomegranate peel. The extract from the pomegranate peel is developed as a natural organic inhibitor. The performance of inhibitor was studied and obtained with the help of the test like weight loss measurement, impressed voltage current test, and the pH. The results of a Pomegranate Peel Extract (PPE) coated rebar were compared with conventional steel.

Materials

Pomegranate. The anti-oxidant Nature of the Pomegranate consist of the chemicals which comprises of the structure that has effective corrosion inhibition performance [6]. The waste peel of the pomegranate fruit was collected from the juice shops. The peel was oven dried and powdered. The peel powder was sieved in 75 microns. The powder that passes the 75-micron sieve was taken for the extract preparation.

Fig 1. Pomegranate Peel Powder

Pomegranate Peel Extract Preparation. PPE is prepared with 20 grams of sieved powder. 20 gram of powder was dissolved in 100 ml of distilled water. The solution was boiled for 100 degrees Celsius and the solution was cooled for 24 hours at the room temperature. The solution has been filtered many times and the boiling process should be carried out till the solution changes to the form of extract. Finally, the obtained extract will be collected [7].

Test on specimens

Carbonation test. The two major phenomena in concrete which leads the path to corrosion are carbonation & chlorination. The protective layer of the concrete cover will be destroyed if carbonation happens in the concrete. in this an atmospheric CO_2 will reacts with the calcium hydroxide that present in the concrete. The strength and the quality of the concrete will depend upon the rate of the carbonation. Carbon attack will first destroy the passive film of concrete cover and it would break down the steel concrete bond, when it reaches the rebar. Thus, it leads to rust which finally ends in corrosion if sufficient moisture is present within the concrete. The carbonation process was explained with expression.

$$CO_2 + H_2O \rightarrow H_2CO_3 \quad (1)$$

$$H_2CO_3 + Ca(OH)_2 \rightarrow CaCO_3 + 2H_2O \quad (2)$$

The depth of the carbonation in the concrete can be easily obtained with the help of the simple chemical test. The test can be carried out by spraying phenolphthalein indicator on the freshly broken concrete [2].

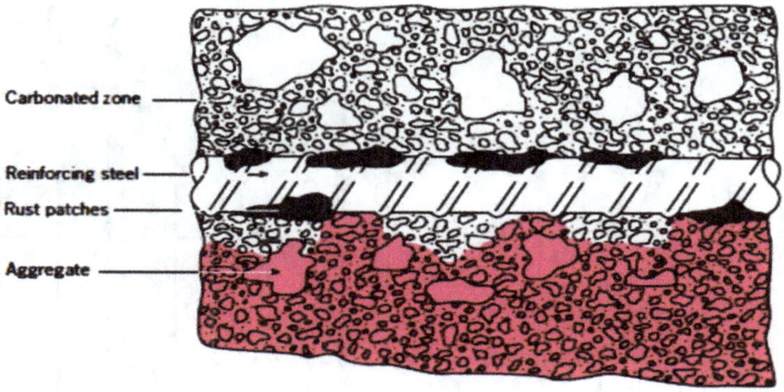

Fig.2. Carbonation test

pH & Chloride content test. The pH will increase rapidly with the presence of chloride. The test was conducted in a simple way by dipping the pH meter in the solution. The solution is prepared by dissolving the powder obtained by abrading the corroded steel in corroded region in the distilled water. Now the pH of the solution was observed by immersing the pH meter. If the value obtained in higher than 11.5, it shows that the solution is acidic and the rod is said to be corrosive.

Impressed voltage test. The corrosion is the long run process and it is not so easy to determine the corrosion rate of the steel in normal laboratory test and thus accelerated corrosion method is preferred to find the corrosion efficiency of the rebar. The impressed voltage current test used in this study is one of the accelerated corrosion methods. For this test PPE coated and uncoated conventional rebars of 200 mm length and 12 mm diameter were taken as the specimens. The rods were reinforced in the concrete such that 50 mm of the specimen to be exposed above the concrete cubes. The concrete with the rebar is made to be partially immersed in 3 % Na-Cl solution. The potential of 10 volt has been applied on the specimen as shown in fig 4. This set up has been kept without disturbed till the rod were attaining a weight loss of 20 %. The variation in time to weight loss for both rods has been observed and noted [8].

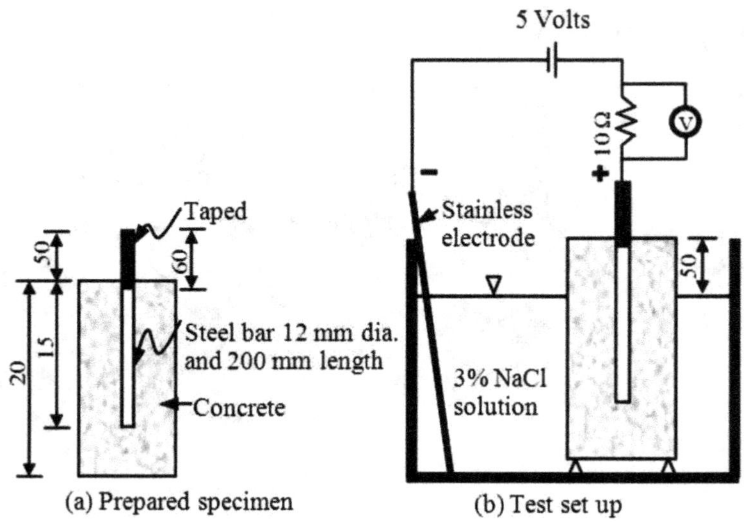

Fig 3. Impressed Voltage current Test

Weight loss measurement test. The corrosion rate of the concrete can be evaluated by the reduction the mass of the steel. In this test the steel samples opted for the test were tested in the two different acid solution HCl and H_2SO_4. The samples were cleaned dried and immersed in the 5% HCl and H_2SO_4 concentrations in 40 mg/litre at room temperature. The initial weights of the samples were noted as W1 before immersed into the solution. The samples were kept undisturbed for different exposure of time, 7, 14 and 28 days. The specimens were taken cleaned and weighed as W2. The loss of weight in the samples were obtained and estimated by using the following equation.

$$W = \frac{(w2-w1)}{A} \tag{3}$$

Were,

W – Weight loss in mg/cm^2

W1 & W2 are the weight of the samples before and after corrosion.

The inhibitor efficiency can be estimated by using the following expression

$$\eta = \frac{w-wp}{w} \times 100\,\% \tag{4}$$

Were,

η % = efficiency in percentage of the PPE Inhibitor,

W = weight loss in mg/cm^2 without pomegranate peel and

Wp = weight loss in mg/cm^2 with pomegranate peel added for the same exposure time for W.

Results & Discussion

Carbonation and pH test

Table1. Carbonation test

Specimen	Result	pH
Conventional steel	Dark pink colour	12.7
PPE coated steel	No colour change	6.2

Table. 1 shows the carbonated prone zone in the concrete and the pH value of the concrete. The phenolphthalein test explains that dark pink colour change observed in the conventional concrete shows that it was carbonated zone. The acidic nature of the solution which contains the dissolved constituents of the corroded samples. The pH value of the conventional steel with 12.7 clearly describes that it is prone to corrosion.

Impressed Voltage test

Table. 2. Accelerated corrosion Test

Specimen	Result
Conventional steel	20 %Weight loss in 8 hours
PPE coated steel	20%Weight loss within 11hours

Table 2 shows the weight loss comparison among the PPE coated and conventional steel sample. The constant potential was applied by the voltmeter to the samples and the weight loss of the samples were observed periodically. It seems that 20 % loss in weight of the samples occurs in 8 hours for the coated rebars whereas for the PPE coated rebars it took 11 hours. Thus, from the test it is observed clearly that the PPE as an inhibitor holds good against the corrosion.

Weight loss measurement test

Table.3. Weight loss measurement test

η in HCL		η inH_2SO_4	
Exposure in days	**η %**	**Exposure in days**	**η %**
7	81.20	7	83
14	88.21	14	86.23
21	94.23	21	95.21
28	95.10	28	94.60

Table 3 explains the inhibitor efficiency of the developed organic inhibitor. The Performance of PPE was studied and observed in two different acidic media HCl and H_2SO_4. Inhibition Efficiency of steel with PPE against corrosion was estimated. The table shows that PPE acts as a best inhibitor against corrosion.

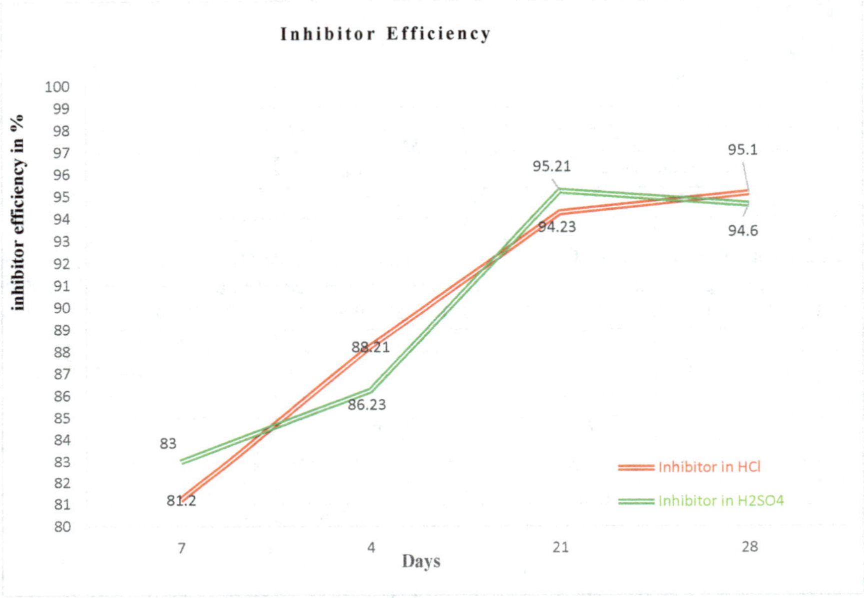

Fig 6. η in HCl Vs η inH_2SO_4

The comparative study on efficiency of the inhibitor in HCl and H_2SO_4 is shown in the graph figure. 6. The figure describes that the efficiency rate of the inhibitor in acidic medium is strong enough to fight against the corrosion.

Conclusion

The observation on carbonation visualize that PPE coated rebars was not carbonated since there is no colour modification in the concrete contains the PPE coated rebars. The pink colour on the concrete with conventional rebars confirms the carbonation on it. The pH value of the concrete confirms the ingression of chlorine which leads to the chloride attack and ends in the corrosion. It was clearly explained with pH value greater than 11.5. The corrosion rate was studied by accelerated corrosion method with the 20 % weight loss of the samples. Inhibitor efficiency rate was evaluated and it proves that the Organic Inhibitor PPE hold good against corrosion the efficiency of the inhibitor reaches 95% as the time of exposure increased. Thus, PPE is suitable to act as an organic inhibitor.

Acknowledgements

Fig.4 the diagrammatic illustration of the carbonation test was taken from BRE part 1

References

[1] Rakesh Verma & Amit Singhal, A Review on Study of CPCC, FBEC & CRSD as Reinforcing Bar, International Journal of Engineering Research & Technology (IJERT), 2017; ISSN: 2278-0181

[2] Corrosion Testing of Concrete Structures, David Simpson, Peter Robery, John Broomfield and Simon Bladon, January 2016

[3] D.B.McDonald, Corrosion Protection for Concrete Structures in Marine Environments, DOI, 10.1061/41190(422)8; August 2011. https://doi.org/10.1061/41190(422)8

[4] Pandian Bothi Raja, Seyedmojtaba Ghoreishiamiri and Mohammad Ismail, Natural Corrosion inhibitors for Steel Reinforcement in Concrete - A Review, Surface Review and Letters, Vol. 22, No. 3 (2015) 1550040. https://doi.org/10.1142/S0218625X15500407

[5] Han-Seung Lee, Hwa-Sung Ryu, Won-Jun Park and Mohamed A. Ismail, Comparative Study on Corrosion Protection of Reinforcing Steel by Using Amino Alcohol and Lithium Nitrite Inhibitors, Materials 2015; *8*, 251-269. https://doi.org/10.3390/ma8010251

[6] Habib Ashassi-Sorkhabi, ShojaMirzaee, Taghi Rostamikia, and Robabeh Bagheri, Pomegranate (Punica granatum) Peel Extract as a Green Corrosion Inhibitor for Mild Steel in Hydrochloric Acid Solution, Hindawi Publishing Corporation International Journal of Corrosion Volume 2015, Article ID 197587. https://doi.org/10.1155/2015/197587

[7] Mustafa Sabri Cheyad, Characterization of Red Apple and Pomegranate Peels Extracts as Corrosion Inhibitors for α-Brass in Acidic Media,

[8] Suresh Bhalgamiya, Govind Tivadi, Mehul Jethva, Techniques for Accelerated Corrosion Test of Steel Concrete for Determine Durability, International Research Journal of Engineering and Technology (IRJET), Volume: 05, Issue: 04, Apr-2018

[9] Sahib Mohammed Mahdi, Study the Pomegranate's Peel Powder as a Natural Inhibitor for Mild Steel Corrosion, International Journal of Materials Chemistry and Physics Vol. 1, No. 1, 2015, pp. 74-81. https://doi.org/10.1155/2015/197587

Comprehensive Review on the Influence of Natural Materials in Soil Stabilization

S.C. Boobalan[1,a,*], M. Dhanabharathi[1,b], S. Dineshkumar[1,c] and M. Gokuldas[1,d]

[1]Civil Engineering, Sri Krishna College of Engineering and Technology, Coimbatore, Tamil Nadu, India

[a]boobalansc@skcet.ac.in, [b]18eucv009@skcet.ac.in, [c]18eucv011@skcet.ac.in, [d]18eucv014@skcet.ac.in

Keywords: Rice Husk Ash, Egg Shell Powder (ESP), Tamarind Kernel Powder (TKP), Lime, Soil Stabilization, Sandy Soils

Abstract. This comprehensive review insists on the impacts of various natural materials in the conventional ground improvement techniques. This review paper focuses on the utilization of natural materials such as Eggshell powder, Rice husk ash, Wheat husk ask, Tamarind Kernel Powder, Jaggery, Chebula, Lime and coir fibers as soil stabilizers. The properties of Unconfined Compressive Strength (UCC), California Bearing Ratio (CBR), index and consolidation characteristics had been compared with existing conventional strength of soils. Out of the materials used for soil stabilizing agents, Rice Husk Ash, Eggshell powder and Tamarind Kernel Powder showed better ground improvement properties. The eggshell powder with 5% optimum replacement by weight of dry soil sample showed an improvement in strength. The properties of the soil sample achieved the improvement in strength with rice husk ash of 6-8% optimum level. TKP of 10% was added in the soil sample showed the soil liquid limit increase to 117% from 67%. TKP of 2% and 8% were added in the soil sample showed the shrinkage limit decrease to 15.4% and 11.4% respectively. Meanwhile, TKP of 8% in the soil sample resulted in a decrease of dry density to 14 kN/m^2 from 17.1 kN/m^2. Similarly, improved strength for achieved from the following combinations of jaggery and eggshell powder, lime and Chebula.

Introduction

In this advanced age of science and technology, man has made rapid strides in all fields since the ancient ages. Right from the dark ages till the present information age, man has grown by leaps and bounds and has tried to come around all kinds of problems encountered by him. From an engineering point of view, especially from a geotechnical perspective, land use for development work has brought to the fore the problem of acute land shortage. This has led to the reclamation of unusable land for development activities.

Ground improvement technology has been the driving force that has brought about this revolution in the reclamation of unusable land, which has led to a sudden spurt in developmental activities. However, all human technologies, since times immemorial have proven to have their drawbacks in most cases. Similarly, this technology, though beneficial in several ways, had environmental impacts. With rapid urbanization and industrialization, land reclamation works have been done with ground improvement technologies. With the rising use of some of these techniques such as cement grouting, lime columns etc., the impact on the environment came to be slowly recognized. In this review study, an effort to see the insights and options of utilizing natural materials like Rice Husk Ash, Jaggery, TKP, ESP, etc. for the beneficial improvement of problematic clays [5].

Need for this study

Ground improvement may be well-defined as the process of refining the soil engineering properties and creating firmer ones. The need for ground improvement arises under the following circumstances: i. Current industrial and urban development insisting demand land renovation. ii. Consumption of unbalanced and environmentally affected ground. iii. With the use of ground improvement techniques, the safe disposal of waste materials is easy. Even though better results were achieved in soil stabilization by cement grouting, lime columns, chemical stabilization, but their effect on the environment is still available. Hence in this review discussion, an effort has been made for the improvement of problematic clays by use of using natural materials. The current comprehensive review is also envisaged due to the environmental impacts of certain popular methods of ground improvement which have resulted in the need for environment-friendly methods to be adopted [14].

Objective of this study

The primary objective of this comprehensive review is to attain improvement of poor soils with negligible influence on the environment, to attain sustainable growth in utilizing unfeasible land without spoiling the environment. Their main objective can be disaggregated into the following stages. i. To document and evaluate the environmental consequences of recent ground improvement methods. ii. To adopt natural materials in part or whole in ground improvement techniques, after validating their positive results in the improvement of problematic clays [14].

Natural Materials with Soil

Improvement of soil by the recent methods have proved to be of great use in improving the soil and providing the scope for use of otherwise unusable land but on the other hand, have had an impact on the quality of the environment. In recent years more emphasis has gone towards protecting the environment by reducing activities of pollution and using alternative methods that minimize the impact on the environment. In the field of ground improvement, this can be achieved slowly by phasing out polluting methods of ground improvement and use of natural materials that have minimal impact on the environment, in this case, soil and groundwater. Several natural materials can be used in the process of natural materials.

Tamarind Kernel Powder

The performance of Tamarind Kernel Powder (TKP) as an additive agent has been identified recently by various researchers. TKP can be utilized as hydrocolloids in cement and lime to improve their performance. The hydrocolloids function as a thickener, water retention binder and agent, lubricant, suspending agent and friction-reducing agent [5].

Egg Shell Powder

The main aim of using Egg Shell Powder (ESP) is to make the soil well stabilized at a low cost. ESP, as the name suggests, is a powdered form of the outer shell of eggs. The use of ESP in soil stabilisation has been documented to give moderate results in improving problematic soils. The performance of ESP has been documented as lesser compared to that of albumin powder. The use of ESP as a replacement in lime stabilization was studied to find out the optimal percentage of lime for stabilization followed by the optimum quantum of ESP that can replace lime in the stabilisation. The results indicated that the performance of the ESP replaced lime stabilisation gave results marginally lesser than pure lime stabilisation. In this context, ESP can be used as a replacement without much loss in strength, thereby resulting in a decrease in the cost of stabilisation. Albumin on the other hand has been long used in some Middle East countries since ancient times as a waterproofing agent against humidity under paints. Furthermore, it was used in

Materials Research Proceedings 23 (2022) 276-283 https://doi.org/10.21741/9781644901953-31

Japan as mortar in building masonry – arch bridges, some of which are still in service now. Albumin on one hand has been documented to be good material in its ability to improve the strength of soil, durability and artificially manufactured albumin is costly. However, the prospect of using albumin in small quantities along with ESP is very good [1, 16, 17].

Jaggery and Chebula

Jaggery and Chebula have long been utilized in infrastructure development since so many years ago in the southern part of India. There have been several instances of its use in the arena of Civil Engineering. Brick Jelly Concrete, made with broken bricks, lime, kadukkai (Chebula), and jaggery and so on, was preferred in South India for roofing works since the olden days for thermal insulation of roofs (The Hindu, 2004). One of the oldest churches in India, St. Mary's Church situated in Fort St. George, Chennai underwent major restoration works. Moisture content availability in the walls was the major issue over there, so de-plastering the solution was suggested by many of the researchers. The de-plastering work was done by using the jaggery and lime along with kadukkai (Chebula), and for smooth marble, finishing was done by using lime along with egg yolks (The Hindu, 2005). The facelift to the historical fort in Thirumayam, about 20 km from Pudukkottai on the Pudukkottai–Karaikudi highway, is in the third phase of the renovation of the structure. In the restoration work, a paste of conventional materials such as lime, mortar, jaggery and Chebula ('kadukkai') was used to strengthen the walls (The Hindu, 2008). Work on the renovation of an ancient temple (Sri Uthamadhaneeswarar Temple at Keezhadhaniyam) taking to the utilization of natural materials as renovation materials in the 9th century AD by ASI. The actual condition of the temple was in a dilapidated form. Every pillar and stone were cautiously detached and arranged in the same order for its re-alignment. Conventional natural materials such as jaggery, gall-nut, ('kadukkai') and lime were utilized (The Hindu, 2008). Kadukkai (Chebula) has been long used as a mortar in infrastructure development. A mortar made by using kadukkai, lime grind and egg-white is utilized as a binding agent for bricks in the architecture of Chettinad. Sri Brihadeeswara temple at Gangaikondachozhapuram, Perambalur district built by King Rajendra Chola, has been beneath the purview of the Archeological Survey of India since 1946. This temple has been completely renovated by using a mix of earth, panai vellam, kadukkai, lime and yellow oxide [5].

Rice Husk Ash

Sample of rice husk ash utilized for testing was obtained from Kamakshi rice mills in the location of Chennai, Tamil Nadu. Normally ash is disposed of in the empty barren land which would be causing environmental issues. It is obtained as a byproduct during the manufacturing process of puffed rice, which would be containing silicate and iron oxide in larger amounts. In India, nearly 60,000 tons of rice husk has been obtained after harvesting paddy fields. Its physical properties showed similar to the naturally available sand and also chemically stable ones. Its friction angle and high angularity property exhibit excellent load-carrying capacity and efficient stability [2, 3, 6, 7, 12]. The ingredients of Rice Husk Ash are listed in Table 1, general soil index properties were given in Table 2 and Table 3 shows the soil sample particle size distribution [3].

Table 1. Rice Husk Ash (RHA) Composition

Constituent	Composition (%)
SiO_2	67.3
Al_2O_3	4.9
Fe_2O_3	0.95
CaO	1.36
MgO	1.81
Loss on Ignition (LOI)	17.78

Table 2. Index Properties of Soil.

Soil Property Description	Value (%)
Liquid Limit	67
Plastic Limit	28
Shrinkage Limit	10.2
Free Swell Index	60
Max. Dry Density	1.71 g / cc
OMC	17.5
UCS	58.5 kN/m^2

Table 3. Results of Particle Size Distribution.

Fractions	Particle Size (mm)	Column B (*t*)
Fine Gravel	20 - 4.75	1.4
Coarse	4.75 – 2	5.1
Medium	2 – 0.425	9.3
Fine	0.425 – 0.075	10.6
Silt	0.075 – 0.002	7.0
Clay	<0.002	66.5

Natural Materials with Sand

As with the expansive soils, silty sandy soils are problematic if they are in contact with water. For sandy soil stabilization techniques, pozzolanic materials, cement, lime and nanoparticles are used. Lime and pozzolan used in their study were taken from the locally available. For understanding the minerals available in pozzolan and lime, an x-ray diffraction test was performed [22]. Nowadays during the stabilization of soil in addition to the strength, stability also plays a major concern. In order the improve that, fibres had been introduced to increase the ductile behaviour during stabilization techniques. Prosopis juliflora fibres were used in combination with cement, to improve the geotechnical properties of sandy soils [23].

Methodology

Preparation of Sample

The testing sample would be prepared as per the requirement of Indian Standards. The soil sample used for testing is to be sieved properly according to the testing procedures. The soil was weighed for the testing process. The alternative natural materials were also sieved and weighed as per the replacement percentages. In the dry state, selected natural materials and soil was mixed thoroughly for preparing the sample. TKP materials obtained from the market was directly utilized along with soil, eggshells were thoroughly washed and dried before utilizing as a replacement. Jaggery and lime available in the market were utilized [11, 18].

Materials

According to the plasticity chart, the soil was plotted in the region of high compressible clays (CH). The various natural materials used for the study include a combination of Chebula and jaggery, Egg Shell Powder, Tamarind Kernel Powder and rice husk ash. Rice husk ash with proportions from 5, 10, 20, 30, 40, 50 and 80% mixed samples were prepared [1, 8, 9]. TKP is a composite mixture encompassing Proteins (18-20%), galactose xyloglucan polysaccharide (55-65%), Lipids (6- 10%) and firm minor constituents as fibres and sugar etc. Egg Shell Powder (ESP), as the name suggests is a powdered form of the outer shell of eggs. The main constituents of eggshells are mineralized Calcium carbonate (95%) and the rest consisting of Calcium phosphate and Magnesium sulphate. Jaggery is a traditional unrefined sugar used throughout South and Southeast Asia [18].

Testing of Samples

Index Tests

The various index tests that were conducted to learn the performance of the natural materials in soil stabilization were the free swell test, shrinkage limit and liquid limit tests. All the index tests were accomplished as per the procedure stipulated in the Indian Standards Code [1, 4].

Compaction Test

The compaction characteristics are generally obtained by performing the standard proctor compaction test. However, in this study, a miniature mould was adopted to perform the compaction test to simplify the procedure. The mould was a scaled-down version of the proctor mould and the compaction energy obtained using the hammer made for this mould was almost the same. The number of blows per layer was modified to achieve similar compaction energy, considering all the other parameters as well. The mould had a dimension of 90 mm height and 38 mm diameter. The weight of the hammer was 1.5 kg. The number of blows per layer was seven. The compaction energy delivered by this hammer was 0.605 J as against the energy delivered by a standard proctor hammer of 0.595 J. The procedure adopted is the same as that of the standard proctor test. About 300g of soil was taken passing through a 2.36 mm sieve and retained in a 710-micron sieve. The test was started with around 10% water content. The mould was weighed before the test. The inside of the mould was oiled before each trial. The soil was divided into three portions and compacted by three layers into the mould. Each layer was compacted by giving three blows of the 1.5 kg hammer. Before filling in the second layer, the surface was scarred using a straight edge, for proper bonding of the layers. The collar was placed before filling in the third layer and sufficient soil was filled in such that the compacted height of the soil was above the height of the mould. After the third layer was compacted the collar was removed and the excess soil was struck off using a straight edge. The weight of the mould plus soil was taken after trimming the excess soil. The soil sample was then extruded and a representative sample from the three layers was placed in the oven for water content determination. The test was then repeated by increasing the water content in the soil in steps of 2%. The trials were continued till the weight of the compacted soil reached a peak value and then decreased. At least two trials after the peak value were performed. The readings were noted after each trial and the maximum dry density and optimum water content were determined by plotting a graph between dry density and water content [3, 19].

Materials Research Proceedings **23** (2022) 276-283 https://doi.org/10.21741/9781644901953-31

Unconfined Compression Strength Test

The soil sample strength, as well as treated soil, was studied by performing unconfined consolidation tests. The unconfined compression test was performed as follows. The density and the water content at which the test was to be performed were fixed initially. The weight of soil required to achieve that density was worked out. The water quantity based on the predetermined water content was worked out. The soil sample was then prepared by filling the mould and applying static compression. After preparation of the sample, it was extruded from the mould. The dimensions of the soil sample were checked and its weight was measured. The sample was placed on the compression testing machine in such a way that the axis of the sample was as close as possible to the loading plate. The dials were adjusted to zero and the proving ring details were noted. The strain rate was set and the motor was started. The load readings and the strain readings were measured at frequent intervals to define the stress-strain relationship. The test was continued till the racks were well-formed. The specimen was then removed and was oven-dried to determine the water content [10, 20].

One Dimensional Consolidation Test

The compressibility characteristics of the soil, as well as the treated soil, were determined by performing the consolidation test. As in the case of the unconfined compression test, the density and the water content at which the compressibility was to be studied was set initially. The height and the diameter of the consolidation ring were measured and the inside surface of the ring was lubricated with a thin film of oil. A sufficient quantity of soil was weighed out to achieve the set density and prepared to the predetermined water content was carefully placed inside the ring, using the spacer. The weight of the soil sample plus the ring was measured. Two filter papers were moistened and placed on either end of the soil sample. A porous stone was soaked in water and placed on the base of the oedometer and the prepared sample was placed on top of it. The second porous stone was placed on top of the prepared sample, after soaking it in water. The loading block was then placed on top of the porous stone. The outer ring was placed on the setup after placing the rubber washer and the entire setup was tightened thoroughly. The sample was then immersed in water completely. The loading platform was adjusted till it touched the loading block. The dials were checked and the initial reading was noted. Then the loading frame was loaded using a load of 0.5 kg/cm^2. After a lapse of 24 hours, the dial reading was noted and the next increment of the load was put. The tests were performed for loads of 0.5, 1, 2, 4 and 8 kg/cm^2. Time-bound readings were taken for any two of the load increments to study the compression versus the time behaviour. After the test, the ring was dismantled and the final weight of the sample plus ring was taken. The sample was ejected from the ring and oven-dried to determine the final water content. The coefficient of vertical consolidation was computed from the time compression tests conducted on the soil as well as soil mixed with various natural materials, using the square root of the 't' method [13, 21].

California Bearing Ratio Test

The California Bearing Ratio Test (CBR) was conducted to determine the strength properties of the soil sample be tested. The increase of rice husk ash replacement in the soil sample reduces the soil strength but adding a small amount of lime will maintain the strength of soil samples [2, 12].

Conclusion

The following are the conclusions made from this comprehensive review paper,

- ✓ While observing the previous works, rice husk ash made a significant effect on the stability properties of soil.

- ✓ By utilizing the ESP and TKP along with clay showed better strength properties in eggshell powder and detrimental effects achieved in tamarind kernel powder.
- ✓ The combination of Jaggery:Clay:Lime would be getting better performance at the lesser percentage of combinations. In addition to that tamarind kernel powder was used to absorb and retain moisture in the areas where the permeability property is required. Hence the huge water retention capability of the soil can be adopted advantageously similar to Bentonite.
- ✓ The utilization of eggshell powder has resulted in the improvement of the characteristics of soil properties and it would be utilized in the soil stabilization technique can be recommended.
- ✓ The combination of J: C: L has improved the index properties of the soil. The use of Chebula can be studied further by avoiding it in combination and studying the results of strength tests. If the results are better than the one studied in this investigation, then the combination sans Chebula can be recommended.
- ✓ With the combination of 15% pozzolan and 3% lime in the stabilization of silty sandy soils, compressive strength has been increased by 16 times in comparison to the natural soil and is considered to be the optimum.
- ✓ The combination of 1% Prosopis juliflora fibres and 9% cement yield higher geotechnical properties in sandy soils.

References

[1] Mohammed N J Alzaidy, Experimental study for stabilizing clayey soil with eggshell powder and plastic wastes, International conference on sustainable engineering techniques, (2019). https://doi.org/10.1088/1757-899X/518/2/022008

[2] Sabyasachi Biswas, Arghadeep Biswas and Amar Dighade, Utilization of rice husk with lime in subgrade soil for a rural road, International conference on emerging frontiers in technology for a rural area, (2012).

[3] R. Rathan Raj, S. Banupriya and R. Dharani, Stabilization of soil using rice husk ash, International journal of computational engineering research, 6, 2, (2016), pp. 43-50.

[4] Deepti V. Cutting and P. L. Naktode, Soil stabilization by using scrap tire rubber, International Journal of Engineering Research & Technology, 9, 6, (2020), pp. 1382-1388. https://doi.org/10.17577/IJERTV9IS060592

[5] Jijo James and P. Kasinatha Pandian, Performance study on soil stabilization using natural materials, International Journal of earth sciences and engineering, 6, 1, (2013), pp. 194-203.

[6] Mohammed Y. Fattah, Falah H. Rahil, Kawther Y. H. Al-Soudany, Improvement of clayey soil characteristics using rice husk ash, Journal of Civil Engineering and Urbanism, 3, 1, (2013), pp. 12-18.

[7] Ankit Patel and Abhijitsinh Parmar, Stabilization of natural soil using rice husk, International Journal of Engineering development and research, (2013), pp. 91-93.

[8] G. Abarajithan, P. Rishab Kumar and R. Srikanth, Feasibility of Soil stabilization using Rice Husk Ash and Coir Fibre, International Journal of Engineering Research & Technology, 6, 4, (2017), pp. 552-556. https://doi.org/10.17577/IJERTV6IS040502

[9] Daryati and M. A. Ramadhan, Improvement of Expansive Soils stabilized with Rice Husk Ash (RHA), Journal of Physics: Conference Series, 1625 (2020), 012006. https://doi.org/10.1088/1742-6596/1625/1/012006

[10] Nafiul Islam, Nayem Ahmed, Sumaiya Akter and A. S. M. Fahad Hossain, Sandy Soil Stabilization using Jute Fiber as Admixture, Journal of Geotechnical Studies, 6, 2, (2021), pp. 30-34.

[11] Vikas Sharma, Pramod Kumar Rathore and Amit Sharma, Soil Stabilization by using Wheat Husk Ash, Journal of Civil Engineering and Environmental Technology, 5, 1, (2018), pp. 31-35.

[12] Syed Aaqib Javed and Sudipta Chakraborty, Effect of Expansive Soil Stabilization using Rice Husk Ash & Marble Dust, Journal of Earth Engineering, 4, 1, (2019), pp. 41-49.

[13] H. Trabelsi M. Chebbi and H. Guiros, Stabilization of clayey soil using fibre reinforcement, International conference on Unsaturated Soils, Hongkong, (2018).

[14] Tehsin Kazi, M. A. Shabiimam, Omar Baig and Feroz Machkuri, Review on comparative study on soil stabilization using natural materials, International Conference on Construction, Real Estate, Infrastructure and Project (CRIP) Management, Pune, (2018).

[15] P. V. Sivapullaiah, K. S. Subba Rao and J. V. Gurumurthy, Stabilization of Rice husk ash as a cushion below foundations on expansive soils, Ground Improvements, 8, 4, (2004), pp. 137-149. https://doi.org/10.1680/grim.2004.8.4.137

[16] O. O. Amu, A. B. Fajobi and B. O. Oke, Effect of eggshell powder on the stabilising potential of lime on expansive clay soil, Journal of Applied Sciences, 5, 8, (2005), pp. 1474-1478. https://doi.org/10.3923/jas.2005.1474.1478

[17] M. S. Chauhan, S. Mittal & B. Mohanty, Performance evaluation of silty sand subgrade reinforced with fly ash and fibre, Geotextiles and Geomembranes, 26, 5, (2008), pp. 429-435. https://doi.org/10.1016/j.geotexmem.2008.02.001

[18] I. T. Jawad, M. R. Taha, Z. H. Majeed & T. A. Khan, Soil stabilization using lime: advantages, disadvantages and proposing a potential alternative, Research Journal of Applied Sciences, Engineering and Technology, 8, 4, (2014), pp. 209-220. https://doi.org/10.19026/rjaset.8.1000

[19] G. L. Sivakumar Babu and A. K. Vasudevan, Strength and Stiffness response of coir fibre reinforced tropical soil, Journal of materials in Civil Engineering, 20, 9, (2008), pp. 571-577. https://doi.org/10.1061/(ASCE)0899-1561(2008)20:9(571)

[20] J. Prabakar, R. S. Sridhar, Effect of random inclusion of sisal fibre on strength behaviour of soil, Construction and building materials, 16, 2, (2002), pp. 123-131. https://doi.org/10.1016/S0950-0618(02)00008-9

[21] E. A. Subaida and S. Chandrasekaran and N. Sankar, Laboratory performance of unpaved roads reinforced with woven coir geotextiles, Geotextiles and Geomembranes, 27, 3, (2009), pp.204-210. https://doi.org/10.1016/j.geotexmem.2008.11.009

[22] Nader Abbasi and Masoud Mahdieh, Improvement of geotechnical properties of silty sand soils using natural pozzolan and lime, International Journal of Geo-Engineering, 9, 4, (2018), pp. 1-12. https://doi.org/10.1186/s40703-018-0072-4

[23] Gopinath Rudramurthy, Poopathi Ramasamy and Arun Rajendran, Geotechnical properties of sandy soil stabilized using cement and Prosopis juliflora fibres, Advances in Science, Technology & Innovation Book Series, (2018), pp.283-285. https://doi.org/10.1007/978-3-030-01665-4_65

Geopolymer Concrete Paver Blocks: A Review

S. Kavipriya[1,a,*], C.G. Deepanraj[2,b], M.P. Iniya[1,c] and B. Jeyanth[1,d]

[1]Kongunadu College of Engineering and Technology (Autonomous), Department of Civil Engineering, Thottiam, Trichy, Tamilnadu – 621215, India

[2]Vivekananda College of Technology for Women, Department of Civil Engineering, Tiruchencode, Namakkal, Tamilnadu – 637205, India

[a*]kavipriyacivil2001@gmail.com, bdeepancdhassini@gmail.com, ciniyaviji@gmail.com and [d]jeyanthbaskaran5890@gmail.com

Keywords: Curing, Geopolymer Concrete, Polymerization, Sodium Hydroxide, Sodium Silicate

Abstract. Geopolymer Concrete is one of the emerging concretes with zero cement. Zero percent usage and hundred percentage utilization of pozzolanic material as source material increases the interest of researchers towards geopolymer concrete day by day. Also, reduction of emission of Co2 rate in environment by the less usage of cement consumption increase the thirst of utilizing geopolymer concrete towards academicians, environmentalists and researches. In geopolymer concrete the main source material which is highly embedded with huge percentage of silica and alumina is utilized instead cement. Stimulator solution, combined with Na_2Si0_3 and NaOH is used for processing and activating polymerization process. As a result of polymerization process, chain links and bonds of Silicon-Oxygen-Aluminum are formed to stimulate the strength properties of concrete. Steam, oven and ambient temperature curing methods are preferred in this concrete. Nowadays, interest of geopolymer concrete turn over to applications of geopolymer products rather than construction of structural elements. Applications of geopolymer concrete include paver blocks, precast slabs, pipes, bricks, tiles, etc. Construction of paver blocks in geopolymer concrete plays an important role. This paper describes the review of GPC paver blocks for the past two to three decades which will help to update the knowledge of GPC paver blocks in all aspects.

Introduction

Ordinary Portland Cement (OPC)-a prime most key ingredient in the manufacture of concrete and construction industry. Production of Portland Cement increases in a higher rate due to modernization and development of infrastructure. A well-known unavoidable fact in the production of cement is that limestone which is found to be greedier in releasing CO_2 into the universe, which highly relates to greater environmental issues such as global warming, green house effects, etc. Reduction in usage of cement is the only solution to reduce environmental hazards.

Utilization of waste materials into the concrete is yet another good solution to rejuvenate the environment to be more eco-friendly. Hence GPC could be mentioned as green concrete since prepared with waste materials and zero cement. Only by these alternate solutions we can build greener environment with quality air for respiration. Such replacement also preserves the natural resources in concrete with the usage of waste material providing a better way for the possibility of replacing concrete. This alternate solution in replacement to cement and concrete helps to develop sustainable environment.

Cement Replacement Materials

Many efforts have been made to choose an alternative material to Portland cement as waste material which overcomes all the above known problems is one of the best remedies to alter normal cement with zero percent cement and hundred percent usage of fly ash or material abundant in silica and alumina. All thermal power plants generate huge quantity of fly ash which is dumped into the soil without any use. This fly ash hinders the interest of researches in the past few decades, which becomes an extra ordinary replacement to cement in the past years. And thirst of eco-friendly environment increases considerable research towards GPC. Since it possesses excellent strength and durability properties, low cost, eco-friendly nature increases its demand day by day. Unlike Portland cement, it has no environmental effects since $CO_2$is not let into the atmosphere.

Application of GPC

GPC can be used as sustainable binders in precast industries in the manufacturing of various products such as bricks, pavers, slabs, hollow block units, pipes, roofing tiles etc. Also, GPC can act as binders in the manufacturing of light weight components in both cement and concrete field. The combination of Silicon-Oxo-Aluminate is called "Sialate" since geopolymer possesses good durability properties so that it can be used in severe environmental applications such as bridges, fiber composites, sealant for industrial ceramics, foundry components etc. Also, GPC is highly resistant to acidic and alkali resistance, hence utilized in mining and sewer systems.

Utility of Geopolymer Technology provides the opportunity for sustainable development by converting waste materials into useful and valuable products and also by reducing greenhouse gas emissions, particularly in the cement industry. Its application also involves the manufacturing of geopolymer pastes and geopolymer coating materials, geopolymer aggregates. Applications of GPC are almost similar to ordinary Portland cement concrete applications.

Paver Blocks

Paver blocks are used to improve the aesthetic looks of paved areas. These pavers are best suited for all areas. Coloured blocks provide more aesthetic beauty to pavers. The paver blocks are cast in different sizes and thickness. These paver blocks act as wearing surface by reducing the stresses acting on subgrade and also help in resisting pavement deformations and elastic deflections similar to base course of flexible pavement. These paver blocks are used for various applications in toll plaza, city streets, bus depots, intersections, road repairs during monsoon, car parks, truck parking areas, roads in high altitude areas, footpaths, cycle tracks, residential streets, sidewalks, level crossings, petroleum bunks, parks, and in many other areas. Blocks are used in light traffic, medium traffic and heavy traffic areas. The blocks are cast into various colours, shape and sizes to suit to the requirement to satisfy the norms of IS: 15658-2006.

Literature Review

Geopolymer Concrete Paver Properties

Mechanical Strength

[1] Illustrate the various methods used to develop mortar specimens for the determination of optimum dry density. Strength properties using various parameters were analyzed using experimental data. It supports to develop a phenomenol model to create with different combinations of ingredients to form geopolymer blocks to reach the desired strength of results. Using set of various experimental data, the validity of models was evaluated. The result reveals the same predicted values compared with experimental data. [2] Investigated composite geopolymer matrix embedded with reinforcement and basalt fibers. To evaluate the improvement

in mechanical properties with respect to material cast with no reinforcement. Tests were conducted with basalt fabrics reinforced geopolymer material. Tests were carried out with the evaluation of strength properties. There was a significant impact between the strength properties of tested composite geopolymer matrix and developed basalt fabric matrix. [3] Revealed the mechanical strength properties with different mixing ratios of fly ash and rice husk ash. Different ratios between these two combinations were considered for casting and testing. Results indicate WA that specimens with more amount of fly ash ratio possess more compressive strength, water absorption, minimum thermal conductivity, bulk density and minimum porosity compared to addition of more percentage of rice husk ash. [4] Discussed manufacturing process of geopolymer cement by mixing with various proportions of calcium carbonate. These combinations of calcium carbonate cast specimens were tested for its compressive strength, water absorption and other durability properties. [5] Investigated with basalt fibre reinforced geopolymer concrete with different ratios of alkaline activator solution, ratio of source material to alkaline solution. [6] It was found that workability of concrete increases with increase in ratio of superplasticizer and rice husk ash. [7] Examined foundry waste sand, fly ash and GGBS as replacement to aggregates and cement in paver blocks. Different tests were performed to determine strength and durability properties of geopolymer concrete with the maximum utilization replacement percentage of waste materials. It was found that by increasing percentage of GGBS strength results increase in excellent percentage compared to increase in foundry sand and fly ash. [8, 9] Investigated the improvement of building envelope using sustainable material. The main aim behind this research is to develop an eco-friendly, low thermal transmittance material. Research was carried out using aluminate silicate source material to develop foam geopolymer concrete. Further it was compared with normal conventional building construction materials. They were tested for its physical, thermal and mechanical performance of glass powder geopolymer blocks.

Durability

[10] Investigated Geopolymer foamed block with more inter connected pores by adding H_2O_2 solution and oleic acid. Adsorption capacity gets increased due to the formation of pores. Also test results revealed that zeolites can be used as low cost replacement material in waste water treatment units. [11] Examined the different combinations of mixture with metakaolin and NaOH solution. Development of bricks was achieved through laterite. Research develops compressed earth bricks with and without stabilizers using 8% of Portland cement. After the process of curing both stabilized and unstabilized bricks were subjected to testing. [12] investigated the utilization of recycled asphalt pavement aggregates in paver blocks. More percentage of recycled waste aggregates were generated from digging existing bituminous roads due to tremendous growth in road development projects. Still more percentage of aggregates were dumped in urban areas without utilization. [13, 14] Investigated utilizing DWTS as replacement of sand in paver blocks. Five different ratios of concrete paver blocks were designed and tested for its compressive strength, leachability, abrasion resistance, and acid and sulphate attack. Micro structure analysis indicates that addition of DWTS result in the formation of ettringite. [15] Describes the paver blocks manufactured using alkali activated and clayey soil binders. The binder was generated from fly ash with more percentage of unburned material. Lime and slag was utilized as calcium sources. Research revealed that the texture of the soil is the most key element for manufacturing compacted soil masonry blocks. Highest sand content, high dry density indicates the lower percentage of pores and water absorption with lowest compressive strength. [16] Explained that increasing alumina content develops properties of GPC whereas decreases its strength properties. Strength of the concrete differs with different ratios of Silica to alumina ratio. [17] Investigated energy efficiency

of the developed material produced by foam alkali activated blend fly ash and glass powder. The energy and thermal efficiency of the material was evaluated. A detailed thermal analysis was carried out using an incompressible conjugate heat. [18] Explained the workability and strength nature of paver blocks in addition of BKRHA particles. It results with increase in strength. Excellent results were found with flexural and tensile strength too. Also, durability properties, resistance to acid and frost attack showed better results in GEOPAV blocks.

Economic Benefits of Geopolymer Concrete Pavers

[19] Investigated the mechanism, typical cooling techniques to develop cool pavements in thermal environment. Cost benefits, policies towards application of cool pavements were analyzed in this research. It was found that the applications of cool pavements remain incomplete in this research. Author suggested that cool pavements towards air temperature is unknown. [20] Explained the technologies suited to connect pathways towards development of Nador village. Also author describes the social- economic activities that influence the rural life. [21] Many difficulties regarding various disagreeable conditions in connecting the pathways were identified. [22] Investigated numerically obtained pertinent quantities with average, normal temperature iso-surface variation through the solid zone and liquid zone and profile of temperature along axis direction. Compressive strength test along with cost analysis and energy demand was carried out on the materials. It could be attained in incorporating glass powder in foamed geopolymer blocks.

Applications of Geopolymer Concrete Pavers

[23] Explained the techniques adopted to reduce the temperature of water retaining blocks in wet and dry conditions at different temperature. Author described the methodology adopted to maintain permeability of water retaining block in heavy rain condition and the process of draining excess water from the blocks. [24] Described the difference between the performance of conventional cement and metakaolin geopolymer blocks. Test results of GPC blocks revealed excellent leaching property, good resistance to acid and sulphate solution. After freeze-thaw cycles and high temperature retention tests geopolymer blocks resulted with low compressive strength. [25] Developed low cost, light weighted non aerated geopolymer blocks using various proportions of GGBS to fly ash ratio. These NAG blocks were durable in nature and resistance to acid. Maximum industrial waste material was utilized for the development of light-weight NAG blocks with minimum concentration of alkaline solution.

Conclusion

From the review, it was found that more researches were carried out with fly ash as source material. In recent articles, basalt fabrics were added into GPC to achieve more strength and durability. More researches were carried out in the recent years in geopolymer blocks and bricks. Researches must be carried out using different alkaline solution and source material.

References

[1] Radhakrishna, S., Design and properties of fly ash, ground granulated blast furnace slag, silica fume and metakaolin geopolymeric based masonry blocks. Eco-efficient Masonry Bricks and Blocks, 2015: p. 329.

[2] Qin, Y., A review on the development of cool pavements to mitigate urban heat island effect. Renewable and sustainable energy reviews, 2015. 52: p. 445-459. https://doi.org/10.1016/j.rser.2015.07.177

[3] Khan, K.A., et al., Mechanical and water transport performance of sustainable geopolymer composite using binary blends of pozzolanic materials. Jurnal Teknologi, 2019. 81(5).

[4] Maruti, S.V. and M. Mounesh, Properties of geopolymer cement mortar and blocks with calcium carbonate. Materials Today: Proceedings, 2020. 24: p. 1518-1524.

[5] Duxson, P., et al., Geopolymer technology: the current state of the art. Journal of materials science, 2007. 42(9): p. 2917-2933. https://doi.org/10.1007/s10853-006-0637-z

[6] Kanu, N.J., et al., An insight into processing and properties of smart carbon nanotubes reinforced nanocomposites. Smart science, 2021: p. 1-16.

[7] Patil, A.R. and S.B. Sathe, Feasibility of sustainable construction materials for concrete paving blocks: A review on waste foundry sand and other materials. Materials Today: Proceedings, 2021. 43: p. 1552-1561. https://doi.org/10.1016/j.matpr.2020.09.402

[8] Khan, K.A., et al., Study on Energy Requirements of Building Envelope Developed from Glass Powder Foamed Geopolymer Blocks, in Recent Advancements in Civil Engineering. 2022, Springer. p. 275-282.

[9] Raut, A., et al., Evaluation of thermal and energy consumption behavior of novel foamed copper slag based geopolymer masonry blocks. Ceramics International, 2022.

[10] Jin, Y., et al., Solvent - Free Syntheses of Hierarchically Porous Aluminophosphate - Based Zeolites with AEL and AFI Structures. Chemistry-A European Journal, 2014. 20(52): p. 17616-17623. https://doi.org/10.1002/chem.201403890

[11] Sore, S.O., et al., Stabilization of compressed earth blocks (CEBs) by geopolymer binder based on local materials from Burkina Faso. Construction and Building Materials, 2018. 165: p. 333-345. https://doi.org/10.1016/j.conbuildmat.2018.01.051

[12] Chen, Y., et al., Application of phase change material in asphalt mixture-A review. Construction and Building Materials, 2020. 263: p. 120219. https://doi.org/10.1016/j.conbuildmat.2020.120219

[13] Liu, Y., et al., Utilization of drinking water treatment sludge in concrete paving blocks: Microstructural analysis, durability and leaching properties. Journal of environmental management, 2020. 262: p. 110352.

[14] Liu, Y., et al., Properties and microstructure of concrete blocks incorporating drinking water treatment sludge exposed to early-age carbonation curing. Journal of Cleaner Production, 2020. 261: p. 121257.

[15] César, J.-S., et al. Analysis of the use of technological tools in the e-learning process. in 2020 IEEE ANDESCON. 2020. IEEE.

[16] Mohammed, A.A., H.U. Ahmed, and A. Mosavi, Survey of Mechanical Properties of Geopolymer Concrete: A Comprehensive Review and Data Analysis. Materials, 2021. 14(16): p. 4690. https://doi.org/10.3390/ma14164690

[17] Singh, R.J., et al., Influence of glass powder incorporated foamed geopolymer blocks on thermal and energy analysis of building envelope. Journal of Building Engineering, 2021. 43: p. 102520.

[18] Mahdi, S.N., N. Hossiney, and M.M.A.B. Abdullah, Strength and durability properties of geopolymer paver blocks made with fly ash and brick kiln rice husk ash. Case Studies in Construction Materials, 2022. 16: p. e00800. https://doi.org/10.1016/j.cscm.2021.e00800

[19] Qin, Y. and H. Yang, Carbonation dominates the acid intake of recycled concrete aggregate subjected to intermittent leaching. Construction and Building Materials, 2015. 89: p. 110-114. https://doi.org/10.1016/j.conbuildmat.2015.04.038

[20] Gupta, R., et al., The requirement of intra village pathways for roadway technology adoption: A rural survey in Nador village, Madhya Pradesh, India. Technology in Society, 2016. 47: p. 101-110. https://doi.org/10.1016/j.techsoc.2016.09.005

[21] Bhardwaj, P., et al., Corrosion and fire protective behavior of advanced phosphatic geopolymeric coating on mild steel substrate. Silicon, 2020. 12(3): p. 487-500. https://doi.org/10.1007/s12633-019-00153-1

[22] Singh, N.P., Black is a Country. 2021: Harvard University Press.

[23] Bao, T., et al., Geothermal energy recovery from deep flooded copper mines for heating. Energy Conversion and Management, 2019. 183: p. 604-616. https://doi.org/10.1016/j.enconman.2019.01.007

[24] Tan, Q., et al., Comparative performance of cement and metakaolin based-geopolymer blocks for strontium immobilization. Journal of the Ceramic Society of Japan, 2019. 127(1): p. 44-49. https://doi.org/10.2109/jcersj2.18130

[25] Kejkar, R.B., et al., Performance Evaluation of Cost-Effective Non-autoclaved Aerated Geopolymer (NAAG) Blocks. Arabian Journal for Science and Engineering, 2020. 45(10): p. 8027-8039. https://doi.org/10.1007/s13369-020-04581-9

Sustainable Materials and Smart Practices - NCSMSP-2021 Materials Research Forum LLC
Materials Research Proceedings **23** (2022) 290-296 https://doi.org/10.21741/9781644901953-33

Experimental Study on Strength of Concrete using Glass Fiber

K. Rajendhiran[1,a,*], S. Gopinath[1,b] and G. Rajasekar[1,c]

[1]Velammal Engineering College, Chennai – 66, India

[a] rajendrancicon@gmail.com*, [b] sgopinathcivil@gmail.com, [c] rajasekar.g122@gmail.com

Keywords: Light Weight Concrete, Glass Fiber, Compressive Strength, Flexural Strength and Spilt Tensile Strength & GFRC

Abstract. Plain concrete retains veritably low tensile strength, limited rigidity and little resistance to cracking. Cracks are innately present in concrete and their poor tensile strength is due to propagation of similar micro-crack filaments when added at certain chance in the concrete to ameliorate the strain parcels as well as crack resistance, to rigidity as flexure strength and durability. Substantly, the studies and exploration of fiber corroborating concrete has been diverted to sword filaments. In recent times, glass filaments have also become available which are free from erosion problems associated with sword filaments. In this design, examination was conducted using glass fiber with concrete. Anti crack, hyperactive dissipation, alkali resistance glass fiber of periphery 14 microns, having an aspect ratio of 857 were employed in probabilities, varying from 0.33 to 1 chance weight in concrete and the parcels of this fiber corroborated concretelike contraction,flexure strength, durability,and tensile strength was studied.

Introduction

Concrete is a structural material which is rich in diversity and vacuity.RFC is used in areas where a demand for high tensile strength is needed, along with precast concrete periods that can be produced in any form according to demand.Large and heavy structure periods, islands, beans, are completed periods for longer period by the involvement of concrete for now has ever been involvement with numerous other structureaccoutrements in order to enhance theits parcels similar as the GFRC.

Types of Fibres

1. Natural fibre
2. Artificial fibre

Natural fibre

Some of the natural fibres include angora fibre, mortal hair, bagasse, bamboo, coire, jute etc. Bagasse Bagasse is the stringy matter that remains after sugarcane is crushed to produce its juice.It's presently used as a biofuel and in the manufacture of pulp.

Literature References

GRFC is a fibre grounded concrete in which fibres are slightly distributed and incompletely acquainted along with other materials similar as cement, summation, beaches,water, fly ash etc. By adding glass fibre it'll increase the compressive strength,tensile strength, resolve tensile strength of the concrete and it'll also reduce the permeability and cracks of the concrete.

Selin Ravikumar and Thandavamoorthy (2004). The study shows a significant increase in the use of fibres in concrete for improving its properties such as tensile strength and ductility. The

Materials Research Proceedings **23** (2022) 290-296 https://doi.org/10.21741/9781644901953-33

fibre concrete is also used in retrofitting existing concrete structures. Among many different types of fibre available today, glass fibre is a recent introduction into the field of concrete technology.

Subramani, Mumtaj (2006), did, in the present investigation, study the behaviour of glass fibres in concrete. In the present investigation, sand has been replaced with glass fibres by 5, 10 and 15% to produce concrete.

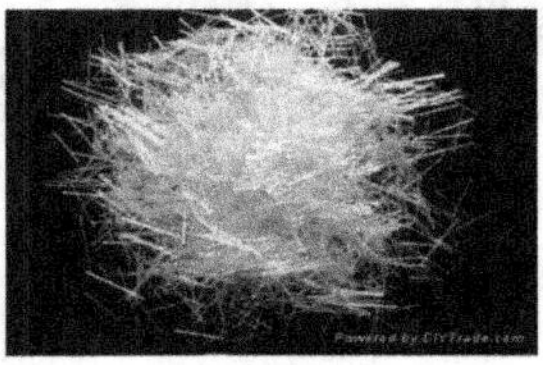

Fig. 1- Sample of Glass fiber

Objectives

- To perform the material test for all the materials worn.
- To determine the strong point of glass FRC using compression, split tensile and flexural test.

Types of Fibre Reinforced Concrete (FRC)

- Steel FRC.
- Natural FRC.
- Polypropylene FRC.
- Carbon FRC.
- Glass FRC.

Applications

- In Africa,sisal fibre corroborated concrete has been used considerably for making roof penstock,corrugated waste.
- Pipes,cilos, gas and water tank Elephant lawn fibre corroborated in mortar and cement is being used in Zambia for low-cost house construction.
- While wood and sisal filaments are being used for making cement compound panel filling, soffits and for sound and fire sequestration.

Need For Study

1. It increases the tensile strength of concrete.
2. It reduces air voids and water voids the essential porosity of strength.
3. The addition of small, nearly spaced and slightly dispersed filaments to concrete would act as a crack arrester.
4. It increases the continuity of concrete.
5. Glass has excellent resistance to creep.

Methodology

In this design, our main ideal is to study the influence of partial relief of cement on glass fibre and to compare it with the compressive strength, flexural strength, complete tensile strength of ordinary M30 concrete.

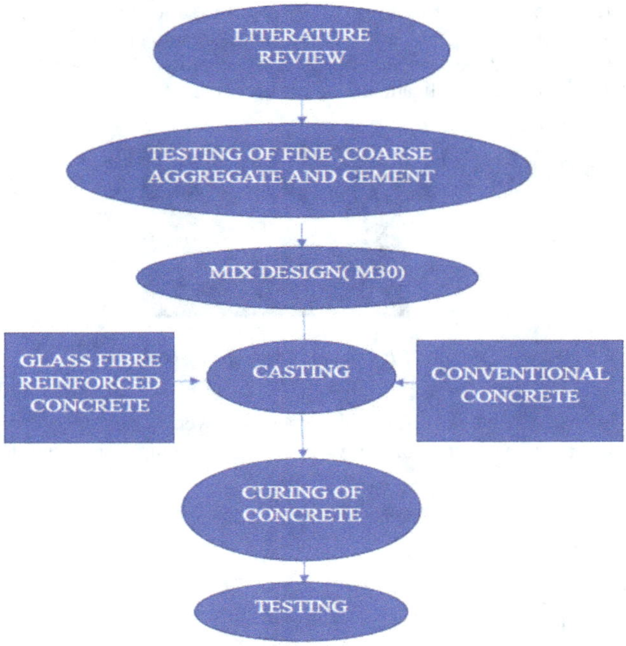

Fig.2 Process flow chart of Glass fibre reinforced concrete

Test on Concrete

a) Compressive Strength Test Cell moulds of 100mm x 100mm x 100mm were used for compressive strength study. Moulds were duly maintained by drawing and slicking before each casting. A wobbling table was used for better contraction and filled in three layers.

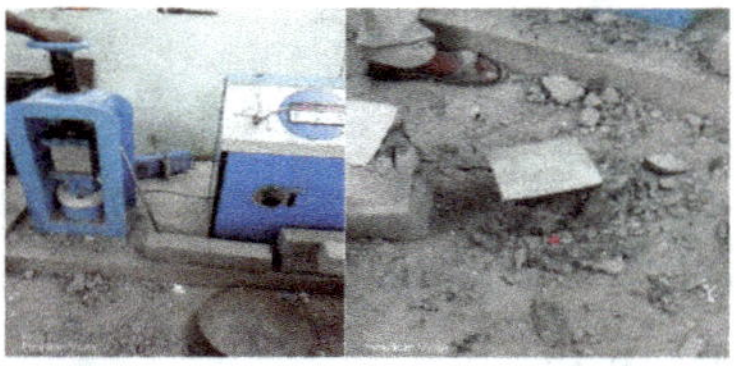

Fig 3: Compressive strength on GF Concrete

b) Split Tensile Test Originally, take the wet instance of size 15cm*30cm after 28 days of curing. Also, wipe out water from the face, for instance. After that, insure that they're on the same axial plane.

Materials Research Forum LLC
https://doi.org/10.21741/9781644901953-33

Fig 4: Split tensile on GF Concrete

c) Flexural Strength Test Flexural strength is the capability of a ray or arbor to repel failure when bending.It's measured by loading-reinforced concrete ray of 500mm x 100mm x 100 mm. Strength value is about 12 to 20 percent of compressive strength.

Fig 5: Flexural strength test

Experimental Investigation

The experimental investigation is divided under 3 headings

I.Compressive strength Test (cube)

Compressive strength for M30 Concrete with various mixed proportions with glass fibre

Table 1: Compressive strength for conventional mixed proportion

Trail	Compressive Load KN	Compressive Strength N/mm^2
1	685	30.30
2	730	32.31
3	705	31.68

II. Split Tensile Test (cylinder)

Split tensile strength for M30 Concrete with various mixed proportions with glass fibre

Table 2: Split tensile strength for conventional mixed proportion

Trail	Tensile Load(KN)	Tensile Strength N/mm^2
1	187	2.6
2	206	2.8
3	182	2.5

III. Flexural strength Test (beam)
Flexural strength for M30 concrete with various mixed proportions with glass fiber

Table 3: Flexural Strength for conventional mixed proportion

Trail	Flexure Load KN	Flexure Strength N/mm^2
1	6	2.67
2	8	3.55
3	9	3.8

Conclusion

- GF corroborated concrete is a recent invention in the field of concrete technology. Glass fibre corroborated concrete is a type of concrete which principally corresponds to cementitious matrix composed of cement, beach, coarse total, water, polymerand cocktails in which short-length glass fibres are dispersed. Fibres similar to photography and glass have excellent resistance to creep.

- The addition of small, nearly spaced and slightly dispersed fibres to concrete would act as a crack arrester and would mainly ameliorate its static and dynamic parcels. The mixed design of the M30 grade is adopted. Material testing is done for fine aggregate, coarse aggregate, cement. The mould of the shapes of cube, cylinder and prism is adopted. The concrete mix without any glass fibre is made and casted. Similarly, concrete mix with 0.3% and 0.6% of glass fibre is made and casted in all three moulds.

- After 28 days of curing compression, a split tensile and flexural test is done and their respective values are noted. From the comparison of strength obtained from conventional concrete and glass FRC, it is seen that the use of glass FRC has higher crack resistance and load bearable than conventional concrete.

- For cube, the test results for compression strength are given below: Conventional -31.4 N/mm^2 0.3% Glass Fibre -30.35 N/mm^2 0.6% Glass Fibre -22.4 N/mm^2

- For cylinder cylinders, the test results for split tensile strength are given below: Conventional -2.66 N/mm^2 0.3% Glass Fibre -1.94 N/mm^2 0.6% Glass Fibre -1.96 N/mm^2

- For prism, the test results for flexure strength are given below: Conventional -3.41 N/mm^2 0.3% Glass Fibre -2.8 N/mm^2 0.6% Glass Fibre -2.5 N/mm^2

- From the results obtained, it is observed that the strength obtained for conventional concrete is noted to be higher as the percentage of glass fibre added is less. So it is concluded that for future studies, glass fibre of 3%, 6% and similar percentages has been decided to be used to obtain greater strength from GRC.

References

[1] Chandramouli, Imran (2010), "cladding of buildings", third edition published 2002, pp.82.

[2] Chawala and Tekwari (2012), "Performance of Glass Fibre Reinforced concrete ", International journal of Inovations in engineering and technology (IJIET), volume I Issue 6 pp.264-248.

[3] Gaurav Tuli, Ishan Garg, (2002), 'Why Alkaline resistance Glass Fibre ", IOSR Journal of civil Engineering , volume 13,issue 3

[4] Jyothikumari (2012) "Manufacture of Glass Fibre Reinforced concrete products", Unit 1, Division of YOGI group UAE, August 2008.

[5] Pitrodaet al. (2015), " Effect of Glass Fibre Reinforced in Rigid Pavements", International Journal of scientific Research and education and research and education (IJSAE), Volume 2 Issue 9,,pp.1797-1804.

[6] Ravikumar and Indess, (2013) "Fibre Reinforced Cementitious Compositities ", second Edition Thandavamoorthy, "Performance of Glass Fibre Reinforced Self Compacting Concrete", international Journal of Engineering Development and Research (IJEDR)

Volume3 Issue 2, pp.755-760.

[7] Subramani. T , A. Mumtaj (2006), "Fibre Reinforced Cement Composites" , Arnon Bentur and Sidney M, Chapter 8, pp.278

[8] Ahmad, Structural behavior of precast lightweight concrete sandwich panel under eccentric loading: an overview. ,2008

[9] Ahmed Fouad Elragi, (2006) Selected Engineering Properties and application of EPS.

[10] Nahro Radi Husein et al., [2013] An experimental study on using lightweight web sandwich panel as a floor and slab. International journal of Innovative Technology and exploring engineering.2013, (IJITEE)

[11] Poluraju., [2014] Journal of sandwich structures and materials.

[12] TR Greeley - 1997 A Review of Expanded Polystyrene (EPS)Properties, Performance and New Applications.

Materials Research Proceedings **23** (2022) 290-296 https://doi.org/10.21741/9781644901953-33

[13] Karthik Jet al (2014), "Experimental Study on Strength Characterisitics on M20 Concrete with Partial Replacement of Fine Aggregate with Copper Slag" International Journal of Advanced Research.

[14] Binaya Patnaik et al (2014), Strength and Durability Properties of Copper Slag Admixed Concrete" International Journal of Research in Engineering and Technology.

[15] Balamurugan et al (2013) "Use of Quarry Dust to Replacing Sand in Concrete – An Experimental Study" International Journal of Science and Research Publications.

Sustainable Materials and Smart Practices - NCSMSP-2021 Materials Research Forum LLC
Materials Research Proceedings **23** (2022) 297-303 https://doi.org/10.21741/9781644901953-34

Comparative Analysis on Mechanical Properties of Polymer Concrete by using Various Lightweight Aggregates

T.S. Mukesh[1,a], P. Kulanthaivel[1,b], A. Naveenkumar[1,c], K. Ramesh[1,d]

[1] Civil Engineering Department, Kongu Engineering College, Perundurai, Erode,638060, Tamil Nadu, India

[a]mukesh.sakthivel@gmail.com, [b] pkulanthaicivil@gmail.com, [c]svnaveen88@gmail.com, [d] rameshramesh102000@gmail.com.

Keywords: Epoxy Resin, Lightweight Aggregates, Density

Abstract. The Polymer concrete is first established in the year of 1950s and gained popularity in 1970s repair works, building cladding and floors, as precasts components. The Polymer concrete has found applications in particularly specialised sectors due to its qualities such as high compression strength, quick curings, high specific strengths, and chemical resist. The objective of this experiment is to use destructive and non-destructive experiments to analyse the mechanical behaviour of Polymer concrete using various lightweight aggregates. Using destructive and non-destructive testing, this experiment evaluates mechanical characteristics of a lightweight polymeric concrete comprising four different types of polymeric ratios. The most important component is to reduce weight. Pumice, perlite, vermiculite, saw dust, and rice husk were utilized as light weight aggregates. Destructive tests revealed that raising the polymer ratio increased the compressive, impact strengths, and splitting-tensile, and the energy absorption of light weight polymer concrete. The properties such as ductility, impact energy, energy absorption shows decrease in efficiency. Pumice was discovered to have the best outcomes among the various lightweight aggregates. These study's findings are significant in understanding a performance of Lightweight polymer concrete and ensuring its safe deployment in a engineering applications which requires a high performance of strength to weight ratio material.

Introduction

General

Polymers concrete is a class of concrete-polymer composite that is manufactured by replacing traditional cement hydrate binders with polymer binders. At room or ambient temperature, most liquid resin, as thermosetting resins, methacrylic resin, and tar-modified resin, polymerize to produce polymeric concrete. The binder phase of polymer concrete contains no cement hydrates and is entirely composed of polymers. The aggregates are firmly held together by polymer binder. The advantage and disadvantage of a polymer binders are immediately transferred to polymer concrete. As a consequence, polymer replacement greatly enhances property such as, adhesion, abrasion, strength, chemical resist, watertightness, freeze-thaw durability, and resist in contradiction to ordinary cement concrete.

The potential of polymer concrete research was introduced as early as 1971, when the American Concrete Institute created Committees 548—Polymer in Concrete. The Committees was in charge of establishing a comprehensive database on the characteristics of polymers concrete. The Committees has also provided up-to-date literature on polymer concrete, as well as user manuals. Polymer concretes is defined as "concrete that organic polymer acts as a binder," according to the 'ACI-CT-13', ACI – Concrete Terminology-13 – An ACI Standard. Polymer concretes, to put it another way, is a complex material made up of aggregates linked together in matrix using a polymeric binder. It improves concrete strength by reducing porosity.

Materials

Polymer

Table.1. Properties of epoxy resin

Properties	Test methods	Units	Values
Epoxy Equivalent Weight	ASTM D1652	g/eq	185-192
Epoxy Viscosity at 25°C	ASTM D445	P	110-150
Colour	ASTM D1544	Gardner	1 max
Pounds/gallon at 25°C		lbs/gal	9.7
Density at 25°C		g/ml	1.16
Physical form			Clear liquid
Vapour pressure at 77°C		Mm Hg	0.03
Refractive index at 25°C			1.573
Specific heat		BTU/lb/ °F	0.5

Table.2. Comparative physical nature of light weight aggregate

Light Weight Aggregate	Specific Gravity	Shape	Density (kg/m$^{3)}$	Bulk Density (kg/m3)	Unit Weight (kg/m3)	Compression Strength (Mpa)
Pumice	0.95	round with open textured	550-1650	350-650	1200-1600	5-15
Perlite	2.4	Round, rough surface	100-400	40-200	400-500	1.2-3
Vermiculite	2.4	cubical	100-400	60-200	300-700	1.2-3
Saw dust	2.29	Arbitrary shape	370-415	190-220	--	--
Rice husk	2.14		--	90-150	83-125	--

From this pumice is found to be more effective than any other light weight aggregates because of having high compressive strength, bulk density, density, unit weight.

Results and Methods

Workability test

Table.3. Slump cone value

Percentage of light weight aggregates	Pumice	Perlite	Vermiculite	Sawdust	Rice Husk
0.5%	52	51	53	52	53.5
1%	53.5	50.5	52.5	51.5	52
1.5%	52.5	50	51	50	51
2%	50	49	50.5	49	50.5
2.5%	49.5	48.5	49	49.5	49

This table shows the slump value of polymer concrete with various light weight light weight aggregates such as perlite, vermiculite, pumice, rice husk, sawdust. Here the slump value of polymer concrete increases when the percentage of light weight aggregates increases.

Mechanical properties

Table.4. Compressive strength of polymer concrete

Percentage of light weight aggregates	Pumice	Perlite	Vermiculite	Sawdust	Rice Husk
0.5%	16.52	16.20	16.38	16.32	16.35
1%	18.89	18.56	18.26	18.35	18.78
1.5%	21.67	21.32	21.59	21.65	21.56
2%	23.45	22.98	22.81	22.92	22.87
2.5%	22.5	22.54	22.35	22.25	22.15

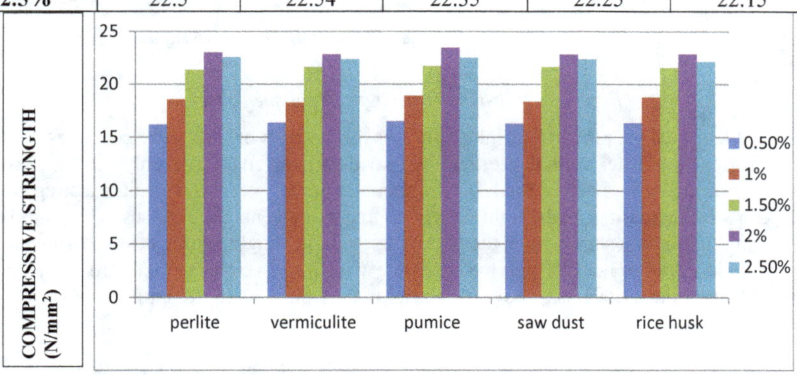

Figure.1. Compression strength of polymer concrete

This table shows the comparative representation of compression strength of a polymeric concrete with light weight aggregate perlite, vermiculite, pumice, sawdust, rice husk. The polymer concrete with pumice aggregates shows higher compressive strength than any other light weight aggregates. The value keeps on increasing as the light weight aggregate percentage increases. The compressive strength of 23.45(N/mm^2) shows when the percentage of light weight aggregates is 2% then it starts slightly to decreases. Then the polymer concrete with light weight aggregates such as perlite, vermiculite, saw dust, rice husk, shows decrease in the compression strength.

Table.5. Split-tensile strength of polymer concrete

Percentage of light weight aggregates	Pumice	Perlite	Vermiculite	Sawdust	Rice Husk
0.5%	2.5	2.45	2.31	2.30	2.25
1%	2.7	2.67	2.51	2.38	2.28
1.5%	3.1	3.1	2.98	2.83	2.75
2%	3.0	2.95	2.79	2.71	2.67
2.5%	2.9	2.88	2.81	2.72	2.66

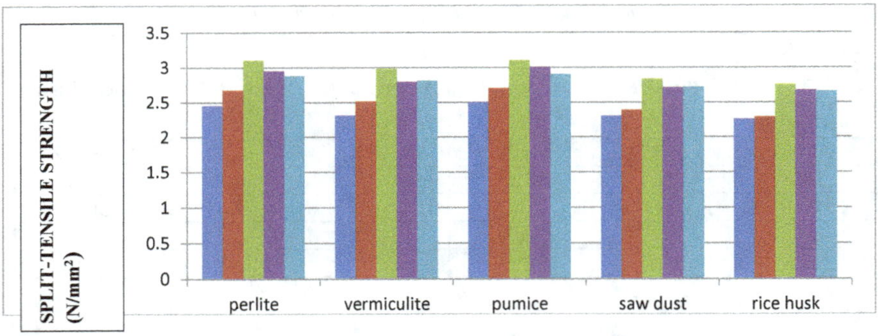

Figure.2. Split-tensile strength of polymer concrete

This table shows the properties of the Split-Tensile strength of a polymer concretes with light weight aggregate perlite, vermiculite, pumice, sawdust, rice husk. The polymer concrete with pumice aggregates shows higher Split-Tensile strength than any other light weight aggregates. The value keeps on increasing as the light weight aggregate percentage increases. The split-Tensile strength of 3.0(N/mm^2) shows when the percentage of light weight aggregates is 2% then it starts slightly to decreases. Then the polymer concrete with light weight aggregates such as perlite, vermiculite, saw dust, rice husk, shows decreases in the Split-tensile strength.

Table.6. Flexural strength of polymer concrete

Percentage of light weight aggregates	Pumice	Perlite	Vermiculite	Sawdust	Rice Husk
0.5%	9.32	9.22	9.15	9.10	9.02
1%	10.57	10.49	10.35	10.28	10.18
1.5%	12.68	12.65	12.58	12.51	12.43
2%	12.65	12.61	12.56	12.47	12.41
2.5%	12.73	12.71	12.65	12.59	12.51

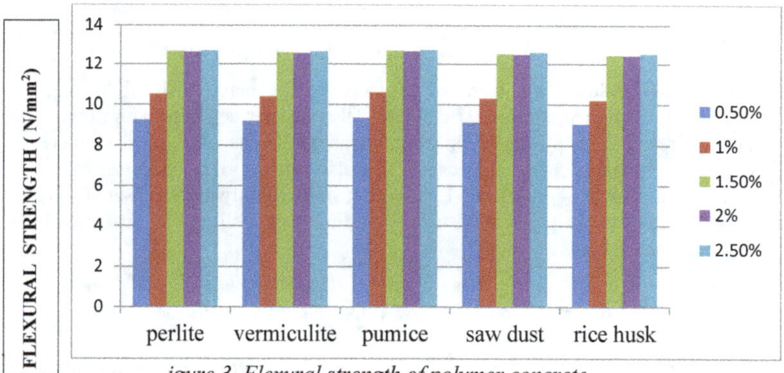

igure.3. Flexural strength of polymer concrete

This table shows a value of flexural strength of a polymeric concrete with the light weight aggregate perlite, vermiculite, pumice, sawdust, rice husk. The polymer concrete with pumice aggregates shows higher flexural strength than any other light weight aggregates. The value keeps on increasing as the light weight aggregate percentage increases. The flexural strength of 12.73(N/mm^2) shows when the percentage of light weight aggregates is 2.5%. Then the polymer concrete with light weight aggregates such as perlite, vermiculite, saw dust, rice husk, shows a gradual decrease in the flexural strength

Table.6. Pulse velocity of polymer concrete

Percentage of light weight aggregates	Pumice	Perlite	Vermiculite	Sawdust	Rice Husk
0.5%	4.23	4.21	4.16	4.11	4.01
1%	6.72	6.67	6.52	6.48	6.42
1.5%	6.78	6.73	6.68	6.61	6.53
2%	5.37	5.35	5.31	5.27	5.21
2.5%	5.29	5.24	5.21	5.18	5.15

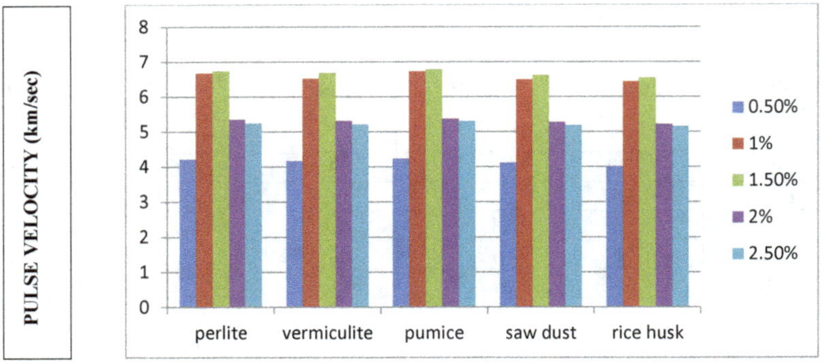

Figure.4. Pulse velocity of polymer concrete

This table shows a value of a pulse velocity of a polymer concrete with a light weight aggregate perlite, vermiculite, pumice, sawdust, rice husk. The polymer concrete with pumice having the pulse velocity greater than any other light weight aggregates. The value keeps on increasing as the light weight aggregate percentage increases. The pulse velocity of 6.78(km/sec) shows when the percentage of light weight aggregates is 1.5% then it starts slightly to decreases. Then the polymer concrete with light weight aggregates such as perlite, vermiculite, saw dust, rice husk, shows decrease in the value of pulse velocity.

Conclusion

- The physical property of pumice shows that the it has better density, bulk density, unit weight and the compression strength.
- The use of pumice in the polymer concrete reduces the density by 20% compared to normal polymer concrete.
- Compression strength is find as maximum for polymer concrete with the pumice aggregate of 1.5% as the substitution of fine aggregate.
- Split-tensile strength is find to be maximum for the polymer concrete with the pumice aggregate of 1.5% as the substitution t of fine aggregate.
- Flexural strength is found to be maximum for the polymer concrete with the pumice aggregate of 1.5% as the substitution t of fine aggregate.
- Pulse velocity is found to be maximum for the polymer concrete with the pumice aggregate of 1.5% as the substitution of fine aggregate.
- When compare to a Portland cement, polymer concretes has high compression and flexural strength. Compressive strengths of 70 to 120 Mpa have been noted.
- LWPC has a lesser density than PC, around 20% lower. This can improve to lower the total cost of construction as well as the structural damage caused by particular types of loads (eg. earthquakes).
- Increase the polymers content of LWPC specimens ranges from 11% to 14% improves mechanical characteristics, ductility, and energy absorption, with improvements of 56% in splitt tensile strength, 34% in compressive strengths, 68% in energy absorptions, and 53% in flexural strengths,
- As a polymer content increases, rate of increase in a compression strength reduces, and the compression strength of 14% polymer content show only a little difference. The porosity

of specimens decreases as polymeric content rises, according to non-destructive examinations.

- It's well known that polymer concretes has considerably superior mechanical characteristics and durabilities than traditional Portland cement concretes. Because of its superior mechanical qualities and a durability, polymer concretes has been prove to be a promising material.
- The mechanical characteristics of LWPC and their fluctuation with temperature were discussed in this work. Although the findings are encouraging, they also highlight the significance of careful LWPC design and execution, as the material's mechanical characteristics can deteriorate dramatically under specific situations.

Reference

[1] Bedi, Raman, Rakesh Chandra, and S. P. Singh. "Mechanical properties of polymer concrete." Journal of Composites 2013 (2013). https://doi.org/10.1155/2013/948745

[2] Heidarnezhad, Fatemeh, Khashayar Jafari, and Togay Ozbakkaloglu. "Effect of polymer content and temperature on mechanical properties of lightweight polymer concrete." Constructiokkn and Building Materials 260 (2020): 119853 https://doi.org/10.1016/j.conbuildmat.2020.119853

[3] Ferdous, Wahid, et al. "Optimal design for epoxy polymer concrete based on mechanical properties and durability aspects." Construction and Building Materials 232 (2020): 117229. https://doi.org/10.1016/j.conbuildmat.2019.117229

[4] Reis, João Marciano Laredo dos. "Effect of temperature on the mechanical properties of polymer mortars." Materials Research 15, no. 4 (2012): 645-649. https://doi.org/10.1590/S1516-14392012005000091

[5] Comparison of mechanical properties for polymer concrete with different types of filler by Marinela Barbuta,Maria Harja and Irina Baran

[6] Lokuge, Weena, and Thiru Aravinthan. "Effect of fly ash on the behaviour of polymer concrete with different types of resin." Materials & Design 51 (2013): 175-181. https://doi.org/10.1016/j.matdes.2013.03.078

[7] Mani, P., Gupta, A. K., & Krishnamoorthy, S. (1987). Comparative study of epoxy and polyester resin-based polymer concretes. International journal of adhesion and adhesives, 7(3), 157-163. https://doi.org/10.1016/0143-7496(87)90071-6

[8] Reis, João Marciano Laredo dos. "Effect of textile waste on the mechanical properties of polymer concrete." Materials Research 12, no. 1 (2009): 63-67.

[9] Assaad, J.J. and El Mir, A., 2020. Durability of polymer-modified lightweight flowable concrete made using expanded polystyrene. Construction and Building Materials, 249, p.118764. https://doi.org/10.1016/j.conbuildmat.2020.118764

[10] Cui, H. Z., Tommy Yiu Lo, Shazim Ali Memon, and Weiting Xu. "Effect of lightweight aggregates on the mechanical properties and brittleness of lightweight aggregate concrete." Construction and Building materials 35 (2012): 149-158 https://doi.org/10.1016/j.conbuildmat.2012.02.053

Study on Physical and Mechanical Properties of Porous Concrete using Recycled Concrete Aggregate

T.S. Mukesh[1,a,*], P. Kulanthaivel[1,b], G. Gowthaman[1,c], J. Hariharan[1,d]

[1]Department of Civil engineering, Kongu engineering college, Perundurai, Erode, Tamil Nadu, India

[a]mukesh.sakthivel@gmail.com, [b]pkulanthaicivil@gmail.com, [c*]gowthaman78836@gmail.com, [d]hariharanjayavelu2001@gmail.com

Keywords: Porous Concrete, Recycled Concrete Aggregates, Permeability, Polyvinyl Alcohol

Abstract. While porous concrete has a poor compressive strength because of its porous structure, concrete will lead to failure when the load applied is high. With recycled concrete aggregate and PVA fibre, an attempt was made to improve porous concrete's physical and mechanical properties. The features of ingredients such as recycled concrete aggregate, cement, Polyvinyl alcohol fibre, and water were first discovered, and then the mix was created and cast. After proper curing, the mechanical properties of concrete were determined by conducting the following tests: compression strength, split tensile strength, and flexural strength, along with physical factors such as permeability. The findings reveal that using the aggregate size of 1012mm improves mechanical characteristics. Permeability was also excellent. On the other hand, strength increased as the percentage of PVA fibre grew and at a certain point strength as well as porosity decreased. The maximum strength was attained at an incorporation % of 1. The mechanical characteristics of the aggregate declined when the size increased, as the surface area available for the cement to bind them together shrank.

Introduction

General

Porous concrete plays an important role in groundwater recharge and stormwater runoff reduction by absorbing and draining rainwater into the ground [10]. It's constructed with cement, coarse particles, and water, much like regular concrete. It is free of sand, resulting in a porous amorphous structure that allows water to flow freely [3]. It can absorb 3 to 5 gpm per sqft of surface area of stormwater which can surpass those required to avoid runoff during typical rainstorms [2].

From 2012 to 2014, the building and demolition industry generated over 370 million tons of concrete trash, according to the US Environmental Protection Agency [4]. If current trends continue, more than 4300 lakh tonnes of concrete debris could be produced in a year by the end of the decade. Despite the fact that concrete waste recycling is becoming more popular, much of it is still dumped in landfills [5]. Furthermore, replacing natural aggregates with recycled aggregates has the potential to cut 15%–20% of CO2 emissions while also protecting a major percentage of limestone resources all around the world. Replacement ratios of two have been found to be effective in studies.[1]

Up to 75% of a city's surface area is impermeable pavement, which prevents groundwater recharge, contributes to abrasion and flooding, transports pollution to adjacent waters, and increases the complexity and cost of storm water treatment [6]. Parking and access spaces benefit from pervious pavement with a compressive strength of up to 4000 psi [7]. Summer ambient temperature can be reduced by 2–4 degrees [8]. This study looked at the research on porous concrete made from recycled concrete aggregate, as well as its physical and mechanical qualities and ways to improve its performance [9].

Scope of our project

- Reducing the use of natural aggregate in concrete and consuming recycled concrete aggregate.
- Reduction of concrete waste and its impact on environment
- Accelerating the ground water table level by reducing the storm water runoff and stagnation of water

Objective

- To make a systematic study on porosity and mechanical properties of porous concrete using recycled concrete aggregates.
- To achieve maximum strength by using poly vinyl alcohol with required porous parameters.
- To give exact conclusion based on tests and observations.

Methodology

Initially the project started with identification and collection of materials, followed by specimen casting for different incorporation's and two types of tests were preferred namely destructive tests and non-destructive tests. Destructive tests include compressive, split tensile and flexural tests. Non-destructive tests include ultrasonic pulse velocity, permeability, porosity tests. The test results were analysed and conclusion was arrived.

Materials

- Cement OPC Grade 43
- Recycled concrete aggregate
- Water
- Poly vinyl alcohol

Cement

Cement is a substance that holds things together. Cement comes in a variety of forms that are utilized in building. Based on its composition, each variety of cement has different features, uses, and advantages. This experiment used Ordinary Portland Cement (OPC) 43 Grade cement. Various tests were done to find the property of cement that we utilized.

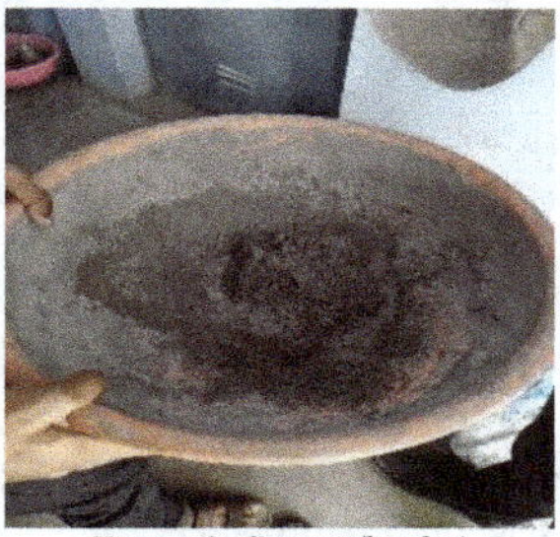

Figure.1. Cement(binder)

Coarse aggregate

As the major coarse aggregate, we employ recycled concrete aggregates that pass a 20mm screen and are retained at 10mm. Crushed cement concrete or asphalt pavement from construction detritus that is reused in other building projects is referred to as recycled aggregates. The utilisation of recycled materials in construction is a long-term strategy in the business. It has a number of advantages, including minimising the requirement for fresh aggregates, lowering energy use, reducing landfill trash, and lowering emissions. These recycled aggregates can be used in the construction of bicycle lanes, pavement shoulders, and many other parts of construction once they have been gathered and processed.

Table.1. Properties of RCA

Specific gravity	2.54
Water absorption	3.44
Impact test	33%
Fineness modulus	7.90

Figure.2. RCA

Polyvinyl alcohol fibre

Polyvinyl alcohol is the full name for PVA. Polyvinyl alcohol is a synthetic or manmade polymer. Vinyl and alcohol groups are present in the PVA polymer molecule. PVA is adsorbed onto cement and moistened goods when a little amount is introduced. The properties of PVA fibre are tabulated below.

Table.2. Characteristics of PVA

Elastic modulus(GPa)	45
Density(g/mm^3)	1.26
Fibre length(mm)	6
Tensile strength(MPa)	1600

Experimental programmes

To determine the strength of permeable concrete, compressive, split tensile, and flexural strength tests were done with 0, .5, 1, 1.5, 2 percent polyvinyl alcohol added to the cement mass. The mix proportion for M30 grade is designed using IS 10262 – 2009[12]. Cube, cylinder and beam were casted and tested for strengths and porous property using the compression testing machine, Flexural testing machine, Ultrasonic pulse velocity etc.

Results

Compressive strength

The cubes were tested by Compression Testing Machine. Compressive strength of the concrete cubes for five types of mixes (one is adding polyvinyl alcohol in the concrete and other is the reference concrete) for 28 days are found. Table 3 shows the results obtained.

Table.3. Compressive strength test

Mix	Fibre incorporation (%)	Compressive strength(N/mm^2)
P0	0%	10
P1	0.5%	12
P2	1%	13.5
P3	1.5%	14
P4	2%	15

Split Tensile strength

The cylinders were tested by Compression Testing Machine. The concrete cylinders' tensile strength was evaluated after 28 days and the findings were tabulated in table 4.

Table.4. Split tensile strength test

Mix	Fibre incorporation (%)	Tensile strength
P0	0%	2.5
P1	0.5%	3.2
P2	1%	3.3
P3	1.5%	3.6
P4	2%	3.8

An increase in strength of 52% was observed for the highest incorporation (2%) from the reference concrete with 0% incorporation.

Flexural strength

The Beams were tested by Flexural Testing Machine (FTM). The Flexural strength of concrete beams for 28-day mixes was discovered. Table 5 shows the outcomes achieved.

Table.5 Flexural strength test

Mix	Fibre incorporation (%)	Flexural strength
P0	0%	1.5
P1	0.5%	1.8
P2	1%	2.1
P3	1.5%	2.3
P4	2%	2.5

Flexural strength for pavements must be between 1 to 4 MPa. Hence These values (between 1.5 and 2.7 MPa) are well within the limit and can be proceeded. Further an increase of strength about

66% was observed for the highest percentage of incorporation (2%) from the reference concrete P0.

Ultrasonic pulse velocity test

The cubes are tested by ultrasonic pulse velocity tester. The ultrasonic values for cubes at 28 days were found out and tabulated in table.

Table.6. Ultrasonic pulse velocity

Mix	Fibre incorporation (%)	Values (km/sec)
P0	0%	3.3
P1	0.5%	3.4
P2	1%	3.6
P3	1.5%	3.8
P4	2%	3.9

Values more than 4.5 are considered as excellent. Values between 3.5 to 4.5 are considered good. They are considered as medium when their values range between 3.0 to 3.5 doubtful when they are below 3. Our test result shown a range of 3.3 to 3.9 which are considered as good.

Permeability

Table.7. Permeability

Mix	Fibre incorporation (%)	Permeability(cm/s)
P0	0%	3.1
P1	0.5%	2.7
P2	1%	2.9
P3	1.5%	2.6
P4	2%	2.4

From the results of permeability tests, it is known that the permeability of the material has decreased, as a result of the increased inclusion of PVA Fibre and decreased gradually and the permeability values are good enough to be used in construction of porous concrete.

Porosity

Porous concrete has a porosity of 15 to 35 percent is considered as nominal in sites. Results from figure 9 shown a range from 25 to 26%. Which is high enough to be considered to be a permeable concrete.

Table.8. Porosity Table

Mix	Fibre incorporation (%)	Porosity (%)
P0	0	25.9
P1	0.5	25.5
P2	1	25.7
P3	1.5	25.4
P4	2%	25.2

Porous concrete has a porosity of 15 to 35 percent is considered as nominal in sites. Result from table 8 shown a range of 25.2%. Which is high enough to be considered to be a permeable concrete.

Conclusion

We specifically conclude from the results of the experiments conducted that

- The optimum range of PVA fibres is at 1%, where all the test results are at the good satisfiable value.
- The mechanical qualities of porous concrete grew in strength as the amount of fibre incorporated rose, however the porosity property declined for the most part as the amount of fibre incorporated increased.
- Because of improved surface porosity, A greater bonding between recycled aggregate and cement paste resulted in increased strength and abrasion resistance.
- The use of recycled concrete generally aggregates specifically showed actually good result in all segments which ensures that it can specifically be used as a substitute for virgin aggregate.
- Incorporation of fibre resulted a 50%, 52%, 66% increase in the property of compressive, tensile and flexural strength respectively from the reference concrete with 0% incorporation of fibre to the highest value of incorporation that is 2%.
- For a better result an incorporation of 1% of PVA fibre is basically recommended for achieving sufficient strength and permeable property of porous definitely concrete at the same time in a subtle way.

Reference

[1] Matthew Upshaw and C.S. Cai, Critical review of recycled aggregate concrete properties, improvements, and numerical models Journal of Materials in Civil Engineering, (2020). https://doi.org/10.1061/(ASCE)MT.1943-5533.0003394

[2] R. Rizvi, S.L Tighe, J. Norris, and V. Henderson, Incorporating recycled concrete aggregate in pervious concrete pavements Annual conference and exhibition of the transportation association of Canada-transportation in a climate of change, (2009).

[3] Z. Yang, W. Ma, W. Shen and M. Zhou, The aggregate gradation for the porous concrete pervious road base material Journal of Wuhan University of Technology-Mater. Sci.Ed., (2008) 23(3), pp.391-394.

[4] R. Sriravindrarajah, N.D.H. Wang, and L.J.W. Ervin, Mix design for pervious recycled aggregate concrete International Journal of Concrete Structures and Materials, (2012), 6(4), pp.239-246.

[5] M. Gupta, L. Dudi, R. Karkhanis and V. Kumar, Determination of Optimum Parameters of Porous Concrete for Adequate Strength and Permeability Urbanization Challenges in Emerging Economies: Resilience and Sustainability of Infrastructure Reston, VA: American Society of Civil Engineers, (2018) (pp. 129-136). https://doi.org/10.1061/9780784482032.014

[6] A.S. Agar-Ozbek, J. Weerheijm, E. Schlangenand K. Van BreugelInvestigating porous concrete with improved strength: Testing at different scales Construction and Building Materials 41, (2013) pp.480-490.

[7] D. Singhand S.P. Singh, Influence of recycled concrete aggregates and blended cements on the mechanical properties of pervious concrete Innovative Infrastructure Solutions 5(3), (2020), pp.1-12. https://doi.org/10.1007/s41062-020-00314-x

[8]E.J. Elizondo-Martinez, V.C. Andres-Valeri, J. Rodriguez-Hernandezand D. Castro-Fresno, Proposal of a New Porous concrete dosage methodology for pavements Materials, (2019) 12(19), p.3100.

[9] J. Yang and G. Jiang, Experimental study on properties of pervious concrete pavement materials Cement and Concrete Research, (2003), 33(3), pp.381-386. https://doi.org/10.1016/S0008-8846(02)00966-3

[10] M.A. Tijani, W.O.Ajagbe, A.A. Ganiyuand O.A. Agbede, Effect of aggregate type on properties of pervious concrete Journal of Modern Technology and Engineering, (2019), 4(1), pp.37-46.

[11] Vignesh Prabu, V., Shanthini, S., Prasath, S.B., Ramakrishnan, S. Mechanical properties of concrete made with partial replacement of fine aggregate by rice husk ash (2019) Indian Journal of Environmental Protection, 39 (2), pp. 185-189.

[12] IS: 10262-2009, Code of Practice for standard and high-performance concrete mix designs.

[13] Ramakrishnan, S, Velrajkumar, G & Ranjith, S, 'Behavior of Cement-Rice Husk Ash Concrete for Pavement', International Journal of Emerging Trends in Engineering and Development, Vol.1, Issue 4, pp 31-41, January (2014).

A Review on Recent Experimental Research on Soil Stabilization

S. Durgadevi[1,*], A. Dinesh[2], S. Saran[1], G. Selvakumar[1]

[1]Department of Civil, Bannari Amman Institute of Technology, Sathyamangalam, Erode, Tamilnadu, India

[2]Department of Civil, Sri Ramakrishna Engineering College, Coimbatore, Tamilnadu, India

*durgadevi@bitsathy.ac.in

Keywords: Liquefaction, Slope Stabilization, Dynamic, Geo Cell

Abstract. The increasing population results in executing the infrastructure and buildings to be constructed also in weak and soft soil. Therefore, the soil must be strengthened to surrender larger settlements, ground subsidence, etc to resist the collapse of building. In such cases, the necessity of ground improvement has been implemented. The ground improvement techniques is majorly performed for reducing settlement in soft soil, improving the soil bearing capacity, slope stabilization and preventing from earthquake liquefaction. This technique is done for stabilizing the properties of soil and attempted to modify the existing ground. Ground improvement can be done by various techniques i.e., vibro-compaction, dynamic-compaction, ground freezing, Vibro-replacement stone columns, Electro kinetic stabilization, grouting, etc. Recent improvement techniques are introducing Geo cell, Geo-textiles and Geo-membranes has been developed in which the soil is reinforced with materials like aluminum, stainless steel, polyester, fibers, polyamides in the form of strips or grids to stabilize the soil. This paper will give some of the recent techniques adopted in site to improve the soil properties.

Introduction

Due to overpopulation the need for requirement is in peak. In this fast growing world the development of urban areas are playing a major role. This leads to the development of railways, electric power plants, sewage systems and so on. Engineers are in a situation to use both soft and weak soil for construction. Hence construction of bridges, multistory building, transmission towers, etc are unobstructed which can be improved by ground improvement techniques. Also to overcome the ground subsidence in earthquake zones ground improvement techniques are also used.

Ground improvement is a technique in which it improves the engineering properties of soil. It also has various objectives like it improving the bearing capacity of soil, reduces the settlement of soft ground, prevents the liquefaction in the ground, etc. The settlement of soft soil occurs in low land areas, for e.g.: place like Chao Phraya in Thailand. Ground subsidence also occurs in earthquake zones and the main phenomenon for ground subsidence is the liquefaction of granular soils. Geo grid, Vibro compaction, Stone column, grouting, mechanically Stabilized Earth (MSE), etc were the some of the ground improvement techniques. Geo grid is a geo synthetic material which is provided as reinforcement to soft soils. Vibro compaction is done by penetrating the vibrator to the granular soil for compaction, which is then followed by filling with granular soil and the final process is ground level finishing. Stone column is a technique which is as similar to vibro compaction which gives high integrity and interlocks the surrounding soils without ground subsidence. Mechanically Stabilized Earth (MSE)is an artificial reinforcement provided to soil, which is constructed for abutments, retaining walls etc. Grouting is a process in which soil and rock is derived through the fluid grout. This method helps in improving the physical property of the soil without any subsidence.

Ground improvement techniques

Ground Improvement Technique followed by using Natural Prefabricated Vertical Drains (NPVD)

Asha.S.et.al (2012) studied the ground improvement technique using the natural pre-fabricated vertical drains (NPVD), geo cell and geo foam. Natural prefabricated vertical drains (NPVD) and polymer prefabricated vertical drains (PPVD) were compared on the basis of discharge capacity, compressive stresses and hydraulic gradient. The report suggested that the measure flow rate per unit width of NPVD was lesser than PPVD.NPVD were categorized according to materials, out of which experimental analysis has been carried out in comparison of woven jute and coir strand with non-woven jute. Out of which, woven jute and coir strands showed better flow rate results than the non-woven jute and coir strands. On the other hand, the reinforced soil (with geo cell system) shows better results i.e., 31.35mm than unreinforced soil (without geo cell system) i.e., 133.4 mm in the aspect of vertical displacement. From these results, it has been concluded that, the geo cell reinforced soil can resist the accumulation of shear stresses than in unreinforced geo cell system which also consequently provides stability in cohesion less soil. The study also focuses on geo foam embankments which are most likely preferred over soft soils, because the magnitude of horizontal and vertical displacement is lesser than unreinforced embankment. The factor of safety in geo foam embankment is relatively higher i.e., 2.96. It was found that, the NPVD, geo cell and geo foam are preferable in stabilizing soils in the developing countries [1]. Figure 1a and 1b shows the natural prefabricated vertical drains made from coir strands which is then wrapped with woven and nonwoven jute materials.

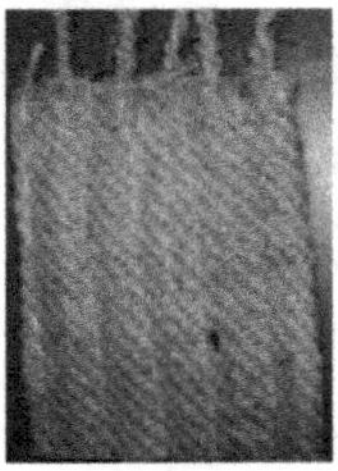

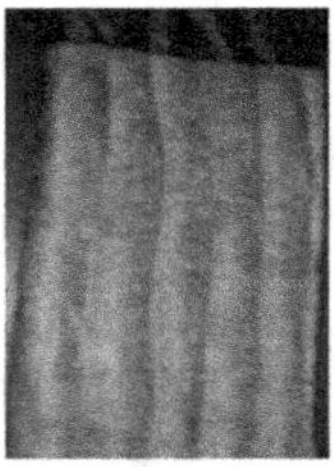

Figure 1a and 1b woven jute and nonwoven jute geotextile.

Test results

Compressive stresses and hydraulic gradient were noted for linear flow rate per unit width. According to ASTM D4716, linear flow rate is measured. The tests were performed at the rates of compressive stresses of 0, 10, 30, 50, 70, 100, 150, 200, 250 and 300kPa at different hydraulic gradients of 0.1, 0.15, 0.25, 0.5, 0.75 and 1.0. Figure 2 shows the test setup for flow rate measurement.

Materials Research Forum LLC
https://doi.org/10.21741/9781644901953-36

Figure 2 Flow rate measurement

A graphical representation is drawn between normal compressive stress (kPa) at x-axis and Flow rate (cubic meter/second) is represented in figure 3.

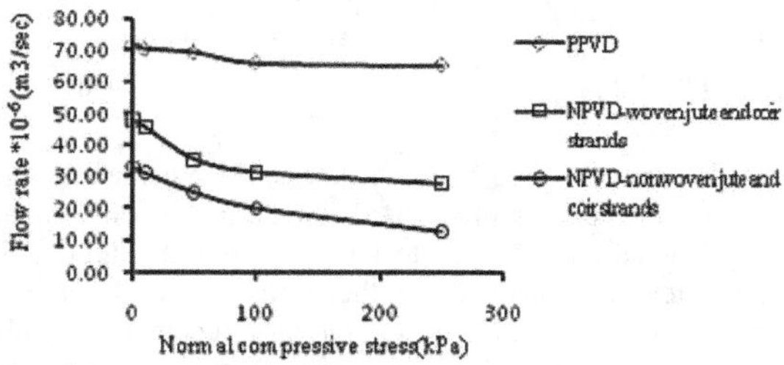

Figure 3 Flow rate versus compressive stress

From the test datas, it showed that NPVD made from woven jute and coir has achieved a better flow rate when compared to nonwoven jute and coir strands. It is found that the flow rates per unit width are comparatively less than PPVD. The reason for less flow rate is due to that compression of core material made of polymer in PPVD is much less than the NPVD coir strands.

Ground Improvement Technique followed by using natural fiber-based Composite-structured geotextiles

Basu.G.et.al (2019) studied the results on Bioengineering of river earth embankment using natural fiber-based Composite-structured geotextiles [2]. Geotextile is made up of jute of HDPE composite structured membrane to hold the soft soils on the river embankments. They have good physical and mechanical properties with a tensile strength of about 10 kN/m. The following are some of the methods involved in the preparation of geotextiles (Jute threads, coconut fibres): Jute threads are prepared using high speed jute spinning machine. Coconut fibers are used to make geotextile net which is done manually by hand knotting method. Testing of geotextile net involves the fiber to place in a humidity of 65+5% and a temperature of 27+2 oc. Tensile strength of geotextile net involves the Instron tensile tester as per BIS 1670 (2002).Mechanical strength of fibre involves the testing of sample in tensile testing machine as per ASTM D4595 (1994). Water

Materials Research Proceedings **23** (2022) 311-320 https://doi.org/10.21741/9781644901953-36

permeability of geotextile sample is tested by flowing water perpendicularly to the geotextile sample with a constant water pressure. From this methodology, hydraulic property of a fiber is known. Table 1 shows the different properties of geotextile polymers

Table 1 shows properties of polymers used as geotextiles.

Type of Fiber	Specific Gravity (kg/m^3)	Modulus of Elasticity (GPa)	Tensile Strength (MPa)	Elongation at break (%)
Polypropylene	910	10	780	55
Polyethylene terephthalate	1400	1.5	25	95
Polyamide	1140	6.5	65	45
Polyethylene	940	65	550	72

Area of the geotextile, size of the mesh and thickness of the geotextile sample was referred as per BIS (2002). Experimental study involves digging a toe-trench longitudinally along both sides of the embankments and then the slopes were leveled to provide a point to point contact to the soil and geotextile net. Geotextiles was laid on the surface from toe-trench to other side of embankment. Coconut fiber is also laid over it. Composite geotextile member is fixed by using matured bamboos and metallic U shaped hooks. Geotextile tubes were placed over the geotextile and coconut net. Then they are covered with aggregates and sand to form a filter and this helps in sealing the flow of soil from embankment. Finally the embankment is covered with sandy soil. Bioengineers concluded that it has two major functions like filtration and reinforcement. Effective production in vegetative growth through the geotextiles is the reason for improvement of river-bank condition. The monitoring process has been done for about 8 years. The result shows there isn't any subsidence and migration of slope even after 8 years. The cost of this project is based on three main things like raw materials, cost of machine and transportation.

Ground Improvement Technique followed by using mechanical stabilization, stabilization by preloading method

Brajesh Mishra (2016) carried out a research in ground improvement techniques and its applications. The analysis which includes mechanical stabilization, stabilization by preloading method, stabilization by sand drain method, physical and chemical modification technique, modification by addition and confinement techniques, sand pile using blasting. Mechanical stabilization is a method of improving the density of granular soils by compaction. The compaction is done by applying a mechanical force which acts on the soil by means of static, vibratory rollers and plate rollers. Stabilization by preloading method provides stability to loose soil, soft soil even to clay and organic silts. This technique involves the compaction and densification of a loose grained soil by the application of loading. The Consolidation process for a compressible layer takes more time as it has to penetrate through the dense layers and reach the surface of the drainage.

This conflict can be solved by introducing sand drains or vertical drains of higher permeability as it helps in reducing the time taken for draining the water present in the soil which easily outreach the drainage surface. Physical and chemical modification technique involves the insertion of waste materials, soils and industrial by-products under the pressure which is also known as grouting. In this modification technique, the soil is compacted by alternate heating and freezing. Modification by addition and confinement techniques involves adding reinforcement to the soil in the form of meshes, fibers and strips. It improves the tensile strength of the soil. The main advantage of this technique is that, it does not involve the migration of soils and keeps the soil as opaque. Sand pile using blasting necessitates the hidden explosions and extracts the soils which then backfill the excavated area. This process enlarges the soil intensity and makes the soil compacted.

Ground Improvement Technique for construction on peat

Duggan A.et.al (2014) performed a study on factors affecting embodied carbon and embodied energy associated with ground improvement techniques for construction on peat. The study focuses on modification or removal of peat and emerging of Embodied Carbon (EC) and Embodied energy (EE) which are taken in consideration in ground improvement techniques. Peat lands were improved as they have high moisture content, low shear strength and creep. This can be achieved by soil-mixing, piling and replacement methods. As it is difficult to predict the settlement in peat type of soil and also the time period for secondary consolidation is excessive, usual ground improvement techniques for soft soil cannot be preferred. Hence, the most reliable method for building roads in peat is considered as Excavate and Replace method. By restoring and rewetting the peat land, CO2 and N2O emission is largely reduced. For stabilizing the peat disposal sites, soft rushes and sphagnum were planted which also helps in binding the peat.

Ground Improvement Technique for construction on peat

Indraratna.B.et.al (2010) studied the system of combined form of vertical drains including vacuum and surcharge preloading and this method is considered as one of the productive methods for the consolidation of soil. Analytical models were used to study the consolidation process and the changes in the properties of the multi layered soil through spectrum method by absorbing vacuum preloading and smear well resistance. The accurate soil behavior was observed when it was exposed to vacuum pressure can be predicted from the demonstration of spectral model. The cylinder specimen were examined containing different percentage of cement and surcharges. The effect of surcharge were compared and is shown in figure 4.

From figure 4, it is evident that, the diffraction intensity of $CaCO_3$ increased to even 4000 a.u, when the carbonation time increased even to about 180days. Due to this, the effect the calcium hydroxide content is slowly diminishing.

Experimental study on Ground improvement using chemical methods.

The recent study shows that Ground improvement plays an vital role in the construction of structures such as, Dams, Petroleum storage yards, large Skyscrapers, Highways and other important structures. The four major polymeric materials used in soil as binders such as Epoxy resins, Polyacrylamide, Polyurethane and Lignosulfonates. The study shows that, hydraulic conductivity decreased from 1.45×10^{-2} to 7.52×10^{-4} cm/s for a Poly Urethane concentration of 4–8%, making the soil type to make stable against water.

Materials Research Proceedings **23** (2022) 311-320 https://doi.org/10.21741/9781644901953-36

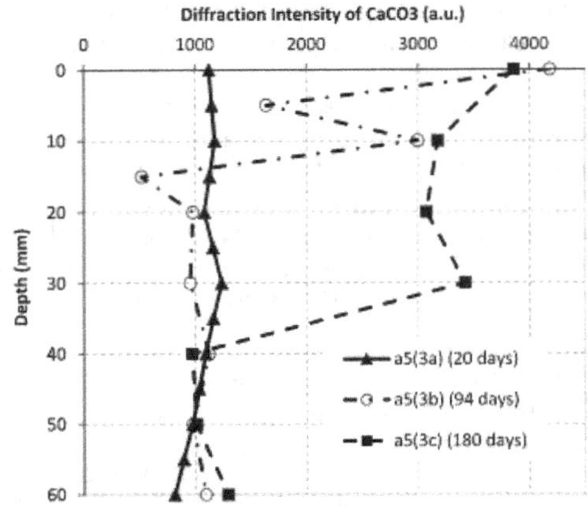

Figure 4 Proportions of CaO diffraction intensity for varying depths and duration.

Another chemical, named, lesser quantity of Lignosulfonate have a greater potential against the soil erosion by improving the tensile strength.

Figure 5 & 6 shows SCM of the kaolinite sample for various percentage of cement. Initially, cement were added in the proportions of 5%, 10%, inaddition, the combination of epoxy resin + kaolinite and 10% cement + kaolinite + epoxy resin is taken as samples.

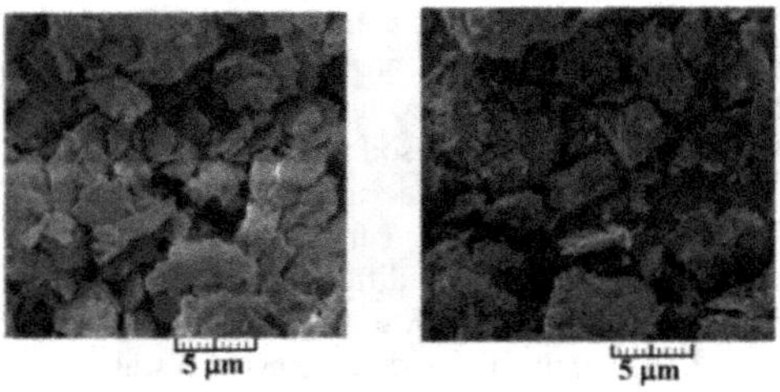

Figure 5a Kaolinite and Figure 5b kaolinite + 10% cement

Figure 6a Epoxy resin and Kaolinite, Figure 6b Kaolinite +10% cement + Epoxy resin

It is seen that, by adding these admixtures, it reduces the development of high intensity of pore pressure. Also the dispersion of particles are gives the better flow capability. From the study, it is observed that, kaolinite is more effective than illite.

Experimental study on bearing capacity of geo cell reinforcement in embankments.
Ling Zhang.et.al (2009) discussed about the experimental study on bearing capacity of geo cell reinforcement in embankments. Geo cell is a three dimensional geo synthetic cellular network formed from thin polymer strips. The beneficial use of geo cell is that it provides a working platform for construction, reduces migration of soil, reduction of ground subsidence and it has high bearing capacity. The study focuses about the comparison of ultimate bearing loads by constructing the embankment with and without geo cell. In first method, concrete tank was installed under a loading frame and it was filled with clayey sand. Then three layers of sub grade soil were filled. Two soil pressure cells were fixed in a clay bed for measuring the vertical soil pressure. Then the embankment is constructed over the soft soils which are then compacted using wooden rod. With the use of hydraulic jack, a surcharge load is applied to the embankment. The load is gradually increased until the surface reaches the settlement of above 55 mm. Second method is as same as first method, but it involves the installation of geo cell between soft soil and the embankment. After leveling the soft clay bed, a layer of crushed stones was filled and geo cell is placed over that. The hollow region of geo cell was in-filled with crushed stones and leveling was done using wooden rod. After that the embankment is constructed and the load is applied over the embankment. The ultimate bearing load carrying capacity was increased by 78.22% when using geo cells are used.

Experimental study on shear strength of recycled construction materials intended for use in vibro ground improvement.
Mckelvey.D.et.al (2002) investigated the shear strength of recycled construction materials intended for use in vibro ground improvement. This paper aims on the replacement of granular materials. They made a comparative study between crushed concrete, building debris, quarry waste. Crushed concrete and building debris were tested by using a shear box of suitable dimension. In this test different vertical pressures were applied and each layer of the respective material was compacted using proctor hammer. To systemize the energy provided to the setup they have used a metal plate of suitable dimension for crushed concrete and building debris. The same procedure was followed for quarry waste with a replacement of wooden block placed over it during the supply of load. As a result it is concluded that the usage of recycled materials instead of naturally available resources has a good impact. Especially high grade recycled concrete wastes possess better strength and are a good refilling material. Usage of these materials does not affect the overall performance in construction. It was also observed that addition of such materials in the construction should be careful as there are possibilities for the materials to get smudged (contamination).

Experimental study on soil improvement using additives.
Mohammad Bilal.et.al (2016) gives a brief study on soil improvement using additives, by mechanical method, soil improvement without using admixtures, preloading and thermal methods. Soil improvement using additives involves the usage of additives like lime, cement and fly ash. Research study shows the improvement in load bearing capacity, workability, stiffness and stability of the soil sample. Soil improvement by mechanical method deals with the compaction of soil by rollers, vibrators, etc. This method is followed by Stone column, Vibro Floatation, Micro Piles and Soil Nailing. Stone column method involves the process of extraction and backfilling the

Materials Research Proceedings **23** (2022) 311-320 https://doi.org/10.21741/9781644901953-36

excavated area by stone filling. Vibro compaction is same as stone column and the granulated soil is used as the backfill material. Micro pile is a process of inserting the steel piles which effectively resist the compressive, tensile or lateral loads. Soil Replacement involves the replacement of soft soil with well compacted suitable materials which involves techniques like Soil Replacement, Vertical Drains, Sand Drains, Pre-Fabricated Vertical Drains (PVDs). Soil improvement using thermal involves heating and freezing. Strength of the soil can be improved by heating and freezing involves the reduction in temperature which causes the water to freeze and it acts as cementing agent.

Experimental study on ground improvement techniques involving vibro-compaction, vibro-flotation.

NimishaKachra.et.al (2016) discussed about the experimental study on ground improvement techniques involving vibro-compaction, vibro-flotation, back-fill material, vibro-displacement, soil-nailing and grouting. Vibro-compaction is mostly suitable for loose soil, which compacts and densifies the soil using vibrator. Vibro-flotation is mainly implemented for the soil which does not have optimum relative density. In this method, oscillator vibrator is used to achieve a densely packed soil particle thereby reducing the displacement. Backfill material is used to refill the same dug out or excavated area of a particular region. Vibro -displacement method carries the process of displacing the soil. Soil nailing is a remedial construction technique which involves the reinforcement bars to treat the unstable soil structures. Grouting is the method of filling the holes of excavated area with closely compacted material i.e., denser material in the place of reinforcement.

Experimental study on bearing capacity of embedded strip foundation on geo grid-reinforced sand.

Patraa.C.R.et.al (2005) studied the bearing capacity of embedded strip foundation on geo grid-reinforced sand. Geo grids are polymeric products which come under geo synthetic materials used to reinforce the soft soil structures. This is considered to be an emerging technology which can be used as an alternative of geotextile. These grids are used to overcome the tension force which pulls the soil aside. The foundation used for this experiment was made out of mild steel plate. The base of the foundation was made rough by applying glue and rolling it on sand. The inner surface walls of mild steel plate were polished to decrease its friction rate. The foundation and the geo grid were placed beneath the soil surface at a desired height. Different loads were applied on this setup and the results were recorded until the setup reached its failure point. These types of experiments can be conducted on weak cohesive soil which is useful for the determination of bearing capacity of the soil, the efficiency of reinforcement.

Experimental study on development of ground improvement techniques being introduced in the recent time and old days.

Dr. Tiwari S.K.et.al (2014) investigated the development of ground improvement techniques being introduced in the recent time and the methods which were used in olden times. The techniques were Mechanical improvement technique, Hydraulic modification, Physical and chemical modification. Mechanical improvement technique involves the application of mechanical force, in which the density of the soil is increased. Static or fixed type vibrators are used in such cases. In Hydraulic modification, the refinements of soil properties are attained by pressurizing the pore water out from the soil by means of wells or drains. The water level is reduced in case of coarse grained soil by laying bores or trenches. For fine grained soil, external or artificial loads are applied. While dealing with Physical and chemical modification methodology, adhesive materials are mixed physically which are used to hold up the constituents together. Natural soil wastes or cementing materials can also be used. In some instances, the adhesives are inserted through grouting.

Experimental study on Ground improvement techniques by compaction piling
Zoltan V.Solymar.et.al (1986) discussed Ground improvement techniques by compaction piling. This experiment is done on loose silty fine sand which is prone to liquefaction caused by seismic disturbances. To prevent liquefaction in such loose soils and to reduce the settlement, following tests were conducted and results were recorded. Cone Penetration Test (CPT) shows the relation between relative density and the volume of soil sample taken. Standard Penetration Test (SPT) gives that resistance against penetration in loose silty fine sand is uniform after compaction. After performing the tests, it showed slight increment in the results. It was also studied that, this type of sand does not liquefy much during seismic occurrence. Buckling or horizontal movement or failure of steel piles may happen due to liquefaction in loose silty fine sand and results in decreasing the lateral support of steel piles. Compaction piling is done to get rid of these difficulties and also stabilize the soil. Compaction piling involves the process of drilling of holes and back filing the hole with layered sand. The usage of compaction pile increases the permeability, densifies the soil against earthquake, and increases the firmness of the soil.

Conclusion
Due to population increase and rapid urbanization there is a high demand for stable and firm sand for construction of buildings as well as highways. Improving the properties of loose soil and soft clay by ground implement techniques prevents the settlement and collapse of the structures built on it. The stability and bearing capacity of the soil has been raised by controlled compaction method or by addition of admixture or stabilizer. The most effective method of stabilization of fine-grained soil is Terra Zyme. On the other hand, high permeable soils are stabilized by grouting. Electrical and chemical stabilization also involves improving the soil properties which are expensive compared to other modes of stabilization. A system of Prefabricated vertical drains (PVDs) combined with vacuum preloading is an effective method of for accelerating soil consolidation. From the test results, it is evident that, Natural vertical drains are comparatively less than polymer prefabricated drains. But, still further more studies are required on Natural polymer vertical drain.

The unstable slopes of soils were economically remediated by soil nailing. Geo-textile and Geo-membrane are widely used nowadays for improving soil characteristics. Although variety of improvement techniques have developed, the suitable method has been adopted based on the economy needed and structure to be constructed. A combination of suitable techniques based on the type of soil, type of project may provide a most sustainable solution for ground improvement solution.

References

[1] B.S. Asha, B. Ram Rathan Lal, A. H. Padade, T. Mandal, J. N. Mandal (2012), "Emerging trends in ground improvement techniques", Journal of Geotechnical and Geo-Environmental Engineering, Geo Congress, ASCE, pp-594-603. https://doi.org/10.1061/9780784412121.062

[2] Basu G (2019), "Bio Engineering of river earth embankment using natural fiber based composite structured geotextiles", Geotextiles and Geo membranes, pp 266-1144.

[3] Brajesh Mishra (2016), "A study on ground improvement techniques and its applications", International Journal of Innovative Research in Science, Engineering and Technology, 5, pp 72-86.

[4] A. Duggan, B.A. McCabe, J. Goggins, E. Clifford (2014), "Factors affecting embodied carbon and embodied energy associated with ground improvement techniques for construction on peat", Journal of Cleaner Production, Volume 271, 122540. https://doi.org/10.1016/j.jclepro.2020.122540

[5] B. Indraratna, C. Rujikiatkamjorn, A.S. Balasubramaniam, G. McIntosh (2010), "Soft ground improvement via vertical drains and vacuum assisted preloading", Geotextiles and Geomembranes, 30, pp 16-23. https://doi.org/10.1016/j.geotexmem.2011.01.004

[6] HarshalVerma, ArunavaRay, RajeshRai, TusharGupta, NeerajMehta, "Ground improvement using chemical methods: A review", Heliyon Volume 7, Issue 7, July 2021, e07678 https://doi.org/10.1016/j.heliyon.2021.e07678

[7] Ling Zhang (2009), "Study on bearing capacity of geo cell reinforcements in embankments", Geotextiles and Geomembranes, 28(5), pp 475-482. https://doi.org/10.1016/j.geotexmem.2009.12.011

[8] Mckelvey D (2002), Shear strength of recycled construction materials intended for use in vibro ground improvement", Proceedings of the Institution of Civil Engineers- Ground Improvement, 6 (2), pp 59-68. https://doi.org/10.1680/grim.2002.6.2.59

[9] Mohammad Bilal, Abdullah Talib (2016), "A study on advances in Ground Improvement techniques and its applications", National conference on Advances in Geotechnical Engineering, at Aligarh.

[10] Nimisha Kachra, Megha k Patel (2016), "Ground Improvement techniques" Global Research and development Journal for Engineering, pp 407-414.

[11] Patraa C R (2005), "Bearing capacity of embedded strip foundation on geo-grid reinforced sand", Geotextiles and Geomembranes, 23(5), pp 454-462. https://doi.org/10.1016/j.geotexmem.2005.02.001

[12] Dr S K Tiwari, N K Kumawat (2014), "Recent development in Ground Improvement Techniques", IJRDET, (2), pp 67-77.

[13] Zoltan V Solymar (1986), "Ground Improvement Techniques by compaction piling", Journal of Geotechnical Engineering, 112(12), pp 1069-1083. https://doi.org/10.1061/(ASCE)0733-9410(1986)112:12(1069)

Investigation on Strength Properties of Concrete using Marble Powder for Ecofriendly Environment

R. Ashwathi[1,a,*], B. Jeyanth[1,b], V. Prabhakaran[1,c], O.P. Deepak[1,d]

[1]Department of Civil Engineering, Bannari Amman Institute of Technology, Erode, India

[a]ashwathisingh@gmail.com, [b]prabhakaranv@bitsathy.ac.in, [c]jeyanthbaskaran5890@gmail.com, [d]deepak@bitsathy.ac.in

Keywords: Concrete, Natural Aggregate, Marble Powder, Eco Friendly, Economical

Abstract. The development of a country is well exhibited through its infrastructure and technology, where civil engineering plays an influencing role in the development of a nation. In the field of construction, concrete is the major material that is used for building purpose. Particularly, natural aggregates are involved in making concrete sound better. The natural aggregates are getting diminished due to its extensive practice. If the same situation prolongs, the natural aggregate gets extinct. This paper majorly focuses on making the environment better and in order to achieve sustainability and to preserve the environment, marble power is employed as partial material in the place of fine aggregate. On the other side, it aims at attaining the strength properties using M20 grade without any compromise. The test results clearly exhibits that incorporation of marble waste in making concrete is eco friendly as well as economical.

Introduction

The main source of making concrete is natural aggregate. There in making concrete, fine aggregate, coarse aggregate and cement are majorly added. In addition to it, admixtures and other materials are added if needed. Owing to the over exploitation of river sand, the natural resources are getting depleted. If the same situation prevails, after certain years there would not be any natural materials. And on the other side, the waste and the byproducts coming out from various industries are dumped into the land even without proper treatment which is causing adverse impact on environment. In order to balance the ecosystem and the depletion of resources, the attempt in incorporating industrial byproducts and municipal byproducts which have become successful [1]. Even though, it is successful, the research is being conducted in finding its optimum percentage of usage and identifying the suitable admixture to achieve the strength and required property [2]. This incorporation of certain raw materials and byproducts provides new dimension to the construction field and promotes the environment towards sustainability [3].

Marble waste is one of the wastes coming out from marble manufacturing industry, which is found in abundant places [4]. These wastes may differ in size. And according to its nature of size and treatment adopted, it may be incorporated in making concrete either as fine aggregate or coarse aggregate as partial replacement [5]. The strength and properties are found to be the same for the marble waste and this makes clear that the strength of concrete does not get affected [6]. By making use of these options we can make protect our environment and the generation of waste also gets decreased. Here lies the emerging concept of wealth from waste which paves way for economy of the nation and provides better sustainability [7]. Also, this helps in building the economy of a nation thereby paving way for energy conservation.

Objective

1. To adopt the concept of wealth from waste for energy consumption.
2. To identify an alternative material for natural aggregate.
3. To indicate the properties of marble waste.
4. To find the optimum percentage of replacement.

Materials and properties

Concrete is a building material composed of homogeneous mixture of natural aggregates with cement, the binding material. In order to bind these materials together and to make it workable, water is added to the mixture [8]. The mixture is prepared and the cube is casted according with mentioned proportions.

Portland Pozzolana cement

Portland Pozzolana Cement conforming to IS standard is adopted and the test is conducted [9]. The cement adopted is free from lumps and dust.

Table 1. Physical properties of OPC cement.

Composition	Percentage
Calcium Oxide, CaO	61-67%
Silicon oxide, SiO2	19-23%
Aluminium Oxide, Al2O3	2.5-6%
Ferric Oxide, Fe2O3	0-6%
Sulphate	1.5-4.5%

Table 2. Chemical properties of OPC cement.

Property	Value
Consistency	30%
Setting time - Initial	30 minutes
Setting time - Final	600 minutes
Fineness Modulus	3.5 %
Specific gravity	3.15

Fine aggregate

The fine aggregate confining to IS: 2386 (Part III) - 1963 of zone-II grading has been successfully adopted and various tests have been conducted [10]. Fine aggregate free from dust and impurities are used.

Table 3. Physical properties of fine aggregate.

Property	Value
Specific gravity	2.43
Fineness modulus	2.88
Grading	Zone –II

Materials Research Proceedings **23** (2022) 321-328 https://doi.org/10.21741/9781644901953-37

Coarse Aggregate
The coarse aggregate confining to IS: 383-1970 of Zone-II grading were used [11]. The coarse aggregate of size 20mm size is adopted.

Table 4. Physical properties of coarse aggregate.

Property	Value
Specific gravity	2.71
Fineness modulus	6.68
Grading	Zone –II

Marble powder
Marble is one of the floor finishing materials used in the construction field. It is generally a metamorphic rock obtained as a transformation of pure limestone [12]. The color of the marble purely depends on its purity. Marble powder is obtained during the time of marble production at the industry [13].

Water
Water is another significant material in the process of concrete preparation. This makes all the materials to mix together and make it workable. It is also important to ensure proper water cement ratio in order attain the required strength [14]. The water used should be clean. It has to be free from dust, impurities and chemical contaminants.

Strength test

Compressive strength
Compressive strength of concrete is one of the prominent characteristics of concrete. The compressive strength of the concrete is determined using cube mould [15]. The mixture is prepared with different proportion of 10, 20, 30, 40 and 50% of marble powder with fine aggregate. The cube is casted and proper curing is done at 7th, 14th and 28th day of interval [16].

The compressive strength of concrete with different proportions of 10, 20, 30, 40 and 50% of marble powder with fine aggregate are shown below.

Table 5. Results of Compressive strength test

Mix	Compressive strength (N/mm²)			Average compressive strength(N/mm²)		
	7 days	14 days	28 days	7 days	14 days	28 days
CM	10.58	16.97	20.08	11.42	17	19.65
	12.35	17.29	19.77			
	11.33	17.09	19.02			
10%	13.37	20.08	22.2	12.9	19.62	23.13
	12.75	19.77	24.04			
	12.59	19.02	23.15			
20%	16.8	22	24.96	16	22.83	25.3
	16.26	22.01	25.91			
	14.96	24.05	25.06			
30%	16.96	22.94	23.95	17.05	24.38	20.73
	16.24	23.82	19.33			
	17.97	26.4	18.93			
40%	14.48	23.83	17.15	15.7	22.02	17.58
	16.8	22.06	18			
	15.8	20.17	17.06			
50%	15.56	11.02	15.37	14.55	15.23	17.07
	12.71	14.51	14.44			
	15.39	20.17	21.42			

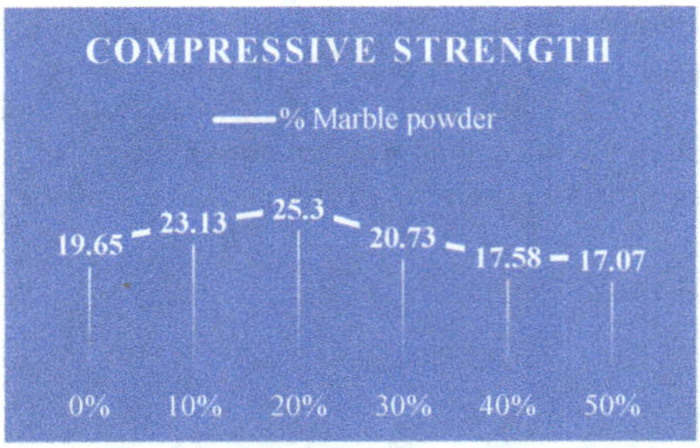

From the examination, it is clearly witnessed that the compressive strength increased gradually and after a stage on additional increase in percentage of marble powder, the compressive strength gets decreased.

Split tensile strength

The split tensile strength of hardened concrete is useful in determining the performance of concrete [17]. Split tensile strength test is obtained by casting and appropriate curing at different ages of 7, 14, 28 days [18].

Table 6. Results of Split tensile strength test

Mix	Split tensile strength (N/mm^2)			Average Split tensile strength(N/mm^2)		
	7 days	14 days	28 days	7 days	14 days	28 days
CM	1.34	1.59	2.02	1.35	1.67	1.92
	1.4	1.71	1.92			
	1.3	1.70	1.82			
10%	1.42	2.01	2.21	1.43	1.95	2.35
	1.42	1.95	2.49			
	1.44	1.90	2.35			
20%	1.48	2.19	2.53	1.51	2.27	2.56
	1.5	2.20	2.62			
	1.54	2.41	2.52			
30%	1.1	2.28	2.45	1.16	2.39	2.12
	1.18	2.38	1.99			
	1.2	2.52	1.92			
40%	1.3	2.41	1.75	1.31	2.21	1.76
	1.32	2.21	1.82			
	1.32	2.02	1.71			
50%	1.18	1.15	1.57	1.15	1.56	1.72
	1.16	1.41	1.41			
	1.1	2.11	2.19			

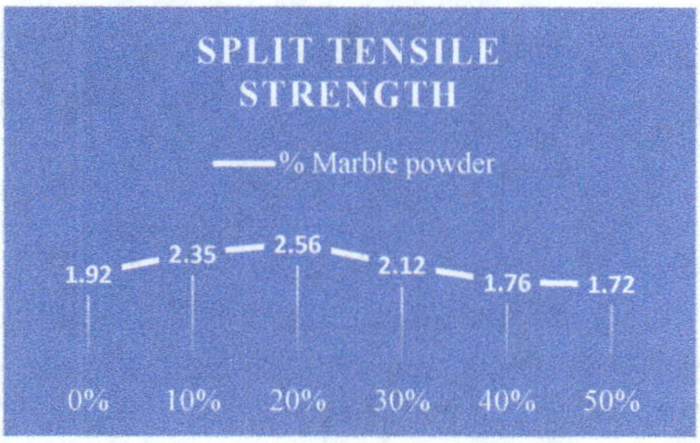

From the test conducted, it is clearly perceived that the split tensile strength increased gradually and after a stage on further increase in percentage of marble powder, the split tensile strength gets decreased.

Flexural strength

The flexural strength of hardened concrete is useful in determining the performance of concrete [19]. Flexural strength test is obtained by casting and appropriate curing at different ages of 7, 14, 28 days [20].

Table 7. Results of Flexural strength test

Mix	Flexural strength (N/mm²)			Average Flexural strength (N/mm²)		
	7 days	14 days	28 days	7 days	14 days	28 days
CM	2.29	2.89	3.14	2.38	2.90	3.12
	2.48	2.91	3.12			
	2.37	2.90	3.09			
10%	2.61	3.15	3.31	2.54	3.11	3.38
	2.51	3.11	3.44			
	2.50	3.07	3.39			
20%	2.90	3.29	3.52	2.83	3.33	3.54
	2.85	3.28	3.58			
	2.73	3.41	3.51			
30%	2.87	3.35	3.45	2.88	3.46	3.22
	2.81	3.42	3.15			
	2.95	3.61	3.07			
40%	2.69	3.42	2.95	2.77	3.29	2.99
	2.84	3.29	3.12			
	2.79	3.16	2.89			
50%	2.78	2.35	2.81	2.69	2.68	2.91
	2.53	2.65	2.72			
	2.75	3.05	3.20			

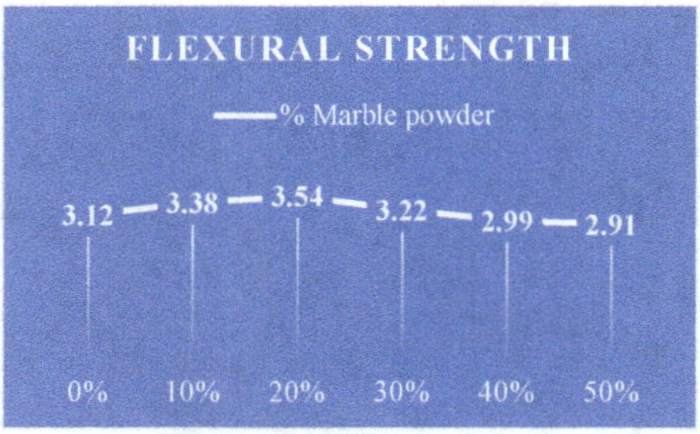

From the test conducted, it is clearly witnessed that the flexural strength increased gradually and after a stage on further increase in percentage of marble powder, the flexural strength gets decreased.

Conclusion

The tests had been conducted on hardened concrete with different proportion of marble powder as partial replacement for fine aggregate and the results have been noted. From the results, the following observations are made.

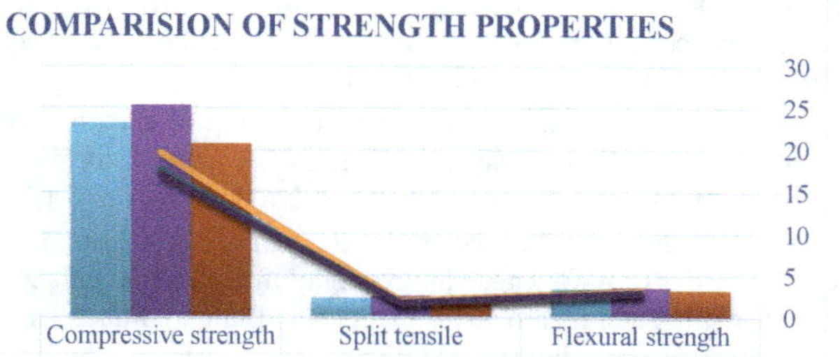

	Compressive strength	Split tensile	Flexural strength
10%	23.13	2.35	3.38
20%	25.3	2.56	3.54
30%	20.73	2.12	3.22
0%	19.65	1.92	3.12
40%	17.58	1.76	2.99
50%	17.07	1.72	2.91

10% 20% 30% 0% 40% 50%

- The compressive strength of concrete increases up to 20% of replacement.
- The split tensile strength of concrete increases up to 20% of replacement.
- The flexural strength of concrete increases up to 20% of replacement.

From these results, it has become evident that, the optimum percentage of replacement for fine aggregate by marble powder is 20. Also, it has been observed that, the decrease in strength beyond 20% is due to the nature and strength of the material. This evidently paves way for attaining better strength and sustainability.

References

[1] Aliabdo, A.A., Abd Elmoaty, A.E.M., & Auda, E.M. (2014). Re-use of waste marble dust in the production of cement and concrete. Construction and Building Materials, 50, 28-41. https://doi.org/10.1016/j.conbuildmat.2013.09.005

[2] Alyamaç, K.E., & İnce, R. (2009). A preliminary concrete mix design for SCC with marble powders. Construction and Building Materials, 23, 1201-1210. https://doi.org/10.1016/j.conbuildmat.2008.08.012

[3] André, A., Brito, J., Rosa, A., & Pedro, D. (2014). Durability performance of concrete incorporating coarse aggregates from marble industry waste. Journal of Cleaner Production, 65, 389-396. https://doi.org/10.1016/j.jclepro.2013.09.037

[4] Belaidi, A.S.E., Azzouz, L., Kadri, E., & Kenai., S. (2012). Effect of natural pozzolana and marble powder on the properties of self-compacting concrete. Construction and Building Materials, 31, 251-257. https://doi.org/10.1016/j.conbuildmat.2011.12.109

[5] Binici, H., Kaplan. H., & Yilmaz, S. (2007). Influence of marble and limestone dusts as additives on some mechanical properties of concrete. Scientific Research and Essay, 2, 372-379.

[6] Ceylan, H., & Manca, S. (2013). Evaluation of Concrete Aggregate Marble Pieces. SDU Journal of Technical Science, 3, 21-25.

[7] Corinaldesi, V., Moriconi, G., & Naik, T.R. (2010). Characterization of marble powder for its use in mortar and concrete. Construction and Building Materials, 24, 113-117. https://doi.org/10.1016/j.conbuildmat.2009.08.013

[8] Ergun, A. (2011). Effects of the usage of diatomite and waste marble powder as partial replacement of cement on the mechanical properties of concrete. Construction and Building Materials, 25, 806-812. https://doi.org/10.1016/j.conbuildmat.2010.07.002

[9] Geseoglu, M., Guneyisi, E., Kocabag, M.E., Bayram, V., & Mermerdas, K. (2012). Fresh and hardened characteristics of self-compacting concretes made with combined use of marble powder, limestone filler, and fly ash. Construction and Building Materials, 37, 160-170. https://doi.org/10.1016/j.conbuildmat.2012.07.092

[10] Hameed, M.S., & Sakar, A.S.S. (2009). Properties of Green Concrete Containing Quarry Rock Dust and Marble Sludge Powder as Fıne Aggregate. ARPN Journal of Engineering and Applied Sciences, 4, 83-89.

[11] Kaseva, M.E., & Gupta, S.K. (1996). Recycling-an environmentally friendly and income generating activity-towards sustainable solid wastes management. Resources, Conservation and Recycling, 17, 299-309. https://doi.org/10.1016/S0921-3449(96)01153-6

[12] Malpani, R., Jegarkal, S.C., Shepur, R., Kiran, R., & Adi, V.K. (2014). Effect of Marble Sludge Powder and Quarry Rock Dust as Partial Replacement for Fine Aggregates on Properties of Concrete. International Journal of Innovative Technology and Exploring Engineering (IJITEE), 4, 39-42.

[13] Omar, O.M., AbdElhameed, G.D., Sherif, M.A., & Mohamadien, H.A. (2012). Influence of limestone waste as partial replacement material for sand and marble powder in concrete properties. Housing and Building National Research Center (HBRC) Journal, 8, 193-203. https://doi.org/10.1016/j.hbrcj.2012.10.005

[14] Baboo R., Khan N.H., Abhishek Kr., Tabin R.S., & Duggal S.K. (2011). Influence of Marble powder/granules in Concrete mix. International Journal of Civil and Structural Engineering, 1, 827-834.

[15] Soliman, N.M. (2013). Effect of using Marble Powder in Concrete Mixes on the Behavior and Strength of R.C. Slabs. International Journal of Current Engineering and Technology, 3, 1863-1870.

[16] Soykan, O., & Ozel, C. (2012). Effects to Polymer Concrete Properties of Particle Size of Marble Dust. SDU International Journal of Technologic Sciences, 4, 102-111.

[17] Tayeb, B., Abdelbaki, B., Madani, B., & Mohamed, L. (2011). Effect of Marble Powder on the Properties of Self-Compacting Sand Concrete. The Open Construction and Building Technology Journal, 5, 25-29. https://doi.org/10.2174/1874836801105010025

[18] Topcu, I. B., Bilir, T., & Uygunoglu, T. (2009). Effect of waste marble dust content as filler on properties of self-compacting concrete. Construction and Building Materials, 23, 1947-1953. https://doi.org/10.1016/j.conbuildmat.2008.09.007

[19] Uysal, M., & Yilmaz, K. (2011). Effect of mineral admixtures on properties of self-compacting concrete. Cement & Concrete Composites, 33, 771-776. https://doi.org/10.1016/j.cemconcomp.2011.04.005

[20] Vaidevi, C. (2013). Study on marble dust as partial replacement of cement in concrete. Indian Journal of Engineering, 4, 14-16.

Effect of Damping on Multistory Frame using Response Spectrum

Preetha Vellaichamy[1,a*], Vishalachi Chakkaravarthi[1], Senthilkumar Veerasamy[2], Aravindhan Chokkan[1]

[1]Department of Civil Engineering, Bannari Amman Institute of Technology, Erode.

[2]Department of Civil Engineering, M.Kumarasamy College of Engineering, Karur.

[a]preethav@bitsathy.ac.in

Keywords: Response Spectrum Method, Story Displacement, Story Drift, Story Shear

Abstract: The absolute response of the multi-storey frame is ensured by elastic and damping forces. The peak structural response of the frame is obtained using response spectrum analysis under linear range to obtain the lateral forces developed in the structure situated at zone V. In this paper the influence of damping ratio on irregular multi-storey frame on elastic response spectra have been investigated. The analytical results show that the impact of damping ratio and modal response combination are significant on the shape of response spectra. The impacts on storey drift and storey displacement parameters are also studied for gradual increase in damping ratios.

Introduction

The substantial elastic strain energy is released by the unexpected slip at the fault so that seismic waves move throughout the body and also along the surface, causing a violent shaking of the Earth. The seismic motions are caused by the generation of seismic waves with violent energy release that ruptures at the fault causes ground shaking on the Earth's surface. These waves arrive at various times, have varying amplitudes, and convey varying amounts of energy. As a result, motion at every point on the ground is random in nature, with amplitude and direction changing at random times [1]. Technology has evolved in order to combat these calamities, such as establishing disaster warning systems and implementing preventative measures. It occurs suddenly for a few seconds, depending on the severity of the earthquake, and causes minor, moderate, or significant loss of life and property. It's manageable, but not entirely eradicated. As a result, today's global worry is a reduction approach. The visual representation and precision of earthquakes have been used to assess the seismic quality of building complexes on several occasions [2]. Structural engineers confront non-convergence and a number of difficulties when endeavouring to assess the seismic quality of building complexes [3]. Adding viscous dampers may be done using a CSM-based design process. Despite its flaws, their technique is useful for determining the optimal damping ratio [4]. In recent years, several papers have been published on analytical approaches for estimating earthquake's physical damage depending on the energy input to the structure. These concepts are, however, only applied in a few real-world structural design applications [5]. In seismic layout and structure study, response spectrum analyses are often utilised. Typically, a nominal critical damping of 5% is used to develop seismic ground motion response spectra [6]. These damping techniques are referred to as viscous damping for the sake of convenience. Both viscous and hysteretic damping is included in the concept of essentially equivalent viscous damping in seismic analysis and design [7]. The direct displacement-based techniques have been developed for focussing a structural seismic layout and analysis research in recent years. The direct displacement method works by replacing a multi degree of freedom structure with an equivalent

Materials Research Proceedings **23** (2022) 329-339 https://doi.org/10.21741/9781644901953-38

single degree of freedom structure, referred as replacement structure. They are distinguished by viscous damping and secant stiffness (elastic and hysteretic damping) that correspond to the optimal displacement response [8]. The viscous damping of replacement structure generally has critical damping greater than 5%. Heat transfer devices have been used to improve the seismic performance of critical structures on several occasions. In the first mode response of a structural system, frictional sliding, metal yielding, viscoelastic material deformation, fluid orificing, and metal phase change can all supply up to 40% critical damping. Despite the fact that various supplemental damping devices' energy dissipation qualities are not perfect viscous, they may be linked to an equal damping ratio [9]. Response spectra for damping levels more than 5% can be calculated using response spectrum prediction methods, which can estimate the spectral ordinate for a variety of damping levels directly. They also can be created by converting current prediction equations or code-based response spectra with different damping ratios utilising response spectrum damping modification factors. The second alternative may have multiple drawbacks, and ground motion prediction equations for varying degrees of damping require a large amount of research [10]. Although there is a little possibility that a structure would be subjected to ground motion from a large earthquake in its lifetime, it will be prone to structural damage if it can respond in the elastic range when exposed to inertia forces induced by a massive earthquake [11].

The design life of a construction might be anything from 50 to 100 years. It may be uneconomical to construct a structure to withstand a greater earthquake that strikes once every 500 years in such a situation. As a result, the structure is designed to withstand base shear forces that are far lower than the actual seismic load. This makes it possible to utilize strength reduction parameters to reduce elastic strength capacity while increasing inelastic drift demand. To obtain lower elastic design force or design base shear, the Indian seismic code divides the seismic reaction by the response reduction factor (R) [12]. The percentage of over strength that can be achieved depends on the construction type and ground motion parameters [13]. When rigid connections are replaced with semi-rigid connections, the over strength factor lowers by roughly 50% while the ductility factor increases greater than 25% [14]. Previously, researchers looked at a variety of factors that influence building performance in a significant ground motion, emphasising the importance of over-strength in structure performance during a severe earthquake [15]. The structure's durability becomes increasingly important as the building's height rises. Tall constructions become more sensitive to a variety of loading events as the number of storeys grows, including extremely high loading values induced by lateral forces such as earthquakes. These guidelines should be followed, and every engineer should make great effort to ensure that the structure is stable and strong enough to bear lateral loads. When the number of floors is reduced, the torsional irregularity coefficient rises [16].

ETABS is used to analyse a G+26 storey RC structure under seismic loads for various damping ratios in Zone-V locations using the response spectrum method. By combining the features of numerous independent ground motion recordings, the response spectrum shows an interaction between ground acceleration and the structural system. In line with IS 1893 (Part 1):2002, a variety of dead and live load combinations have been observed. The earthquake loads on a structure of various heights are calculated. When lateral systems are used, the displacement, shear, and moment of the structure reduces, enhancing the structure's stiffness to resist lateral stresses [17]. This paper focuses on the impacts of various damping ratios and modal combinations for the G+26 storey RC building located in Zone-V under response spectrum analysis.

Analytical Investigation

The response spectrum has been analysed using linear dynamic analysis. Response spectrum calculates vibration in each mode and then combines the responses from numerous mode to get the overall response. Deformation and acceleration are two examples of responses. The response spectrum is determined by plotting the highest response against the natural period. ETABS is used to do the analysis for a G+26 level multi-story structure. The plan of the G+26 RC building with its elevation is shown in Fig. 1 & 2.

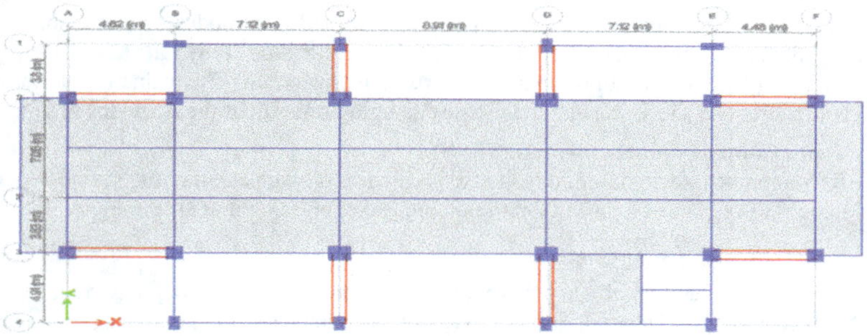

Fig 1. Plan of the G+26 RC building for first floor

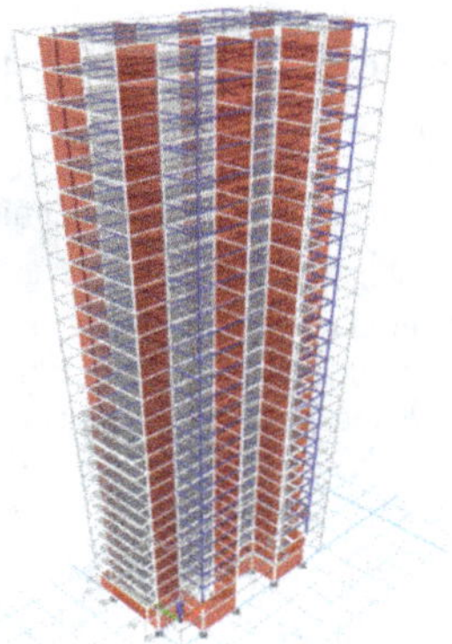

Fig 2. Elevation of the G+26 RC building

Modal Combination

Complete quadratic combination (CQC)

Traditional Response Spectrum Analysis using the CQC rule implies the following steps: (1) solving the real-values eigen problem to produce the modal matrix and accompanying spectral matrix; (2) picking the structure's number of vibrational modes (m); (3) finding the interest's generic structural response y(t) is a combination of the m modal responses, according to Eq. 1.

$$y(t) = \sum_{i=1}^{m} \quad e_{i\,\theta_i}(t) \qquad \text{Eq. 1}$$

where θi(t)= qi(t)/Γi, ei, is the influence coefficient that contribute the i-th mode shape to the response quantity of interest affected by the i-th participation factor [for example ei= ϕri Γi when the rth structural displacement has to be evaluated, that is y(t) = ur(t)]. The term i(t) denotes the solution to the following differential equation, $\ddot{\theta} i(t) + 2\zeta\omega i\dot{\theta}(t) + \omega_i^2\theta i(t) = \ddot{u}_g(t)$ (4) Applying the CQC rule as given in Eq. 2 to the design value of y(t):

$$y^{CQC} = \sqrt{\sum_{i=1}^{m} \sum_{k=1}^{m} \rho_{ik}\, e_{ie_k} \frac{A_{e(T_{i},\xi_i)}}{\omega_i^2} \frac{A_{e(T_{i},\xi_i)}}{\omega_k^2}} \quad \text{Eq. 2}$$

Where ρik is the correlation coefficient between the i-th and k-th modes of vibration, which is calculated assuming zero-mean stationary seismic acceleration [18]. In terms of pseudo-acceleration, Ae(Ti,ζi) is the i-th ordinate of the response spectrum. This value is determined by the vibration durations Ti= 2π/ωi and the viscous damping ratios ζi of the m modes of vibration.

Square root of sum of squares method (SRSS)

In the SRSS approach, the maximum response is calculated by calculating the square root of the sum of squares of responses in each vibration mode and expressing it as shown in Eq. 3

$$r = \sqrt{\sum_{i=1}^{n} r_i^2} \quad \text{Eq. 3}$$

The SRSS technique of combining maximal modal responses is fundamentally sound when the modal frequencies are well spaced. This approach, on the other hand, delivers disappointing results when the frequencies of significant contributing modes are close together.

Absolute Sum Method (ABSSUM)

The ABSSUM approach adds the peak responses of all modes algebraically, assuming that all modal peaks occur at the same time. Eq. 4 calculates the system's maximum response.

$$r = \sum_{i=1}^{n} |r_i| \quad \text{Eq. 4}$$

The ABSSUM method gives a much more conservative estimate of the total response peak value as well as an upper bound on the final response quantity.

Structural configuration

The Commercial building is considered for the analysis using response spectrum analysis to predict the behaviour of structural elements. Various damping ratios and their percentage are shown in Table 1 and the structural parameters taken for study of response spectrum analysis are shown in Table 2 and. In this study, analysis of G + 26 storey RC building for zone V has been carried out. The earthquake forces as specified by code IS: 1893-2002 (Part I) has been followed.

Table 1. Damping Ratio with Percentage

Damping percent (%)	Damping ratio
2	0.02
5	0.05
7	0.07
10	0.1
15	0.15
20	0.2
25	0.25
30	0.3

Table 2. Structural parameters

Description	Parameters
No.of Storeys	G+26
Typical floor height	3.6m
Ground floor height	5.1m
Total height of building	93.85m
Plan dimension	20m
Grades of concrete	M50, M35
Types of sreel	Fe500, Fe250
Dead load	2.5KN/m^2
Live load	3KN/m^2
Thickess of slab	225 mm
Sizes of beam	400mmX750 mm
	450mmX900mm
Sizes of column	1000mmX1000mm
	1000mmX500mm
	900mmX750mm
	500mmX1000mm
Type of zone	V
Zone factor	0.36
Importance factor	1
Type of structure	SMRF
Response reduction factor	5

Results and discussion

A. Storey displacement

When displacement compared to base of the structure is said to be storey displacement. Storey displacement is a critical parameter to consider when seismic force affects the structure. It is determined by the structure's height. The storey displacement for G+ 26 storeys RC building in Zone-V for various damping ratios like 0.02, 0.05, 0.07, 0.1, 0.15, 0.20, 0.25 and 0.30 are obtained for both X and Y direction. Analysis was done using response spectrum using CQC, SRSS and ABSSUM methods. The values of displacement at top storey are higher than at bottom storey. Hence the storey displacement for 26th storey for 5% and 25% of damping ratio using response spectrum analysis in X-direction is shown in Fig.3 and Fig.5. The storey displacement for 26th storey for 5% and 25% of damping ratio using response spectrum analysis in Y-direction is shown in Fig.4 and Fig.6.

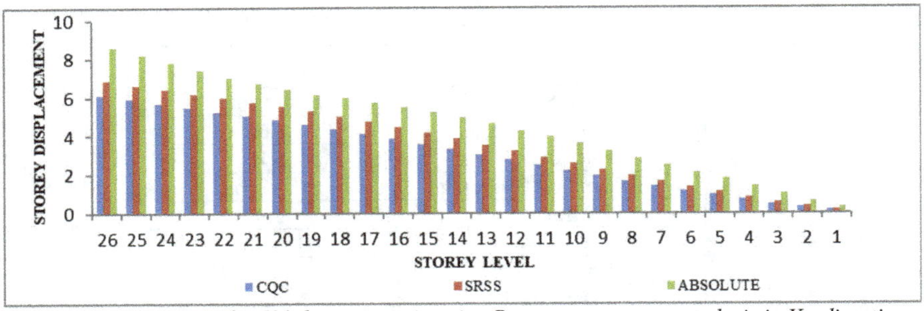

Fig 3. Displacement for 5% damping ratio using Response spectrum analysis in X - direction

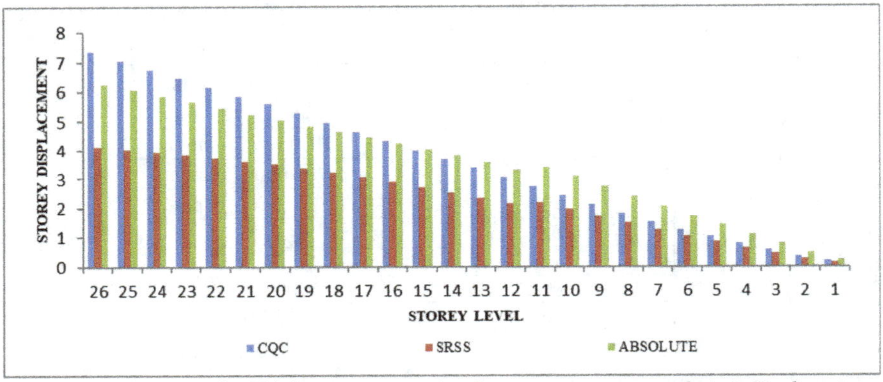

Fig 4. Displacement for 5% damping ratio using Response spectrum analysis in Y - direction

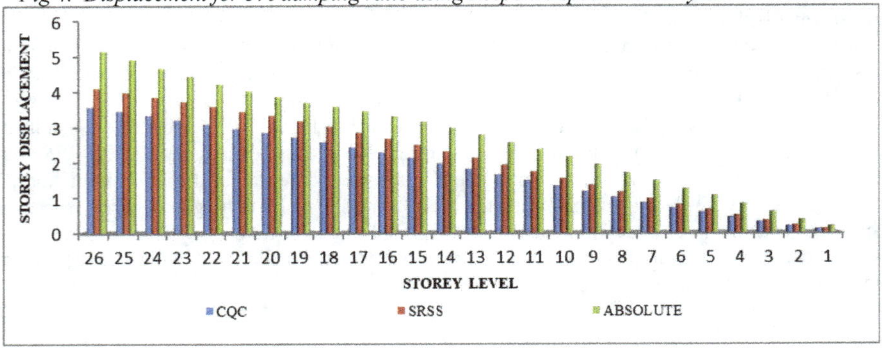

Fig 5. Displacement for 25% damping ratio using Response spectrum analysis in X – direction

 Materials Research Forum LLC
https://doi.org/10.21741/9781644901953-38

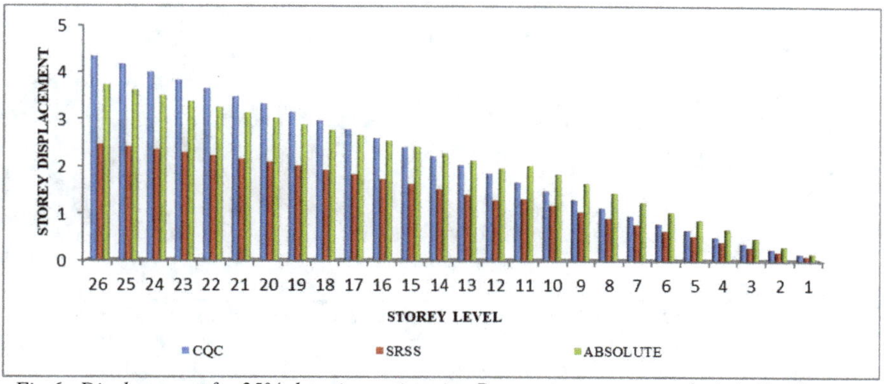

Fig 6. Displacement for 25% damping ratio using Response spectrum analysis in Y - direction

B. Storey Drift

Storey drift is defined as a storey's displacement in relation to the next storey. Its maximum value should be equal to 0.004 of the storey's height. In general, the value of drift is greatest at the building's H/2 height. The storey drift for G+ 26 storeys RC building in Zone-V for various damping ratios like 0.02, 0.05, 0.07, 0.1, 0.15, 0.20, 0.25 and 0.30 are obtained for both X and Y direction. Hence the storey drift for 26th storey for 5% and 25% of damping ratio using response spectrum analysis in X-direction is shown in Fig.7 and Fig.9. The storey displacement for 26th storey for 5% and 25% of damping ratio using response spectrum analysis in Y-direction is shown in Fig.8 and Fig.10.

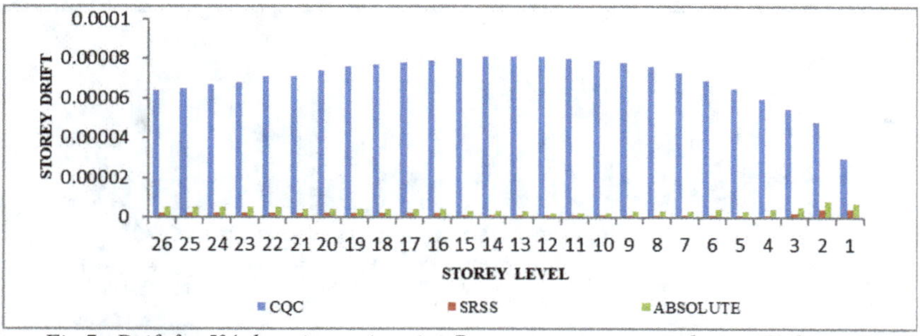

Fig 7. Drift for 5% damping ratio using Response spectrum analysis in X-direction

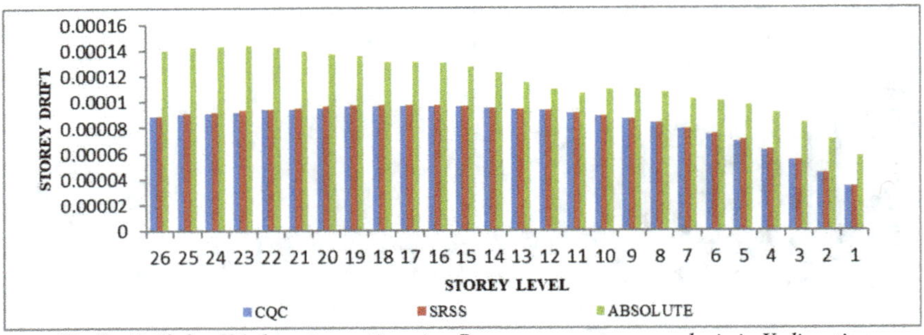

Fig 8. Drift for 5% damping ratio using Response spectrum analysis in Y-direction

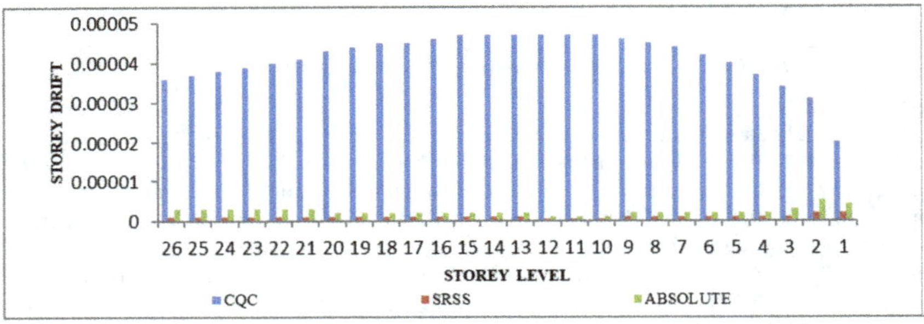

Fig 9. Drift for 25% damping ratio using Response spectrum analysis in X-direction

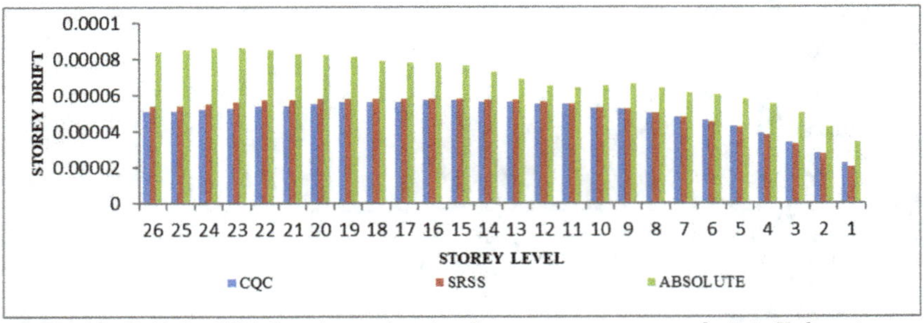

Fig 10. Drift for 25% damping ratio using Response spectrum analysis in Y-direction

C. Storey Shear

Base shear refers to the seismic force at the building's base. Storey shear refers to the lateral forces experienced by different floors as a result of an earthquake. Its value is highest at the bottom storey and lowest at the top storey. The storey shear for G+ 26 storeys RC building in Zone-V for various damping ratios like 0.02, 0.05, 0.07, 0.1, 0.15, 0.20, 0.25 and 0.30 are obtained for both X and Y direction. Analysis was done using response spectrum method. For both directions shear values are same hence storey for 26th storey and Base shear for 5% and 25% of damping ratio are shown in the Fig.11, Fig.12, Fig.13 and Fig.14.

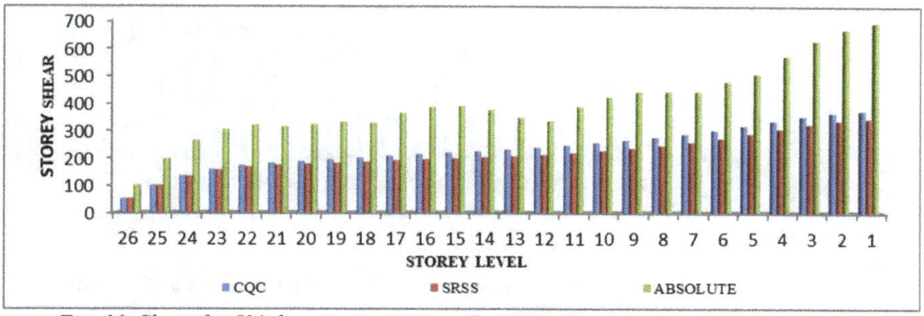

Fig 11. Shear for 5% damping ratio using Response spectrum analysis in X-direction

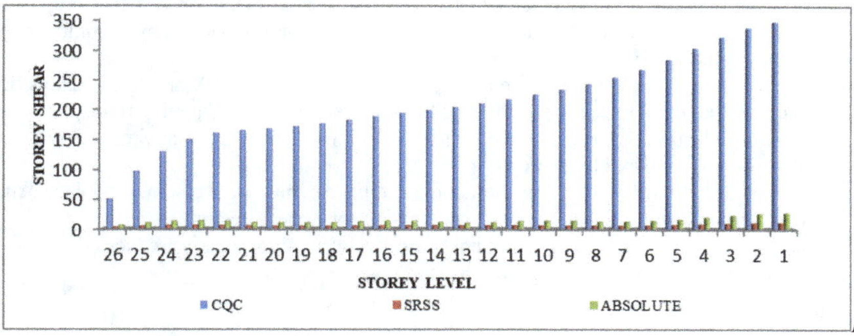

Fig 12. Shear for 5% damping ratio using Response spectrum analysis in Y-direction

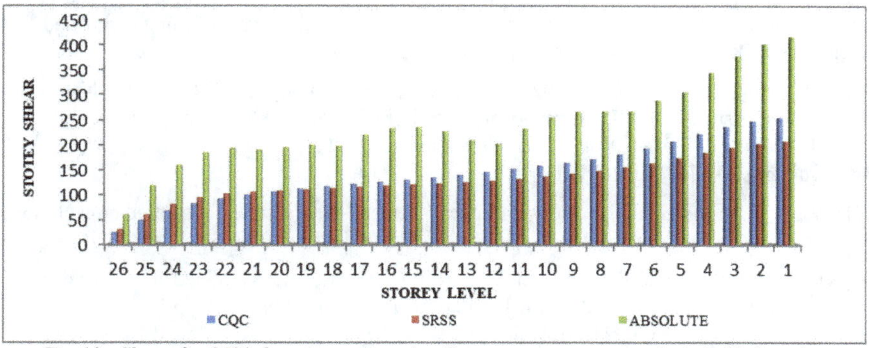

Fig 13. Shear for 25% damping ratio using Response spectrum analysis in X-direction

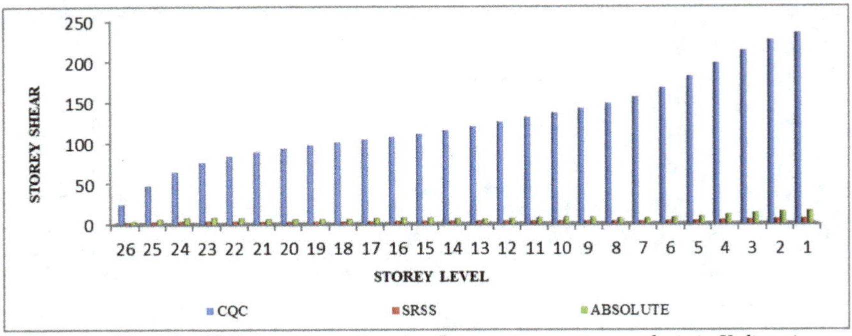

Fig 14. Shear for 25% damping ratio using Response spectrum analysis in Y-direction

Conclusions

1. From the analysis it is concluded that the storey displacement for G+ 26 storeys RC building in Zone-V is maximum for 0.02 damping ratio and in 0.05 damping ratio storey displacement value is average.
2. When analysis is done using Response spectrum method in both X and Y directions the storey drift is found to be maximum at 15th and 17th storey for all damping ratios
3. The storey shear in Response spectrum method for both directions is maximum at the ground when compared top storeys.
4. It is noted that when increasing the damping ratio parameters like storey displacement, storey drift and storey shear values are decreasing for Zone V region.
5. Hence by using the higher damping ratio we can significantly reduce the storey displacement, storey drift and storey shear values.

References

[1] Rashmi Agashe, Marshal Baghele, Vaishanvi Deshmukh, Sharad Khomane, Gaurav Patle, Kushal Yadav (2020) To study Analysis and Design of Multi-Storey Building using staad-Pro and comparing with manual calculations, International Research Journal of Engineering and Technology (IRJET), e-ISSN: 2395-0056, p-ISSN: 2395-0072, Volume: 07, Issue: 04, Apr 2020.

[2] S. R. Kangle, D. S. Yerudkar (2020) Response Spectrum Analysis for Regular Multistory Structure in Seismic Zone III, International Journal of Engineering Research & Technology (IJERT), ISSN: 2278-0181, Vol. 9 Issue 09, September-2020. https://doi.org/10.17577/IJERTV9IS090262

[3] Chopra, A.K. and Goel, R.K. (1999) Capacity-demand-diagram Methods based on inelastic spectrum. Earthquake Spectra. 15(4), pp.637-656. https://doi.org/10.1193/1.1586065

[4] Kim, J., Choi, H. and Min, K.W. (2003) Performance-based design of added viscous dampers using capacity spectrum method. Journal of Earthquake Engineering, 7(1), pp.1-24. https://doi.org/10.1080/13632460309350439

[5] Takahashi, M. and Akiyama, H. (1999) The maximum displacement and energy response of multi-story frames under earthquakes. Journal of Structural and Construction Engineering (AIJ), 515, 59-66 (in Japanese). https://doi.org/10.3130/aijs.64.59_1

[6] Sheikh, M. Neaz; Tsang, Hing-Ho; Yaghmaei-Sabegh, Saman; and Anbazhagan, P., "Evaluation of damping modification factors for seismic response spectra" (2013). Faculty of Engineering and Information Sciences - Papers: Part A. 1985.

[7] Blandon C.A. and Priestley, M.J.N. 2005. Equivalent viscous damping equations for direct displacement-based design. Journal of Earthquake Engineering; 9(2):257-278. https://doi.org/10.1142/S1363246905002390

[8] Priestley M.J.N, Calvi G.M. and Kowalsky M.J. 2007. Displacement-Based Seismic Design of Structures. IUSSPress: Pavia, Italy.

[9] Lee S.H., Min K.W., Hwang J.S. and Kim J. 2004.Evaluation of equivalent damping ratio of a structure with added dampers. Engineering Structures; 26: 335-346. https://doi.org/10.1016/j.engstruct.2003.09.014

[10] Stafford P.J., Mendis R. and Bommer J.J. 2008. Dependence of damping correction factors for response spectra on duration and number of cycles. Journal of the Structural Engineering (ASCE), 134(8):1364-1373. https://doi.org/10.1061/(ASCE)0733-9445(2008)134:8(1364)

[11] S.A. Barakat, A.I. Malkawi, A.S. Shatnawi, A Step towards evaluation of seismic response reduction factor in multi-storey reinforced concrete frames. Nat. Hazards 16, 65-80 (1997). https://doi.org/10.1023/A:1007972616511

[12] IS: 1893 (Part 1): 2002. Criteria for Earthquake Resistant Design of Structures. (Part 1: General Provisions and Buildings) (5th Revision).

[13] R, Meli, Code-prescribed seismic actions and performance of buildings, in Proceedings of the 10th World Conference on Earthquake Engineering, vol. 10 (A. A. Balkema, Rotterdam, 1992), p. 5783-5788.

[14] T. Balendra, X. Huang, Overstrength and ductility factors forsteel frames designed according to BS 5950. J. Struct. Eng. 129(8), 1019-1035 (2003). https://doi.org/10.1061/(ASCE)0733-9445(2003)129:8(1019)

[15] J.A. Blume, Allowable stresses and earthquake performance, in Proceedings of the 6th World Conference Earthquake Engineering, International Association for Earthquake Engineering (1977), p. 165-174

[16] M B Vikram, Roopesh M, Sandeep, Sanjay Kumar, "Comparison and Analysis of Multi-Storey Building in Various Seismic Zones", International Journal of Emerging Trends in Engineering and Development, Issue 7, Vol. 3, 2017, pp.233-238.

[17] AkashPanchal, Ravi Dwivedi, "Analysis and Design of G+6 Building in different Seismic Zones of India", International Journal of Innovative Research in Science, Engineering and Technology, Vol. 6, Issue 7, 2017, pp.14331-14338.

[18] IS: 875 (Part 2):1987 (Reaffirmed 1997). Indian Standard code of practice for design loads (other than earthquake) for buildings and structures, part 2, imposed loads (2ndRevision).

Study on Controlled Low Strength Materials using GGBS with Dredged Soil and M-Sand

S. Loganayagan[1,a], S. Ramakrishnan[2,b], V. Jeevanantham[1,c], M. Gowtham[1,d]

[1]Civil Engineering, Bannari Amman Institute of Technology, Erode 638401, India

[2]Civil Engineering, Sri Krishna college of engineering and technology, Coimbatore 641042, India

[a]loganayaganme@gmail.com, [b]ramakrishnans@skcet.ac.in ,[c]jeeva.civil.tpgit@gmail.com
[d]gowtham.ce19@bitsathy.ac.in

Keywords: Low Strength Material, Dredged Soil, GGBS, M-Sand

Abstract. In general, CLSM mixtures contain common ingredients such as Portland cement, fly ash, good mixing and water. CLSM is forced to fill in the back material and not low-strength concrete, rather it can best be described as property which is designed as concrete and strength flow and strength as per requirement and used as a backfill to avoid soil issues. CLSM can be built with a variety of strengths and sizes, taking into account costs future requirements, low power CLSM will be required to allow future excavation, and if there is no space for future digging the energy can be high on the other hand, furthering the size of CLSM can be adjusted according to the cost and material requirements. However, some industrial products and recycled products are also accepted and promoted as long as they are available, costing a particular use and the necessary characteristics of a combination such as flow, power, extraction, and quantity are acceptable. The aim of this study was to test whether it was possible to apply red mud such as placing a portion of Portland cement in a low-power controlled (CLSM) component made of industrial-grade products. The control mixture was initially made from the Portland cement, fly ash, and water. Bleeding, flow, the initial time for the setting of new CLSM compounds is measured and subsequent complications include compression. Results-They performed well and complied with CLSM requirements at ACI 229 levels in terms of flow, bleeding rate, initial set-up time, uncompressed compression strength. Low power control devices (CLSM) remove the problems of ground receding to provide the strength of the supporting structure.

Introduction

Controlled Low Strength Material (CLSM) mixtures contain common ingredients such as Portland cement, fly ash, good mixing and water. CLSM is forced to fill in the back material and not low-strength concrete, rather it can best be described as property which is designed as concrete and strength flow and strength as per requirement and used as a backfill to avoid soil issues[1]. CLSM can be built with a variety of strengths and sizes, taking into account costs future requirements, low power CLSM will be required to allow future excavation, and if there is no space for future digging the energy can be high on the other hand, furthering the size of CLSM can be adjusted according to the cost and material requirements. However, some industrial products and recycled products are also accepted and promoted as long as they are available, costing a particular use and the necessary characteristics of a combination such as flow, power, extraction, and quantity are acceptable[2]. This project aimed to test whether it was possible to use GGBS with M-sand and dredged soil, as binding and filler materials. GGBS is a waste obtained from quenching molten iron slag from the blast furnace in iron manufacturing industries. CLSM is one of the environmentally friendly materials made using waste materials like GGBS, fly ash etc. and cement

Materials Research Proceedings **23** (2022) 340-347 https://doi.org/10.21741/9781644901953-39

as an activator for pozzolanic reaction. In the present study, the ratio of GGBS to the cement is maintained as 3.5 and 2.8, water to binder ratio varied from 0.5, 0.55, 0.6, 0.65. The sand content of 1350 g and 1425 g were used for GGBS/cement ratios of 3.5 and 2.8 respectively. Fresh properties like density, flow ability, bleeding, and initial setting times are studied. It is observed that an increase in river sand content can lead to improved stability (bleeding rate) in the proposed CLSM operation. Increasing water content greatly influence the fresh properties in the proposed CLSM. The hardened properties like compressive strength and split tensile strength were studied [3]. It is observed that the increase in water content has a specific impact on the fresh and hardened properties of the CLSM using GGBS. The entire CLSM mixtures exhibit required strength as per ACI Committee 229. This research paved way for the usage of GGBS in controlled low strength materials and effective utilization of waste materials in a sustainable manner.

Literature Review

Backfills

The Ease of placing CLSM in restricted places without compaction facilitates the reduction in trench width or excavation.

Traditional methodology of backfilling in layers and compacting will never provide with the uniformity of density as facilitated by CLSM. CLSM can be placed in layers, allowing each layer to harden prior to placing the next layer.

Structural Fills

CLSM with higher strength can be produced to act as structural fills, in case of BC soil it can distribute structures load on greater area.

CLSM can provide a uniform and level surface for uneven sub-grades under foundation footings and slabs.

Utilities Bedding

CLSM gives a great sheet material to pipe, electrical, phone, and different sorts of conductors.

Erosion Control

Lab studies, just as field execution, have shown that CLSM opposes disintegration better than numerous other fill materials.

Void filling

Tunnel shafts and sewers—filling deserted passages and sewers, it is vital to utilize a stream capable combination.

A steady stockpile of CLSM will assist with keeping the material streaming and make it stream more noteworthy distances.

Methodology

Introduction
Literature review
Material collection & Testing on materials
Casting on mortar and concrete
Comparison on both concrete cubes
(Dredged soil & M-sand)

Figure 1. Methodology

Materials Research Forum LLC
https://doi.org/10.21741/9781644901953-39

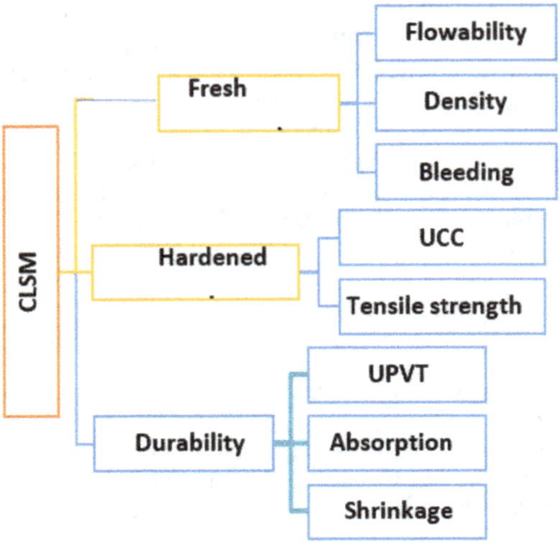

Figure 2. Schematic for the testing program

Materials

Common Portland concrete (solid) of 53 Grade out there in local market is utilized inside the examination. The solid which is utilized had tried for a very long time according to IS: 4031-1988 and situated to be affirming to shifted determinations of IS: 12269-1987 having explicit gravity of 3.0.Manufactured sand (M sand) is used as a substitute for normal stream sand in this audit[4]. Particles going through 4.75 mm sifter are used. Crushed exact sums of size 20mm and 10mm appearance express gravity 2.79 and 2.74 exclusively was used in the survey. Water for relieving and blending affirming to IS 456 was used.

Figure 3. Dredged soil

Gathering Dredged soil from the jungle gym as shown in the figure 3.. Particles going through 4.75 mm strainer are utilized. Squashed precise totals of size 20mm and 10mm appearance explicit gravity 2.79 and 2.74 separately was utilized in the review[5].

Figure 4 Ground Granulated Blast Furnace Slag

Ground Granulated Blast Furnace Slag (GGBS) is a concrete substitute that further develops toughness and furthermore the natural certifications of substantial blends. It is a side-effect of the iron-production industry [6].

Experimental evaluation

The experimental evaluation of all the parameters for the present study is based on the ASTM standard procedure and the IS standard procedures. The details of the test and the codal provisions are given in Table 1. A tentative schematic is also provided for the methodology adopted for the present study [7].

Table 1 Details of the test and corresponding code provisions

S.No	Materials Property	Details of test	Procedure from Standard codes
1	Ash Content	Loss on Ignition	ASTMD1762-84
2	Gradation	Particle size distributions	ASTM D6913 / D6913M - 17
3	pH	pH Test	
4	Elemental Analysis	Atomic absorption spectroscopy	
5	Flowability	Flow Consistency	ASTM D6103 / D6103M - 17e1
6	Density	Density of the materials at fresh and hardened state	ASTM D6023 - 16
7	Unconfined Compressive strength	Compressive strength of the CLSM	ASTM D2166 / D2166M – 16
8	Ultra-Sonic Pulse Velocity	Homogeneity of the CLSM Materials	ASTM C597

Mix design for M sand & Dredged soil

The proportioning of the ingredients was as per IS10262:2009, with fly ash as the main ingredient and cement as supplementary material, further the focus of the mix design was to develop substitute for backfill which matches its rate but consist the advantages of CLSM, along with that GGBS as partial replacement of cement [8].Dredged soil sand as complete replacement of M sand. The design mix proportions were given in table 2 and 3.

Table 2 Mix design for M sand

S.No	Cement (g)	GGBS (g)	M sand (g)	Water (ml)
1	200	700	1350	450
2	200	700	1350	495
3	200	700	1350	540
4	200	700	1350	585
5	200	700	1425	475

Table 3 Mix design for Dredged soil

S.No	Cement (g)	GGBS (g)	Dredged soil (g)	Water (ml)
1	200	700	1350	450
2	200	700	1350	495
3	200	700	1350	540
4	200	700	1350	585
5	200	700	1425	475

The above design mix were casted into cubes for various tests which includes basic testing of materials, setting time, California Bearing Ratio (CBR), Slump Cone Test and Ultra sonic pulse velocity [9].

Figure.5 CLSM mould casting

Test on setting time

For deciding the typical consistency and the hour of setting of our blend, that comprise of bar gauging, having a needle in each end, and backing in outline with graduated scale to quantify the distance to which the needle infiltrates our blend as shown in the figure 5[10].

Materials Research Proceedings **23** (2022) 340-347 https://doi.org/10.21741/9781644901953-39

Figure 6 Initial setting time and final setting time

California Bearing Ratio (CBR)

CBR test decides the relative bearing proportion and development qualities under realized additional charge weight of base, sub base and sub level soils for the plan of streets, asphalts and runways [11]. The CBR test is utilized broadly in determination of materials and control of sub levels [12].

Slump Cone Test

The concrete slump test estimates the consistency of new cement before it sets. A different test, known as the flow table, or slump-flow, test, is utilized for substantial that is excessively liquid (non-workable) to be estimated utilizing the standard slump test, in light of the fact that the substantial won't hold its shape when the cone is eliminated [13].

Figure 7 Flow-ability tests on CLSM

Ultrasonic pulse velocity

It is used to measure the transit time, velocity and parameter of evaluating concrete. To check the quality of concrete and to measure the velocity of an ultrasonic pulse passing through a concrete [14]. The test has been performed for both the samples such as M-sand and dredged soil. The comparison value were tabulated in table 4

Table 4 Comparison Result of UPV test

S.No	Pulse Velocity by Cross Probing for M sand (Km / Sec)	Pulse Velocity by Cross Probing for Dredged soil (Km / Sec)
1	2.61	1.72
2	2.72	1.79
3	2.78	1.86
4	2.96	1.92
5	3.2	1.98

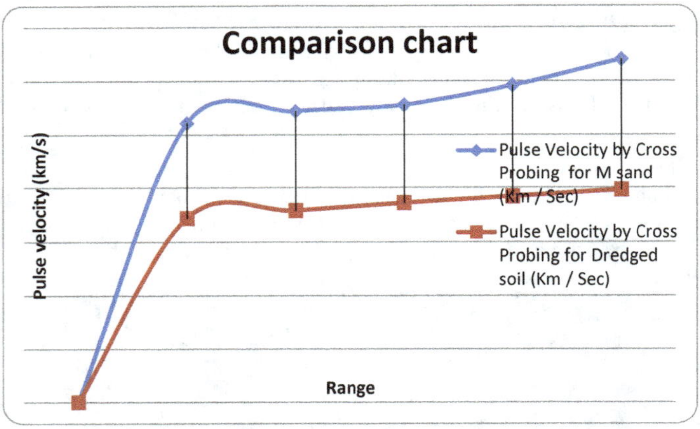

Figure 8.Comparison Result of UPV test

The figure 8 shows the values of ultra sonic pulse velocity ratio for different ratios with different samples [15]. The result may vary according to the strength of the soil and

Conclusion

From the observation of various test results, it clearly shows that increase in M-sand & Dredged soil content can lead to improved stability (bleeding rate) and speed up the setup process in the proposed CLSM operation. Finally, in terms of strength within 28 days, the most powerful CLSM can be produced. All in all, it is important to know that M sand & Dredged soil can be potential in CLSM production. soil or soils developed / compacted with pre-defined gradation (GSB, WMM) are in the past it was used for recycling, (both completed / non-completed structures), mortar foundations, empty filling etc., with its natural limits of air, durability, need for congestion, inclination of water, tendency to settle.

References

[1] M. A. Fauzi, M. F. Arshad, and N. Md Nor, "Statistical models to develop optimised controlled low-strength materials with wastepaper sludge ash," Constr. Build. Mater., vol. 286, p. 122816,2021. https://doi.org/10.1016/j.conbuildmat.2021.122816

[2] A. BarghiKhezerloo, S. M. Dehghan, and M. Sepasdar, "Using Dune Sand in Controlled Low-Strength Material for Backfilling: Field Installation and Numerical Simulation," J. Pipeline Syst. Eng. Pract., vol. 12, no. 1, p. 04020055,2021. https://doi.org/10.1061/(ASCE)PS.1949-1204.0000508

[3] D. AhadzadehGhanad, A. Soliman, S. Godbout, and J. Palacios, "Properties of bio- based controlled low strength materials," Constr. Build. Mater., vol. 262, p. 120742, 2020. https://doi.org/10.1016/j.conbuildmat.2020.120742

[4] S. P. Singh, K. Bhagya, and M. Mishra, Properties of Fly Ash-Based Controlled Low Strength Material, vol. 85.2020. https://doi.org/10.1007/978-981-15-6086-6_19

[5] C. P. Ramesh, H. P. Vageesh, T. Raghavendra, B. C. Udayashankar, and A. Shashishankar, Proportioning of CLSM Containing Class C Fly Ash, GGBS, Quarry Dust and Processed Slag Sand.2019. https://doi.org/10.1007/978-3-030-01929-7_6

[6] J. Zhang, J. Wang, X. Li, T. Zhou, and Y. Guo, "Rapid-hardening controlled low strength materials made of recycled fine aggregate from construction and demolition waste," Constr. Build. Mater., vol. 173, pp. 81-89,2018. https://doi.org/10.1016/j.conbuildmat.2018.04.023

[7] T. Manh Do, G. O. Kang, and Y. sang Kim, "Development of a new cementless binder for controlled low strength material (CLSM) using entirely by-products," Constr. Build. Mater., vol. 206, pp. 576-589,2019. https://doi.org/10.1016/j.conbuildmat.2019.02.088

[8] Chandra S., (2005), "Implications of using recycled construction and demolition waste as aggregate in concrete", Session lead paper, International Conference on Sustainable Waste Management and Recycling, Cement Concrete Composite, 27(6), 738-741.

[9] Ishimoto H., Origuchi T., Yasuda M., (2000), "Use of papermaking sludge as new material", Journal Material Civil Engineering 12(1), pp.310-313. https://doi.org/10.1061/(ASCE)0899-1561(2000)12:4(310)

[10] Dhir RK, Leelawat T, Limbachiya MC., (2000), "Use of recycled concrete aggregate in high-strength concrete", Materials Structure, vol. 33, 574-580. https://doi.org/10.1007/BF02480538

[11] JianzhuangXiaoa, YuedongSuna, H. Falknerb, (2005), "Seismic performance of frame structures with recycled aggregate concrete",Department of Building Engineering, Institute of Building Materials.

[12] Khaldoun Rahal, Ã, (2007), "Mechanical properties of concrete with recycled coarse aggregate Building and Environment", Department of Civil Engineering, Kuwait University, 42, 407-415. https://doi.org/10.1016/j.buildenv.2005.07.033

[13] Khalaf FM, DeVenny AlanS, (2004), "Recycling of demolished masonry rubble as coarse aggregate in Concrete", ASCE J Mater Civil Eng, 331-340. https://doi.org/10.1061/(ASCE)0899-1561(2004)16:4(331)

[14] Mozaffari E., O'Farrell M., Kinuthia J.M., Wild S., (2006), "Improving strength development of waste paper sludge ash by wet-milling", Cement and Concrete Composites, 28, 144-152. https://doi.org/10.1016/j.cemconcomp.2005.10.007

[15] Ramachandran VS., (1981), "Waste and by-products as concrete aggregates", Canadian Building Digest Report, CBD-215, NRC,http://www.irc.nrc-cnrc.gc.ca/pubs/cbd/cbd215.

Sustainable Materials and Smart Practices - NCSMSP-2021 Materials Research Forum LLC
Materials Research Proceedings **23** (2022) 348-361 https://doi.org/10.21741/9781644901953-40

Utilisation of Sustainable Materials in Geopolymer Composites– A Review

Preetha Vellaichamy[1,a,*], Vishalachi Chakkaravarthi[1], Senthilkumar Veerasamy[2], V.M. Gnanasundar[1]

[1]Department of Civil Engineering, Bannari Amman Institute of Technology, Erode. India

[2]Department of Civil Engineering, M.Kumarasamy College of Engineering ,Karur, India

*preethav@bitsathy.ac.in

Keywords: Binder Materials, Matrix, Geopolymer, Concrete, Strength

Abstract. Geopolymer composites made from sustainable ingredients which are used to make ecofriendly concrete in the infrastructure sector. The dramatic increase in infrastructure growth around the world demonstrates the daily demand for cement production. This study provides an overall view of research on the use of materials and the performance of geopolymer matrix based on strength and durability. Unlike cement, the reutilization of industrial by-products reduces greenhouse gas emissions during manufacture. Hence geopolymers can contribute to a better alternative to Portland cement. Natural raw materials, agricultural waste, and industrial waste by products from diverse industries are used as composite filler / binder materials in geopolymer matrix to improve workability , durability and reducing geopolymer concrete manufacturing costs. With the help of various curing procedures, the compressive strength of geopolymer concrete can be increased in a short amount of time. It has also been discovered that adding fibres to geopolymer concrete improves tensile strength, lowering the cost of structural maintenance.

Introduction

Concrete which is a commonly used global building material made with Portland cement. This increases cement production due to the enormous escalation in infrastructure and mechanization development. Geopolymer concrete is one made use of environmentally friendly materials to reduce CO_2 emission by extreme use of OPC. Initially geopolymer concrete research was constrained with the usage of natural source materials like calcined clays, kaolin, metakaolin and silica fumes. However, in order to solve the problem of industrial waste disposal, items includes fly ash, GGBS and palm oil ash etc., are commonly used to make the environment more sustainable while emitting less CO_2 than cement. The cement replacement materials should possess enough pozzolanic property [1]. He also stated that geopolymer concrete has an earlier strength, is more durable, and does not have the harmful alkali–aggregate interaction that portland cement has. Geopolymer concrete with low calcium fly ash and treated with various alkaline solutions has demonstrated great resistance to harsh environments.

Geopolymerisation process

Natural/alternative source materials and alkaline solutions are the two most significant components of geopolymer. Geopolymers are a polymer network made up of silicon and aluminium atoms connected jointly through oxygen atoms. As a source material for geopolymer binder, fly ash, ground granulated blast furnace slag (GGBS), metakoalin, Silica fume, and other amorphous minerals rich in Si and Al can be used. Geopolymer, like Portland cement, does not use a calcium-silicate-hydrate (CSHs) gel to improve structural stability; instead, it uses a poly condensation process with silica and alumina precursors. In the geopolymerization process, aluminosilicate oxides react chemically with alkali polysilicates to generate polymeric Si – O – Al links [2]. In the geopolymerization process, the amount of alkali solution, type of binder, Si/Al ratio, temperature, and curing conditions are all important variables. The key concept behind the fly ash-based geopolymer matrix is the alkali breakdown of aluminoslicate followed by the polycondensation process. Condensation between flyash and alkali solution produces Si4+ and Al3+ species, which are then nucleated, oligomerized, and polymerized to produce aluminoslicate matrix. The energetic Al3+ and Si4+ species react to form nuclei and aluminosilicate oligomers, as shown in Fig. 1. The aluminosilicate matrix network is balanced by alkali cations. The final structure of geopolymer materials is determined by the appropriate Si/Al ratio. Different aluminosilicate networks form as aluminosilicate oligomers, which eventually form geopolymer concrete/mortar, depending on the Si/Al molar ratio.

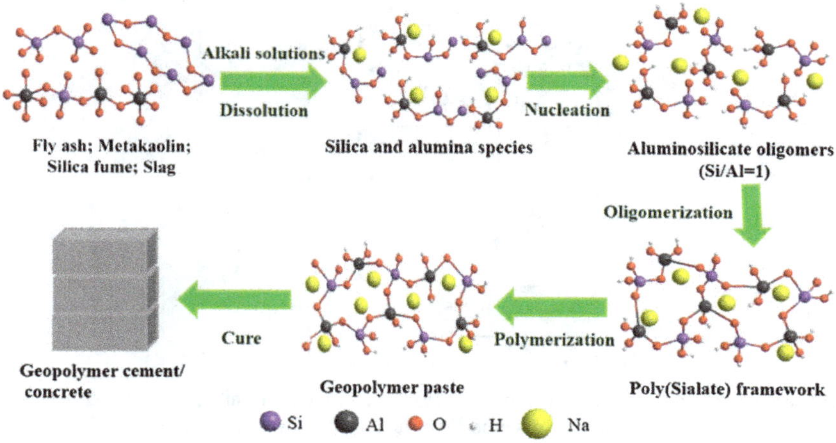

Fig. 1. Geo polymerisation Process [2,3]

The setting behaviour, microstructure, and property development of fly ash-based geopolymers are fundamentally different from metakaolin-based geopolymers. The Gluhhovsky model divides the process into three stages: destruction–coagulation, coagulation–condensation, and condensation–crystallization. The essential process in geopolymerization is depicted in Fig. 2 as a very simplified reaction mechanism. A supersaturated alumina silicate solution is generated when aluminosilicate sources are dissolved in water. When concentrated liquids are present, a gel forms, which subsequently condenses to form huge networks, allowing the water to escape. The reaction media contains water, but it is trapped inside the pores of the gel [4, 5].

Effect of Filler materials on Geopolymerisation

To hydrate partial replacements of highly mixed cement with flyash, several alkali solutions are utilised. In terms of performance, this is preferable to normal Portland cement. The formation of gel, which consists of C-S-H and N-A-S-H, causes the setting and hardening of alkaline-hydrated (fly ash-rich) super-blended cement [6]. The fineness and surface qualities of the raw materials are primarily responsible for geopolymerization. The geopolymerization phase, which consists of a mixture of two or three source minerals, such as kaolinite, albite, and fly ash, activates and speeds up the reaction of the necessary gel phase, resulting in a high compressive strength [7].

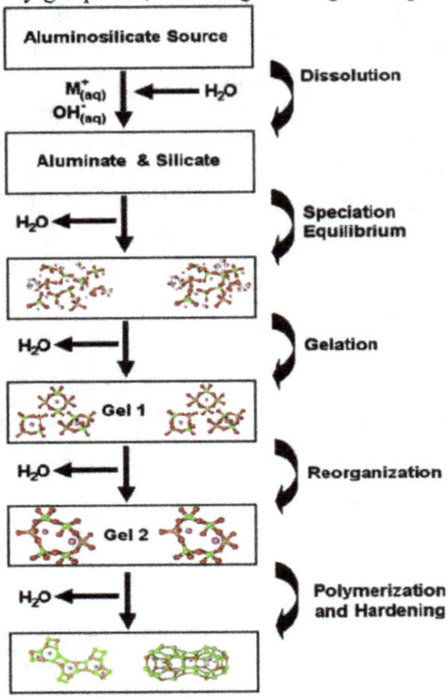

Fig. 2. Conceptual model of Geopolymerisation [3]

Based on curing temperature and methodology, the influence of alkali activator concentration enhances the process of geopolymerisation reaction and, moreover, leads to strength improvement [8]. Increased alkali activator solution proportions speed up the setting time, which reduces strength, thermal stresses, reaction rate, and geopolymerization peak time [9]. The amount of dissolving in NaOH solutions was greater than in KOH solutions when the most reactive components were utilised. In both alkaline and acidic conditions, Si and Al behave in a coordinated manner [10].

Effect of combination of binder materials with their performance

Reference	Waste Materials utilized	alkaline medium	Remarks
(Bhardwaj and Kumar 2019) [11]	GGBS+ flyash + waste foundary sand	NaoH, Na_2SiO_3 (14M)	Drop in workability and rise in strength with 60% foundary sand
(Ali, Naje, and Nasr 2020) [12]	Flyash +Chopped tyre rubber	Na_2SiO_3/ NaOH by w.t of FA (45%)	Compressive strength increases by 10%

(Republic and Republic 2006) [13]	Flyash + Ground Slag	SiO_2 /Na_2O (6-10%)	Microstructure phases are studied with co-existence of C–S–H phase. Maximal geopolymer strength was reached.
(Hawa, Tonnayopas, and Prachasaree 2013) [14]	Metakaolin + palm oil ash	SiO_2 /Al2O3 , Cao/Sio2	Highest compressive strength obtained from (SiO2 :Al2O3 =2.88 : 1)
(Chen and Id n.d.) [15]	Metakaolin + Ca	CaO:Na2O:Al2O3:SiO2:H2O (0.4:1:1:4:12.1)	Faster gel formation by calcium enhanced by faster metakolin dissolution
(Shi et al. 2020) [16]	Flyash + Metakaolin (Foamed Geopolymer)	calcium stearate, H2O2, NaOH, sodium silicate and expanded polysterene (EPS)	H2O2 causes expansion of foamed geopolymer, Metakaolin reduce pore diameter
(Oyebisi et al. 2020) [17]	Harnessed corncob ash (CCA) + (GGBFS)+ Portland lime cement	Sodium Silicate / Sodium Hydroxide – 2.5, Al / B - 0.54 (12M,14M,16M)	14 M of SH solution achieved highest compressive, flexural, and splitting tensile strengths for both M 30 and M 40.
(Panizza et al. 2020) [18]	Metakaolin + GGBS + flyash + furnace slag	Si/Al, K/Al	Geopolymer mortar produces better shrinkage and long term behaviour

Effect of curing temperature on alkali activator

The effect of curing procedures, as well as the inclusion of binder/filler ingredients in various alkaline activator dosages, on geopolymer concrete's strength qualities, may be beneficial. Curing temperature has a significant impact on geopolymeric processes and material characteristics [19]. The effect of the Si/Al ratio in the range of 1.7 to 2.9 on weight losses, residual strength, volumetric shrinkage, and water sorptivity vary dramatically with temperature, as illustrated in Fig. 3 and 4. The colour of the specimens changes from grey to bright crimson as the temperature rises, accompanied by the formation of microscopic cracks. At higher temperatures, the lower Si/Al ratio causes more weight loss and volumetric strain. The higher Si/Al ratio exhibits best results in terms of residual compressive strength after exposure to elevated temperatures as shown in Fig. 5 & 6.

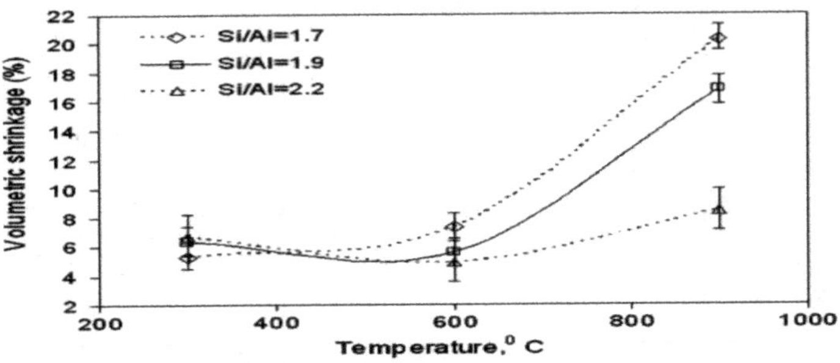

Fig. 3. Variation of volumetric shrinkage

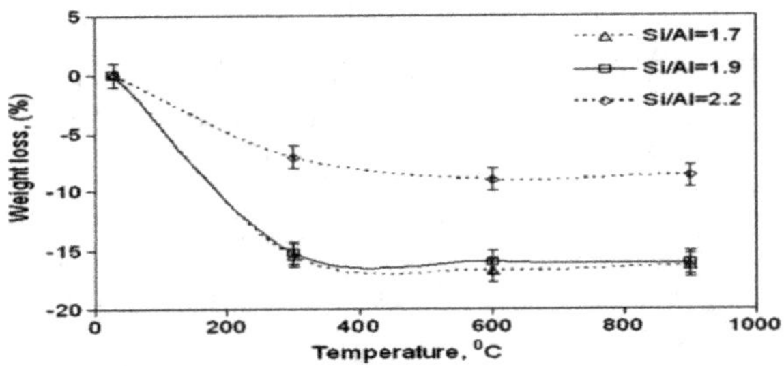

Fig. 4. weight loss of geopolymer with respect to temperature [20]

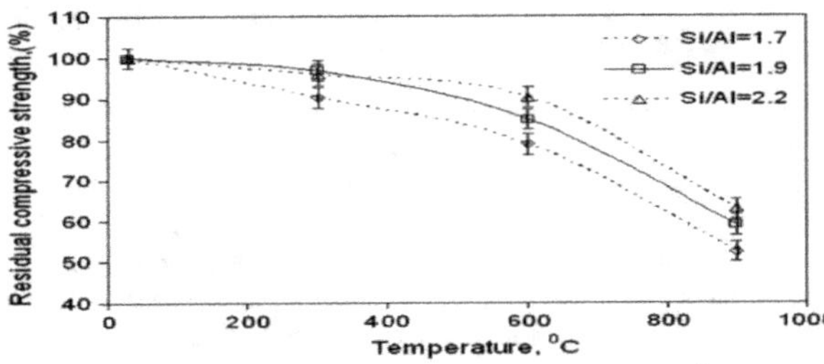

Fig.5. Variation of compressive strength

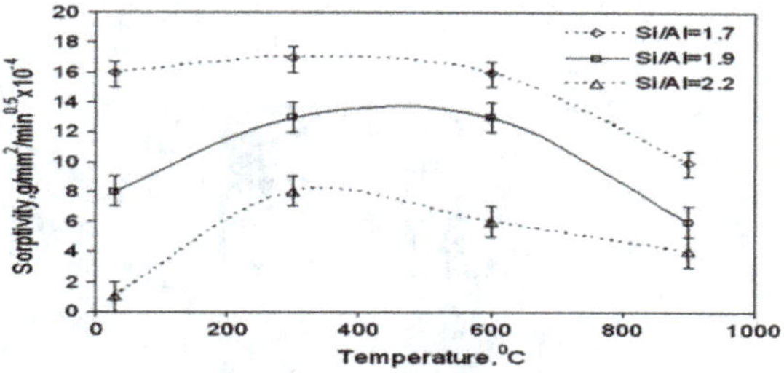

Fig. 6. Variation of sorptivity against temperature of geopolymer concrete [20]

Specimen	Colour Name	HEX Code	Shades
Unexposed	Alice Blue	#F0F8FF	
300°C	Light Grey	#D3D3D3	
600°C	Rosy Brown	#BC8F8F	
900°C	Chocolate	#D2691E	

Fig. 7. Specimens at different temperature exposure and standard colour indices [20]

When subjected to high temperatures, higher Si/Al ratios, on the other hand, have proved to have significant strength as shown in Fig. 7. As a result, increasing the Si concentration in geopolymer composites increases their performance when exposed to higher temperatures, demonstrating that they are temperature resistant materials [20]. The compressive strength of concrete was improved by raising the SiO3/OH ratio from 0.5 to 1.5 after the geopolymerization procedure was successful. As the temperature rises, the strength of geopolymer concrete gradually increases [21].

Environmental sustainability
The severe environmental impact of carbon dioxide emissions, as well as the predicted ongoing expansion of industrial development, needs retooling the construction industry with sustainable and environmentally acceptable materials in order to develop alternate binder systems. The impacts of estimating the carbon footprint of geopolymer and OPC concrete, including energy consumption from all activities required to source raw ingredients, create concrete, and complete the construction process. The CO2 footprint of geopolymer concrete was roughly 9% lower than that of OPC concrete binder, as shown in fig. 8. Greater OPC emission was linked to cement and sodium silicate in geopolymer concrete [22].

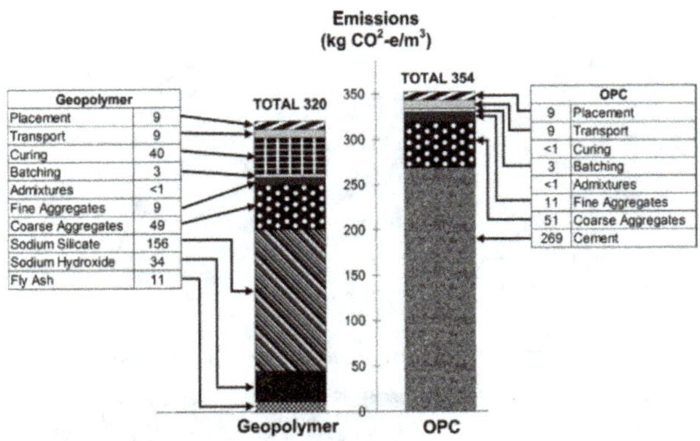

Fig. 8. Summary of CO_2 emission for OPC and geopolymer binders [22]

The energy consumed in the production of ordinary Portland cement (OPC) and supplemental cementitious materials (SCM'S) is depicted in Fig. 9. Cement clinker calcination absorbs a significant quantity of fossil fuel energy and releases additional CO2 via a decarbonation reaction. The extensive use of basic materials such as FA and GGBS as geopolymer binder in power plants and steel manufacturing has no detrimental impact on CO2 emissions. Geopolymer binders [23] provide a game-changing answer to the OPC industry's environmental issues. Green concrete's growth is aided by the use of SCMs in its manufacture, such as high-volume fly ash concrete, ultra-high performance concrete, geopolymer concrete, lightweight concrete, and so on. Increased strength, durability, workability, low permeability, controlled bleeding, reduced shrinkage, and greater resilience to severe environments are just a few of the long-term and cost-effective benefits of green concrete. Green concrete also promotes the creative and long-term use of waste and alternative materials in the construction industry [24].

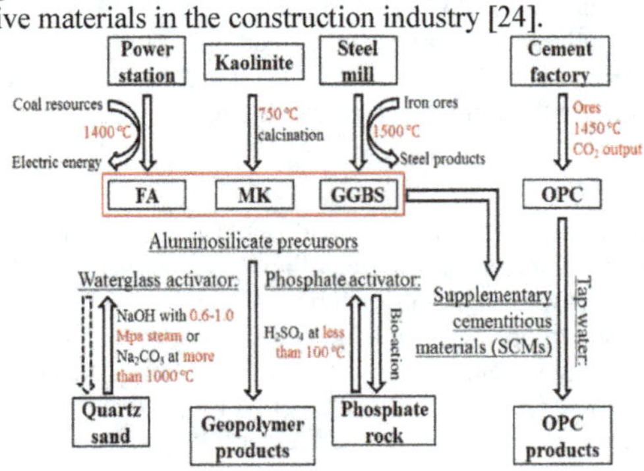

Fig. 9. Energy Expenditure diagrams of Geopolymer and OPC products [23]

Utilization of Nanomaterials in Geopolymer concrete

Geopolymers have been successfully made from fly ash, silica fume, rice husk ash, red mud, and crushed granulated blast furnace slag. The use of nanoparticles improves the structural properties of geopolymer concrete and mortars. Nanomaterials such as nanosilica (NS), nanotitania (NT), nanoalumina (NA), nano clay (NC), geothermal nanosilica waste (GSW), and carbon nanotubes considerably improve the mechanical and durability properties of geopolymers, according to various publications. Silicon dioxide (SiO2) nanoparticles form a dense concrete mix that increases the strength and endurance of the materials [25, 26]. The workability of nanoparticles in geopolymer concrete is limited due to their larger surface area. The beginning and final setting durations of geopolymer matrix are lowered due to the acceleration effect of nanoparticles, making the optimum content of nanoparticles with varying percentages appropriate for geopolymer applications. A larger concentration of nanoparticles diminishes mechanical strength and increases water absorption and porosity by forming more voids in the matrix due to particle agglomeration and pore dispersion. In geopolymer concrete, nano-silica with a particle size of 10 nm was used up to 8%, resulting in increased compressive strength, decreased permeability, and a dense geopolymer concrete matrix. In GPC-N cubes containing 8% nano silica, chloride and sulphate incursions were found to be low. Geopolymer concrete containing nano silica, according to preliminary study, is a feasible alternative to OPC. The GPC-N also performed well in a number of situations [27]. Adding 2% weight of nano-Al2O3 particles to a fly ash-based geopolymer matrix improves mechanical properties, as seen in figures. 10 and 11 [28]. As a result, nanoalumina, like nanosilica, speeds up the geopolymerisation procedure, resulting in a denser geopolymeric gel.

Fig. 10. SEM image of nano- Al_2O_3 powder [28]

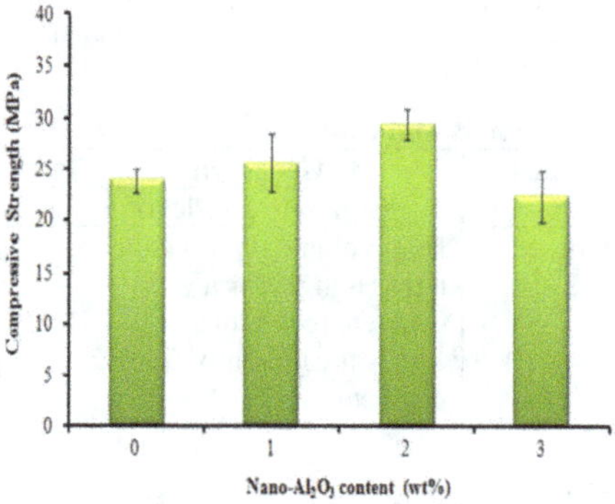

Fig. 11. Compressive strength of geopolymer nanocomposites as a function of nano-Al_2O_3 content [28]

As shown in fig. 12, adding nanotitania (TiO2) to flyash-based geopolymer concrete with varied percentages improves mechanical properties in later ages. Increased geopolymerisation processes triggered the strength [29].

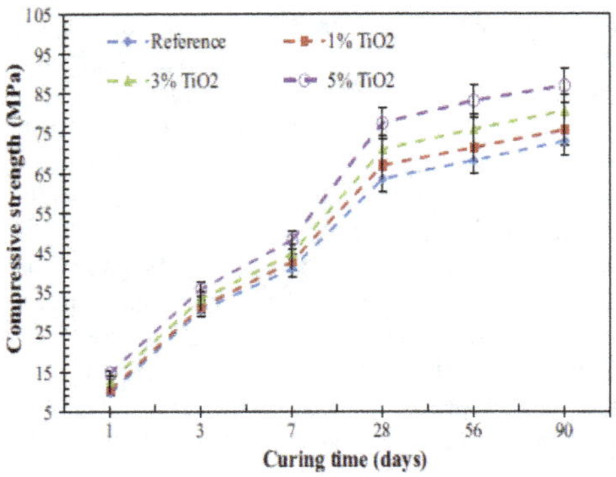

Fig. 12. Effects of nano-TiO2 on the compressive strength of geopolymer at 1, 3, 7, 28, 56 and 90 days [29]

Effect of fibres in geopolymer concrete

Fibres, in combination with a geopolymer binder, are used in varying proportions to improve strength and durability, as well as to reduce building costs. Glass, polypropylene, and steel fibres have all been utilised in studies and have proven to have superior durability. Increased rupture modulus, bond strength, compressive strength, and flexural characteristics are all benefits of steel fibre. When fibres are added to concrete, its workability is reduced [30, 31, 32, and 33]. The application of cellulose fibre geopolymer matrix composite enhances specimen strength and fracture resilience via cellulose modification [34]. The brittleness of geopolymer concrete is reduced when glass and polypropylene fibres are mixed with hybrid fibres. The presence of hybrid fibres resulted in greater performance in the post-cracking stage, while the presence of polypropylene fibre resulted in a delay in the formation of the first crack with an increase in toughness index [35, 36, and 37].

Applications of geopolymer matrix and concrete

Source	Materials	Geopolymer	Remarks
		Concrete applications	
(Liew et al. 2017) [24]	Flyash based concrete	High volume fly ash concrete, ultra High performance concrete (containing admixtures), lightweight concrete	High strength, Increase durability, improved workability, Reduced permeability and Shrinkage.
(Laskar and Talukdar 2017) [38]	Blast furnace slag + flyash	As Concrete repairing agent	Improvement in strength and workability aspects.

(Deb, Nath, and Sarker 2015) [39]	GGBS+ flyash	Heat cured methodology is employed for concrete strength enhancement	Improvement in strength and higher GGBS contents leads to lower workability aspects.
(Lee et al. 2019) [40]	GGBS+ flyash	Excellent chloride resistance concrete	Durability test – Nine months of curing period. Better compressive strength.
(Deb et al. 2015) [39]	Low calcium Flyash+ Blast furnace slag	Early strength development with 10-20% GGBS	Reduction in shrinkage aspects
Marine applications			
(Hassan, Arif, and Shariq 2020) [41]	Flyash based mix	Amorphous aluminosilicate geopolymeric gel formation are chemically stable in sea water	Low Permeability. Inhibit penetration of sea water
(Arbi et al. 2016) [42]	Alkali activated materials (Flyash + slag)	Durability enhancement by chemistry of pore solution	Better resistance to chemical attacks. Improvement in durability aspects.
(Mahmood 2019) [43]	Steel furnace slag aggregate +fly ash + blast furnace slag	Size reduction in break water structures	Repair works , Delayed hydration process
(Fan et al. 2018) [44]	Flyash based mix	High performance geopolymer mortar for repair works	Better for aggressive environments such as seawater and acidic environments

Conclusion

The review summarises the important findings of several investigations on the utilisation of materials and the strength of geopolymer concrete. The geopolymer binder's source material should have a strong pozzolanic activity, with a suitable alkali solution ratio activating the geopolymerisation reaction. A greater grasp of the source materials' chemical and physical properties can help you decide which binders to use and in what ratio. A combination of materials, curing temperature, curing procedure, and alkali activator ratio improve the strength of the Geopolymer. The longevity of geopolymer concrete is affected by the component and design mix ratios used in its preparation. Pavements, maintenance work, coastal area constructions, and bridges are where geopolymer concrete is most commonly used. The addition of fibres to geopolymer concrete improves its resiliency after it has cracked.

References

[1] Davidovits, Joseph., Geopolymer Chemistry and Applications, 5-Th Edition (2020).

[2] Zhuang, Xiao Yu, Liang Chen, Sridhar Komarneni, Chun Hui Zhou, Dong Shen Tong, Hui Min Yang, Wei Hua Yu, and Hao Wang, Fly Ash-Based Geopolymer: Clean Production, Properties and Applications, Journal of Cleaner Production 125:253-67 (2016). https://doi.org/10.1016/j.jclepro.2016.03.019

[3] Shamsaei, Ezzatollah, Owen Bolt, Felipe Basquiroto de Souza, Emad Benhelal, Kwesi Sagoe-Crentsil, and Jay Sanjayan., Pathways to Commercialisation for Brown Coal Fly Ash-Based Geopolymer Concrete in Australia, Sustainability (Switzerland) 13(8) (2021). https://doi.org/10.3390/su13084350

[4] Duxson, P., A. Fernández-Jiménez, J. L. Provis, G. C. Lukey, A. Palomo, and J. S. J. Van Deventer, Geopolymer Technology: The Current State of the Art, Journal of Materials Science 42(9):2917-33 (2007). https://doi.org/10.1007/s10853-006-0637-z

[5] Xu, Hua and J. S. J. Van Deventer, The Geopolymerisation of Alumino-Silicate Minerals, (2000). https://doi.org/10.1016/S0301-7516(99)00074-5

[6] Palomo, A., A. Fernández-Jiménez, G. Kovalchuk, L. M. Ordoñez, and M. C. Naranjo, Opc-Fly Ash Cementitious Systems: Study of Gel Binders Produced during Alkaline Hydration, Journal of Materials Science 42(9):2958-66 (2007). https://doi.org/10.1007/s10853-006-0585-7

[7] Xu, Hua and Jannie S. J. Van Deventer, Effect of Source Materials on Geopolymerization, Industrial and Engineering Chemistry Research 42(8):1698-1706 (2003). https://doi.org/10.1021/ie0206958

[8] Görür, E. B., O. Karahan, C. Bilim, S. Ilkentapar, and E. Luga. 2015, Very High Strength (120 MPa) Class F Fly Ash Geopolymer Mortar Activated at Different NaOH Amount , Heat Curing Temperature and Heat Curing Duration, 96:673-78 (2003). https://doi.org/10.1016/j.conbuildmat.2015.08.089

[9] Ling, Yifeng and Ashley Buss, Proportion and Performance Evaluation of Fly Ash- Based Geopolymer and Its Application in Engineered Composites, Iowa State University Captones, Theses and Dissertations 140 (2018).

[10] Panagiotopoulou, Ch, E. Kontori, Th Perraki, and G. Kakali, Dissolution of Aluminosilicate Minerals and By-Products in Alkaline Media, Journal of Materials Science 42(9):2967-73 (2007). https://doi.org/10.1007/s10853-006-0531-8

[11] Bhardwaj, Bavita and Pardeep Kumar, Comparative Study of Geopolymer and Alkali Activated Slag Concrete Comprising Waste Foundry Sand, Construction and Building Materials 209:555-65 (2019). https://doi.org/10.1016/j.conbuildmat.2019.03.107

[12] Ali, Isam Mohamad, Ahmed Samir Naje, and Mohammed Salah Nasr, Eco-Friendly Chopped Tire Rubber as Reinforcements in Fly Ash Based Geopolymer Concrete, Global Nest Journal 22(3):342-47 (2020).

[13] Republic, Czech and Czech Republic, Microstructure Of Geopolymer Materials Based On Fly Ash, 50(4):208-15 (2006).

[14] Hawa, Abideng, Danupon Tonnayopas, and Woraphot Prachasaree, Performance Evaluation and Microstructure Characterization of Metakaolin-Based Geopolymer Containing Oil Palm Ash, The Scientific World Journal (2013). https://doi.org/10.1155/2013/857586

[15] Chen Xu, Andre Sutrisno and Leslie J. Struble, Effects of Calcium on Setting Mechanism of Metakaolin -based Geopolymer, Orcid ID : 0000-0002-6813-2849, 0-2.

[16] Shi, Jinyan, Baoju Liu, Yuanchun Liu, Enliang Wang, Zhihai He, Huijie Xu, and Xiaodong Ren, Preparation and Characterization of Lightweight Aggregate Foamed Geopolymer Concretes Aerated Using Hydrogen Peroxide, Construction and Building Materials 256:119442 (2020). https://doi.org/10.1016/j.conbuildmat.2020.119442

[17] Oyebisi, Solomon, Anthony Ede, Festus Olutoge, and David Omole, Geopolymer Concrete Incorporating Agro-Industrial Wastes: Effects on Mechanical Properties, Microstructural Behaviour and Mineralogical Phases, Construction and Building Materials 256:119390 (2020). https://doi.org/10.1016/j.conbuildmat.2020.119390

[18] Panizza, Matteo, Marco Natali, Enrico Garbin, Vilma Ducman, and Sergio Tamburini, Optimization and Mechanical-Physical Characterization of Geopolymers with Construction and Demolition Waste (CDW) Aggregates for Construction Products, Construction and Building Materials 264:120158 (2020). https://doi.org/10.1016/j.conbuildmat.2020.120158

[19] Shin, S., G. Goh, and C. Lee., Predictions of Compressive Strength of GPC Blended with GGBFS Developed at Varying Temperatures, Construction and Building Materials 206:1-9 (2019). https://doi.org/10.1016/j.conbuildmat.2019.01.181

[20] Thokchom, Suresh, Kalyan Kr Mandal, and Somnath Ghosh., Effect of Si/Al Ratio on Performance of Fly Ash Geopolymers at Elevated Temperature, Arabian Journal for Science and Engineering 37(4):977-89 (2012). https://doi.org/10.1007/s13369-012-0230-5

[21] Rajarajeswari, A. and G. Dhinakaran., Compressive Strength of GGBFS Based GPC under Thermal Curing, Construction and Building Materials 126:552-59 (2016). https://doi.org/10.1016/j.conbuildmat.2016.09.076

[22] Turner, Louise K. and Frank G. Collins., Carbon Dioxide Equivalent (CO2 -e) Emissions : A Comparison between Geopolymer and OPC Cement Concrete, Construction and Building Materials 43:125-30 (2013). https://doi.org/10.1016/j.conbuildmat.2013.01.023

[23] Wang, Yan Shuai, Yazan Alrefaei, and Jian Guo Dai., Silico-Aluminophosphate and Alkali-Aluminosilicate Geopolymers: A Comparative Review, Frontiers in Materials 6(May 2019):1-17.

[24] Liew, K. M., A. O. Sojobi, and L. W. Zhang.,Green Concrete: Prospects and Challenges, Construction and Building Materials 156:1063-95 (2017). https://doi.org/10.1016/j.conbuildmat.2017.09.008

[25] Jindal, Bharat Bhushan and Rahul Sharma., The Effect of Nanomaterials on Properties of Geopolymers Derived from Industrial By-Products: A State-of-the-Art Review, Construction and Building Materials 252:119028 (2020). https://doi.org/10.1016/j.conbuildmat.2020.119028

[26] Rashad, Alaa M., Effect of Nanoparticles on the Properties of Geopolymer Materials, Magazine of Concrete Research 71(24):1283-1301 (2019). https://doi.org/10.1680/jmacr.18.00289

[27] Vyas, Sameer, Sameer Mohammad, Shilpa Pal, and Neetu Singh., Strength and Durability Performance of Fly Ash Based Geopolymer Concrete Using Nano Silica, International Journal of

Engineering Science Technologies 4(2):1-12 (2020). https://doi.org/10.29121/ijoest.v4.i2.2020.73

[28] Alomayri, Thamer., Experimental Study of the Microstructural and Mechanical Properties of Geopolymer Paste with Nano Material (Al2O3), Journal of Building Engineering 25(February 2019):100788. https://doi.org/10.1016/j.jobe.2019.100788

[29] Duan, Ping, Chunjie Yan, Wenjun Luo, and Wei Zhou., Effects of Adding Nano-TiO2 on Compressive Strength, Drying Shrinkage, Carbonation and Microstructure of Fluidized Bed Fly Ash Based Geopolymer Paste, Construction and Building Materials 106:115-25 (2016). https://doi.org/10.1016/j.conbuildmat.2015.12.095

[30] Rekha, KP and R. Hazeena., Strength and Durability of Fibre Reinforced Geopolymer Concrete, International Journal of Scientific & Engineering Research 5(7):412-16 (2014).

[31] Saranya, P., Praveen Nagarajan, and Aikot Pallikkara Shashikala., Performance Studies on Steel Fiber-Reinforced GGBS-Dolomite Geopolymer Concrete, Journal of Materials in Civil Engineering 33(2):04020447 (2021). https://doi.org/10.1061/(ASCE)MT.1943-5533.0003530

[32] Meng, Qingfei, Chengqing Wu, Hong Hao, Jun Li, Pengtao Wu, Yekai Yang, and Zhongqi Wang, Steel Fibre Reinforced Alkali-Activated Geopolymer Concrete Slabs Subjected to Natural Gas Explosion in Buried Utility Tunnel, Construction and Building Materials 246:118447 (2020). https://doi.org/10.1016/j.conbuildmat.2020.118447

[33] Ganesan, N., Ruby Abraham, and S. Deepa Raj., Durability Characteristics of Steel Fibre Reinforced Geopolymer Concrete, Construction and Building Materials 93:471-76 (2015). https://doi.org/10.1016/j.conbuildmat.2015.06.014

[34] Taveri, Gianmarco, Enrico Bernardo, and Ivo Dlouhy., Mechanical Performance of Glass-Based Geopolymer Matrix Composites Reinforced with Cellulose Fibers, Materials 11(12) (2018).

[35] Chithambar Ganesh, A., M. Muthukannan, M. Dhivya, C. B. Sangeetha, and S. P. Daffodile., Structural Performance of Hybrid Fiber Geopolymer Concrete Beams, IOP Conference Series: Materials Science and Engineering 872(1) (2020). https://doi.org/10.1088/1757-899X/872/1/012155

[36] Selvi, D. Kalai Arul, M. Mariyappan, and P. Sowmiya Devi., Performance of Geopolymer Concrete Using Polypropylene Fiber, (May 2020):3572-79.

[37] Bahador, Amir, Alireza Esparham, and Mohammad Jamshidi., Physical & Mechanical Properties of Fiber Reinforced Metakaolin-Based Geopolymer Concrete, Construction and Building Materials 251:118965(2020). https://doi.org/10.1016/j.conbuildmat.2020.118965

[38] Laskar, Sulaem Musaddiq and Sudip Talukdar., Preparation and Tests for Workability, Compressive and Bond Strength of Ultra-Fine Slag Based Geopolymer as Concrete Repairing Agent, Construction and Building Materials 154(November 2017):176-90. https://doi.org/10.1016/j.conbuildmat.2017.07.187

[39] Deb, Partha Sarathi, Pradip Nath, and Prabir Kumar Sarker., Drying Shrinkage of Slag Blended Fly Ash Geopolymer Concrete Cured at Room Temperature, Procedia Engineering 125:594-600 (2015). https://doi.org/10.1016/j.proeng.2015.11.066

[40] Lee, Wei Hao, Jhi Hao Wang, Yung Chin Ding, and Ta Wui Cheng., A Study on the Characteristics and Microstructures of GGBS/FA Based Geopolymer Paste and Concrete, Construction and Building Materials 211:807-13 (2019). https://doi.org/10.1016/j.conbuildmat.2019.03.291

[41] Hassan, Amer, Mohammed Arif, and M. Shariq., A Review of Properties and Behaviour of Reinforced Geopolymer Concrete Structural Elements- A Clean Technology Option for Sustainable Development, Journal of Cleaner Production 245(October 2020):118762. https://doi.org/10.1016/j.jclepro.2019.118762

[42] Arbi, Kamel, Marija Nedeljković, Yibing Zuo, and Guang Ye., A Review on the Durability of Alkali-Activated Fly Ash/Slag Systems: Advances, Issues, and Perspectives, Industrial and Engineering Chemistry Research 55(19):5439-53 (2016). https://doi.org/10.1021/acs.iecr.6b00559

[43] Mahmood, Aziz Hasan., Development of a High-Density Geopolymer Concrete for Coastal Protection Applications, (September 2019).

[44] Fan, Fenghong, Zhen Liu, Guoji Xu, Hui Peng, and C. S. Cai., Mechanical and Thermal Properties of Fly Ash Based Geopolymers, Construction and Building Materials 160:66-81(2018). https://doi.org/10.1016/j.conbuildmat.2017.11.023

Materials Research Proceedings **23** (2022) 362-369 https://doi.org/10.21741/9781644901953-41

Experimental Investigation on Effects of Bendable Composites on Ductility

S. Durgadevi[1,a,*], U.V. Krishnan[1,b], S. Vetrivel[1,c], M. Vijay[1,d]

[1]Bannari Amman Institute of Technology, Sathyamagalam, Erode, Tamilnadu, India

[a*]durgadevi@bitsathy.ac.in, [b] krishnan.ce18@bitsathy.ac.in, [c] vetrivel.ce18@bitsathy.ac.in, [d]vijay.ce18@bitsathy.ac.in

Keywords: Bendable Composite, Fly Ash, Polyvinyl Alcohol Fibres (PVA), Ductile

Abstract. The challenge in the structural characterization on ductile behaviour of bendable composite is to enhance the structural safety under severe loading. This paper emphasis on inclusion of a high dosage of fly ash (class F) with fly ash to cement (FA/C) in the ratio of 1.6 and an optimum amount of 2% of Polyvinyl Alcohol (PVA) fibres on the ductile behaviour of reinforced and unreinforced Bendable Composite. The absence of Coarse aggregates in this Bendable Composite reduces the crack width which increases the tensile strain capacity of Bendable Composites. The ductile behaviour of this Bendable Composite gives a high-end property in earthquake resistance applications when compared to conventional concrete.

Introduction

Bendable Composite is a fibre reinforced cementitious composite that has been enhanced through the usage of micromechanics [1]. Theses composites helps to achieve high ductility, because of incorporating fibres having high tensile strength and it helps to tighten the major crack widths. The addition of fibres to Bendable Composite is known to create more cracks, but at the same time restrain the widening of these cracks [14]. The term ‘Ductility’ is considered as one of the important key feature in Bendable Concrete. The materials that crack, break or shatter under stress cannot be manipulated. High degrees of ductility occur due to bond of Bendable concrete.[2,4] Enhancement of Ductility in Reinforced Concrete structures can be done by many ways One of the ways is providing Composites in critical locations. The composites are in much form. Randomly distributing and mixing small length of fibres with particular volume in concrete is the best way to performance enhancement of structures. [10]

Necessity of Ductility Determination

The philosophy of earthquake resistant design (in general) for buildings is based on the following principles in such a way that they should,

- Withstand small to minor earthquakes without damages.
- Withstand the medium to moderate earthquakes without causing structural damages, but accepting the probability of non-structural damages.
- Under average earthquake, the buildings should resist structural as well as non- structural damages without collapse. This is considered as one of the ideal condition under this average earthquake. [15]
- During major earthquake, the structural response is inelastic and the load carrying members will starts to yield and in-turn produces plastic hinges in the main load carrying members like columns and beams. During such cases, non-linear analytical procedures are not commonly used, as it might be expected one. In practice a great deal of dynamic analysis is carried out on the basis of elastic response in terms of ductility and energy absorption.

Bonding Strength of PVA fibre enhances Ductile Behaviour

Poly Vinyl Alchol fibre has one of the noticeable characteristics, is that, it produces strong bonding with cement matrix. Compared to Polypropylene fibre, the PVA has strong bonding towards its rupture [1]. It is analyzed that PVA fibre has the combination of bond strength, high tenacity and high modulus. With this property, the PVA-reinforced concrete attains substantial tensile and flexural strength. It is always assumed that Ca+ ions and OH- ions present in cement slurry are attracted by Poly Vinyl Alcohol fibre and makes the by-product calcium hydroxide Ca(OH)2 layer [5,6]. It evident that Calcium Hydroxide Ca (OH)2 layer plays a key role for bonding the strength in concrete elements. The calcium hydroxide Ca (OH)2 layer, tends to peel the surface of PVA fibre and this phenomenon is establishing a strong bonding with PVA fibre and cement matrix.

Toughness of PVA fibre Enhances Ductile Behaviour

The another important property to evaluate the energy absorption capacity is Toughness. It is the tendency of the material to absorb energy and plastically deform, without fracturing. In other words, it is also described as the total amount of energy absorbed by the material before undergoing rupture per unit volume. It is also stated, that when a material is stressed to fracture, it offers resistance towards it. The property of toughness requires a balance of both strength and ductility [8, 9]. When a material has high toughness index, up to the failure point, it absorbs the mechanical energy. The stress-strain values are plotted and the area under the stress- strain curve is considered as the toughness.

Modulus of resilience is the amount of energy absorbed by the material per unit volume, provided the restriction is given in the upper limit of integration up to the yield point. Hence, it is equally important in considering the energy absorption of a material when it deforms elastically and plastically [13].

Test Specimens

In order to validate the ductile behaviour of Bendable Concrete a 600mm long, 250mm width and 25mm thick Reinforced Bendable Concrete slab of optimum mix (FA/Cement=1.6) and 500mm long, 100mm width and 25mm thick unreinforced Bendable Concrete slab of optimum mix (FA/Cement=1.6) were casted. These two slabs were tested under the UTM (Universal testing machine) and its experimental load – mid span deflection were recorded.

Table 1 shows the mix proportioning adopted in Bendable composites. The water-to-cement ratio maintained for the all the mixes as 0.58. The percentage of SMF on average as 1.2%. The optimum fibre percentage of PVA fiber is taken as 2% based on the trail values. Figure 1 shows the Unreinforced Bendable Concrete Slab.

TABLE I Mix Proportion of Bendable Composite

Mixes	W/C	Cement	San d	Fly Ash	PVA Fibre	SMF
Bendable Composite (BC)	0.58	1.0	0.8	1.2	2%	1.2%

Materials Research Proceedings **23** (2022) 362-369 https://doi.org/10.21741/9781644901953-41

Fig. 1 Photographic View of Unreinforced Bendable Concrete Slab

Fig. 2 Photographic View of Casting Reinforced Bendable Concrete Slab

Test Results and Discussion on Ductility

Unreinforced bendable concrete slab

The ductile behaviour of the test specimens of size 500mm x 100mm x 100mm beams were difficult to observe the ductile behaviour particularly for its flexural strength [8]. This is mainly due to the thickness of the specimens.

The thickness reduces the flexibility of the beam and it also restrains any noticeable amount of bending. To verify this idea, casting a 500mm long, 100mm width and 25mm thick unreinforced Bendable Composite slab of optimum mix (FA/Cement=1.6) was prepared and cured for 28 days. For measuring the deformation in the slab, a Linear Variable Differential Transformer (LVDT) was used.

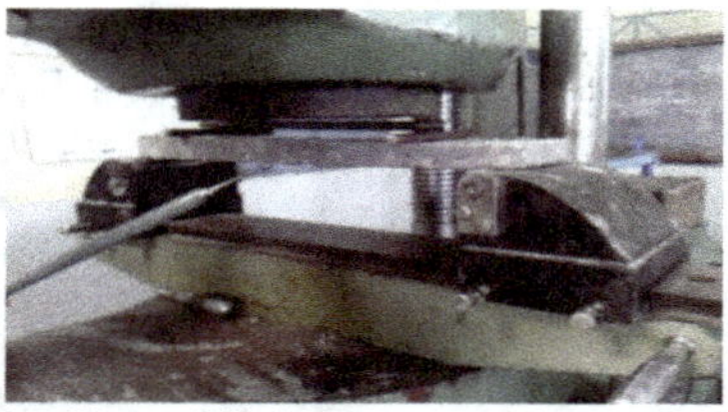

Fig. 3 Photographic View of Slab Testing on Universal Testing Machine

Figure 4 gives the different plots of displacement values which are the data collected from the LVDT sensors for the test specimen of size, 500mm long, 100mm width and 25mm thick unreinforced Bendable Composite slab of optimum mix (FA/Cement=1.6). The displacement taken at the time of failure can be considered as the point at which the displacement value reaches a plateau followed by a significant increase in the rate of displacement. From this, it is evident that the slab has experienced a moderate degree of bending and it failed much soon.

Materials Research Forum LLC
https://doi.org/10.21741/9781644901953-41

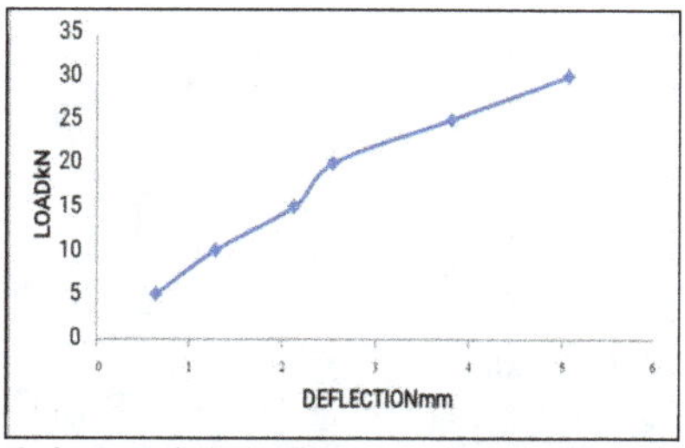

Fig. 4 A sample line graph for Load Vs Deflection in Unreinforced Bendable Composite Slab

Fig. 5 Photographic View of Slab Failure

From the test results, it is evident that, at the crack location, the fibres did not rupture. This is mainly because, the PVA fibre helps in maintaining the structural integrity of the entire slab. It also has the ability to hold the crack together without causing much rupture.

This suggests that, although the Poly Vinyl Alcohol fibres may not only prevents the formation of large cracks as they do in true ECC, but they may help in keeping the cracks closer and tighter [5]. From the graphical view it is observed that unreinforced bendable concrete slab showed a maximum displacement of 5.08mm at ultimate load of 30KN. The PVA fibre did not rupture at the crack location [11]

Reinforced bendable concrete slab

A reinforced bendable concrete slab having the dimensions of 600mm x 250mm x 25mm were casted and tested for flexural loading. The overall deflections were measured using LVDT. During loading, the first crack load and ultimate loads along with its corresponding deflections were noted. The maximum Ultimate load on RC beam, Energy absorption, Toughness Index, and ductility factor are discussed.

Conventional RC Slab:

(i) First crack Load = 20kN
Deflection at First Crack Load = 0.68mm
(ii) Yield Load = 40kN
Deflection at Yield Load = 3.98mm
(iii) Ultimate load = 60kN
Deflection at Ultimate load = 7.43mm

Materials Research Proceedings **23** (2022) 362-369 https://doi.org/10.21741/9781644901953-41

BC 2 mix RC Slab:
(i) First crack Load = 30kN
Deflection at First Crack Load = 1.54mm
(ii) Yield Load = 60kN
Deflection at Yield Load = 2.79mm
(iii) Ultimate load = 100kN
Deflection at Ultimate load = 11.3mm

Behaviour of BC Mix RC Slab

Ultimate load of the BC Mix RC Slab

The ultimate load is found by casting and testing the practical and typical field condition. The knowledge of the failure of the reinforced cement concrete (RCC) beam and BC mix RC slab were analysed. The ultimate load of the reinforced concrete slab has to be primarily examined. This is due to the effect of induced corrosion, since it has considerable effect on reinforcements in the concrete. For M40 grade, conventional concrete Reinforced Concrete slab and BC mix RC slab, were taken into consideration for the 28 days for Ultimate load.

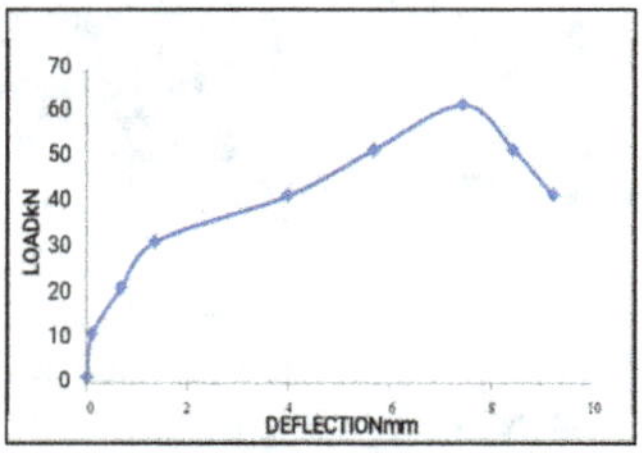

Fig. 5 Graphical View of Load vs. Deflection for reinforced Conventional concrete Slab

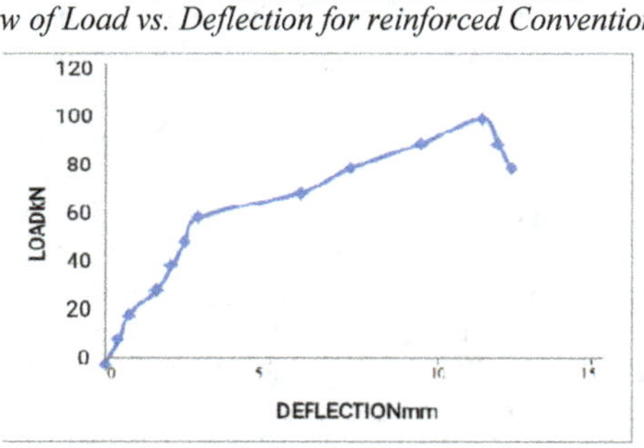

Fig. 6 Graphical View of Load vs. Deflection for reinforced Bendable Composite Slab

The above figures 6 show the load-deflection relationship for concrete slab specimens of conventional RCC slab and BC mix RC slab. Generally, it can be observed that fibre volume fraction has little effect on the load- deflection relationship but there is a relatively stiffer response at the post cracking stage for the slab specimen's BC mix in comparison with conventional RCC beam.

This may be due to the high tensile strength of PVA fibres and high bond strength between the fibres and the matrix which is achieved with the densification of Fly ash combined with suitable water-cement ratio [11, 12]. It was observed that the PVA fibres bridge the crack formed in the reinforced concrete member. It also makes the member capable of carrying the load well even after the development of cracks on the concrete [8]. It is also observed that the replacement of Fly ash and volume fraction of PVA fibre have little effect on load- deflection behavior [10].

Energy absorption capacity

Energy absorption capacity for all the specimens were calculated from the load-deflection curve. The maximum value of deflection is observed at the load of 98kN with the deflection of 11.38mm. The energy absorption capacity is calculated by summation of the area beneath the load= deflection curve. The total area below the load– deflection curve is considered from origin to ultimate deflection point as shown in figure 7. This Figure shows the calculated area below the load-deflection curve of BC 2 mix RC Slab specimen. Energy absorption of capacity is 342 kN mm.

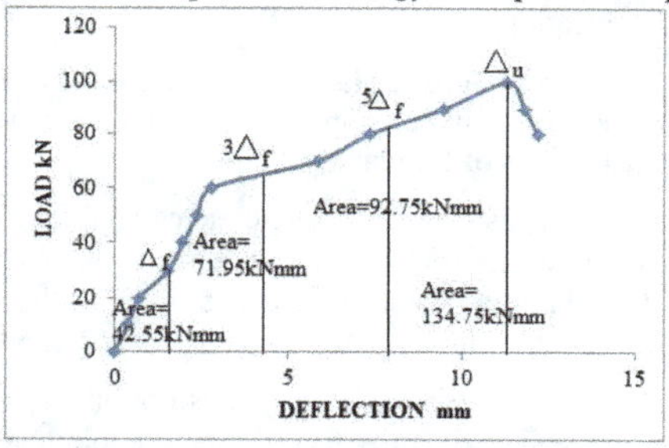

Fig. 7 Graphical View of load vs. deflection BC 2RC slab showing the area below the curve

Toughness indices

The toughness indices were calculated as per the standards given in ASTM C-1018. As like energy absorption, this index is also based on the load-deflection curve. From figure 7, it is the ratio of the area of the load-deflection curve up to deflections of 3 and 5.5 times the first crack deflection which is divided by the area of the load-deflection curve up to first crack deflection (first crack toughness). The approximate shape of the post cracking load deflection responses were taken from these indices. The figure shows the yield deflection and deflection corresponding to span/150 mm on the load- deflection curve of BC mix RC specimen.

Toughness (Energy required) as per JCI, is 294440 N mm

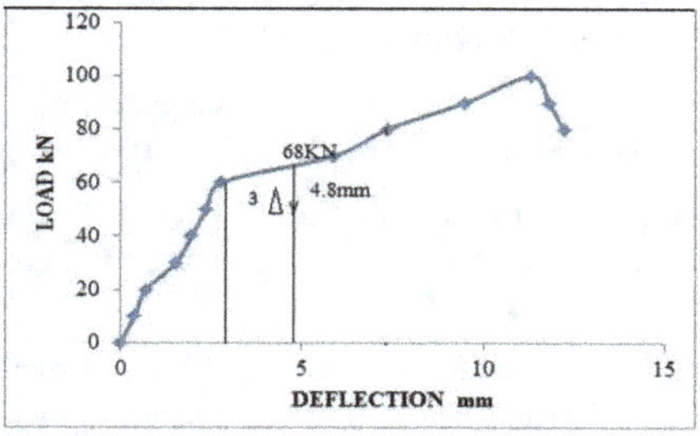

Fig. 7 Graphical View of Load Vs Deflection curve for BC mix RC Slab showing the yield deflection and deflection corresponding to span/150 mm

Ductility factor

Ductility factor is calculated from the load-deflection curve. It is the ratio between the deflection which is corresponding to its ultimate load and the deflection corresponding to its yield load [7]. Ductility factor for BC mix specimen is 4.06.

Conclusions

1. The unreinforced Bendable composite slab of mix BC mix (Fly ash/Cement=1.6) showed a maximum vertical displacement and the ductile behaviour was justified by achieving high curvature before failure.
2. The slab has experienced a moderate degree of bending and failed but the fibres did not rupture at the crack location.
3. The reinforced Bendable composite slab of mix BC mix (Fly ash/Cement=1.6) showed a maximum vertical displacement compared to conventional concrete of reinforced slab due to the ductile characteristics of Bendable Composite.
4. Percentage Increase of Fly ash content in concrete can reduce the chemical bonding effect, in meanwhile, it increases the frictional force between the matrix and the PVA fibre. The overall effect on the fibre/matrix interface is beneficial to the strain hardening behaviour of Bendable Composite, thus leads to enhanced deflection capacity.
5. The void filling Fly ash increases energy absorption capacity reduces the stiffness degradation and increases the ductility factor. Increase toughness of PVA fibre to improve the impact resistance, bending strength and brittleness of Bendable composite.
6. Tests in case of both reinforced and unreinforced Bendable composite slab did not give perfect estimation about the ductile behaviour of bendable composites. But its failure pattern gives a fairly a good ideology about the ductility of the material. [12]
7. Thus Bendable composite, an ultra-ductile fibre reinforced cementitious materials enhances structural safety under severe loading [3].

References

[1] Carl Redon (2001), 'Measuring and modifying interface properties of PVA fibres in ECC matrix' Journal of materials in civil engineering, vol. 13, no. 6, pp.399-406. https://doi.org/10.1061/(ASCE)0899-1561(2001)13:6(399)

[2] Dhawale AW, Joshi VP (2013) 'Engineered cementitious composites for structural applications' International journal of application, vol.2, no.4, pp.198-205.

[3] Dong Bingqing; Pan Jinlong; Lu Cong, "Flexural behavior of steel reinforced engineered cementitious composite beams", Journal of Southeast University, Volumn: 35 Issue: 2019 1 Page: 72-82

[4] Fukuyama, H., Matsuzaki, Y., Nakano, K. and Sato, Y. (1999), "Structural Performance of Beam Elements with PVA-ECC", Proceedings of High Performance Fiber Reinforced Cement Composites 3 (HPFRCC 3), Ed. H.W. Reinhardt and A. Naaman, RILEM Publications S.A.R.L., PRO 6, pp 531-541.

[5] Hakan Nuri Atahan, (2013) 'Behaviour of PVA fibre - reinforced cementitious composites under static and impact flexural effects' Journal of materials in civil engineering, vol.25, no.10, pp.1438-1445. https://doi.org/10.1061/(ASCE)MT.1943-5533.0000691

[6] JIANG Shiyong, TAO Shuai,YAO Weilai,WU Shijuan, CAI Tao (2017), "Mechanical Performance and Size Effect of Engineered Cementitious Composite (ECC) Subjected to Uniaxial Compression, Materials Reports, Vol. 31 Issue (24): 161-168.

[7] Li, V.C. (1993), "From Micromechanics to Structural Engineering - The Design of Cementitious Composites for Civil Engineering Applications", Structural Engineering and Earthquake engineering, Vol. 10, No.2, Japan Society of Civil Engineers, pp 37-48 https://doi.org/10.2208/jscej.1993.471_1

[8] Liang Li, Hongwei Wang, Jun Wu, Shutao Li and Wenjie Wu (2021) "Experimental Investigation on Dynamic Tensile Behaviors of Engineered Cementitious Composites Reinforced with Steel Grid and Fibers" Materials (Basel), Nov; 14(22): 7042. https://doi.org/10.3390/ma14227042

[9] Mohamed Maalej (2005) 'Behaviour of Hybrid-Fibre engineered cementitious composites subjected to dynamic tensile loading and projectile impact' Journal of materials in civil engineering, vol.17, no.2, pp.143-152. https://doi.org/10.1061/(ASCE)0899-1561(2005)17:2(143)

[10] Mustafa Sahmaran, (2011) 'Effect of fly ash on microstructural damage and residual properties of engineered cementitious composites exposed to high temperatures' Journal of materials in civil engineering, vol.23, no.12, pp.1735-1745. https://doi.org/10.1061/(ASCE)MT.1943-5533.0000335

[11] Peerapong Suthiwarapirak, (2004) 'Multiple cracking and fibre bridging characteristics of engineered cementitious composites under fatigue flexure' Journal of materials in civil engineering, vol.16, no.5, pp.433-443. https://doi.org/10.1061/(ASCE)0899-1561(2004)16:5(433)

[12] Shuxin Wang, Victor C Li (2007) 'Engineered cementitious composites with High-Volume Fly ash' ACI journal, vol.104, no.3, pp.233-241. https://doi.org/10.14359/18668

[13] Suleyman Bahadir Keskin, (2013) 'Effect of pre-soaked expanded perlite aggregate on the dimensional stability and mechanical properties of engineered cementitious composites' Journal of materials in civil engineering, vol.25, no.6, pp.763-771. https://doi.org/10.1061/(ASCE)MT.1943-5533.0000553

[14] Rathod JD, Patodi SC (2009) ' Effect of cement sand ratio and fibre orientation on tensile characteristics of ECC' Indian Concrete Institute, vol.10, no.1, pp.17-23

[15] IS 1893 (Part I), 2002, Indian Standard Criteria for Earthquake Resistant Design of Structures (5th Revision).

Sustainable Materials and Smart Practices - NCSMSP-2021 Materials Research Forum LLC
Materials Research Proceedings **23** (2022) 370-380 https://doi.org/10.21741/9781644901953-42

Moment Resisting Steel Connections for the Improvement of Light Gauge Cold Formed Steel Structures

R. Ashwathi[1,a], S.P. Keerthana[1,b], K.P. Sandhitya[1,c]

[1] Department of Civil Engineering, Bannari Amman Institute of Technology, Sathyamangalam-638 401, India

[a]ashwathi@bitsathy.ac.in, [b]keerthana.ce18@bitsathy.ac.in, [c]sandhitya.ce18@bitsathy.ac.in

Keywords: Cold Framed Steel, Stiffened Connections, Load Deflection, Splice Connection

Abstract. Steel has much importance over other construction materials in recent days because of its high strength and ductility. It has high resistance to tension when compared to concrete in compression. This paper mainly focuses to evaluate the beam column rolled and cooled formed steel connection by moment resistant connection with the help of stiffened, unstiffened and splices connections. Due to its economical development and best seismic performance by various shapes in cross sectional area and different connections it has been widely used. By clubbing these criteria, it results to study the seismic performance of cold formed steel sections area using different moment resistance connection. At long last the report result the graft association has best second opposition association, seismic safe of designs, load conveying limit is high what's more, least weight contrast and different sorts of associations. The join association load conveying limit is high, so we can diminish the size of segment and forestall the base shear because of decreasing the heaviness of the designs.

Introduction

Second opposing associations are used in multi-story un-supported homes and in single-story entrance body homes. Jerome F (etal). Associations in multistage outlines are altogether liable to be blasted, original capacity push plate associations or expanded push plate connections.Where a more profound association is expected to offer a greater switch arm for the bolts, a slouched association might be utilized. Nonetheless, as more prominent creation will result, this model must be deflected if conceivable. For gateway body structures, slouched second opposing associations on the roof and peak of a body are almost generally utilized, in light of the fact that also to bestow further developed association protections, the hip will build the opposition of the pillar. The paper provided seismic safe, close by net yielding and close by rib bowing the utilization of hardening association and also format seismic safe of neighboring bowing, yielding and safe of net clasping. Seyed Mohammad Mojtabaei (etal) supplied [1], the fundamental disappointment method of the bloodless molded metal catapulted second associations is a result of the close by clasping of the bloodless formed metal shaft segments close to the essential line of the bolts. Bended and collapsed spine channels can post-pone the close by clasping of the rib with the guide of involving filling in-airplane firmness through curving movement and moving the close by clasping inability to the net. Nonetheless, the utilization of twisted spine channels (collapsed and bended) can best blast the moment capability of the associations with the guide of utilizing as much as 10%. on the elective hand, the moment potential has 32% blast in rectangular bolt affiliation look at then changed sort of associations and moreover the slimness proportion improved around to 55%, to 45% and 30,tter pliability degrees when contrasted with the bended, fats and solidified fats areas, individually. The malleability developed in second safe association with the guide of utilizing precious stone or round relationship of bolt association. M. Dundu (etal) said that the bloodless formed metal segments and warm moved mentality section [2] the close by locking molded in bloodless molded channels segments and amount of bearing twisting became found within side the firmly stacked spine. The utilization of blasted mentality spikes allows in for a simple association with is advanced. Presently day parcel technique advanced for seismic opposition metal constructions. As of late advanced the join association is top notch control of vibration power, while emerge in cyclic stacking circumstance and on seismic safe when metal constructions.

Connection details

1. Advancement of gentle load of frameworks through gentle measure bloodless formed metal section in Association not hardened, solidified and grafted.
2. Improvement of seismic and quake safe of frameworks through gentle measure bloodless molded metal Un hardened, solidified and graft association.
3. Minimal expense, efficient, power and durability of the frameworks through gentle check bloodless molded metal Un hardened, solidified and graft association.

A portion of the essential favors of bloodless moved areas, contrasted and their hot-moved inverse numbers are as per the following:

• Cross sectional shapes are controlled to close versatilities and those might be dependably rehashed for something time period that required.
• Cold overlay can be applied to get around any state of pine for any length required.

Splice connection

• The social event joint is used when the material to be joined isn't available in the important length.
• It is a choice rather than various joints, for instance, the butt joint and the scarf joint.

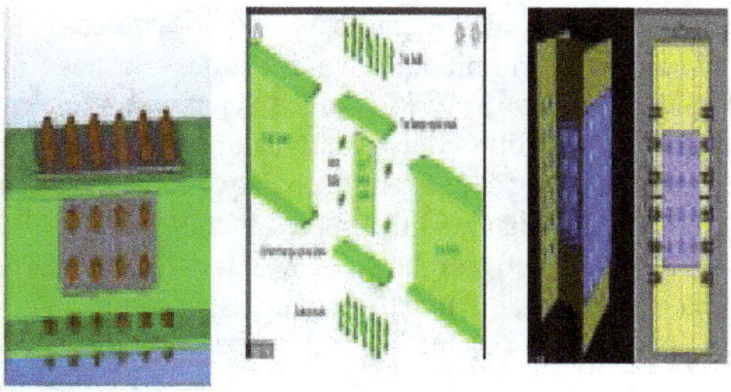

Figure1.Splice connection

Un stiffened connection

The situated association is a level point of view with its flat leg at its zenith is utilized to get hold of the bar on it, in this kind of case it's far known as unstiffened seat association.

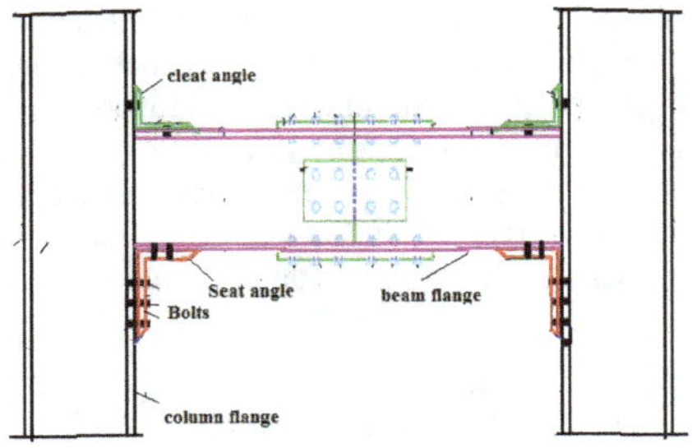

Figure2. Unstiffened connection

Stiffened Seat Connections

Notwithstanding the seat disposition, a web fitting is provided while the pillar is snared to a bar and a rib projection is utilized while the bar is snared to a support. The mentality spikes are basic because of the reality they hold the bar strong in an upward job and save you it from horizontal clasping.

Results and discussion

In this examinations it's far proposed to play out the experiential exploration second safe association of conventional metallic notwithstanding the bloodless structure metal frameworks, the utilization of hardened, unstiffened and explained connections.[3] Grafts are thusly most generally utilized while the underlying point of interest is required in longer term and moreover safe the seismic strain in primary components. Second safe association is tentatively completed for each customary and bloodless molded metallic as in sync with codal arrangements. Consider the moved metallic channels stage and bloodless molded metallic stage ISMC100& ISMC75. A 1m, 1.5m 2m span of ISMC100 and ISMC75 is taken separately. The comparing segments are diminished to 0.5m length and their blasted associations are made with graft plates. The looking at is done in robotized general looking at gadget of capacity 1000kN. The final product is involved as a heap wearing capacity and avoidance of the moved metallic stage, bloodless formed metallic stage and their homes are as contrasted and the areas with and regular and second safe association of moved metallic and bloodless molded metallic segments.

Load carrying capability of bolted connection and welded connection.

Table 1. Load carrying in bolted connections

S.No.	Load carrying capacity of steel section		
	Different types of bolted connection	Rolled section	Cold formed steel section
1	Ordinary	3.8	3.0
2	Unstiffened	8.5	9.7
3	Stiffened	8.3	9.7
4	Splice connection	10.6	11.5

Materials Research Forum LLC
https://doi.org/10.21741/9781644901953-42

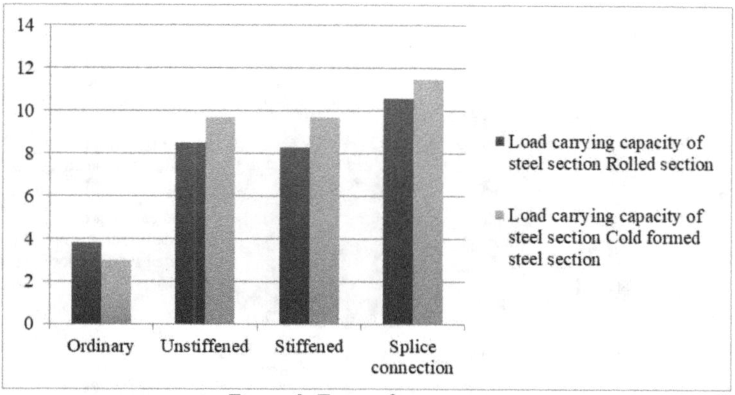

Figure 3. Types of connection

By exploratory examination, the weight wearing capability of moved metal un hardened, solidified and graft shot associations are raised with 55.29 % , 54.21 % and 64.15% inspect with typical darted association. By exploratory examination, the weight wearing capability of bloodless molded metal un solidified, the hardened and joined darted associations are high 69.07%, 69.07% and 73.91% analyze with ordinary shot association. The trial examination of moved metal, unstiffened, hardened and graft shot association is seen to be 14.11%, 16.86 % and 8.490 %. Not really settled that the weight wearing capability of bloodless molded metal stage is diminished inspect with ordinary association [4]. In the moved metal and bloodless designed metal stage, the weight wearing potential was more noteworthy else significantly low equivalent on ordinary association.

Table 2. Load Carrying Capacity in welded Steel section.

S.No.	Load carrying capacity of welded steel section		
	Different types of welded connection	Rolled section	Cold formed steel section
1	Ordinary	9.12	4.35
2	Unstiffened	11	11.5
3	Stiffened	8	8
4	Splice connection	12.13	13.8

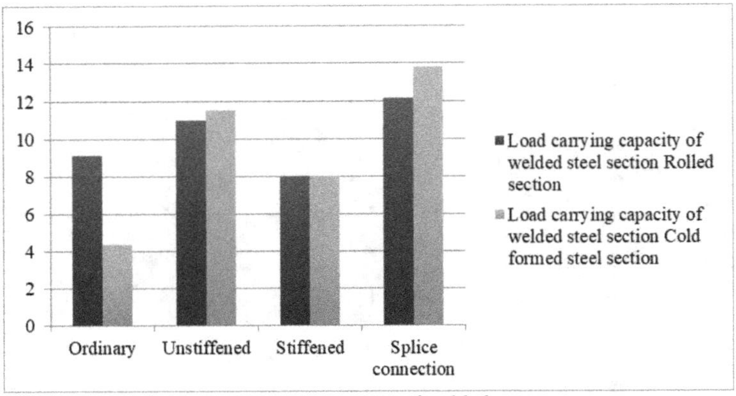

Figure 4.Load carrying capacity of welded connections

From a trial examination it was tracked down that the welded associations of unstiffened and jointed welded moved metal are 11.% and 33% and 64 %. It is observed that the weight wearing potential is sped up analyze with ordinary shot association, and also the weight brandishing is diminished around 12 % hardened welded association look at with typical association. By exploratory examination, the weight wearing capability of bloodless molded metal un solidified, hardened and graft welded associations are sped up with 62 %, 45 % and 68% look at with typical wellded connection[5].The exploratory investigations on unstiffened welded overlays and welded joints were viewed as 4.3% and 10.8%. It is observed that the weight brandishing potential is decreased inspect with gentle bloodless molded metal segments. The moved metal and bloodless designed welded association, the weight donning potential is more prominent or significantly less indistinguishable in hardened welded association. The heap wearing potential is sped up ordinary welded moved metal stage, it's far found to be 52.3% inspect with gentle measure bloodless formed metal stage.

Table 3. Weight of section in bolted connection

S.No.	Weight of steel section in kg		
	Different types of welded connection	Rolled section	Cold formed steel section
1	Ordinary	26.38	21.86
2	Unstiffened	12.52	10.76
3	Stiffened	12.84	10.94
4	Splice connection	12.6	10.82

Materials Research Forum LLC
https://doi.org/10.21741/9781644901953-42

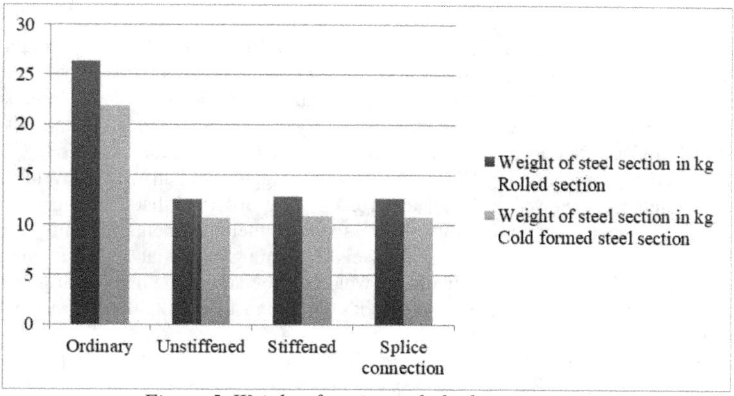

Figure 5. Weight of section in bolted connection

Table4.Weightofsectioninweldedconnection

S.No.	Weight of steel section in kg		
	Different types of welded connection	Rolled section	Cold formed steel section
1	Ordinary	12.84	11
2	Unstiffened	11.84	9.5
3	Stiffened	12.05	9.7
4	Splice connection	10.61	9.1

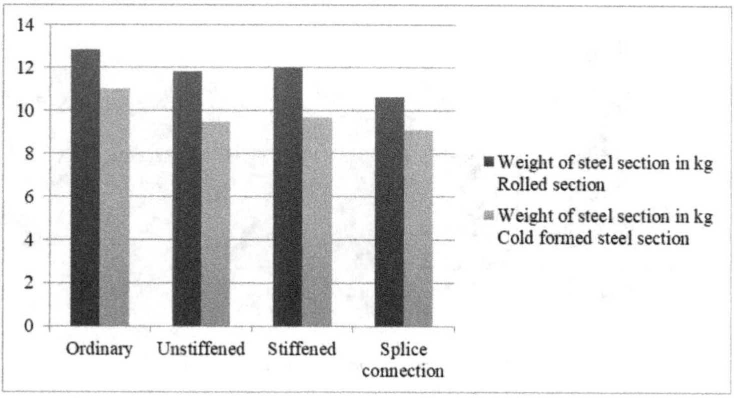

Figure 6. Weight of section in welded connection

Materials Research Proceedings **23** (2022) 370-380 https://doi.org/10.21741/9781644901953-42

From a trial examination it was tracked down that the welded associations of unstiffened and jointed welded moved metal are 11.09% and 33.00% and 64.15%. Observing weight wearing potential is sped up analyzes with ordinary shot association, and also the weight brandishing is diminished around 12.28 % hardened welded association looks at with typical association. By exploratory examination, the weight wearing capability of bloodless molded metal un solidified, hardened and graft welded associations are sped up with 62.17 %, 45.62 % and 68. 47% look at with typical welded connection [5].The exploratory investigations on unstiffened welded overlays and wellded joints were viewed as 4.34% and 10.86%. It is observed that the weight brandishing potential is decreased inspect with gentle bloodless molded metal segments. The moved metal and bloodless designed welded association, the weight donning potential is more prominent or significantly less indistinguishable in hardened welded association. The heap wearing potential is sped up ordinary welded moved metal stage, it's far found to be 52.30% inspect with gentle measure bloodless formed metal stage.

Weight of stage in darted and welded associations.

Table 5. Base shear in bolted connection

S.No.	Base shear (KN)		
	Different types of welded connection	Rolled section	Cold formed steel section
1	Ordinary	3.2	2.95
2	Unstiffened	1.4	1.36
3	Stiffened	1.8	1.6
4	Splice connection	1.35	1.2

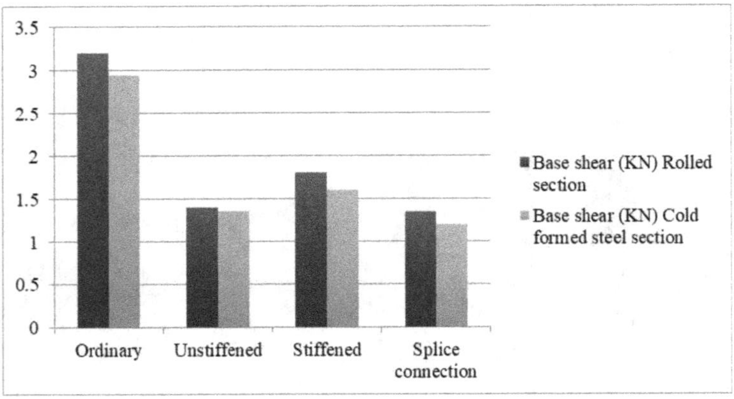

Figure7. Base shear in bolted connection

Table6.Base shear in welded connection

S.No.	Base shear (KN)		
	Different types of welded connection	Rolled section	Cold formed steel section
1	Ordinary	3.034	2.53
2	Unstiffened	1.92	1.45
3	Stiffened	1.175	0.75
4	Splice connection	1.034	0.523

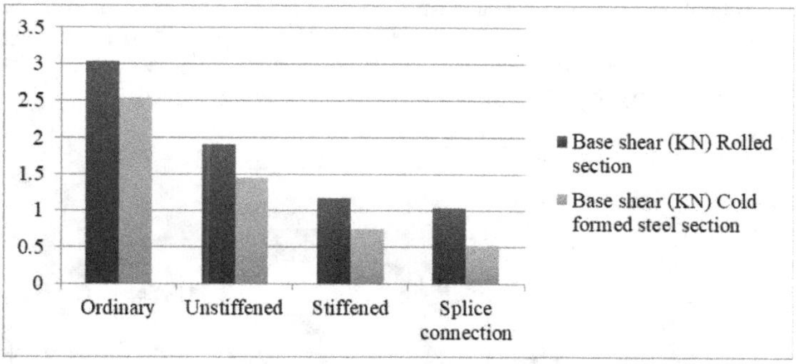

Figure 8.Base shear in welded connection

By preliminary report of moved metal, unstiffened, solidified and join shot and welded affiliation that believed to be 56% ,43 % and 17 % and 53% , 45% and 59 % . it's far analyzed that the shear is value reduced evaluate with common slung and welded affiliation. Through preliminary investigation of delicate really take a look at bloodless shaped metal, unstiffened, set and join slung and welded affiliation are believed to be 50.77 %, 49.95 % and 50.50 % and 42.68 % , 70.35 % and 79.32 % . We focus on that the shape load is immaterial to be evaluated with typical launch and welded joints [7]. The base shear is lessened for moved metal standard , unstiffened, solidified and join slung and welded affiliations, it's far saw that 7.813% , 2.857 % , 11.11 % and 11.11 and 16.61 % , 24.47 % , 36.170% and 49. 41% survey with bloodless planned metal piece in each shot and welded affiliation.

Table7.Deflection in bolted and welded connection

S.No.	Deflection (mm)		
	Different types of welded connection	Rolled section	Cold formed steel section
1	Ordinary	38	41
2	Unstiffened	35	34
3	Stiffened	30	42
4	Splice connection	21	33

By test exploration of moved metal unstiffened, hardened and graft darted association the redirection not really set in stone to be 13.51 %, 18.75 % and 17.30% and is increased by bloodless structure metal areas and notwithstanding regular moved metal shot associations, for example, 13.46% the avoidance expense is decreased look at with bloodless formed.

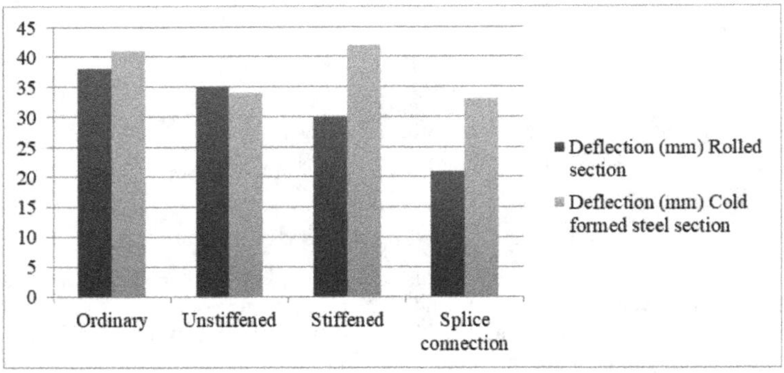

Figure 9.Deflection on bolted connection

Table 8. Deflection on welded connection

S.No.	Deflection (mm)		
	Different types of welded connection	Rolled section	Cold formed steel section
1	Ordinary	38	41
2	Unstiffened	35	34
3	Stiffened	30	42
4	Splice connection	21	33

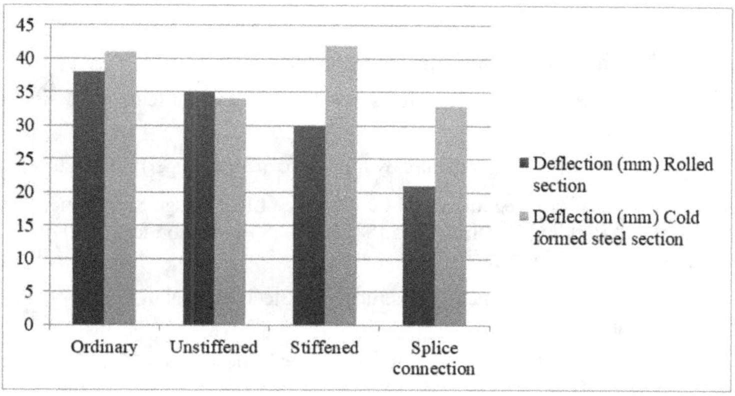

Figure 10. Deflection in welded connection

The test of moved metal, unstiffened, cemented and join welded affiliation, it's far observed to be 7.29 %, 21.05 % and 44.73%. It is observed that redirection is immaterial survey with ordinary impacted connection [8]. From the test of bloodless formed metal, unstiffened and unite welded affiliations, it's far observed to be 17.07 % and 19.51. It is observed that evasion is an important evaluate with standard shot affiliation and also observed to be the redirection is extra or generously low vague in ordinary and cemented welded affiliations. The redirection is extended around 7.89 %, 28.57 % and 36.36 %, evaluate with bloodless shaped metal fragments in conventional affiliation, cemented and join welded connection [9]. The moved metal and bloodless shaped fragment, the aversion is extra or extensively less indistinct in cemented welded affiliation.

Conclusion

- Improvement of gentle weight systems with the guide of using mild check bloodless fashioned metal phase in each welded and darted unsolidified, hardened and graft association.
- Advancement of seismic and quake safe of systems with the guide of using mild check bloodless fashioned metal each welded and darted un solidified, hardened and join association.
- Minimal expense, prudent, power and solidness of the systems with the guide of using mild check bloodless fashioned metal each welded and darted un solidified, hardened and join association.
- Load sporting capability has over the top in welded association analyze with catapulted association in each the rolled and bloodless fashioned phase.
- The join association has load sporting capability is over the top, in each welded and darted association analyze with hardened and unsolidified association.
- The combination of declare five in expansion including, the graft association has load sporting capability is over the top, so we will seen the aspects of stage and save you the base shear on the grounds that of lessen the heap of systems[10].
- The graft association has magnificent seismic generally speaking execution in each bloodless fashioned metal and rolled metal phase examine with different varieties of associations.

Reference

[1] S.K.Duggal "Limit State Design of Steel Structures".

[2] N. Subramanian "Design of Steel Structures based on Limit state Method of design as per standardized code IS:800:2007.

[3] S.S Bhavikatti "Design of steel structure by limit state method as perIS: 800:2007".

[4] IS: 801:1975 Code of Practice for Use of Cold Formed Light Gauge Steel Structural Members in General Building Construction.5.IS811:1987 Specification for Cold Formed Light Gauge Structural Steel Section.

[5] IS 800 Code of Practice for General Construction in Steel , Bureau of Indian Standards,NewDelhi,2007

[6] IS 1893 -2002 " Criteria for Earthquake Resistance and Construction of Buildings "Bureau of Indian Standards, New Delhi

[7] A Jayaraman (etal) "Seismic Performance of Cold Formed Steel and Conventional Steel of Industrial Structures Using Splice Connections" International Journal of Recent Technology and Engineering" ISSN: 2277-3878, Volume-7 Issue-4S2,December2018.

[8] JEROMEF(etal)"ContinuityPlateDetailingforSteelMoment-ResistingConnections "ENGINEERING JOURNAL", September-2003.

[9] Seyed Mohammad Mojtabaei (etal) "Efficient design of cold-formed steel bolted-moment connections for earthquake resistant frames" Thin-Walled StructuresISSN0263-8231 January2019. https://doi.org/10.1016/j.tws.2018.12.015

[10] M.Dundu (etal) "Angle Cleat Base Connections"20th International Specialty Conference on Cold-Formed Steel structures November2010

[11] B. A. Ali, Structural performance of cold-formed steel frame with bolted moment connection [Ph.D. thesis], Universiti Teknologi Malaysia, 2012.

[12] D. Dubina and R. Zaharia, "Experimental evidence of semi-rigid behaviour of some cold-formed steel truss bolted joints," in Proceedings of the International Conference on Experimental Model Research and Testing of Thin-Walled Steel Structures, Prague, Czech Republic, 1997. https://doi.org/10.1016/S0263-8231(97)00028-1

[13] V. F. de Paula, L. M. Bezerra, and W. T. Matias, "Efficiency reduction due to shear lag on bolted cold-formed steel angles," Journal of Constructional Steel Research, vol. 64, no. 5, pp. 571-583, 2008. https://doi.org/10.1016/j.jcsr.2007.10.008

[14] J. R. U. Mujagic and W. S. Easterling, "Connections in cold-formed steel framing-designing with AISI, 2007: specification and application overview," STRUCTURE Magazine, pp. 9-11, 2009.

[15] G. J. Hancock, "Cold-formed steel structures," Journal of Constructional Steel Research, vol. 59, no. 4, pp. 473-487, 2003. https://doi.org/10.1016/S0143-974X(02)00103-7

Review on Utilizing E-Waste Plastic in Bitumen for Better Strength and Sustainable Environment

P. Krithiga[1*], M. Vishnu Preethi[1], K. Samritha[1], K Senthil Kumar Mena[1]

[1]Kongu Engineering College, Perundurai, Erode, Tamil Nadu, India

* krithipalanisamy@gmail.com

Keywords: E-Waste, Bitumen, Composition, Plastic, Recycling

Abstract. E-Waste or Electronic waste is the broken pieces or junk or which is not used in present that is thrown out at the end of their lives. Generation of Electronic Waste is rising year by year due to the demand for newer electronic products which made the public to upgrade their technologies. The composition of plastics in Electronic Waste is high which is non degradable may cause consequential reaction. These wastes would contaminate in water, air, soil and also severely affect the humans and environment. Managing the Electronic Waste with tactical approach may create a way for sustainable waste management. For effective waste management process it is essential to adopt the 4R methods of Reduce, Recovery, Reuse and Recycle. Because it is significant to contemplate the health of the people and also by generating jobs in e-recycling field. In recent times the research is underway to examine the possibilities of using E-Waste in construction field. By adding the Electronic Waste as an alternative material to conventional material in bitumen for various percentages like 5%, 10%, 15%, 20% and 25%. Reusing the E-Waste plastic in aggregate form as certain or diverse forms probably low-budget and it is feasible in technical manner for disposing the huge E-waste. Replacing Electronic Waste in various forms in bitumen gives better strength than conventional bitumen.

Introduction

In recent times, Electrical and electronic industries are burgeoning towards a greater step. E-Waste or Electronic waste is the broken pieces or not useable at now or junk that is produced at the end of their lives. The plastic waste dismantled by E-Waste is rising at a greater pace around the earth. Every year, lot tones of Electronic Wastes are needed to discard in lands contribute to pollution in the surroundings. [1]. Managing of these wastes with tactical approach requires a systematic effort to create a method for reaching sustainable waste management. E-Waste Management is a booming global issue, which showcase the consequence of such E-Wastes on humans and environment. For effective waste management it is quite essential to execute the recycling options. The origination of E-Waste and the pollution possibilities relates the problems with electrical and electronic industries. [2]. Electronic products contains dangerous and harmful toxins in the components that leads to environmental problems when they are throw away in the lands. Owing to fast development, the duration of computers have been draw back to small period of time. Every latest unearthing have potential of twice the antiquation wage, also by advertising and giving offers to find its affordable and agreeable to get a newer one rather upgrading the older appliances. Cathode ray tubes (CRT's) are present in computer monitors and TV's that contains more amount of lead oxide. In inclusion most of the electronic components have batteries that consists of toxic materials like cadmium (Cd),

nickel and many ferrous metals. These toxic materials might affect the workers staff and people who are living on that side. [3]. The Electrical and electronic appliances will pursue to be manufactured, reused, recycled, recovered and disposed because it is significant to contemplate the health of the people and also by generating jobs in e-recycling field owing to the huge chemicals are intrinsic in the E-Waste flow. In e-recycling, the formal type offers services to civilization and propose limitless offers to the people. If we done recycling in applicable method it will be the best method and it will recover the materials from E-Waste and also reduces the wastes from landfills [4]. The disposing of E-Waste plastic is contemplated to exist as huge problem because of its little bit of biodegradability and large quantities in generation. In recent times the research is underway to examine the possibilities of using E-Waste in construction field. Reusing the E-Waste plastic in aggregate form as certain or diverse forms probably low-budget and it is feasible in technical manner for disposing the huge E-waste. Contingent on size of particle and chemical composition, E-waste practiced as fine aggregate, coarse aggregate and fine filler in field of construction. [5].

Classification of e-waste

E-Waste is enlarging epidemically in proportions by reason of more demand of newer versions in all equipment. Diverse government sectors, private sectors, hospitals and public are fast feeding the old appliances [3]. The Electronic items considered as E-Waste:Large domestic appliances (clothes dryer, air coolers, refrigerator, Dish washers), Small domestic appliances (blender, electric fry pan), Information Technology devices (Monitors, keyboards, computers, telephones, mobile phones, laptops), User electronics (Flat screen TV's, MP3 players, DVD players), Lighting devices (LED downlights, track light, floor lamp), Electrical tools and equipment (electric drills, saws, screwdrivers), Toys, sports equipment (Tennis baseball automatic server), Medical equipment in hospitals (ventilator, ECG machines) Control and monitoring instruments (Thermostats, smoke detectors) [3, 6]. The figure 1 shows the sources of E-Waste.

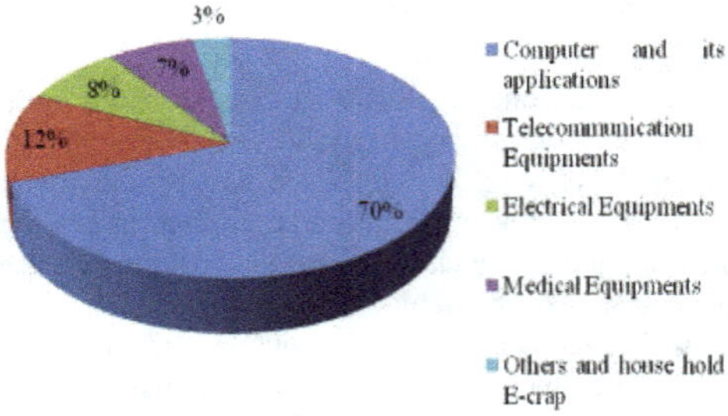

Fig. 1, Source of E-Waste
(SOURCE: ASSOCHAM-KPMG Report, 2018)

Generation of e-waste

Electrical and electronic waste is enlarging towards a fast and unmanageable allowance and also rapidest increasing proportion in MSW. It can be presumed that the disposal of outdated electronic products is mainly driven by the production of new electronic products. This signifies that the growth in the global production will result in alike growth in E-Waste generation [7]. Nowadays, human beings coming with technological development that each second they attached with the electronic equipment like mobile phones, laptops, computers etc., but the life span of these electronic products is 3-5 years. The lesser life duration of equipment, the higher its proportion. For example, computers have shorter life span up-to three years so it comprises greater proportion than other electronic waste products like ovens and refrigerators these have duration of life up-to 10-12 years [8]. In 1990 after the Economic liberalisation of first phase, the difficulties in E-Waste have found exhibiting. As stated in TRAI, In 2008, 113.26 million newer consumers have been added by India with 9.5 million average consumers of every month. The growth market of cellular from (2007-2008) is 168.11 million to 261.97 million [9]. In 2007 A GIZ-MAIT study had put the approximate of 333,000 tonnes and other different estimates to 420,000 tonnes.

As per approximation UNEP report 2010, the generation of E-Waste in equipment wise are: Refrigerators of 100,000 tonnes; Mobile phones of 1700 tonnes; Personal computer of 56,300 tonnes ; Televisions of 275,000 tonnes ; of 4700 tonnes [10].

In 2014, the overall electronic waste generated by the world is 41.8 million tonnes and it is anticipated to increase averagely by 2018 is 50 million tonnes. 3-5% is the approximated growth rate of E-Waste stream. The rate will be thrice rapidest rather than the all wastes [11]. The components of electronic wastes are categorized as lighting equipment, large household equipment, small equipment, Computer monitors and screens, temperature exchange equipment, Information Technology and telecommunication equipment. [12]. The below table 1 shows the different categories of E-Waste.

Table 1. DIFFERENT CATEGORIES OF E-WASTE
(SOURCE: AMIT KUMAR et al, E-WASTE [12])

CATEGORIES	AMOUNT (in million tonnes)
Lighting equipment	1.0
Large household Equipment	11.80
Small household Equipment	12.80
Information Technology and Telecommunication equipment	3.0
Cooling and freezing equipment	7.0
Computer monitors and screens	6.3

In sequence of tallest allowance to electrical and electronic equipment of the top states involve Tamil Nadu, Madhya Pradesh, Uttar Pradesh, Maharashtra, West Bengal, Andhra Pradesh, Delhi, Punjab, Karnataka and Gujarat [13]. The table 2 gives the quantity of WEEE generated in Indian states.

Table2. QUANTITY OF WEEE GENERATED IN INDIAN STATES
(SOURCE: E-WASTE SCENARIO IN INDIA [13])

SI.NO	INDIAN STATES	WEEE (TONNES)
1	Andhra Pradesh	12,780.3
2	Andaman and Nicobar Islands	92.2
3	Nagaland	145.1
4	Maharashtra	20,270.6
5	Kerala	6,171.8
6	Jammu and Kashmir	1,521.5
7	Gujarat	8,994.3
8	Daman and Diu	40.8
9	Chandigarh	359.7
10	Assam	2176.7
11	West Bengal	10,059.4
12	Uttar Pradesh	10,381.1
13	Tamil Nadu	13,486.2
14	Rajasthan	6,326.9
15	Puducherry	284.2
16	Utarakhand	1,641.1
17	Tripura	378.3
18	Sikkim	78.1
19	Punjab	6,958.5
20	Orissa	2,937.8
21	Chhattisgarh	2149.9
22	Goa	427.4
23	Himachal Pradesh	1,595.1
24	Bihar	3055.6
25	Dadra and Nagar Haveli	29.4
26	Haryana	4,506.9
27	Delhi	9,729.2
28	Meghalaya	211.6
29	Madhya Pradesh	7,800.6
30	Jharkhand	2,021.6
31	Mizoram	79.6
32	Karnataka	9,118.7
33	Lakshadweep	7.4
34	Manipur	231.7
35	Arunachal Pradesh	131.7
	TOTAL	146,180.7

In 2014, as stated in United Nations Environment Program 62 million tons of electronic wastes are produced around the world. Out of that 8 lakhs tones have been produced by India. Three equipment is taken for this report that is cellular phone, computers and television [14]. As reported

Materials Research Proceedings **23** (2022) 381-400 https://doi.org/10.21741/9781644901953-43

by Global E-Waste Monitor 2017. In 2016 India produced 2 million metric tonnes of Electronic waste, flows next to China, United States and Japan as 7.2, 6.3 and 2.1 million metric tonnes respectively in the top 10 countries which produces more E-waste among the global nations [15]. The figure 2 shows the total E-Waste generated up-to year 2021.

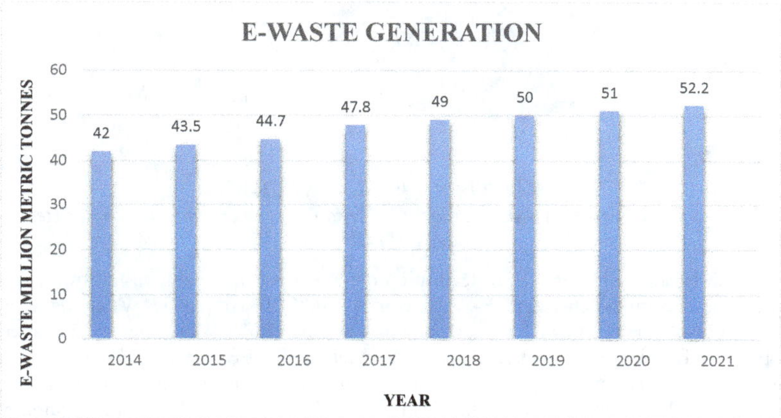

Fig. 2, TOTAL E-WASTE GENERATED
(SOURCE: THE GLOBAL E-WASTE MONITOR 2017 [15])

Resultant to the requirement of new appliances, more performance and effectual technology leads to the more generation of electronic waste. [16]

Composition of e-waste

At the end of life electronic waste provides a precious raw material for the production of new electronic products. In E-Waste boundless substances are found, there are number of batteries, thermostats, condenser comprising elements namely mercury, Aluminum, lead, Poly chlorinated biphenyls, Tin, PVC (Poly Vinyl Chlorides), Copper, thermosetting plastics, fibreglass, Carbon, Silicon, Beryllium, and Iron [17]. Large domestic equipment like machines for dish washers, washing machines, for vegetable and fruit storage refrigerants, air conditioners etc., devices using in homes viz., mobile phones, MP3 players, FM's, personal computer etc., Lighting equipment like bulbs, Compact Fluorescent Lamp (CFL), tube light, fluorescent lamps, entertainment equipment, toys etc., Equipment used in medical applications like machines for sonography, MRI, Blood pressure monitor all these equipment are abundant in Mercury (Hg), Barium (Ba), Cadmium (Cd) [18]. Information Technology and Telecommunication Equipment like telephones, computer, laptops, tablets etc., and the systems used for calling services at Business Process Outsourcing having more toxic composition viz., Ni, Cd, Copper (Cu) [19]. E-Waste consists of abundant number of toxic substances there are broadly categorized into hazardous material, non-hazardous, ferrous metals, non-ferrous metals, glass, plastic, , precious metals, printed circuit boards etc. [20]. The figure 3 shows the material fractions in E-Waste.

Materials Research Proceedings **23** (2022) 381-400
https://doi.org/10.21741/9781644901953-43

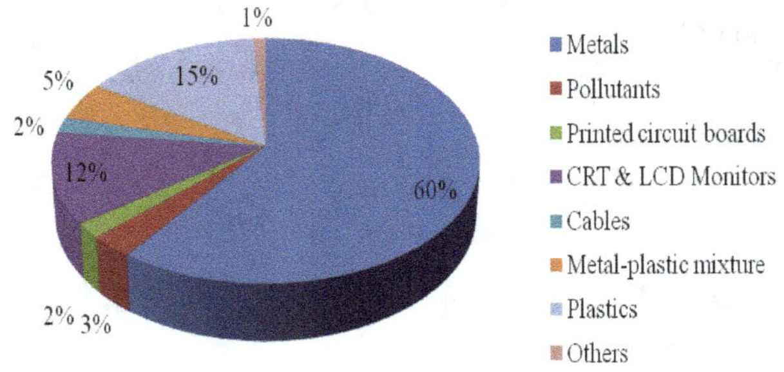

Fig. 3, MATERIAL FRACTIONS IN E-WASTE
(SOURCE : The Generation, Composition, Collection, Treatment and Disposal system and Impact of E-Waste)

In the last decades, Electrical and Electronic Equipment was the main waste flows and also it will be pursue in coming generation. E-Waste contains 10%-30% of plastics which makes it more valuable to recycle the plastic in the value point of view and also as WEEE target. The percentage of plastics in WEEE is generally high it is very dependent on WEEE category. Control instruments, toys and monitoring equipment have more than 50% of plastics while medical devices and lighting equipment have less than 5% weight of plastics [21]. Electrical and Electronic Equipment are categorized under 15 various types of engineering plastic including high-impact polystyrene (HIPS), polystyrene (PS), polyamide (PA), polypropylene (PP), styrene-acrylonitrile (SAN), acrylonitrile–butadiene–styrene (ABS), blends of polycarbonate (PC)/ABS, polyurethane (PU), blends of HIPS/poly (Poly Phenylene oxide) (PPO) and polyesters.[22]

The elements of E-Waste are Americium (AM), Barium (BA), Cobalt (CO), Manganese (MN), Palladium (PD), Sb, Pt, Gallium (Ga), Bi, Eu, Au, Nickel (Ni), Se, Ta, Ti,Yttrium(Y),Tb, Arsenic (As), B, Indium (In), Ru, Li, Th, Vanadium (V), Rh, Nb, Silver (Ag) [23]. The figure 4 shows the composition of E-Waste.

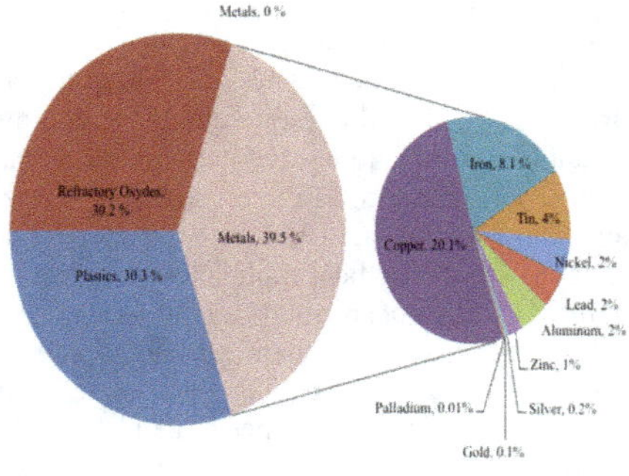

Fig. 4, COMPOSITION OF E-WASTE
(SOURCE: PERCENTAGE OF CHEMICAL CONSTITUTES IN THE TOTAL E-WASTE [23])

Table 3. COMPONENTS AND COMPOSITION OF EEE (SOURCE: ELECTRONIC WASTE, AN OVERVIEW OF TREATMENT TECHNOLOGIES OF E-WASTE [24, 25, 26])

COMPOSITION	COMPONENTS OF ELECTRICAL AND ELECTRONIC EQUIPMENT
Aluminum (Al)	Mobile phones, connectors, cathode ray tubes, computer chips, hard drives
Barium (Ba)	Getters in cathode ray tubes, fluorescent lambs, phones
Cadmium (Cd)	Solder, SMD chip resistor, detectors, batteries, toners, printed circuit boards
Copper (Cu)	Cathode ray tubes, heat sinks, computer chips, cables, central processing unit, Wirings
Polychlorinated Biphenyls (PCB)	Condenser, transformers, coolants in generators, dielectric fluids, ceiling fans, capacitor and transformer, dish washers
Mercury (Hg)	Gas discharge lamps, fluorescent lamps, thermostats, monitors, sensors, cells, housing
Tin (Sn)	Liquid crystal display screens, computer chips, solder, printed wiring board
Scandium(Sc)	Aerospace, laser and lightings
Nickel (Ni)	printed wiring board, mobile phones, housing, batteries
Chromium (Cr)	Data tapes, anticorrosion coatings, floppy disks, decorative hardener
Indium (In)	LCD displays, Printed wiring board
Beryllium (Be)	Power supply boxes, silicon controlled rectifiers, x-ray machines
Tantalum (Ta)	capacitors, Printed wiring board and power supply

Effects of e-waste

Many cities in India doesn't consider the effects of Electronic waste problems. Mostly the electronic wastes are collected, handled, disassembled and recycled by nonunionized or non-formal sections they are shortage of knowledge in the systematic process. Environmental problem and human health correlated with electronic wastes be hypercritical and contemplative within nature foremost to pollution and degradation of natural resources and causes a dreadful diseases like cancer, skin disease etc., in human beings [27]. In disposing the electronic wastes there are two types, i.e. incineration and landfill, both of these methods will produce substantial pollution problems. Even the recycling of electronic waste hand over the poisonous substances to the environment and perhaps influence the health of human [28]. There are two ways that affect the health of human while disposal of electronic wastes: a)OBSTACLE IN FOOD CYCLE:

Materials Research Proceedings **23** (2022) 381-400 https://doi.org/10.21741/9781644901953-43

defilement by hazardous matters beginning with process of recycling and disposal wastes leads to infiltrate in food web and it shifting to the human health. b) UNMEDIATED CONSEQUENCES ON LABOUR: labor who are working in the elementary of recycling process will lead to the hazardous matters exposure to their health. [29]. The figure 5 shows the effects of E-Waste on Human and Environment.

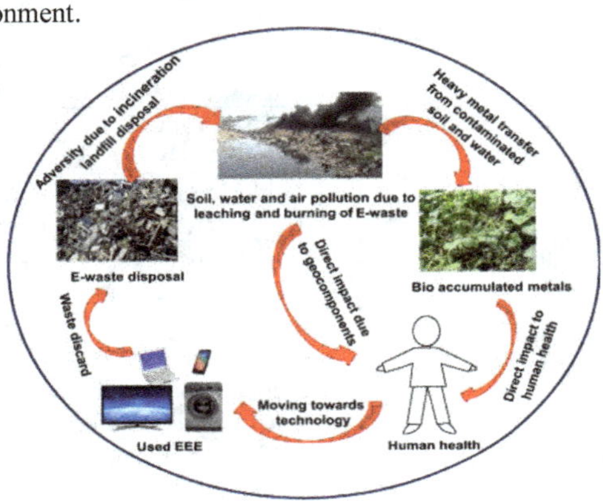

Fig. 5, EFFECTS OF E-WASTE ON HUMAN AND ENVIRONMENT
(SOURCE: MANAGE AND RECYCLE E-WASTE IN INDIA)

In the course of divergent steps of recycling operations (mincing, extracting and diverge process) have a chance of toxic substances to enhance as airborne having the form of aerosol matters that size ranges from nanoparticles to tens of microns. Among the most important, industry is mainly responsible for the health hazards of humans in form of aerosol nanoparticles which may travel and deeply stay into the tract of respiratory and lungs [30]. Electronic components contains dreadful toxic which may intense exertion on environment and human health. Polyvinyl Chloride (PVC), Computer monitors, Brominated flame retardants (BFRs) contains more quantity of lead. Repeated exposure of this component affect the internal organs like reproductive system, kidney, endocrine, nervous system. Due to this affects, electronic goods are differentiated into three:BROWN GOODS: radios, computers, WHITE GOODS: washing machine, refrigerators, GREY GOODS: color printer, scanners

During recycling process, brown goods and white goods have fewer toxics rather grey goods [31]. Nowadays mobile phones production are more due to the new innovative in technology makes them to use only 2 years but duration of devices is 7 years. So discarding those plastic from mobile phones require an incineration burning process to remove priceless metal. The burning of electronic waste plastic leads to the pollution of environment and also affects human health [32]. In innovation of new technology the metals usage are in long term so production and demand for toxic goods also raised. Electronic appliances are supplied mainly by the mining can't be disregard. Large scale lands are needed for mining activities also that toxic substances may end up in air, land and water. Utilization of natural resources leads to the release of radioactive substances and it may affect the environment and human health. Also the extraction of priceless components from electronic waste will release carbon-dioxide. Production of valuable components in tons like silver, gold will emit 10,000 tons of carbon-dioxide ($CO2$) [33]. Main origin for groundwater is from water bodies and it transfers to voids in underground soil. The water which is moving is in upper surface will contaminate with toxics present in soil and also contaminate with water. These contamination in water will affect the health of human being by drinking and using of these toxic water cause a chronic diseases [34]. One of the components in electrical and waste i.e. exposure of mercury have more chances to affect the pregnant women and also influence the health of fetal. The contiguity of mercury with water interchange them into furthermore poisonous form methyl mercury and corrupt the food cycle. The human beings who are interlinked with recycling operations will definitely exposure to poisonous chemicals. Among human beings, kids and female are hugely affected to the electronic waste contamination [35]. In an electronic waste, the second hugest component is plastics it have a weight description to come near to 20-35% of total weight. Part of the plastic are summed with many poisonous substances like flame retardant, if it doesn't disposed in sequence manner it will affect the environment leads to pollution [36]. Some other poisonous substances in electronic waste comprises of polychlorinated Biphenyls (PCBs), hydro chlorofluorocarbons (HCFCs), arsenic, asbestos, chlorofluorocarbons (CFCs). Presence of few quantity of these chemicals may cause huge pollution problem, further evaporation of di methylene mercury and metallic components cause dangerous to environment [37]. Plastic which disposing in lands guide to biotic and abiotic downgrade of plastic and supplements present in plastic (EXAMPLE: plasticizers, stabilizers, heavy metals) could discharge into the environment. During downgrade the plastics release methane leads to global warming [38, 39]. Pollution in environment and human health problems will be reduced when the process of landfills are performed well [40].

Management of e-waste
In India, conventional managing of electronic waste should have Centre of attention about collecting, extraction and also upraise having knowledge of health problems and environmental affect due to the inappropriate management methods. Have to gather the details about the chemical components added in the electronic waste from zonal research so it will be easy to find the fault existed and also to bring a standard solution for environmental problems and health issues [41, 42]. Ministry of Environment, Forest and Climate Change (MOEF) have formatted specification for managing electronic waste in eco-friendly manner. The specification consists recommendation for disposing and treating electronic waste and also codification of electronic waste [43]. Reasonable procedure for managing electronic waste is coming from various ways to solve the problems by creating stable policies, requisition for systematic approaches of discarding and extracting components from electronic waste [44]. For ecological management of electronic waste, variance modifying technique is to be implemented like the amount and standard component of electronic waste to be improved. From profitable countries, producers could adopt a greatest method for managing electronic waste [45]. Systematic recycling process is important one so the profitable countries must enhance their way of managing electronic waste rather by transporting to emerging countries. By the fundamental idea of Extended Producer Responsibility (EPR), the creators must ensure the extended lifespan of component till they are fully recycled [46].

Primarily, the substantial working of managing electronic waste comprises: Collecting electronic waste plastics, Disposing of plastics, Rehabilitate its energy, Reprocessing it to newer products [47].

Collecting electronic waste plastics
For recouping the electronic waste plastic it have two methods. Firstly, collection of waste plastic later they go in for Municipal solid waste and Secondly, collecting it prior to go in for Municipal solid waste. Electronic waste plastics are the type of 'non-visual' plastics. They are collected outer of the Municipal solid waste for the further process recycling operations in ecological manner [48].

The flow diagram of electronic waste plastic management process is given below:

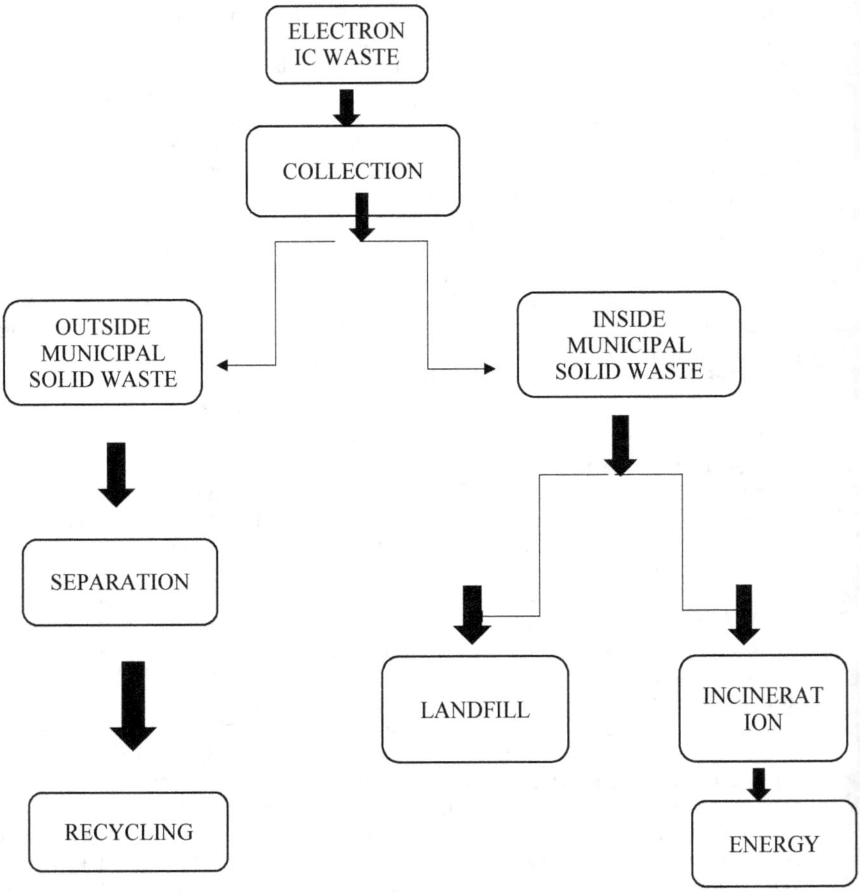

(SOURCE: PLASTIC WASTE MANAGEMENT IN CONSTRUCTION: TECHNOLOGICAL AND INSTITUTIONAL ISSUES [49])

Disposing of plastics

Demeaning of electronic waste plastics and further more waste plastics didn't have a disposing problems on lands. Even though it has slight effect of disposing problems it won't develop more problems in landfilling process and also won't grant harmful leachate on lands. Using the following two techniques in disposing makes environment more better i.e. Photo degradation (by integrating photo sensitive supplement in plastic) and bio-degradable plastic (by integrating starch supplement in plastics). Degradation of plastic would enhance their preferable characteristics in diverse approach [50, 51]

Rehabilitate its energy
Heat of electronic waste plastic could be retrieved beside incineration process. Electronic waste plastics are great combustible origin due to the presence of more resins. It have a warming desirability just equivalent to the coal value. It also gives more benefit by conserving natural etymology, lowering the dependence of energy from foreign and contributing fascinating etymology for replacement of vivacity. 90-95% mass of electronic waste plastic is reduced by incineration method. Many researches are going on to implement these wastes in construction sites [52, 53]

Recycling
Recycling operation is one of the best method to manage the electronic waste plastics because it saves the energy, reduces the major plastics in landfills and also by conserving the natural petrochemical by- products. Various newer technologies makes the electronic waste plastic convert to newer products in many ways. The method which is used in recycling must workable to various composition. Further growing of recycling method is enhanced by giving more knowledge about health and environmental problems [54, 55]

TABLE 4. TECHNIQUES FOR SEPARATION PROCESS

S.NO	METHODS	SEGREGATION TESTS	CONCEPT OF SEPARATION	CATEGORISATION PROCESS
1.	Cleansing, Grating, Screening	**Screening:** Mechanical forces **Grating:** Vibratory	**Cleaning:** To pull out adhering contamination **Grating**: Utilizing graters combined to vibratory revolutionary	**Cleansing:** Segregating contamination out of beneficial component [56] **Grating:** Pulverizing and powdering [57] **Screening:** Fragment dimension segregation [58]

2.	Magnetic Separation	Magnetic field	Utilizing magnets to segregate components from wastes	Segregating non-magnetic and non-ferrous materials from ferromagnetic metals [59, 60]
3.	Eddy current Separation	Electrical density and electrical conduction	Utilizing a strong field of magnet to segregate magnetic composition from components	Separating non-metal and non-ferrous metals [61, 62]
4.	Electrostatic Separation	Electrical conduction	Molecules interconnecting gives various charges molecules repercussion in various intensity	Separating metals and non-metals from components [63, 64]
5.	Triboelectrostatic separation	Dielectric constant	Categorizing positive and negative ions downward an field of electric depending on plan levying features	Separating plastics [65, 66]

Materials

Coarse aggregate

Aggregates are the most vital segment in road construction. They are worn for sub base, granular base in mix proportion of bitumen. It is also used as chief substance in comparatively low-cost roads i.e. water bound macadam. Various sizes like 20mm, 10mm, 4.75mm, 2.36 mm are used. Being a construction material for pavement it must be assessed in consonance with recommendations of Ministry of Road Transport & Highways (MORTH) [67, 68, 69]. The table 5 gives the properties of coarse aggregate.

TABLE 5. PROPERTIES OF COARSE AGGREGATE

S.NO	NAME OF TEST	MORTH ACCEPTABLE VALUE
1	Flakiness Index	30% maximum
2	Water Absorption	2% maximum
3	Specific Gravity	2.0-3.0
4	Impact Test	24% maximum
5	Crushing Test	30% maximum
6	Elongation Index	30% maximum

Bitumen

Bitumen is one of the material in road construction which obtained along with fractional distillation of petroleum products. It consists of 11% hydrogen, 87% carbon, 2% oxygen. The bitumen 60/70 is used. The color is black and it holds adhesive and water resistant properties. It also has a great binding property and holds the aggregate together [70, 71, 72]. The table 6 gives the properties of bitumen.

TABLE 6. PROPERTIES OF BITUMEN

S.NO	NAME OF TEST	MORTH ACCEPTABLE VALUE
1	Ductility	75cm minimum
2	Softening Point	40-50 c
3	Penetration	60/70

E-waste

Using E-Waste as an alternative for conventional is an effectual method for disposing wastes in road construction because it has impact resistance, high tensile strength and stiffness. Polyethylene plastic is used. Here toxics are removed from components and made into e-waste powder [73, 74, 75].

Test analysis

Ductility test

Here the elastic nature of bitumen is compared with conventional and alternative mix design. E-Waste powder is added to bitumen in various proportions like 5%, 10%, 15%, 20% and 25%. From the test results, it exhibits that the value of ductility is reduced when Electronic Waste powder is included supplementary. But the mix design of alternative method of 5% E-Waste powder and 95% bitumen gives more ductility. [76, 77]

Penetration test

In this test, the foremost aim is to notice the distance in mm. For the trials, firstly heating the bitumen at 160°C then permitting the bitumen to dry for 1 hour and keeping it in water for 30 minutes. Obtaining the values by popping the needle into bitumen for thrice times. From the trial mix, when there is no adding of E-Waste powder, there is low penetration. By slightly adding E-Waste powder makes it to have high value of penetration [78, 79].

Softening point
The objective is to examine at which temperature the material melts and enhance soften. This test is to managing the various softening point of bitumen which is used in national highways, village roads and state highways. By using ring ball apparatus the bitumen is heated for 15 min and keeping it in water for 10 min to know at which temperature the bitumen softens and falls from ring. From trial mix, when there is no adding of E-Waste powder, there is minimum of temperature but gradually adding E-Waste powder the temperature is maximum [80, 81]

Viscosity test
The goal of this test is to examine the bitumen smoothness. Smoothness of bitumen is determined by examining the capacity of bitumen flow at required temperature. By adding E-Waste in various percentages like 5%, 10%, 15%, 20%, 25%. From the test results, when there is no adding of E-Waste, there is minimum viscosity value but gradual adding increases the value of viscosity [82, 83].

Marshall stability test
Marshall Test is to determine the extreme load fetched by material at a quality temperature of 60°C. When the specimen is fetching to extreme load there is deformation of flow value. This method is of low amount and simple, it is used to characterize the bitumen mixes in India. This test is practical in nature. The load bear on specimen is perpendicular to the cylinder axis having deformation rate of 51mm per minute. Flow value and Stability is low when no E-Waste powder is added but gradual increase of E-Waste increases the flow value and stability [84, 85].

Conclusion
Based on the study, the Electronic Waste contamination leads to a major problems in humans and environment. By adopting 4R methods of Recycle, Reduce, Recovery, Reuse, the Electronic wastes are utilized in a great manner. Because it saves the energy, reduces the major plastics in landfills and also by conserving the natural petrochemical by- products. Using Electronic waste plastic as an alternative material to conventional bitumen increases the properties of bitumen by adding it in various percentages. It is also cost effective and minimizes the usage of natural resources like bitumen and aggregate. Managing the Electronic Waste is the new technique, if the same method is followed the impacts on environment and humans can be minimized.

References
[1] L. Salmabanu., L. Ismail, Potential application of E-wastes in construction industry, Construction and Building Materials, 203 (2019) 222-240. https://doi.org/10.1016/j.conbuildmat.2019.01.080

[2] M. Lakshmi, Use of e-plastic waste in bituminous pavements, Gradevinar, 70,7 (2018) 607-615. https://doi.org/10.14256/JCE.1375.2015

[3] M. N. Mundada, K. Sunil, A. V. Shekdar, E - waste: a new challenge for waste management in India, International Journal of Environmental Studies, 61,4 (2001) 265-279. https://doi.org/10.1080/0020723042000176060

[4] C. Diana Maria, D. Zhao, The formal electronic recycling industry: Challenges and opportunities in occupational and environmental health research, Environment International, 95 (2016) 157-166. https://doi.org/10.1016/j.envint.2016.07.010

[5] S. Needhidasan, B. Ramesh, S. Gorab Agarwal, Experimental investigation of bituminous pavement (VG30) using E-waste plastics for better strength and sustainable environment, Materials Today: Proceedings, 2214-7853.

[6] S. Sivakumaran, E-Waste Management, Disposal and Its Impacts on the Environment, Universal Journal of Environmental Research and Technology, 3,5 (2013) 531-537.

[7] Y.Richardson, J.B. Walker, A.K. Youn, TCLP heavy metal leaching of personal computer components, Journal of Environmental Engineering, 132,4 (2006) 497-504. https://doi.org/10.1061/(ASCE)0733-9372(2006)132:4(497)

[8] S.k. Ajim Ali, E-Waste Generation and Its possible Impacts on Environment and Human Health: A Study on Kolkata, Asian Profile, 45,1 (2017) 78-94.

[9] Sushant B. Wath, P. S. Dutt, T. Chakrabarti, E-waste scenario in India, its management and implications, Environmental Monitoring and Assessment, 2010.

[10] C. Hicks, R. Dietmar, M. Eugster, The recycling and disposal of electrical and electronic waste in China- legislative and market responses, Environ. Monit. Assess. Journal of Environmental Impact Assessment Review, 25 (2005) 459-471. https://doi.org/10.1016/j.eiar.2005.04.007

[11] K. Amit, H. Maria, Denise Crocce Romano Espinosa, E-waste: An overview on generation, collection, legislation and recycling practices, Resources, Conservation and Recycling, 122 (2017) 32-42. https://doi.org/10.1016/j.resconrec.2017.01.018

[12] G. Cucchiella, D. Adamo, I. Lenny Koh, S.C. Rosa, Recycling of WEEEs: An economic assessment of present and future e-waste streams. Renew, Sustain. Energy Rev, 51 (2015) 263-272. https://doi.org/10.1016/j.rser.2015.06.010

[13] M. Turner, D. Callaghan, Waste electrical and electronic equipment directive, UK to finally implementthe WEE directive, Computer Law and Security Report, 23 (2017) 73-76. https://doi.org/10.1016/j.clsr.2006.11.007

[14] J. Senophiyah Mary, T. Meenambal, Inventorisation of E-Waste and Developing a Policy - Bulk Consumer Perspective,. Procedia Environmental Sciences, 35 (2016) 643 - 655. https://doi.org/10.1016/j.proenv.2016.07.058

[15] Dwivedi, Amrita, V. K. Kumar, Solid Waste Management and Sanitation in Varanasi City, National Geographical Journal of India, 55,3 (2009) 1-12.

[16] J. Ladou, S. Lovegrove, Export of Electronics Equipment Waste, International Journal of Occupation and Environmental Health, 14,1 (2008) 1-10. https://doi.org/10.1179/oeh.2008.14.1.1

[17] S. Badur., R. Chaudhary., 2008. Utilization of hazardous wastes and by-products as agreen concrete material through S/S process: a review, Rev. Adv. Mater. Sci 17 (1-2), 42-61. https://doi.org/10.1002/prs.680170112

[18] W. Rolf, K. Heidi Oswald, K. Deepali Sinha, Max Schnellmann, Heinz Bo, Global Perspectives on E-Waste, Environmental Impact Assessment Review, 25 (2005) 436-458. https://doi.org/10.1016/j.eiar.2005.04.001

[19] P. Vanegas, J. R. Peeters, D. Cattrysse, W. Dewulf, J. R. Duflou, Improvement potential of today's WEEE recycling performance: the case of LCD TVs in Belgium Front. Environ. Sci. Eng., 11,5 (2017) 13. https://doi.org/10.1007/s11783-017-1000-0

[20] Y. Abdollahi, S. B. M. Said, N. A. Sairi, Enhancement of electronic protection to reduce e-waste, J. Indust. Eng. Chem., 29 (2015) 400-407. https://doi.org/10.1016/j.jiec.2015.04.021

[21] M. Grano, P. Ana, S. Luanha, R. Rita, Composition of plastics from waste electrical and electronic equipment (WEEE) by direct sampling, Waste Management, 32 (2012) 1213-1217. https://doi.org/10.1016/j.wasman.2012.02.010

[22] F. Vilaplana, S. Karlson, Quality concept of the improved use of recycled polymeric materials, Macromal Materials Engineering, 293 (2008) 274-297. https://doi.org/10.1002/mame.200700393

[23] A. Pariatamby, D. Victor, Policy trends of e-waste management in Asia, J. Mater. Cycles Waste Manage., 15,4 (2013) 411-419. https://doi.org/10.1007/s10163-013-0136-7

[24] V. Dissanayake, Electronic Waste, Encyclopedia of Toxicology, 2 (2014) 568-572. https://doi.org/10.1016/B978-0-12-386454-3.00565-0

[25] C. Jirang, R. Hans Hogen, Electronic Waste, Management of Electronic Waste, (2011) 281-296. https://doi.org/10.1016/B978-0-12-381475-3.10020-8

[26] K. Peeranart, P. Jatindra Kumar, M. Sanchita, B. Jayanta Kumar, S. Binoy, An overview of treatment technologies of E-waste, Waste Management, 57 (2016) 113-120. https://doi.org/10.1016/j.wasman.2016.01.043

[27] J. Gregory, A. Kirchain, A comparison of North American electronic recycling systems, In Proceedings of the 2007 IEEE international symposium on electronics and the environment, (2007) 227-232. https://doi.org/10.1109/ISEE.2007.369399

[28] S. Zhang, E. Forssberg, Mechanical separation-oriented characterization of electronic scrap. Resources, Conservation and Recycling, 21 (1997) 247-269. https://doi.org/10.1016/S0921-3449(97)00039-6

[29] J. P. Wang, X. K. Guo, Impact of electronic wastes recycling on environmental quality, Biomedical and Environmental Sciences, 19 (2006) 137-142.

[30] T. Ourania, L. Michael, Environmental Risks Associated with Waste Electrical and Electronic Equipment Recycling Plants, Encyclopedia of Environmental Health, (2018) 1-10.

[31] F. J. Barbosa, J. E. Tanus-Santos, R. F. Gerlach, P. J. Parsons, A critical review of biomarkers used for monitoring human exposure to lead: advantages, limitations, and future needs, Environmental Health Perspect, 113 (2005) 1669-1674. https://doi.org/10.1289/ehp.7917

[32] O. Tsydenova, M. Bengtsson, Chemical hazards associated with treatment of waste electrical and electronic equipment, Waste Management, 31,1 (2011) 45-58. https://doi.org/10.1016/j.wasman.2010.08.014

[33] J. H. Rademaker, R. Kleijn, Y. Yang, Recycling as a strategy against rare earth element criticality: a systemic evaluation of the potentialyield of NdFeB magnet recycling, Environmental Science Technology, 47 (2013) 10129-10136. https://doi.org/10.1021/es305007w

[34] S. Fahad, S. Hussain, S. Saud, F.Khan, S. Hassan, W. Nasim, M. Arif, F. Wang, J. Huang, Exogenously applied plant growth regulators affect heat-stressed rice pollens, J Agron Crop Science, 202 (2016) 139-150. https://doi.org/10.1111/jac.12148

[35] P. Hennebert, M. Filella, WEEE plastic sorting for bromine essential to enforce EU regulation, Waste Management, 71 (2018) 390-399. https://doi.org/10.1016/j.wasman.2017.09.031

[36] H. Alter, Environmentally sound management of the recycling of hazardous wastes in the context of the Basel Convention, Resources, Conservation and Recycling, 29 (2011) 111-129. https://doi.org/10.1016/S0921-3449(99)00061-0

[37] S. Zhang, E. Forssberg, Intelligent liberation and classification of electronic scrap, Powder Technology, 105 (1999) 295-301. https://doi.org/10.1016/S0032-5910(99)00151-5

[38] S. M. Ogilvie, WEEE and Hazardous Waste, AEA Technology Environmental, (2004).

[39] R. Hischier, P. Waüger, J. Gauglhofer, Does WEEE recycling make sense from an environmental perspective? The environmental impacts of the Swiss take-back and recycling systems for waste electrical and Electronic Equipment, Environmental Impact Assessment, 25 (2005) 525-539. https://doi.org/10.1016/j.eiar.2005.04.003

[40] T. Puckett, T. Smith, Exporting harm: the high-tech trashing of Asia, The Basel Action Network, Silicon Valley Toxics Coalition, Seattle, (2002).

[41] M. Dwidwey, R. K. Mittal, An investigation into E-Waste flows in India, J.clean production 37 (2012) 229-242. https://doi.org/10.1016/j.jclepro.2012.07.017

[42] A. Borthakur, P. Singh, Electronic waste in India: problems and policies, Int. J. Environmental Science 3,1 (2012) 353-362.

[43] D. Sinha-Khetriwal, P. Kraeuchi, M. Schwaninger, A comparison of electronic waste recycling in Switzerland and in India, Environmental Impact Assessment Rev 25,5 (2005) 492-504. https://doi.org/10.1016/j.eiar.2005.04.006

[44] P. Agamuthu, D.Victor, Policy trends of extended producer responsibility in Malaysia, Waste Manage Res 29,9 (2011) 945-953. https://doi.org/10.1177/0734242X11413332

[45] R. Afroz, M. M. Masud, R.Akhta, J. B. Duasa, Survey and analysis of public knowledge, awareness and willingness to pay in Kuala Lumpur, Malaysia-a case study on household waste management, J. Clean Production. 53(2013) 185-193. https://doi.org/10.1016/j.jclepro.2013.02.004

[46] V. N. Pinto, E-waste hazard: the impending challenge, Indian J. Occup. Environmental Med. 12,2 (2008) 65. https://doi.org/10.4103/0019-5278.43263

[47] V. S. Rotter, P. Chancerel, W. P. Schill, Practicalities of individual producer responsibility under the WEEE directive, Waste Management Resources 29,9 (2011) 931-944. https://doi.org/10.1177/0734242X11415753

[48] T.R. Carlee, The Economic Feasibility of Recycling: A Case Study of Plastic Wastes, Prager, New York, 1986.

[49] H. Alter., The origins of municipal solid waste: II. Policy Options for plastics waste management, Waste Management Resources. 11 (1993) 319-332. https://doi.org/10.1006/wmre.1993.1034

[50] F. P. Boettcher, Environmental Compatibility of Polymers, Emerging Technologies in Plastics Recycling, ACS Symposium Series 513, American Chemical Society (1992) 16-25. https://doi.org/10.1021/bk-1992-0513.ch002

[51] H. Mankowitz, Incineration of municipal solid waste: Scientific and technical evaluation of the state of- the art by an expert panel, Resources, Conservation and Recycling, 4 (1990) 241-252. https://doi.org/10.1016/0921-3449(90)90005-O

[52] J. Gourmands, H. van der Slot, T. Alberts, Waste Materials in Construction. Studies in Environmental Science, 48, Elsevier Science Publishing Company Inc., New York, 1991.

[53] J. W. Barlow, D. R. Paul, The Compatibility of Mixed Plastic Scrap. Secondary Reclamation of Plastic Waste; Research Report-Phase I, Plastics Institute of America (1987) 137-148.

[54] R. S. Stein, Miscibility in Polymer Recycling, Emerging Technologies in Plastics Recycling. American Chemical Society, (1992) 39-48. https://doi.org/10.1021/bk-1992-0513.ch004

[55] Bennett, R.A, Recycled Plastics: Product Applications and Potential, Emerging Technologies in Plastics Recycling. American Chemical Society (1992) 26-38. https://doi.org/10.1021/bk-1992-0513.ch003

[56] F. Fenella, I. de Michalis, A. Scorcher, M. Pelion, F. Velia, Extraction of metals from automotive shredder residue: Preliminary results of deferent leaching systems, Chin. J. Chemical Engineering 23 (2015) 417-424. https://doi.org/10.1016/j.cjche.2014.11.014

[57] J. Cui, E. Forsberg, Mechanical Recycling of Waste Electric and Electronic Equipment: A Review, Hazardous Materials, 99 (2015) 243-263. https://doi.org/10.1016/S0304-3894(03)00061-X

[58] A. Gung, S. M. Gupta, Disassembly Sequence Planning for Products with Defective Parts in Product Recovery,Computer Engineering 35 (1998) 161-164. https://doi.org/10.1016/S0360-8352(98)00047-3

[59] M. Sethuraman, E. D. van Hullebusch, D. Fontana., A.Akcil, H. Devec., B. Batinic, J. P. Leal, T. A. Gasche, M. A. Kucuker, K. Kuchta, Recent advances on hydrometallurgical recovery of critical and precious elements from end of life electronic wastes-A review, Crit. Rev. Environmental Science Technology. 49 (2019) 212-275. https://doi.org/10.1080/10643389.2018.1540760

[60] X. N. Zhu, C. C. Nie, S.S. Wang, Y. Xie, H. Zhang, X. J. Lyu, J. Qiu, Cleaner approach to the recycling of metals in waste printed circuit boards by magnetic and gravity separation, J. Cleaner production (2019). https://doi.org/10.1016/j.jclepro.2019.119235

[61] F. Settimo, P. Bevilacqua, P. Rem, Eddy Current Separation of Fine Non-Ferrous Particles from Bulk Streams, Physical Separation Science Engineering. 13 (2009)15-23. https://doi.org/10.1080/00207390410001710726

[62] Z. Schlett, F. Claici, I. Mihalca, M. Lungu, A New Static Separator for Metallic Particles from Metal-Plastic Mixtures, Using Eddy Currents, Mineral Engineering 15 (2002) 111-113. https://doi.org/10.1016/S0892-6875(01)00215-1

[63] F. Hamersk, A. Krummenauer, A. M. Bernardes, H. M. Veit, Improved settings of a corona-electrostatic separator for copper concentration from waste printed circuit boards. J, Environmental Chemical Engineering. 7 (2019). https://doi.org/10.1016/j.jece.2019.102896

[64] J. Cui, E. Forssberg, Mechanical Recycling of Waste Electric and Electronic Equipment: A Review, J. Hazardous Materials. 99 (2003) 243-263. https://doi.org/10.1016/S0304-3894(03)00061-X

[65] Y. Higashiyama, K. Asano, Recent Progress in Electrostatic Separation Technology, Partical Science Technology 16 (1998) 77-90. https://doi.org/10.1080/02726359808906786

[66] C. W. Kiewiet, M. A. Bergougnou, J. D. Brown, I. I. Inculet, Electrostatic Separation of Fine Particles in Vibrated Fluidized Beds, IEEE Transport appliance IA-14 (1978) 526-530. https://doi.org/10.1109/TIA.1978.4503586

[67] R. Chinapu, R. Aswani kumar, M. Jarupala, The study on use of recycled materials on highway construction, International Journal of Advance Research in Science and Engineering, 5,10 (2016) 331-338.

[68] M. Surya, O. K. C. Bavithran, A. R. Nandhagopal, Stability study on eco-friendly Flexible pavement using E-waste and Hips, International Journal of Civil Engineering and Technology (IJCIET), 8,10 (2017) 956-965.

[69] K. Rajat, C. Rajesh, Use of e-waste and fly ash as a filler replacement in the bituminous concrete pavement, International Journal of Advance Research, Ideas and Innovations in Technology. 4,2(2018) 1283-1288.

[70] M. S. Ranadive, S. Maheshkumar Krishna, Performance Evaluation of E-Waste In Flexible Pavement an Experimental Approach, International Journal of Civil, Structural, Environmental And Infrastructure Engineering Research And Development. 2,3 (2012) 1-11.

[71] S. Prashant, K. Abishek, S. Sushanth, R. Raajev, Use of Plastic waste in flexible pavement, International Journal Of Engineering Research And Technology, 9,9 (2020). https://doi.org/10.17577/IJERTV9IS090423

[72] S. Amit Kumar, R. K. Singh, Application of waste materials in road construction, Non-Conventional Energy Sources for Sustainable Development of Rural areas, (2016) 1-5.

[73] D. Javiya, A. Yogesh, G. Himanshu, A Review on Performance of Bituminous Mix using E-waste and Fly-ash for the Flexible pavement, International Journal of Advance Engineering Research and Development, 4,2 (2017) 11-15. https://doi.org/10.21090/IJAERD.41658

[74] R. Himani, J. Rajesh, Utilization of Mobile Waste in Construction Industry in Preparation of Flexible Pavement, International Journal of Scientific Research in Civil Engineering, 2,6 (2018) 6-10.

[75] C. Vaidevi, D. Sahana, E. Sripriya, T. Tamilselvi, D. S. Vijayanand, Utilization of e-waste in flexible pavement, International Conference on Mechanical, (2020) 030001-030006. https://doi.org/10.1063/5.0024752

[76] P. Pankaj, H. Nikhil, Experimental Study of Bituminous Concrete Containing Plastic Waste, Journal of Mechanical and Civil Engineering, 11,3 (2014) 47-54. https://doi.org/10.9790/1684-11323745

[77] Dr. R. Vasudevan, S.K. Nigam, R. Velkennedy, A. Ramalinga Chandra Sekar, B. Sundarakannan, Utilization of Waste Polymers coated aggregates for flexible pavement and easy disposal of waste Polymers, Proceedings of the International conference on sustainable solid waste management, 5 (2007) 105-111.

[78] S. Needhidasan, S. Gorab Agarwal, A review on properties evaluation of bituminous addition with E-waste plastic powder, Materials Today: Proceedings, (2019). https://doi.org/10.1016/j.matpr.2019.12.127

[79] Sangita, Tabrez Alam Khan, Sabina, D. K. Sharma, Effect of waste polymer modifier on the properties of bituminous concrete mixes, Construction and Building Materials, 25 (2011) 3841-3848. https://doi.org/10.1016/j.conbuildmat.2011.04.003

[80] R. L. Schroder, The use of recycled materials in highway construction, Transportation Research Board, 58,2 (1994) 32-41.

[81] A. Shankar, K. Koushik, G. Sarang, Performance studies on bituminous concrete mixes using waste plastics, Highway Research Journal, 6,1 (2013).

[82] M. Brajesh, Use of Plastic Waste in Bituminous Mixes of Flexible Pavements by Wet and Dry Methods: A Comparative Study, International Journal of Modern Engineering Research, 6,3 (2016) 41-45.

[83] S. E. Zoorab, I. B. Superma, Laboratory design and Performance of Improved Bituminous Composites Utilizing Recycled Plastic Packaging Waste, 5,6 (2006) 203-209.

[84] W. Baron, Y. Zaping, Properties of Modified Asphalt Binders Blended with Electronic Waste Powders, Journal of Materials in civil Engineering, 24,10 (2012) 1261-1267. https://doi.org/10.1061/(ASCE)MT.1943-5533.0000504

[85] S. Manjay kumar, M. Prathiksha, An Experimental Study on Partial Replacement of Aggregate by E-Waste for bitumen Pavement, International journal of trend in scientific research and development, 3,3 (2019) 382-384. https://doi.org/10.31142/ijtsrd22865

Sustainable Materials and Smart Practices - NCSMSP-2021 Materials Research Forum LLC
Materials Research Proceedings **23** (2022) 401-418 https://doi.org/10.21741/9781644901953-44

Investigation on Strength Properties of Concrete using Steel Slag as a Partial Replacement for Fine Aggregate

R. Ashwathi[1,a,*], G.T. Amudhan Vetrivel[1,b], M. Abishek[1,c]

[1]Department of Civil Engineering, Bannari Amman Institute of Technology, Sathyamangalam, Erode, TamilNadu, India

[a]ashwathi@bitsathy.ac.in, [b]amudhanvetrivel.ce18@bitsathy.ac.in, [c]abishek.ce18@bitsathy.ac.in

Keywords: Fine Aggregate Replacement, Mechanical Strength Properties, Steel Slag, Eco-Safety, Concrete Conglomerates

Abstract. The economic and ecological performance of normal concrete can be increased by modulating fine aggregate content with a series of combinations of steel slag which possess different physical and chemical behaviour based on its cooling method, heating processes. Many research shows that magnetic separating of steel slag increases its efficiency towards integrating with clinker by 50% compared to integrating of clinker with non-magnetically separated steel slag. It has proved to be an impressive replacement material in concrete surfaces where high skid resistance is essential. Steel slag produced in various types of furnaces have different characteristic nature, blast furnace slag (BFS) has hydraulic and cement properties when used in water-bound macadam roads and also in flexible and rigid pavements. Issue in steel slag is energy consumed in its production and transportation where it is utilized, if energy consumed in slag grinding/magnetic separation is high when compared to cement calcining and grinding, then it would not be economical in replacement criteria. This study exhibit mechanical properties of concrete with partial replacement of fine aggregate with steel slag in distinct proportions. The optimum amount of replacement in fine aggregate is found to be 20% giving a strength increment of 8% in the compressive strength category. In split tensile and flexural strength criteria strength increment of 7.5% and 40.625% is observed. There are many practical implications of steel slag in the construction industry, road constructions, and clinker substitutes as granulated BF slag, water treatment plants, evidently many researches have proved slag as productive coarse aggregates replacement.

Introduction

A roundabout economy is a right now acknowledged fundamental, where the conventional direct economy is changed into a roundabout economy, where each movement is imagined as a cycle, where squander materials are considered as expected new assets, rather than side-effects to discard.[1]. Innumerable hills and valleys are carved to get the most utilised material in the construction field, the aggregate, 75% of concrete volume is constituted by aggregate or metal. There is a developing interest in involving waste materials as elective total materials and critical exploration is utilized a wide range of materials as total substitutes, for example, coal debris, impact heater slag and steel slag. [1]. More recently through awareness on sustainable development paved the way for extensive research and development to use waste by-products from diverse industries in construction. Slag, which is an expensive waste material, came to light in the last century by growing awareness on environmental issues in our society that had an impact on the effective utilisation of slag. To maintain the new aspiration of a circular economy, engineers and contractors pose a challenge to exploit waste by-products from the building industry to be integrated as raw material in the same building cycle. Utilisation of slag products started with the first ore smelting process in the 19th century and compensation in using by-products and also space to store these products were lacking. Lightweight cement is less ecologically strong than typical weight concrete on a cubic meter reason for two reasons, first and foremost, it is less compelling for warm mass purposes, and also, the typified natural effects will generally be higher due to the energy required in making the totals and the higher concretes substance utilized in the mixes. It is suggested that during the plan and obtainment of concrete, some thought be given to the related ecological effects. [7] This might well show that gross decreases in effects can be made without modification to the nature of the design delivered. It is additionally suggested that singular organizations develop information about the effects related with their strategies, for instance how much waste, to guarantee that the information utilized in any investigation are both explicit and current. This will work on the precision of the computations and the objectivity of the appraisal prompting better educated plan and obtainment choices. [7]

Slag

Slag is a by-product of the pig-iron and steel industries, made up of Calcium, Magnesium, and Aluminium Silicates in various combinations; different kinds of slag are created by changing the cooling process, but the fundamental compositions stay the same. Slag is produced in metallurgical and combustion processes when impurities in metals or ores are removed. During the smelting and refining process, slag floats to the surface, protecting it from oxidation and reduction and keeping it clean. The slag phase is used in the iron and steel industries to purify the liquid metal by eliminating impurities. Iron or ferrous slag is formed in the blast furnace process, non-ferrous slag is produced during the manufacture of non-ferrous metals (Pb,Cu,Ni), and steel slag is produced. The material that results in the solidification of molten BF slag under atmospheric conditions is air cooled slag, which is widely used as a base or sub-base in pavement construction, filler or concrete aggregate, and also in scour protection layers, depending on the cooling conditions in the blast furnace. The blistering and high porosity expanded slag produced by controlled quick cooling of molten slag in water or with a mix of steam and compressed air is utilised in light weight concrete. High pressure water jets are used to produce granulated slag with quenching which gives out glass like granules. Steel slag production process is shown below Fig.1.

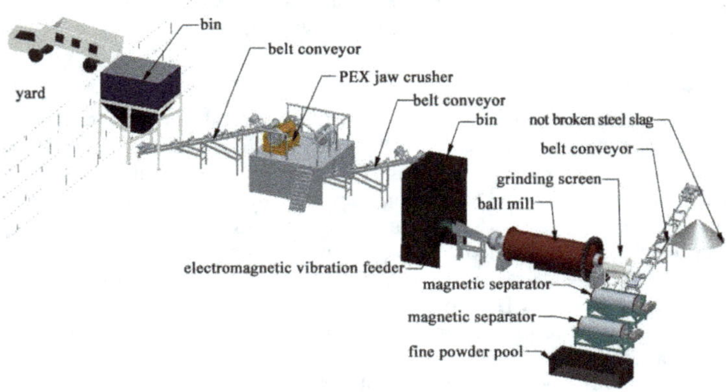

Fig.1: Steel slag production process.

Water glass- activated slag concrete has a stronger mechanical strength and durability than traditional Portland cement concrete with the same dosage. Additionally, adding extra cementitious material to concrete mixes enhances mechanical characteristics and increases structural density, reducing susceptibility to aggressive agent penetration in both OPC and AAS concretes.[13] Granulated slag is then crushed, pulverised and screened to desired sizes, used in cement production to produce slag cement with enhanced heat of hydration, resistance to sulphate and chloride properties. Another well-known and cement-like material called GGBFS (ground granulated blast furnace slag) is produced which when integrated in concrete gives durable and high strength mixes in a longer time-scale. For a long time, several countries throughout the world, including Japan, Australia, Russia, and the European Union, have employed GBFS to produce artificial sand.[12] The GBFS sources in these nations are typically consistent in composition and engineering qualities, and the output slag quality is rigorously monitored. Despite the fact that a large volume of GBFS is generated in Vietnam each year, the countries focus on function is still lacking. As a result, their quality is unregulated, and their composition and qualities are insecure. [12] An alternative material for clinker production reducing energy consumption and overall reduction in carbon footprint of cement industry, after slag drying clinker with 40-50% slag in conveyed into ball mill with 6% gypsum resulting in Portland slag concrete which have low hotness of hydration and low salt total response properties. Portland slag concrete enjoys the benefit of better execution, toughness and ideal expenses likewise being eco-accommodating. Compound organizations of slag is given in Table.1.

Table 1: Steel slag composition.

CONSTITUENTS AS OXIDES	SYMBOL	BF SLAG [%]
Calcium oxide	CaO	41
Free lime	-	0
Silicon oxide	SiO_2	35
Iron oxide	Fe_2O_3	0.7
Magnesium oxide	MgO	6.5
Manganese oxide	MnO	0.45
Aluminum oxide	Al_2O_3	14
Titanium oxide	TiO_2	1
Potassium oxide	K_2O	0.3
Chromium oxide	Cr_2O_3	<0.005
Vanadium oxide	V_2O_5	<0.05
Sulphur	S	<0.6

Steel slag

Steel slag as a road building material might minimise the cost of obtaining and processing naturally existing aggregates from an economic standpoint. The steel industry may also be able to lower the cost of treating and disposing of massive stockpiles of steel slag. [16]. Oxidized elements which remain after removal of silicon and carbon as carbon-di-oxide in basic oxygen furnace (BOF) and electric arc furnace method (EAF) when combined with lime forms steel slag, hard, dense and similar to iron slag. The properties and composition of steel slag depends on type of steel production, Black slag is mostly produced in melting scrap metal in EAF method. Non-resting steel production gives out slag called white slag, which is also produced in small amounts in the secondary metallurgy process in vacuum oxygen decarburisation furnaces. These slags differ in chemical and mineral composition and possibilities of their use will also vary. Another type of slag is used in bituminous mixes in roads called non-ferrous steel slag produced in recovery of smelting process in Non-ferrous metals. It usually contains residual metal substitutes. Steel slag proved to be a barrier remedy in waste dumpsites where leaching of toxic heavy metals to nearby land bodies is evident. Steel slag presents permeable design and huge surface region; likewise, it is not difficult to isolate from water because of its high thickness. In this way, the use of steel slag in modern waste water therapy has gotten serious consideration as of late [5]. Development pace of SMA blend with steel slag is beneath 1% following 7 days, which guarantees the security of steel slag in SMA mixtures. In short, the fruitful use of steel slag as total in asphalt development can give a new and more financially savvy approach for total assets, and abatement the dangers of strong squanders to climate. In any case, more examinations ought to be completed on its reusing interaction and wide application in future [6].20% GGBFS replacement had a considerably positive effect on the cube and cylinder compressive strength but 40% and 60% GGBFS replacements reduced the strength at the age of 28 days. • The maximum loss of strength of 0%, 20%, 40%, and 60% GGBFS concrete is due to the effect of magnesium chloride followed by magnesium sulphate. [9]. the characteristics of water quenched slag are found to be quite similar to those required for use in the Portland slag cement making process. Air-cooled slag, on the other hand, is acceptable for use as an aggregate in the building sector. [14]. Steel slag processing is critical for steel recovery because it produces an angular, typically well-graded material that is largely free of metallic contaminants, and the recovered steel (2 to 4% of raw steel production) is a valuable scrap. [15]. the asphalt concrete has a regular heating time and a prolonged cooling time once steel slag is replaced. Steel slag induction of healing asphalt concretes is given more time to mend. [17]

Literature review

Kunal K. Das and Eddie S. S. Lam (2020), "Partial replacement of cement by ground granulated blast furnace slag and silica fume in two-stage concrete (preplaced aggregate concrete)", to improve the qualities of two staged concrete or preplaced aggregate concrete, researchers tested partial substitution of cement with granulated blast furnace slag (GGBS) and silica fume (SF) (PAC).This concrete is produced by gravity method by placing aggregates in the formwork and injecting grout into it to fill voids between the aggregates. Optimum proportion of replacements is found to be GGBS of 30% and SF of 10% to give the suitable mix. Although adding these replacements increases the shrinkage potential, the amount of shrinkage is significantly lower than that of conventional concrete. [3]

M.Gouthampriya et al (2020), "Partial Replacement of Cement by Ground Granulated Blast-Furnace Slag in Concrete", inspection attempts to find the possibility of replacement of cement with granulated blast furnace slag (GGFS), varying proportions of percentage of binding material is replaced with GGFS and compressive, split tensile and flexural strength properties are observed and found that 20 % replacement of cement is enhanced in all mechanical criteria compared to control concrete. Compressive Strength results are presented below in Table 2.

Materials Research Forum LLC
https://doi.org/10.21741/9781644901953-44

Table 2: Compressive strength results

S.No	Percentage of Replacement [%]	Compressive strength [N/mm^2]		
		7 Days	14 Days	28 Days
1	0%	18.43	22.61	25.82
2	10%	19.48	23.60	26.73
3	20%	21.33	24.93	28.12
4	30%	16.65	18.93	22.65

Split tensile strength for cylinders are as below in Table 3.

Table 3: Split tensile strength

S.No	Percentage of Replacement	Split Tensile Strength [N/mm2]
28 Days		
1	0%	2.86
2	10%	3.28
3	20%	4.34
4	30%	2.20

Flexural strength of beams are as below in Table 4.

Table 4: Flexural strength.

S.No	% of Replacement	Flexural Strength (N/mm2)
28 Days		
1	0%	6.28
2	10%	7.45
3	20%	8.36
4	30%	5.69

María Eugenia Parron-Rubio at al, "Slag Substitution as a Cementing Material in Concrete: Mechanical, Physical and Environmental Properties", to achieve a circular economy by utilizing waste materials as raw materials in construction field, cement substitution by different types of slag from steel mills, behaviour of concrete with slag from blast furnace and ladle furnace in terms of physical and technical properties. For an environmental aspect, leachate test is conducted, it is found that blast slag has excellent substitute capability providing properties above reference concrete and ladle slag can be used for non-structural applications. For physical properties density and porosity are studied. The proportion of slag substitution studied are 30%, 40% and 50%. Totally 10 mixes are prepared with three types of slags of 3 proportions. Results shows for density property, ladle furnace slag 1(LFS 1) makes the mix aerated and porous more than mixes with other slags. Density of other slag compared to control mix is very identical and substitution doesn't pose any threat to this property. GGBFS and LFS2 slags reduced the porosity to 25% and a 43% increase in performance compared to reference concrete which will have good performance in long run preventing the external agents that deteriorate the material's

Materials Research Proceedings **23** (2022) 401-418 https://doi.org/10.21741/9781644901953-44

microstructure, on other hand LFS1 give rise to high porosity and threat of propagation of crack and corrosion of steel frame. In the compressive strength category, GGBFS on the first day of hardening shows less compressive strength, but strength increases even after 30 days. Other slags lose strength giving up chances of structural material. Also this work proves if SiO_2 is high then higher pozzolanic activity, LFS1 attains less strength due to less SiO_2 and also of greater porosity. Optimum proportion in this study is found to be 25% of GGBS based on a previous study. A similar trend is observed for flexural strength where 50% substitution gives 4% increase in strength compared to no substitution mix. Other slag are worst performing, losing strength even after 90 days, owing to their chemical and physical properties.

Mani et al, "Replacement of Coarse Aggregate using Steel Slag in Concrete", this study focuses on mechanical property analysis of concrete where its aggregate part is replaced with steel slag in different proportions, it is found that more stable mix with high durable and resisting nature is formed and optimum quantity of slag replacement is 70 % of aggregates giving a 7-8% increase of strength when compared to control concrete.

Bilim et al.(2008) "Artificial neural networks and predicted the compressive strength of GGBFS concrete", The substantial blends were made with three diverse water-concrete proportions (0.3,0.4 and 0.5) three distinctive concrete measurements (350,400 and 450 kg/m3) and four fractional slag substitution proportion (20%,40%,60% and 80%). Compressive strength of soggy relieved examples were estimated at 3, 7, 28, 90 and 360 days. Results from the review are introduced beneath in Table 5.

Table 5: Results from above study for varying w/c ratios in consequent curing periods, Compressive strength is in [N/mm²]

W/C Ratio	GGBFS (%)	Age [Days]				
		3	7	28	90	360
0.30	0	63.8	75.7	80.3	85.7	92.8
	20	58	72.1	81.8	90.1	95.9
	40	49.8	66.4	83.8	91.4	96.8
	60	36.8	63.2	80.6	92.5	101.3
	80	26.3	50.2	66.3	77.4	78.4
0.40	0	53	59.2	64.3	71	79.4
	20	45.8	60.3	73.5	82.3	86.6
	40	35.9	56.3	66.4	81	84.6
	60	26.9	38.6	61.8	73.4	80.6
	80	22.6	36.7	46.8	54.6	58.9
0.50	0	25.8	36.6	48.7	50.5	60.2
	20	21.3	33.2	50.4	56.2	66.7
	40	16.7	28.3	49.3	53.4	65.1
	60	13	20.8	39.5	49.1	57.4
	80	9	17.5	27.7	35	39.6

For all slag substitution proportion, strength commitment of slag to concrete is low upto 7 days old enough. Notwithstanding, compressive strength of cement containing slag concrete is higher than that of control ordinary Portland concrete cement for 20% and 40% substitution proportions at 28 days, 90 days and one year. Counterfeit neural organization (ANN) model is made with 45 substantial combinations. Information that is utilized in the model is organized in organization of six info boundaries that covers concrete, GGBFS, water, hyper plasticizer, total and time of tests

Materials Research Proceedings **23** (2022) 401-418 https://doi.org/10.21741/9781644901953-44

and a result boundary which is compressive strength of cement. The outcomes showed that ANN can be an elective methodology for anticipating the compressive strength of GGBFS substantial involving fixings as information boundaries.

Pelligrino and gaddo (2009)"Natural aggregates of traditional concrete with black/ oxidizing electric Arc Furnace (EFA) slag", The substantial made with EFA slag as total showed great strength qualities in typical condition and strength properties of the combination containing EFA slag are absolutely tantamount than those noticed for conventional cement, the compressive strength of solid shapes examples with customary and EFA slag total later 7,28and 74 days displayed in table as beneath. It is important to see that compressive strength settles later the initial 28 days for customary combination though it keeps on improving for the EFA slag aggregate, maturing diminishes the strength of conventional cement wherein expanding for EFA aggregate. Aftereffects of compressive strength utilizing EFA slag are given underneath in Table 6.

Table 6: Results with EFA slag as aggregate.

Mix	7 days	28days	74days
Traditional	25.3	32.5	30.4
With EFA slag	37.2	42.3	44.4

Ismail and Hashmi (2008), "Report that the waste iron were reused to partially replaced sand at 10%, 15% and 20% in a concrete mixtures", droop, fresh thickness, dry thickness, compressive strength, and flexural strength tests were done to assess waste - iron significant quality. This technique is useful for restoring ages for the considerable mixes for 3,7,14, and 28 days. The results reveal that considerable mixes including waste iron have better compressive properties. For the 3, 7, and 28 days restoration periods, the compressive strength of the significant blends formed of 20% waste - iron total grew by 22.60 percent, 15.90 percent, and 17.40 percent, respectively. Aftereffects of solidarity increase rate is given in underneath Table 7[10].

Table 7: strength in percentage rise or fall.

% waste iron Aggregates	**Increasing (+) or decreasing (-) (%)**			
	3 days	**7 days**	**14 days**	**28 days**
0	0.00	0.00	0.00	0.00
10	8.20	11.43	-6.60	+1.80
15	15.20	14.90	2.97	1.2.95
20	22.60	15.90	+0.46	17.40

Madhavi et al, "Investigations on Sulphate Attack on Copper Slag Concrete", effect of sulphate on concrete can be as far as substance and actual qualities, whitish appearance in surface of cement is reaction of sulphate assault, sulphate changes hydrations items, for example, calcium hydroxide into calcium sulphate or gypsum and Tri-calcium aluminate hydrates are changed over to ettringite. These inside regulations cause turning and pushing, at last loss of joining between particles, vulnerability increments and water streams easily to the internal part. 2.2-2.3 huge loads of copper slag are delivered for each one ton of copper in the matte purifying cycle. Copper slag is fused at 20%, 40% and 60% in concrete, presented to magnesium sulphate answer for 30 days sway loss of solidarity. Substantial blend in with 40 % substitution of copper slag expands the strength from 60 MPa-64.65Mpa. Compressive Strength following 30 Days Curing of MgSo4 Exposure is displayed in underneath table: Results are displayed beneath in table 8 impact of sulphate on strength qualities. [4]

Materials Research Proceedings **23** (2022) 401-418 https://doi.org/10.21741/9781644901953-44

Table 8: Strength reduction or increase after $MgSO_4$ exposure.

Proportion	Compressive load, kN	Compressive Strength, 30 days	Increase/ Decrease in Compressive Strength %
CS 0	950	42.22	-
CS 20	1000	44.44	21.97
CS 40	1350	60.00	64.65
CS 60	1250	55.56	52.46

Materials

As per codal requirements and standard procedures of experiments materials that are used should be tested for compliance and of shear quality for their correctness or credibility of results. In addition to steel slag, cement, fine aggregate, coarse aggregates were used in experiments. The purpose of these preliminary experiments are to enable an engineer to design a mix for desired strength. Descriptions of each materials is given below:

Portland cement

A scientifically controlled material whose compositions can be varied based on requirement of concrete mix, the function of cement is to fill the voids between sand and aggregates and also to bind them forming a interfacial layer to form a compact mass, although its constitutes less than 20% in whole concrete volume but it is an active binding medium with inevitable place in concrete. Change in proportion of cement increases the strength up to a level then strength remains the same. Portland cement is the most important and widely used type of cement, classified into three types based on grade which signifies the strength of cement. Higher grade cements are finer and costlier than lower grade cements. Lower grade cement gives many advantages such as faster rate of strength attainment, cost saving of 10-15%. In this study Shankar cement of grade 43 is used, it was cool, without lumps or adulteration, various tests conferring to IS 8112:1989 were conducted. It was safely stacked in moisture free condition to prevent hydration. Various tests done are initial and final setting time, specific gravity, fineness and consistency. Properties are presented in table 9 below. In comparison to PSC concrete, replacement of OPC with GGBFS in concrete resulted in comparatively low strength development for the majority of the cement content in the concrete. For most of the cement content, using PSC as cement directly in concrete mix resulted in considerable strength growth for a curing period of up to 28 days. [11] However, compared to OPC and GGBFS concrete, the rate of strength growth for PSC concrete cured at 3 days showed somewhat lower strength gain at low cement content (300 and 320 kg/m3). The compressive strength growth at all ages of curing shows considerable strength development over OPC plus GGBFS concrete at higher levels of cement concentration, between 340 and 380 kg/m3. This is due to a reduction in variation in cement quality and an optimised amount of gypsum in the resulting concrete for improved strength development. [11]

Materials Research Proceedings **23** (2022) 401-418 https://doi.org/10.21741/9781644901953-44

Table 9: Cement properties after conducting respective tests.

SI.NO	CHARACTERISTICS	VALUES OBTAINED EXPERIMENTALLY.
1	Specific Gravity	3.12
2	Initial Setting Time	32 min
3	Final Setting Time	10.03 hours
4	Consistency	28%
5	Fineness	7.31%

Aggregates

Bulk of concrete is covered by aggregates, which is of nearly 75% of whole volume, fine aggregate is essential for giving required workability and binding cement paste and coarse aggregates, they should possess certain requirements so that concrete is workable, strong, durable and economical. They should be strong, clean, durable and well graded. Coarse aggregates are added in combination to get uniform distribution of particles.

Coarse aggregate

Metals held over 4.75 mm sifter are called coarse aggregates , they are delivered by squashing enormous stones or rock, one more way is uncrushed rock that are gotten by normal breaking down. In view of sort of cement and grade aggregates utilization shifts. The sizes shift from 10-20 mm and accessible as rocks up to 40mm. lower sizes of 8 and 6 mm are utilized in self compacting concrete. Locally accessible 20mm size total is utilized in this review. The totals are washed to eliminate residue, soil and a SSD condition is accomplished. The totals are tried as per IS383-1970. Properties of coarse totals are given beneath in Table 10.

Table 10: Properties of coarse aggregate.

SI.NO	CHARACTERISTICS	VALUES OBTAINED EXPERIMENTALLY
1	Maximum size	20 mm
2	Specific gravity	2.65
3	Water absorption ratio	0.51%
4	Shape test(Elongation Index)	11.38%
5	Shape test(Flakiness Index)	16.10%
6	Crushing value	21.33%
7	Impact value	10.94%

Sieve analysis results for coarse aggregates are given below in Table 1.

Table 11: Sieve analysis report for coarse aggregate

Weight of sample taken =3000gm					
SI.NO	**IS Sieve (mm)**	**Wt. Retained (gm)**	**% Retained**	**% Passing**	**Cumulative % Retained**
1	80	0.00	0.00	100	0.00
2	40	0.00	0.00	100	0.00
3	20	68.5	2.28	97.72	2.28
4	10	2776.5	92.55	5.17	94.83
5	4.75	113.5	3.78	1.38	98.62
6	Pan	0.00	0.00	0.00	
	Total	**3000**		**sum**	**695.73**
				FM	**6.95**

5

Fine aggregate

Aggregates which pass over 4.75mm are fine aggregates. They are formed by natural disintegration of rocks and also by crushing stones or gravel. Based on size distribution fine aggregates are categorized into three types. Fine, medium and coarse are types, IS 383-1970 presents 4 grading zones (I to IV). The grading zones become progressively finer from grading zones I to IV. Properties and sieve analysis results are given in below table 12.

Table 12: Sieve analysis report for fine aggregate.

Weight of sample taken =1000gm					
SI.NO	**IS Sieve (mm)**	**Wt. Retained (gm)**	**% Retained**	**% Passing**	**Cumulative % Retained**
1	4.75	14.5	1.45	98.55	1.45
2	2.36	37	3.70	94.85	5.15
3	1.18	246.5	24.65	70.20	29.80
4	600µ	205.5	20.55	49.65	50.35
5	300µ	287.5	28.75	20.90	79.10
6	150µ	177	17.70	3.20	96.20
7	Pan	32	3.20		
	Total	**1000**		**sum**	**262.65**
				FM	**2.62**

Water

Consumable water is thought of as good for blending and relieving of cement. Water ought to be liberated from sewage and any unfavourable foreign substances. Water obtained from lakes and streams are not should have been tried or tested. However, water when suspected to have sewage or taken from modern plants or canneries, it shouldn't be utilized in concrete except if tests are done and palatable outcomes are noticed. Normally water obligated for drinking is took into account utilization in concrete. Water utilized for this review was liberated from any toxins and reasonable for utilization.

Steel slag

For this study steel slag was bought from “Agni steel plant” Ingur in Tamilnadu. Sieve analysis of steel slag is given below in Table 13. Image of steel slag used is given in Fig 2.

Table 13: Sieve analysis for steel slag.

Weight of sample taken =1000gm					
SI.NO	**IS Sieve (mm)**	**Wt. Retained (gm)**	**% Retained**	**% Passing**	**Cumulative % Retained**
1	4.75	14	1.4	98.6	1.4
2	2.36	28	2.8	95.8	4.2
3	1.18	94.5	9.45	86.34	13.65
4	600µ	189.5	18.95	67.8	32.2
5	300µ	329.5	32.95	34.95	65.05
6	150µ	291.5	29.15	5.8	94.2
7	Pan	58	5.8		
	Total	**1000**		**sum**	**210.6**
				FM	**2.10**

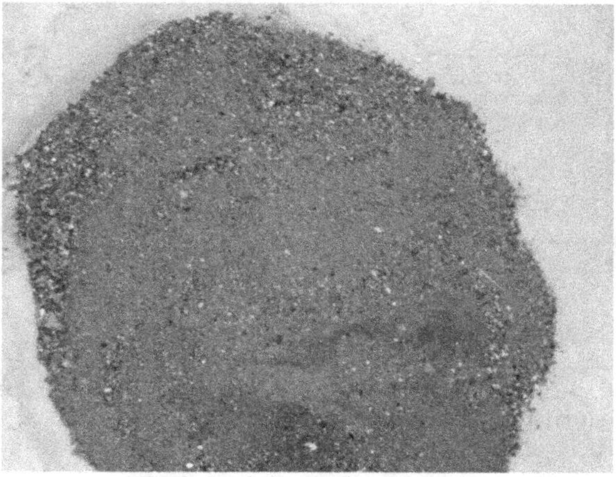

Fig.2: Steel slag used in this study.

Methodology

Mix design

To find the relative proportion of all materials in concrete is mix design. Among various methods to design the concrete mix, a method proposed in Indian standard code was selected and calculations were made. The design mix of grade M20 as designed is given below:

Design stipulation

Characteristic compressive strength

Strength that should be attained on 28 days : 20 N/mm^2
Maximum size of aggregate : 20mm (angular)
Degree of workability : 0.90compaction
Degree of quantity control : Good

Exposure conditions : Mild

Test data for materials
Cement used : OPC
Specific gravity:
Cement : 3.12
Coarse aggregate : 2.69
Fine aggregate : 2.54
Water absorption:
Fine aggregate : 0.89%
Coarse aggregate : 0.43%
Free moisture:
Coarse aggregate : Nil
Fine aggregate : Nil
Sieve analysis in fine aggregate : conforming to grading

Mix proportion of various mixes that are designed based on steel slag percentage is given in table below Table 14:

Table 14:Mix proportions for various SS replacements.

Water(W) Kg/m^3	Cement (C) Kg/m^3	Fine Aggregate(FA) Kg/m^3	Coarse Aggregate(CA) Kg/m^3	Steel Slag (IS) Kg/m^3
180	360	573.86	1233.54	0
180	360	516.48	1233.54	57.38
180	360	459.088	1233.54	114.772
180	360	401.702	1233.54	172.158

Compressive strength testing
In most underlying components, concrete is fundamentally used to oppose compressive burdens. At the point when cement is exposed to pressure, disappointment happens in the upward plane of the part along its slanting, the break shows up because of parallel malleable strains. Miniature break shows up along the upward hub of the part in hub pressure because of horizontal elastic strains.

Test procedure
For compression testing, 150X150X150 mm test specimen cubes were casted. Concrete mixes containing varied percentages of steel slag (0, 10%, 20%, 30%) as a partial substitute for fine aggregate were cast into cylinders and cubes for further testing. To achieve a homogenous mixture, fine aggregate and steel slag were dry mixed first, followed by coarse aggregate and cement mixture. After achieving the desired colour, the needed amount of water is added, and concrete is formed. Before pouring concrete into the mould, the interior surfaces of the mould and the base plate were lubricated. The cubes were taken from the mould after 24 hours and put in clean fresh water at 27+2 degrees Celsius. The samples were evaluated after 7, 14, and 28 days of curing. No cushioning was provided while testing in a compressive testing machine. Load was applied axially till failure or crushing happens. Results of compressive strength varying in the curing period are presented in the next section.

Flexural strength testing

Flexural strength is one of the important parameters considered to design a member to take loads and also play a role in limiting the state of serviceability criterion. It comes into play when an inadequate subgrade support is subjected to wheel loads. IS 516-1959 gives procedure and apparatus required, standards for doing this test.

Test procedure

Specimens of size 15X15X75 mm are prepared since aggregate size is less than 40mm and greater than 19mm. steel rollers of 4 numbers are taken for loading and supporting. Length of rods are taken 10mm longer than specimen width and placed at 3d distance of outer rollers, distance between centres of inner rollers is d, such that all is systematic. The test specimen is tested immediately after storing it in water, whilst it is still wet, the example is set in the machine accurately focused with the longitudinal hub of the example at right points to the rollers. Load is applied at a pace of 400kg/min as the example is of width 15cm. Flexural strength of examples in various relieving periods is given in the following segment.

Split tensile strength testing

The split tensile strength is one of the important parameters that has its extent toward size of crack and cracking load. Usually concrete is not subjected to direct tension owing to its low tensile strength property, but these parameters help us to know where the concrete cracks. This method of test gives more uniform results than the ring tension test and double punch test. IS 5816-1999 presents a procedure for this test with tolerances and mould dimensions.

Test procedure

Each considerable mix is tested in a barrel-shaped form with a 150mm diameter and a 300mm length, in accordance with IS 10086-1982. Cement is laid down in 50mm layers throughout the projecting process. The test instances are stored in a vibration-free environment with a clammy character of about 90% relative dampness and at a temperature of 25 °C to 29 °C for 24 hours (12 hours from the hour of water expansion to the dry fixes). The samples are verified for layering correctness till the nearest 0.2m and then removed from the moulds and submerged in clear water until taken out before the test. Any sand particles on the testing machine's bearing surfaces are removed. Two headings pieces of ostensible (1/8 inch for example 3.175 mm) thick compressed wood liberated from blemishes around (25 mm) wide and of length equivalent to or somewhat longer than that of the example are set from every example, these two bearing strips are put among top and base bearing plates of machine. Polar lines are drawn at each finish of the example with a gadget so they are in a similar pivotal plane. One of the bearing strips is put along the focal point of the lower bearing square. Example is set in a strip and adjusted so the line drawn toward one side is vertical and loped over the compressed wood strip. Second bearing strip is put longwise along the chamber, loped over lines set apart on the finishes of the chamber. Load is applied at a uniform pace of 689-1380 kPa/min till disappointment. Separating two times of most extreme burden applied with result of breadth, phi (22/7) and length gives the split rigidity of an example. Aftereffects of examples are given in the following segment.

Results and discussion.

Consequences of tests led on concrete are examined here. As the target of this review is to observe the mechanical properties of cement with fractional substitution of steel slag as fine total, to accomplish that outcome a test review was intended to explore the impact of steel slag in compressive, split malleable and flexural strength of cement. This included, blend configuration, projecting and relieving of solid shapes likewise with starter trial of material utilized for this review.

Slump test

To determine the consistency and amount of water-cement ratio needed, slump test is conducted for each mix proportions, this gives the workability of concrete and not suitable for very wet and dry mix.IS 456 gives workability values for different converting works in construction. It involves finding loss of height when a concrete filled frustum cone is lifted and behaviour of each mix shows the real picture of consistency. Results obtained from the slump test are given below in Table 15.

Table 15:Slump Test results.

SI.NO	% REPLACEMENT OF STEEL SLAG	W/C RATIO	SLUMP(mm)
1	0	0.50	30
2	10	0.55	30
3	20	0.60	27.5
4	30	0.65	24

Compressive strength

Below table shows mean compressive strength of specimens tested in 7, 14, and 28 days of curing from time it is mixed with water. Strength values are in MPa or N/mm^2 as given in below Table 16. Ultra fine slag replaces concrete with three distinct rates viz 5, 10, and 15% to concentrate on the compressive strength, porosity and sorptivity. Execution of monetarily accessible ultra-fine slag was viewed as substandard as ground one and cement with 10% ultra-fine slag was viewed as ideal. [8]

Table 16: Compressive test results of our study.

Tested day	0%	10%	20%	30%
7th day	16.25	17.55	18.60	17.20
14th day	18.56	19.20	20.50	19.26
28th day	20.98	21.56	22.66	21.10

The compressive strength increases as the percentage of steel slag is increasing, if 10% replaced in fine aggregate, on 7th day compressive strength is 8% more than control mix, correspondingly 14.46% more strength for 20% replacement and 5.8% increase for 30 % replacement. It also shows early strength development for all three replacement mixes compared to reference mixes. Optimum amount of replacement of steel slag is found to be 20 % giving more durability and strength with considerable cost benefits. 28th day strength for 20% replacement is 8% more than control concrete.

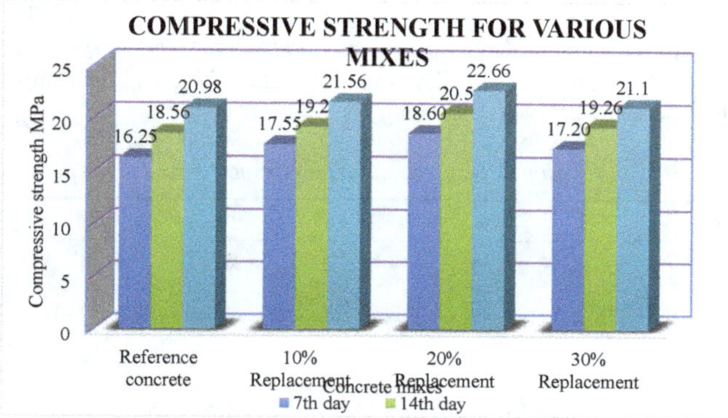

Fig 3: Graph comparing compressive strength for various design mix.

Flexural strength

Below Table 17 presents the flexural test observations for various mixes. From these values (in MPa) we infer that the same trend is continued as before, that increasing proportion ultimately improves flexural strength of the mix compared to CM. for 20 % replacement a 40.62% improvement of strength is observed.

Table 17: Flexural strength outcome after required test.

	0%	10%	20%	30%
7th day	1.24	1.56	1.80	1.62
14th day	1.96	2.24	2.68	2.10
28th day	2.56	3.34	3.60	3.10

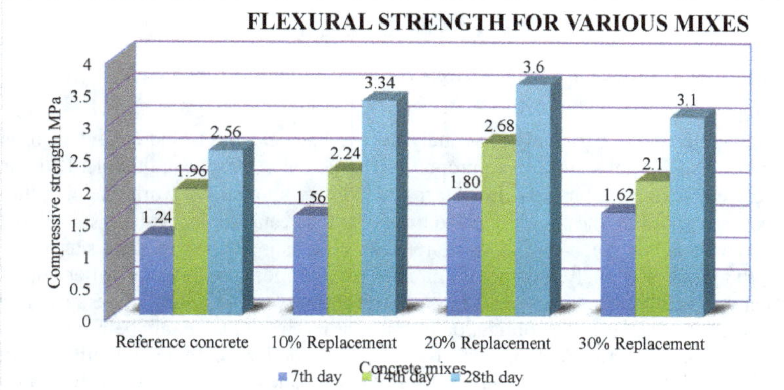

Fig 4: Flexural strength comparison for different mixes.

Split tensile strength

Below 18 tables exhibits strength improvement of 11.667% for 20 % replacement compared to reference concrete, more support for the limit state of cracking is given by this mix than nominal concrete which is more susceptible for cracking.

Table 18: Split tensile strength development in successive curing periods.

	0%	10%	20%	30%
7th day	1.36	1.50	1.62	1.78
14th day	1.80	1.96	2.10	1.98
28th day	2.40	2.46	2.68	2.10

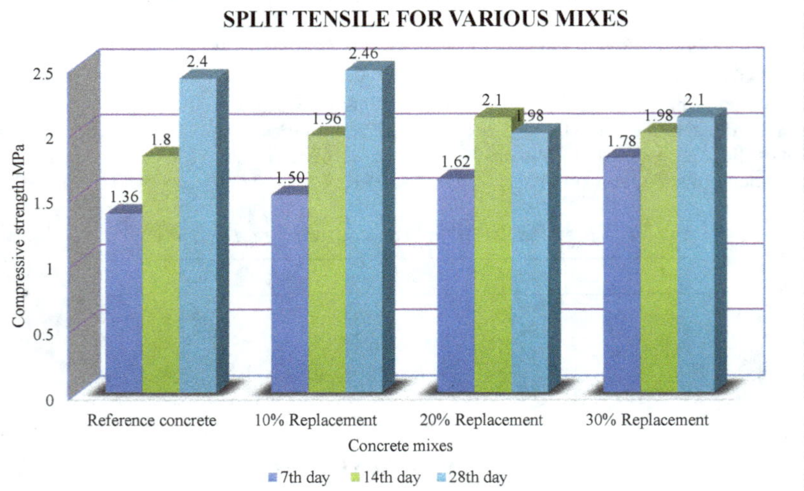

Fig 5: Split tensile strength comparison for different mixes.

Conclusion

The Primary objective of this study is to study the mechanical aspects and behaviour of various mixes replaced with steel slag in three proportions to get an economical, durable, and compact, eco-friendly concrete conglomerate. Infrastructure industry all over the world is going through a rapid phase, so exploiting the remaining resources would be catastrophic and a sustainable way of economic development. Need for aggregate related products is increasing, steel slag would be a suitable product to effectively incorporate in concrete, the most used element after aqua. Many research and development in this field is going on to tackle inadequacy of resources and the engineering field should have the obligation to safeguard a circular economy befriending ecologic wellbeing as a salient factor. A comprehensive study is carried out in reviewing other studies and experimental analysis is carried out. Further study can be made on different detrimental attacks on concrete such as chemical and environmental influence in long term behaviour of slag aggregate concrete.

References

[1] Parron-Rubio, M.E.; Perez-Garcia, F.; Gonzalez-Herrera, A.; Oliveira, M.J.; Rubio-Cintas, M.D. Slag Substitution as a Cementing Material in Concrete: Mechanical, Physical and Environmental Properties. Materials 2019, 12, 2845. https://doi.org/10.3390/ma12182845 https://doi.org/10.3390/ma12182845

[2] Mani V, Dinesh Kumar S, Gaayathri Priya N, Poornima, 2016, Replacement of Coarse Aggregate using Steel Slag in Concrete, international journal of engineering research & technology (IJERT) SNIPE - 2016 (Volume 4 - Issue 25),

[3] Das, KK, Lam, ESS, Tang, HH. Partial replacement of cement by ground granulated blast furnace slag and silica fume in two-stage concrete (preplaced aggregate concrete). Structural Concrete. 2021; 22 (Suppl. 1): E466- E473. https://doi.org/10.1002/suco.201900494 https://doi.org/10.1002/suco.201900494

[4] Irfan, H & Anil, K & Rose, Gowtham & Velan, Senthil. (2018). Investigations on Sulphate Attack on Copper Slag Concrete. Journal of Advanced Research in Dynamical and Control Systems. Vol. 10. 1272-1275.

[5] Huang Yi, Guoping Xu, Huigao Cheng, Junshi Wang, Yinfeng Wan, Hui Chen,An Overview of Utilization of Steel Slag,Procedia Environmental Sciences,Volume 16,2012,Pages 791-801,ISSN 1878-0296,https://doi.org/10.1016/j.proenv.2012.10.108. https://doi.org/10.1016/j.proenv.2012.10.108

[6] Shaopeng Wu, Yongjie Xue, Qunshan Ye, Yongchun Chen, Utilization of steel slag as aggregates for stone mastic asphalt (SMA) mixtures, Building and Environment,Volume 42, Issue 7,2007,Pages 2580-2585,ISSN 0360-1323,https://doi.org/10.1016/j.buildenv.2006.06.008. https://doi.org/10.1016/j.buildenv.2006.06.008

[7] M.W Brocklesby, J.B Davison,The environmental impacts of concrete design, procurement and on-site use in structures,Construction and Building Materials, Volume 14, Issue 4, 2000, Pages 179-188,ISSN 0950-0618, https://doi.org/10.1016/S0950-0618(00)00010-6. https://doi.org/10.1016/S0950-0618(00)00010-6

[8] P. Sharmila, G. Dhinakaran, Compressive strength, porosity and sorptivity of ultra fine slag based high strength concrete, Construction and Building Materials, Volume 120, 2016, Pages 48-53, ISSN 0950-0618, https://doi.org/10.1016/j.conbuildmat.2016.05.090. https://doi.org/10.1016/j.conbuildmat.2016.05.090

[9] Deepankar K. Ashish, Bhupinder Singh and Surender K.Verma, The effect of attack of chloride and sulphate on ground granulated blast furnace slag concrete, Advances in Concrete Construction Volume 4, Number 2, June 2016, pages 107-121, DOI: https://doi.org/10.12989/acc.2016.4.2.107 https://doi.org/10.12989/acc.2016.4.2.107

[10] Ismail, Zainab & Al-Hashmi, Enas. (2007). Reuse of waste iron as a partial replacement of sand in concrete. Waste management (New York, N.Y.). 28. 2048-53. 10.1016/j.wasman.2007.07.009. https://doi.org/10.1016/j.wasman.2007.07.009

[11] Chokkalingam, Ramesh & Rajakannu, Manikandan. (2018). Performance Evaluation of Ordinary Portland Cement with GGBFS and Portland Slag Cement at Same GGBFS

Replacement Level in Concrete. MATEC Web of Conferences. 203. 06004. 10.1051/matecconf/201820306004. https://doi.org/10.1051/matecconf/201820306004

[12] Truc, Nguyen Ngoc. (2018). PARTIAL REPLACEMENT OF NATURAL SAND BY GRANULATED BLAST FURNACE SLAG (GBFS) IN FINE AGGREGATE FOR CONCRETE: PRACTICAL APPLICATION IN VIETNAM. 10.5593/sgem2018/1.2/S02.066. https://doi.org/10.5593/sgem2018/1.2/S02.066

[13] Yadav, Tarun. (2019). Analysis of Effect of Ground Granulated Blast Furnace Slag (GGBFS) on the Mechanical Properties of Concrete using Destructive and Non-destructive Tests. Journal of mechanics of continua and mathematical sciences. 14. 10.26782/jmcms.2019.02.00006. https://doi.org/10.26782/jmcms.2019.02.00006

[14] Tripathy, Sunil & Dasu, Jayalaxmi & Yanamandra, Rama Murthy & Kapure, Gajanan & Pal, Atanu. (2020). Utilisation perspective on water quenched and air-cooled blast furnace slags. Journal of Cleaner Production. 262. 121354. 10.1016/j.jclepro.2020.121354. https://doi.org/10.1016/j.jclepro.2020.121354

[15] J. Emery, Steel Slag Utilization in Asphalt Mixes. National Slag Association, MF 186-1, 1984, www.nationalslagassoc.org

[16] Zumrawi, Magdi. (2015). Experimental Study of Steel Slag Used As Aggregate in Asphalt Mixture. International Journal of Civil, Environmental, Structural, Construction and Architectural Engineering. 9.

[17] Xu, Haiqin & Wu, Shaopeng & Hechuan, Li & Zhao, Yuechao & Lv, Yang. (2020). Study on Recycling of Steel Slags Used as Coarse and Fine Aggregates in Induction Healing Asphalt Concretes. Materials. 13. 889. 10.3390/ma13040889 https://doi.org/10.3390/ma13040889

Experimental Analysis on Utilizing Sugarcane Bagasse Ash as a Replacement for Cement

R. Hemavathi[1,a,*], A. Vijay Vignesh[1,b], R.V.K. Vigneshwar[2,c]

[1]Department of Civil Engineering, Sri Ramakrishna Institute of Technology, Coimbatore, India

[2]Department of Civil Engineering, Coimbatore Institute of Technology, Coimbatore, India

[a]hemavathi859@gmail.com, [b]avijayvignesh@gmail.com, [c]vigneshwar.2902@gmail.com

Keywords: Sugarcane Bagasse Ash, Cement Kiln Dust, Consistency Test, Compressive Strength Test

Abstract. Construction industry is estimated to consume nearly 30-35% of natural resources and associated to major global pollutions. It is essential to move our path into a sustainable way of construction practices, to conserve natural resources for the future generation. Several researchers had investigated the various possibilities of alternatives to produce sustainable green building materials from industrial by-products, for example, blast furnace slag, fly ash and silica (SiO_2). Presently, there has been an attempt to use the large amount of Sugarcane Bagasse Ash (SBA) which is formed when using the bagasse-biomass fuel in electric generation industry. Due to availability of Cement kiln dust (CKD) and Sugarcane Bagasse Ash (SBA) in large amounts, it appears to be cost effective and eco-friendly by the applications for these industrial by-products. The overall goal of this research study was to replace cement by certain percentage through potential applications for the Cement kiln dust (CKD) and Sugarcane Bagasse Ash (SBA) based on a detailed characterization of the materials. The analyses are discussed in terms of Setting time, Consistency test and Compressive Strength. As a result, Sugarcane Bagasse Ash (SBA) for a certain percentage replaced in cement increases the strength of mortar cube but if it exceeds 20%, then the strength starts to decrease.

Introduction

Concrete is an essential product since it is a vital part in development of construction industry. It is one of the most broadly utilized materials on the earth. Roughly 3.6 billion metric tonnes [10] of concrete are manufactured globally each year, with volume anticipated to ascend to in excess of 5 billion metric tons by 2030. The business is filling especially quickly in agricultural nations having an appeal for framework and lodging. Concrete [9] is made by mixing clinker with gypsum under controlled environment. The concrete creation process requires nuclear power for delivering clinker in revolving furnaces. Each huge load of concrete deliveries 0.9 to 1.1 ton of carbon dioxide (CO_2), a significant ozone harming substance, into the air, which unfavorably influences earth's environment. Then again, huge amounts of residue, generally known as cement kiln dust (CKD), is formed during the manufacture of Portland concrete clinker by the dry cycle. Cement kiln dust [4], by nature is a fine-grained, strong, profoundly soluble material, the particles width range between few μm and 50 μm that eliminated from concrete furnace fumes gas via air pollution control equipments. A portion of the produced CKD could be reused into concrete furnace as crude feed. However, this is restricted by the alkali's concentration in CKD, which cause the salts content in concrete surpass as far as possible (i.e., not over 0.6%) [8] as well as diminishing the efficiency of the kiln and creating equipment malfunction.

Research Significance

When Cement Kiln Dust [4] and Sugarcane Bagasse Ash [7][2] are used as the engineering materials, it will reduce the generation of waste which will create a better environment. Due to its vast range of applications, it can be utilized as a pozzolanic material in the construction industry.

Literature review

Waqas ahmad et. al (2021) [7] explored the effect of sugarcane bagasse ash in cement-based composites. In his research, it is found out that SBA contains amorphous silica which makes it a good pozzolanic material. Also, SBA needs less amount of water to achieve its consistency. The strength of the cement-based composites increases when added upto optimum amount (10%), while further addition has reduced the strength of the composites. Banger et al. (2017) [2] utilized bagasse ash for the research work in crushed and sieved through sieve of size 150 micron and passing out portion is utilized in concrete as a partial substitution of concrete in the proportion of 2% 4%, 6%, 8% and 10% by weight of the cement. The test results shows that the strength of concrete is increased when adding Sugarcane Bagasse Ash (SBA) at greater percentage. Sadek et al. (2017) [6] incorporated cement kiln dust in paving blocks and results shows that upto 10% of CKD can be used as a partial replacement of cement. On replacing upto 40% CKD in paving blocks, it reduces the quality of the blocks considerably.

Abdulkadir et al. (2014) [1] evaluates the suitability of Sugarcane Bagasse Ash as a partial replacement of cement in concrete production. They obtained the SBA and burnt at 700° C and to attain the standard size of particles, the SBA is made to pass through the 45μm sieve. It was then used to replace OPC by weight in ratio of 0%, 10%, 20% and 30% respectively. The result shows that there is decrease in the concrete density with increase in percentage replacement of SBA. Kunal et al. (2014) [4] analyzed the effect of bacterial treated cement kiln dust mortar. It is found out that the strength of the mortar increase when adding CKD upto 10% but further addition reduces the strength due to lower cement content.

From the study of Relan & Saxena (2015) [5], it is concluded that there is an increase in workability whenever adding an appropriate percentage of SBA in concrete. It is presumed that finely grounded SBA can be effectively utilized in concrete as a replacement of cement and is answerable for higher compressive qualities than ordinary Portland cement (keeping amount of cement as constant). The results shows that the cement could be beneficially replaced with SBA up to 12.5 % for M35 concrete. The review also uncovers that the compressive strength increased up to 10 % replacement and when its beyond 15 % replacement, the strength was found to be decreasing.

We have gone through several literatures and journals related to our research. Those literatures helped us to study about the previous research studies about SBA and CKD. It helped us to understand the various studies for utilizing this waste material in our research.

Methodology

This part includes the procurement and grinding of materials, study of properties of both CKD and SBA, Casting and testing of mortar cube. The SBA was freshly burnt and collected from the sugar production industry in Namakkal. Ball Mill is a type of grinder used to grind and blend materials. It works on the principle of impact and attrition. It is used to reduce the size of the materials to finer particles through impact process. We used this equipment to make the CKD and SBA to finer particles. The grinded materials are collected in larger quantities and those materials are sieved through 150 micron and 75-micron sieves to make it finer to equalize the fineness of the cement. The casting and testing [3] of mortar cubes for various mix proportions (60:40, 70:30, 85:15) %

Materials Research Proceedings **23** (2022) 419-424 https://doi.org/10.21741/9781644901953-45

are carried out in a controlled environment. The results from various tests for different mix proportions are determined and analyzed.

Laboratory Investigation

Preliminary actions carried out, initially, the procured materials are completely dried under the open sunlight to remove the water content from the collected materials. After drying of the procured materials, the grain sizes of the materials are reduced by grinding the materials in the ball mill to make it finer to equalize its fineness to the cement particles [3]. Then the grinded materials are sieved using the 75μm sieve. The laboratory investigation includes consistency test [3], compressive strength on 7th day for cement kiln dust (CKD) and Sugarcane Bagasse Ash (SBA) on various mix proportion.

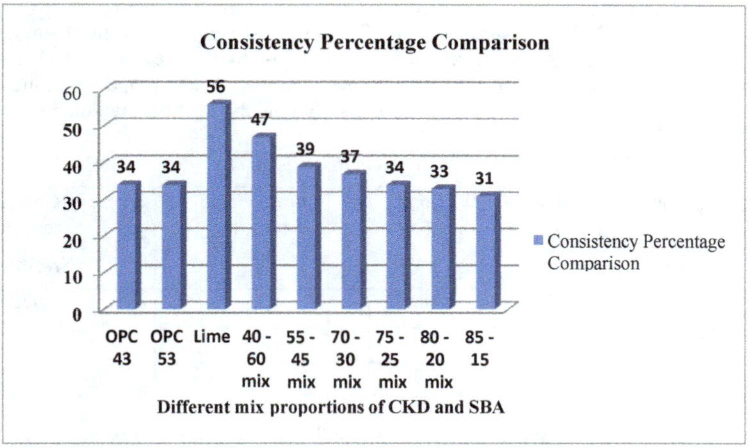

Fig 1. *Comparison of Consistency test for OPC 43, OPC 53 grade of cement, lime and different mix proportions of CKD and SBA.*

The above chart (refer fig.1) shows the comparison of consistency value of OPC 43, OPC 53 grade of cement, lime and different mix proportions of CKD and SBA. It shows that the lime is having higher consistency percentage than others due to higher amount of heat of hydration process. The two grade of cement gives the standard consistency value. In the different mix proportions of CKD and SBA the consistency percentage gets reduced along with increase in the CKD content. The silica in the CKD reacts well and forms paste by consuming lesser amount of water itself.

Table 1. 28th day compressive strength of 85:15 mix of CKD and SBA

Sample	Dial Gauge Reading (kN)	Load (N)	Compressive strength (N/mm²)	Average Compressive Strength (N/mm²)
1	6.5	6500	1.33	1.54
2	8	8000	1.63	
3	8.2	8200	1.67	

There is minimum strength is increased on 28th day compressive strength comparing (refer Table 1) with 7th and 21st day results. But still this strength is not sufficient for the mortar cube. The strength is very less even compared with the third-class brick. The pozzolonic properties in this new blended mix of CKD and SBA are very less and the bonding properties reduced due to this insufficient binding paste. The particle size distribution also will be a reason behind the low strength.

Table 2. 28th day compressive strength of 70:30 mix of CKD and SBA

Sample	Dial Gauge Reading (kN)	Load (N)	Compressive Strength (N/mm²)	Average Compressive Strength (N/mm²)
1	4.8	4800	0.97	1.07
2	5	5000	1.02	
3	6	6000	1.22	

There is no change in the compressive strength even the mix proportion are changed. Comparing with 85:15% mix proportion (refer Table 2), the compressive strength of 70:30 % mix gets reduced due to reduction in the silica content by reducing the CKD percentage.

From the compressive strength value (refer Table 3), it shows that there is decrease in the strength due to reduction in the CKD % in the mix proportion. It is because of reduction of silica content which is insufficient to form the C-S-H gel in the mortar cube and the pores inside the mortar cube also plays vital role in the reduction of the compressive strength.

Table 3. 28th day compressive strength of 60:40 mix of CKD and SBA

Sample	Dial Gauge Reading (kN)	Load (N)	Compressive Strength (N/mm²)	Average Compressive Strength (N/mm²)
1	4	4000	0.81	1.08
2	6	6000	1.22	
3	6	6000	1.22	

Materials Research Forum LLC
https://doi.org/10.21741/9781644901953-45

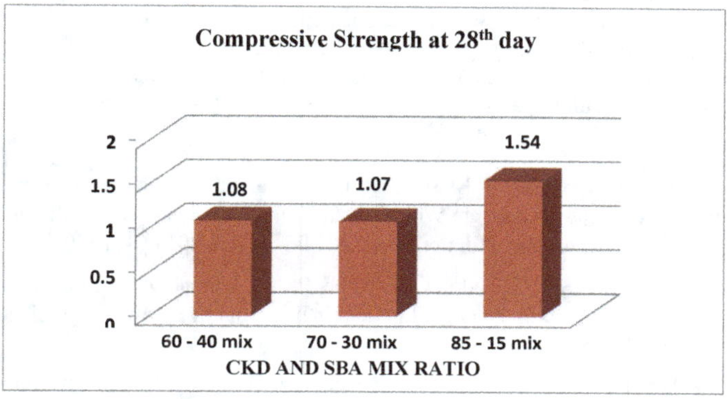

***Fig 2.** Comparison Chart of Compressive Strength for Different Mix Proportions of CKD and SBA*

The compressive strength of 85:15% mix ratio of CKD and SBA at 7th, 21st and 28th day are 1.26, 1.44 and 1.54 N/mm^2. The 28th day compressive strength (refer fig.2) of 70:30 and 60:40% mix proportion of CKD and SBA are 1.07 and 1.08 N/mm^2. It shows that all the compressive strength values are very less compared to cement mortar strength. Even the compressive strength is lesser than the third-class brick. The strength gets reduced because of large number of pores distribution, insufficient pozzolonic reaction, low bonding, and reduction in C-S-H gel formation and may be of increased carbon content in SBA.

Results and Discussion

Comparing the consistency values of OPC 43, OPC 53 grade of cement, Lime and different mix proportions of CKD and SBA (85:15, 70:30, 60:40) the OPC 43 and 53 grade of cement attained their standard consistency value. The fineness of cement results that the permissible percentage of particle size above 90 micron is only 10%. So, the mix proportion of CKD and SBA sieved using 75-micron sieve to make it more finer particles. Then consistency test is carried out for the mix proportion of (60:40, 70:30, 85:15) % of CKD and SBA sieved through the 75 microns. The results of the consistency value lie between the standard consistency value (47%, 34% and 31% respectively) and shows that there is decrease in the consistency value by increase in the CKD percentage in mix proportion of the blended cement of CKD and SBA.

By comparing the results consistency obtained the water cement ratio increases with the decrease in the cement kiln dust (CKD).The strength is not attained up to the expected level. The compressive strength (refer fig.2) at 28th day of (85:15), (70:30), (60:40) % mix of CKD and SBA are 1.54, 1.07 and 1.08 N/mm^2 respectively. The reason for low strength is the formation of C-S-H Gel is very less compared to cement. And there is more porous formation while using this blended cement. It was likewise presumed that not just the pores size is deciding the mortar strength yet in addition its dissemination through fractal aspect D, where the compressive strength diminished with D value increased. Likewise, incorporation of SBA for a certain percentage replaced in cement increases the strength of mortar cube but if it exceeds 20 % the strength starts to decrease. In any case, research should be carried to analyze the results of this study and furthermore evaluate the durability characteristics of the produced mortars.

Using cement kiln dust and sugarcane bagasse ash as a replacement of cement is effective utilization of waste generated in cement and sugar industry. Also, it reduces the harmful environmental and health impacts thus creating an eco-friendly environment.

References

[1] Abdulkadir, T., Oyejobi, D., & Lawal, A. (2014). Evaluation of Sugarcane Bagasse Ash As a Replacement for Cement in Concrete Works. Acta Tehnica Corviniensis - Bulletin of Engineering, 3, 71-76. http://acta.fih.upt.ro/pdf/2014-3/ACTA-2014-3-11.pdf

[2] Banger, S. ., Phalke, S. ., Gawade, A. ., Tambe, R. ., & Rahane, A. . (2017). A review paper on replacement of cement with bagasse. International Journal of Engineering Sciences & Management, 7 (March), 127-131.

[3] IS 4031- Part (1-6). (1996). Method of physical tests for hydraulic cement. Bureau of Indian Standards, New Delhi, Reaffirmed in 2005.

[4] Kunal, Siddique, R., & Rajor, A. (2014). Strength and microstructure analysis of bacterial treated cement kiln dust mortar. Construction and Building Materials, 63, 49-55. https://doi.org/10.1016/j.conbuildmat.2014.04.011 https://doi.org/10.1016/j.conbuildmat.2014.04.011

[5] Relan, N., & Saxena, A. K. (2015). Experimental Study of Replacement of Cement by SCBA in Concrete. International Journal of Science and Research (IJSR), 4(12), 15-18. https://doi.org/10.21275/v4i12.nov151840 https://doi.org/10.21275/v4i12.NOV151840

[6] Sadek, D. M., El-Attar, M. M., & Ali, A. M. (2017). Physico-mechanical and durability characteristics of concrete paving blocks incorporating cement kiln dust. Construction and Building Materials, 157, 300-312. https://doi.org/10.1016/j.conbuildmat.2017.09.107 https://doi.org/10.1016/j.conbuildmat.2017.09.107

[7] Waqas ahmad, Ayaz ahmad, Panuwat Joyklad, Fahid Aslam. (2021)Sustainable approach of using sugarcane bagasse ash in cement-based composites: A systematic review. Case Studies in Construction Materials, Volume 15. https://doi.org/10.1016/j.cscm.2021.e00698 https://doi.org/10.1016/j.cscm.2021.e00698

[8] M. Maslehuddin , O.S.B. Al-Amoudi, M.K. Rahman, M.R. Ali, M.S. Barry Properties of cement kiln dust concrete. Construction and Building Materials 23 (2009) 2357-2361 https://doi:10.1016/j.conbuildmat.2008.11.002 https://doi.org/10.1016/j.conbuildmat.2008.11.002

[9] Bhatty MSY. Use of cement kiln dust in blended cements - alkali-aggregate reaction expansion. World Cem 1985;16(10):386-91.

[10] Oss HG. Minerals information - cement. Retrieved January 2010, from United States Geological Survey; 2010. http://minerals.usgs.gov/minerals/pubs/commodity/cement/

Sustainable Materials and Smart Practices - NCSMSP-2021 Materials Research Forum LLC
Materials Research Proceedings **23** (2022) 425-431 https://doi.org/10.21741/9781644901953-46

Empirical Modeling on Swell Pressure of Clay using Index Properties

V. Jeevanantham[1,a,*], J. Jayashree[2,b], S. Loganayagan[1,c]

[1] Department of Civil Engineering, Bannari Amman Institute of Technology, Sathyamangalam, INDIA

[2] Department of Civil Engineering, Sri Ramakrishna Institute of Technology, Coimbatore, INDIA

[*a]jeeva.civil.tpgit@gmail.com, [b]jjayashree16@gmail.com, [c]loganayaganme@gmail.com

Keywords: Expansive Soil, Montmorillonite, Swell Pressure, Plasticity Index, Matlab

Abstract. Expansive soils undergo volume expansion, when it gains moisture content. Light structures constructed on this type of soil will be lifted by the upward swell pressure. Swelling characteristics decides the degree of safety of structures resting on expansive soil strata. Predicting the swell pressure of the soil consumes nearly 5 days of time (variable with respect to soil potential) in the laboratory as well as needs expensive testing setup. In our study, a correlation is proposed to develop for swell pressure using the index properties of soils namely liquid limit and plastic limit, which shall be assessed at the laboratory relatively short period of time. Swelling Pressure tests by Free Swell Method are performed on dynamically compacted 20 remolded soil samples collected within Coimbatore Corporation limit. The study area is between the four coordinates of 11^{0}08'49.25" N 76^{0}53'36.28" E, 11^{0}12'05.58'' N76^{0}55'57.84" E, 10^{0}59'16.52" N 76^{0}52'17.47" E, 10^{0}57'00.59" N 76^{0}57'43.71" E. Laboratory experimental data given as input in MATLAB gives satisfactory results and correlation is extracted from curve fitting method.

Introduction

Building foundations bears the entire super structure load. The soils beneath the foundations hold play a crucial part in holding the foundation. But Expansive soils, such as montmorillonite deforms more provides lesser resistance against the foundation load. One-fifth of the Indian land is covered by these expansive soils, produces difficulty to building & Road construction activities.

Black cotton soil, which is rich in montmorillonite clayey mineral and exhibits higher degree of shrinkage, swelling and compressibility nature under loading due to its chemical composition. Higher the montmorillonite presence, it becomes black or blackish grey in colour. Montmorillonite has the structural composition of 1 octahedral alumina sheet and 2 tetrahedral silica sheets in the structural term defined as 2:1 composition. Due to the Isomorphous substitution they are highly adsorbs the water and forms diffusible double layer around the surface. It becomes more expansive in nature with increase in moisture content. This water imbibing phenomena is occurring in higher degree, comparatively with kaolinite and bentonite clay mineral.

Coimbatore, Tamilnadu, India is highly covered by the clay deposits, especially montmorillonite mineral. As observed, many of the light structures in Coimbatore lost its stability due to this swelling nature. Hence it is taken as study area.

Sample Collection

The sample used for the conducting various experiments in our project were collected at a depth more than 1.5 meter from the Existing ground level.

The representative soil samples were collected from 20 locations within the Coimbatore city. The study area is between the four coordinates of A (11^{0}08'49.25" N 76^{0}53'36.28" E), B(11^{0}12'05.58" N 76^{0}55'57.84" E), C(10^{0}59'16.52" N 76^{0}52'17.47" E), D(10^{0}57'00.59" N 76^{0}57'43.71" E). The locations of twenty soil samples are shown in Figure1

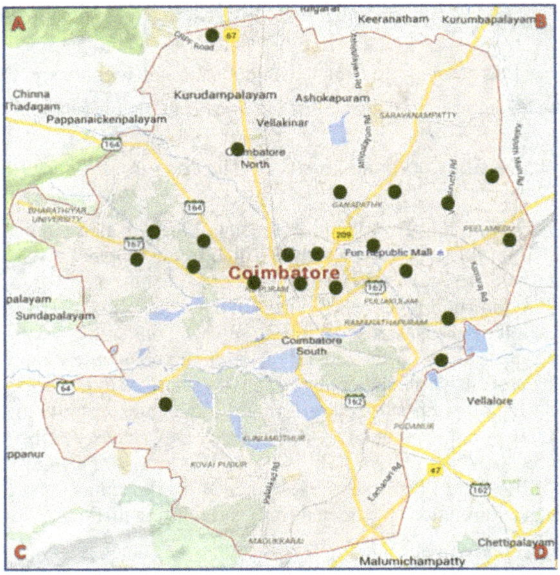

Fig.1 Locations of sample collection within Coimbatore

The locations of twenty soil samples are collected within the Coimbatore limit are shown in Table 1.

Table 1- Locations of sample collection

Sample Identification	Location	Latitude	Longitude
S1	GCT campus	11^{0}01'20.20" N	76^{0}56'09.87" E
S2	SOWRIPALAYAM	11^{0}00'42.42" N	77^{0}00'32.27" E
S3	SINGANALLUR	11^{0}00'06.55" N	77^{0}01'30.03" E
S4	KARUPPARAYAN KOIL	11^{0}01'26.19" N	76^{0}54'56.60" E
S5	CHERAN MANAGAR	11^{0}03'07.34" N	77^{0}01'10.53" E
S6	KALAPATTI ROAD	11^{0}02'29.74" N	77^{0}02'13.25" E
S7	AGRI UNIVERSITY	11^{0}01'04.82" N	76^{0}56'17.04" E

S8	VADA KOVAI	$11^0$01'05.64" N	$76^0$57'15.24" E
S9	RATHINAPURI	$11^0$01'37.78" N	$76^0$58'05.49" E
S10	MULLAI NAGAR	$11^0$01'39.28" N	$76^0$55'03.59" E
S11	GANDHI PARK	$10^0$59'59.21" N	$76^0$56'55.98" E
S12	NARASIMANAICKENPALAYAM	$11^0$07'55.34" N	$76^0$56'05.91" E
S13	KRISHNARAYAPURAM	$11^0$01'55.26" N	$76^0$59'31.55" E
S14	SAIBABA COLONY	$11^0$01'19.84" N	$76^0$56'36.41" E
S15	THUDIYALUR	$11^0$04'57.54" N	$76^0$56'27.75" E
S16	NEELAMBUR	$11^0$04'26.59" N	$77^0$05'05.44" E
S17	SIVANANTHA COLONY	$11^0$01'38.54" N	$76^0$57'18.09" E
S18	PEELAMEDU	$11^0$01'59.00" N	$77^0$01'38.31" E
S19	VELLALORE	$10^0$56'36.09" N	$76^0$55'59.51" E
S20	PERUR	$10^0$56'00.62" N	$76^0$58'52.58" E

Literature review

Various statistical/empirical methods have been attempted to predict the swell pressure based on index properties of soil [1-9]. A review of previous investigations is presented is paramount to appraise the state of art and to get idea of the areas that are already covered by investigators. It also helps to understand various correlations between index properties and engineering properties. The literature review is presented author wise and summary of the review given below as a Table.

Table 2- List of Correlations

AUTHOR	EQUATIONS
Seed et al	$P_s = 21.6 \times 10^{-5} (I_P)^{2.44}$ $P_s = 3.6 \times 10^{-5} (A_c)^{2.44} x (C)^{3.44.}$
Ranganatham and Satyanarayana	$P_s = 21.6 \times 10^{-5} (I_P)^{2.67}$
Komornik and David	For marl, $P_s = 6.7+2.4(I_P)$ For clay, $P_s = 0.9+2.1(I_P)$
Nayak and Christensen	$P_s = 3.6 \times 10^{-2} (I_P)^{1.12} [\frac{C}{Wn}]^2+3.8$
Vijayvergiya and Ghazzaly	$\log P_s = 0.033 (W_L) - 0.083 (W_n) + 0.458$
Guiras-Skandaji	$P_s = -117.59 + 3.0501(W_i)$

Where,

P_s – Swell Pressure
I_P – Plasticity index
A_c – Activity of Clay

Materials Research Proceedings **23** (2022) 425-431 https://doi.org/10.21741/9781644901953-46

C – Clay Fraction
W_L – Liquid Limit
W_n and W_i – Natural water content

Laboratory Investigation
Preliminary tests are conducted as per the IS 2720 code covering various parts on the representative clay soil in order to determine the properties of the soil. The experimental laboratory tests results mainly includes Atterberg's limits and Swell pressure test, which are going to be the Base data for this study. In addition to that, Sieve analysis is conducted to know about the proportion. Standard Proctor test is conducted to get the optimum moisture content to prepare the sample for soaking.

Swell pressure test
Swell pressure, defined as the pressure which is required to return a swelled specimen back to its original state prior to swelling. Swell pressure test is obtained by conducting Free Swell method as per IS 2720 (Part41) – 1997.

The remolded representative soil samples for swell pressure test were prepared under optimum moisture content to achieve maximum dry density. The sample preparation for swell pressure test and swell pressure apparatus is shown in figures 2 and 3. The results obtained from swell pressure test for 20 samples with minimum test time of 4 days each are shown in table 3.

Fig.2 Sample preparation

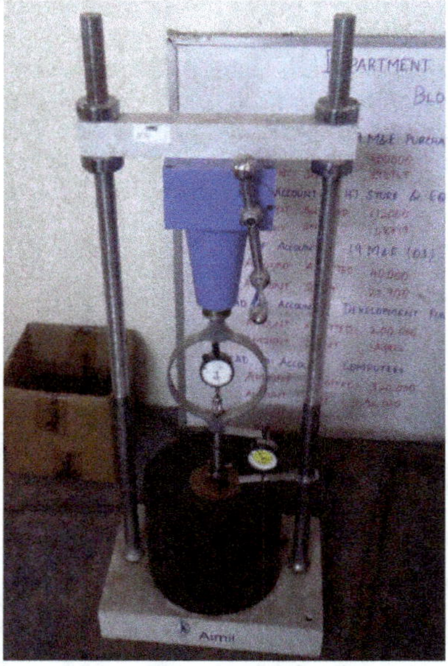

Fig.3 Loading Setup

Table 3- Consolidated results

Sample No.	Soil Type	Liquid Limit, W_L (%)	Plastic Limit, W_P (%)	Plasticity Index, I_P(%)	OMC (%)	MDD (g/cc)	Swell Pressure, P_S (kPa)
S1	CH	57.5	16.67	40.83	20.1	1.67	137
S2	CH	55.2	22.22	32.98	23.4	1.526	94.7
S3	CH	63	22.5	40.5	25.1	1.727	131.2
S4	CH	58.4	11.22	47.18	17.45	1.725	145.7
S5	CI	49	18	31	23.5	1.634	99.5
S6	CH	60.3	15	45.3	23.8	1.52	143.6
S7	CH	65.75	22.48	43.27	21.8	1.59	139.3
S8	CI	54.08	21.05	33.03	25.1	1.47	91.5
S9	CL	34	11.3	22.7	15.8	1.36	71.7
S10	CI	48	14.7	33.3	22.5	1.6	96.7
S11	CH	65.3	12.9	52.4	28	1.487	175.3
S12	CH	68.7	15.3	53.4	23.2	1.49	167
S13	CH	60.12	23.09	37.03	27.1	1.43	115.6
S14	CH	70.1	20	50.1	22.2	1.49	162
S15	CH	63.3	24.1	39.2	30	1.58	97.65
S16	CH	61.2	22.5	38.7	24	1.45	101.45
S17	CH	58.2	19.4	38.8	16.2	1.7	125.4
S18	CH	64.4	22.6	41.8	23	1.542	137.4
S19	CH	66.4	24.3	42.1	25.1	1.595	138.5
S20	CH	68.6	19	49.6	23.5	1.62	162.7

Where,

CL – Low Compressible Clay
CI – Intermediate Compressible Clay
CH – Highly Compressible Clay
OMC – Optimum Moisture Content
MDD – Maximum Dry Density

Plasticity Index (I_P) = Liquid Limit – Plastic Limit

MATLAB - Curve Fitting Tool

Experimental results are need to be given as input parameters and plotted against X and Y axis as per the requirement. Plot will show the special distribution of data given. Under the curve fitting tool, Polynomial order shall be selected to connect the points distributed in the plot. On selecting the different orders, Goodness of fit will be analyzed and displayed in terms of R-Square, Adjusted R-Square and RMSE. The R Square is a measure that is used to determine the relative correlation and the goodness of fit.

R Square >0.8 Strong correlation exists between two sets of variables;
0.2 < *R Square* < 0.8 Correlation exists between the two sets of variables; and
R Square <0.2 Weak correlation exists between the two sets of variables.

Figure 4 shows the input data given in the command window for curve fitting method, where Plasticity index is loaded as X data and Swell pressure is loaded as Y data.

```
clc;
close all;
clear all;
x = [40.83,32.98,40.5,47.18,31,45.3,43.27,33.03,22.7,33.3,52.4,53.4,37.03,50.1,39.2,38.7,38.8,
y = [137,94.7,131.2,145.7,99.5,143.6,139.3,91.5,71.7,96.7,175.3,167,115.6,162,97.65,101.45,125.
ft = fittype ('poly1');
[xdata, ydata] = prepareCurveData(x,y);
c = fit(xdata,ydata,ft);
```

Fig. 4 Data Input (Command Window of MATLAB)

Figure 5 shows the data plotted under curve fitting tool, where polynomial order is given as one.

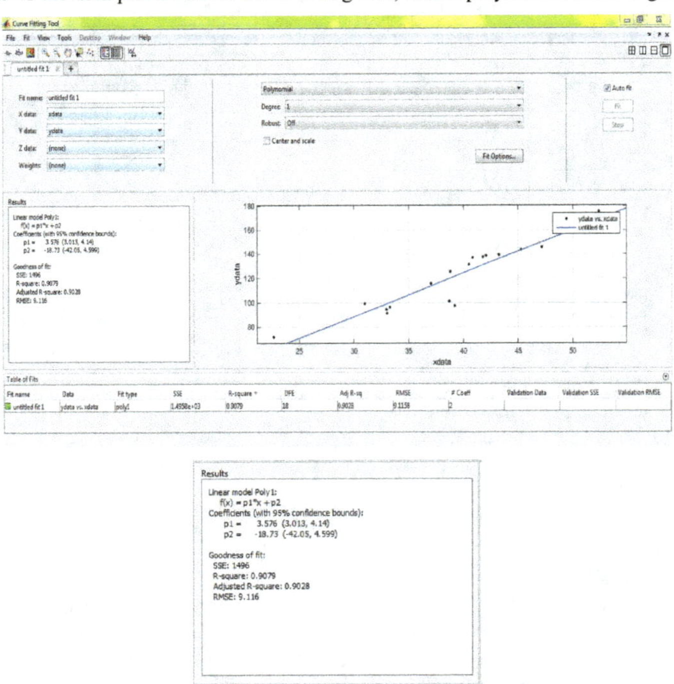

Fig.5 Curve Fitting Window

Results shows that strong relation existing between the plasticity index and Swell Pressure Analysis gives the R Square value of 0.90. Coefficients are with 95 % confidence bounds for this linear curve developed. The correlation given by the Matlab as linear line is shown below.

$$\mathbf{P_S = 3.576\ I_P - 18.73}$$

Where,

P_S -Swell pressure in kN/m^2
I_P -Plasticity Index in %

Conclusion

Plasticity Index can be calculated from quick results of Liquid limit and Plastic limit. As plasticity index is highly influencing the swell pressure, it is selected as the variable for the Matlab modeling. Despite, Amount of Clay, Activity number of clay and type of clay mineral will also influence the Swell pressure, but they are not given as input parameter for this study, since hydrometer analysis involves tedious testing procedures and calculations.

Under the Matlab Modeling, Polynomial order 1 is selected for this analysis.

When the polynomial order is increased, R Square value becomes stronger as observed. The correlation given by the Matlab is shown below.

$$\mathbf{P_S = 3.576\, I_P - 18.73}$$

Where,

P_S Swell pressure in kN/m^2
I_P Plasticity Index (%)

References

[1] Yusuf Erzin , Orhan Erol (Turkey), Correlations for quick prediction of swell pressures, Electronic journal of geotechnical engineering, 14 (2004) 78-87.

[2] Iman Ashayeri - shahabodin Yasrebi, Free swell and swell pressure of unsaturated compacted clays; experiments and neural network modeling, Journal of Geotech Geol Eng. 27 (2009) 137-153 https://doi.org/10.1007/s10706-008-9219-y

[3] Dr. Ch. Sudha Rani, Prediction of swell pressure of expansive soils using compositional and environmental factor, Int. Journal of Civil Engineering and Technology , 4 (2013) 134-142.

[4] V. NAYAK and R. W. CHRISTENSEN, Swelling characteristics of compacted, expansive soils, Journal of Clays and Clay Minerals, 19 (1971) 251-261 https://doi.org/10.1346/CCMN.1971.0190406

[5] Abdel kader Djedid, Naouel Ouadah, Indirect estimation of swelling clay soils parameters, Electronic journal of geotechnical engineering, 18 (2013).

[6] Yusuf Erzin and Nazmi Güneş, The prediction of swell percent and swell pressure by using neural networks, Journal of Mathematical and Computational Applications, 16 (2011) 425-436. https://doi.org/10.3390/mca16020425

[7] Nihat SinanI ýk, Estimation of swell index of fine grained soils using regression equations and artificial neural networks, Scientific Research and Essay 4 (2009) 1047-1056.

[8] Berawala, K. S. Solanki, C. H, Empirical Correlations of Expansive Soils Parameters for the Surat Region, Electronic journal of geotechnical engineering 17 (2012).

[9] Vikas K. S. Thakur and Devendra N. Singh, Rapid Determination of Swell pressure of Clay Minerals, Journal of Testing and Evaluation, 33 (2005). https://doi.org/10.1520/JTE11866

[10] Indian standard code of practice IS 2720 (Part1 to Part41) & IS 1498-1970

Performance Study on Partial Replacement of Bitumen using Waste Polypropylene Cement Bags for Pavement Construction

R. Malathi[1,a*], R. Geethamani[2,b]

[1]Department of Civil Engineering, Sri Ramakrishna, Institute of Technology, Coimbatore, Tamilnadu, India

[2]Department of Civil Engineering, Bannari Amman Institute of Technology, Alathukombai, Sathyamangalam, Tamilnadu, India

[a]malub18@gmail.com, [b]geethamanir@bitsathy.ac.in

Keywords: Waste Polypropylene Bag, Bitumen, Pavement Construction, Conventional Materials

Abstract. Flexible pavement is most commonly used in the bituminous material mix. It consists of different binders such as asphalt or bitumen and other minerals are added and compacted and used together in the lower layer of the pavement. Bituminous pavement surface property changes due to varying weather conditions. In hot climatic conditions, the bitumen becomes soft and during cold conditions it becomes too stiff and brittle in nature. In this study, the main concept is to reuse the waste materials and to promote cost-effective solution and reduce environmental pollution. In this project, waste cement bags (polypropylene) are used to replace a portion of bitumen which is used in road construction. Effect of addition of additive materials on the strength of the pavement was evaluated. Previous studies state that polypropylene emits 60ppm less pollution than bitumen while burning. The used cement waste bag when mixed with bitumen improves the behaviour and mechanical properties as per standard specification. In this experimental work, cement bags are shredded into smaller fragments and heated to 90°C for 30 minutes in a heating mantle till the plastic pieces turned into powder form. On heating, 5 gm of powder was obtained from 20 gm of the used cement bags. This powder was added subsequently to bitumen in 5%, 7%, 9% & 11% for partial replacement of bitumen. Finally, it is inferred that the optimum percentagewas 8.2% without the addition of binders or oxidizing agents, for better efficiency.

Introduction

Our country has initiated improved road development depending on the connectivity of various places with an adequate road network. Roadways are more flexible for travel with the choice of time-saving. The movement passengers are easy to select the major channel of transportation. They play a significant role in improving the socio-economic and living conditions of the peoples as well as goods. Roads are planned and designed to provide a good road network connecting all interconnecting road lines. Additional significance that was achieved through road development is increase in employability to a highly potential level. The government sector invested funds to upgrade the existing road by adopting excellent maintenance. Out of the total length of highways available 98% are of flexible type. A lot of researches experimented utilizing waste materials in road construction, but not experimented in the field. Hence the lab scale study has to be transformed to field model, to promote sustainable development.

Plastic is one of the non-biodegradable materials. Plastics now form part of our day today life. The population is changing and drastically increasing every year and also development activities are improved day by day to meet the current standards. As a result, the production of various plastic waste materials has seen a tremendous increase every day throughout the year all over the world. Increasing the usage of non-biodegradable materials which are not easy to handle and rate of decomposition is too long [8]. Lot of problems are created due to improper disposal of wastes. Though this waste plastic material is useful, improper disposal can create lot of problems. There is no proper method for the disposal of plastics; therefore proper recycling methods should be adopted and encouraged to reduce the plastic waste disposal. Proper stepsare to be initiated to apply in the research area with respect to the addition of such wastes materials to mix with bitumen to improve their quality, increasing life and reducing the cost of road construction work [9].The use of polymeric waste in bitumen to improve the properties of binders proves to be very promising. Polypropylene is an excellent material for partial replacement of bitumen. Polypropylene waste originally prepared from Low-Density Polypropylene (LDPP) has been identified to be studied in modified bitumen [10]. When polypropylene is mixed with hot bitumen, it forms an oily surface on the aggregates and the mixture is laid on the road surface just like a normal tar road.

The main objectives of this study area. To analyze the important properties of polypropylene, bitumen and aggregates, b. To prepare liquid and powdered polypropylene materials and compare their characteristics. c. To conduct laboratory tests on molten and powdered polypropylene and compare their properties with bitumen. d. To evaluate the optimum polypropylene percentage, to reduce the quantity of bitumen and thereby reduce the cost of road construction. e. To collect, clean, shred, reprocess by melting and blending with bitumen and test it for efficiency.

Review of Literature

Waste Plastic

Waste Plastic is used in partial replacement of bitumen in the form of aggregate or binder. The stability of pavement construction was developed using the addition of waste plastic materials. It enhanced the strength and life of the structures. Similarly, if a bitumen concrete blend is mixed with a waste of polypropylene materials it gives a better result when compared to plain bitumen. At the same time, the waste reduction also to be accounted for reduction in pollution. Utilization of waste plastic used in the road construction provides better resistance of water and in turn reduces the failure of bitumen segregation from pavement structures [8]. Newly the road network patterns were introduced in many places for the integrated development of Rural and Arterial road network for socio-economic growth. In the current scenario, reuse or recycling of waste materials is growing among the public. Utilization of polythene or polypropylene bags for replacement in bitumen blended work is promoted, since good results are obtained with replacement. Marshall Stability tests were found to be the formation of deformation variations to study the failure occurring in the road pavement due to poor material mixing and binder occurs. This in turn decreases the possibilities of potholes formation in the pavement structure. The result shows that the use of a higher percentage of plastic waste to reduce the quantity level of bitumen is 10% [9]. Studies were made to analyse the strength performance by using replacement materials with different mixing percentage.Plastic replaced by tar road showed good performance for heavy traffic and showed better binding results in various climatic conditions [10]. Asphalt type roads were also used in the waste plastic materials by both wet and dry methods. Bitumen replacement percentage varied between 5-10% by weight of bitumen. In addition waste plastic in the modified bitumen methods, improved the endurance limit of pavement performance with marginal saving in bitumen usage [11]. Different types of plastic materials also were studied in research work namely High-density polyethylene (HDPE), Low-Density Polypropylene (LDPP) and

Polypropylene (PP). The above specified materials were added in a conventional type of AC-20 graded bitumen and other grades. Generally, the temperature variation ranged between 160°C–170°C. Because of that when temperature increased after setting the road pavement is hardened or shrank and blended with the bitumen 'in-situ', with a shear mixer [12].

Road Construction Materials:
The research methodology studied has to adopt various tests to investigate results on mixing of aggregate, bitumen and plastic. The test result was done in water absorption, aggregate impact, loss Angeles and aggregate crushing test [IS: 2386 (part 4)-1963] for aggregates and softening point, penetration test and ductility test [IS: 1203-1978] for bitumen. Alternative ways of using waste plastic materials are mixing with the aggregates and a small proportionate ratio blended with polymer in the bitumen. Prepared mixing materials are later tested in the laboratory and the results are obtained. To determine the physical properties and flow characteristics of bitumen, tests such as penetration ring, ball softening point and viscosity tests were performed. When compared to bitumen mix with HDPE PMBs up to 3% concentration of polymer in bitumen softening temperature point has been increased gradually. Likewise, the viscosity test showed that the results were increment in concentration of polymer with increase in the viscosity variation [13]. Proposed durability of the roads laid with shredded plastic waste is much more compared asphalted with the conventional mix proportion. While considering a National highway road in the past 4 to 5 years it's claimed that plastic-bitumen roads were used in replacement of conventional roads in the last 10 years. It shows that waste plastics such as high-density polyethylene are made in the form of powder; 3 to 4 % is used to increase the durability of the road. In addition hot bitumen was mixed with varying proportions of 0 to 12%. It was inferred that the values of penetration and ductility have decreased with an increase in the proportion of plastic additives upto 12 % by weight. The softening point was increased in the proportion of plastic additives upto 8 % by weight [14].

Range of Material Testing Values:
Asphalt pavement construction, laminated in hot bitumen is coated over stone aggregates and is laid and rolled in the pavement layer. Mixing of bituminous and recycled plastic waste is mainly done with the low-density polyethylene for the replacement of 30%. Aggregates are chosen based on a standard specification of 2.36 - 5mm aggregates and showed 250% increase in the strength of Marshall Stability. At the same time mix density was reduced to 16% [15]. Waste plastic is heated, grounded and made into powder form and mixed with 4 % plastics in bitumen [16]. The cost of the material was reduced by 7.99% with the addition of plastic to bitumen between the ranges of 5% to 10% [17]. Two percent of polymer composition added with AC-10 bitumen and it's produced AC-20 bitumen. Properties studies that helped to improve the marshal stability, design stability and other desirable properties were observed [18]. Alternative methods of waste disposal packing materials were collected from domestics and industrial. This technology has been used by many researchers because they reviewed variety of research studiesto choose the most effective and efficient technology.Collection and utilisation of waste materials created a pathway to eco-friendly environment. The usage of polymer depends on the equivalent amount of bitumen quantity to reduce the road construction cost [19]. In India, Network systems are adopted more efficiently implanted in a large level of the movement mostly inroads. Apart from the passenger, the freight moves on the roads network. Studied based 65% of freight and 85% of passenger traffic used as roads for their movement. Strength parameter increased in plastic waste when compared to conventional and replacement of bituminous mixes in plastic studies. Good results were produced in the replacement of plastic in the bituminous mixes [20]. According to reviewed result the

marshal stability test obtained optimum use of plastic up to 10% by maintaining a softening temperature of around 130 - 180°C and because of good binding properties, it is used as a good binder for road construction work[21]. Suggestions were made to use the bitumen with optimum replacement waste bag materials.

Methods and Materials

The recycling process involves five steps namely collection, cleaning, shredding, reprocessing by melting and blending with bitumen.

The first step includes collection of waste polypropylene cement bags from construction site. The waste cement bags are then cleansed and immersed in detergent dissolved water to remove dirt and suspended impurities. Then these bags were dried to remove moisture from them. These bags were cut down into small pieces. The shredded pieces were taken in appropriate amount and melted till liquid form is attained with the help of heating mantle. The molten liquid was added to bitumen in desired percentage to determine the optimum polypropylene content by conduct various test such as Penetration test, Ductility test, Marshall Stability test & Centrifugal Extraction test.On heating polypropylene bags in the heating mantle, it is converted into a molten state. But if heated further, it gets converted into powdered form. Liquid forms of polypropylene bags were used in many researches to identify the possibility of partial replacement using additive materials of oxidant agent and others. Powered form of polypropylene bags without using additive agents is concentrated more in this study.

Bitumen

Bitumen is collected from petroleum products and is a more consistent viscous material. Different grades of bitumen were used for road work. Before using bitumen, the grade has to be analysed and the best has to be used for road construction. Bitumen is the heaviest fraction obtained from the fractional distillation of crude oil. The primary use of bitumen is for road construction. Bitumen is highly adhesive, which keeps the aggregate in the road mix bound together greater than 50mm wide and several deep conditions are considered. Pavement cracks are formed due to changes in weathering seasons, thus forming improper bitumen mixing and uneven soil surface nature. Finally, the results occurred in irregular swelling and heaving. Soil stability is one of the criteria to decide to mix bituminous materials. The instability of soils maydevelop more damage in the pavement structure.

Polypropylene

Polypropylene, a thermoplastic, is formed by the polymerization of propylene using an appropriate catalyst. It belongs to a group of polyolefins and is partially crystalline and non-polar. It can resist the actions of organic solvents, salts, acids and alkalis. Due to its mineral composition, it is highly resistant to water.

Materials Testing

Waste Polypropylene

The first step includes the collection of waste polypropylene cement bags from the construction site. The waste cement bags are then cleansed by immersing them in detergent dissolved water to remove dirt and suspended impurities. Then these bags were dried to remove moisture from them. These bags were cut down into small pieces. The shredded pieces were taken in an appropriate amount and melted till liquid form is attained with the help of a heating mantle. The molten liquid was added to bitumen in the desired percentage to determine the optimum polypropylene content by conducting various tests such as Penetration test, Ductility test, Marshall Stability test & Centrifugal Extraction test.

Ductility Test is used to determine the extent to which a material can withstand plastic deformation without rupture. It gives a measure of the adhesive property of bitumen and its ability to stretch. The bitumen is added with waste polypropylene in successive ratios to obtain the optimum value. Penetration Test, in this test, a standard penetration needle is allowed to penetrate a standard bitumen sample for 5 seconds and the value is expressed as 1/10 of mm. It determines the hardness or softness of bitumen by measuring the depth in mm.

Marshall Stability Test
Marshall Stability is an important testing for pavement construction to evaluate the performance prediction measure conducted on the bituminous mix. It measures the maximum load sustained by the bituminous material at a standard loading rate 50.8mm/minute. It is related to the resistance of bituminous materials to distortion, displacement, rutting and shearing stresses. The procedure consists of the determination of properties of mix, flow analysis and optimum bitumen content. The experiment was carried out using Marshall Stability Apparatus [1].

Los Angeles Abrasion Test
Los Angeles Abrasion test on aggregate is used to measure toughness and abrasion resistance and also studied the degradation of aggregate value which is subjected to analyzed graded percentage. Increasing the wheel load or rubbing actions that are affecting pavement structure. This experiment results to evaluate the suitability of stone aggregate in pavement work which relates to the quality of aggregates. This test is carried out by AASHTO T 96 or ASTM C 131: A steel ball is used to measure the strengthening of aggregate crushing values. Quality of aggregate acceptable limits for the bituminous road should not be less than 30 percent [4][5].

Aggregate Impact Test
The aggregate impact test is used to measure resistance to the impact of fracture under the heavy wheel load action. Aggregates percentage losses determined passing through 2.36mm sieve and 15 blows of standard hammer drop. The test performance can be determined within a short duration. It is easy to work and handle equipment and transfer to the construction field. Find properties of aggregates that could be considered the resistance to crushing of aggregates. The percentage of crushing is increased due to varying the load. Toughness values of aggregate are measured to find indirect information of the compressive strength [5][6][7].

Addition of Polypropylene to Bitumen and other Aggregates

Polypropylene in liquid form
After collecting the samples during heating process on every sample 20gms of polypropylene is collected 10ml in molten state materials in a single process. To increase the binding property between polypropylene and bitumen, a strong oxidizing agent and binder agent is needed. The trial proved effective but on re-heating for further addition of liquid polypropylene, the plastic began to segregate from bitumen. This led to the use of fresh bitumen and resulted in the wastage of bitumen. Heating time should not be less than 20 minutes.

Polypropylene in powdered form without using an oxidizing and binding agent
On heating polypropylene bags in the heating mantle, it is converted into a molten state. But if it is heated further, it gets converted into powdered form. Thus powdered waste binds well with bitumen and can be re-used for successive addition for conducting a test to find out the optimum bitumen content. But this resulted in a decrease in quantity. On heating 20gms of polypropylene

wastes, get collects 5gms of powder polypropylene. The heating time of the polypropylene wastes is not exceeding 35 minutes.
Experiment was carried in two trials, since it was easier, more economical and it does not require the use of strong oxidizing agents and binder agents for proper blending. The bitumen was added with 5%, 7%, 9% & 11% of polypropylene powder to determine the optimum polypropylene ratio.Powered form of Waste polypropylene bag materials is shown in Fig.1.

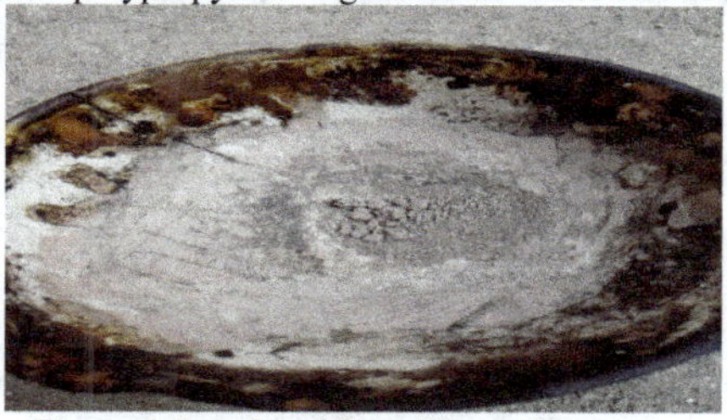

Figure 1.Powdered form Waste Cement bag

Result and Discussion

Material Properties

Properties of Polypropylene
Polypropylene is a tough, rigid and crystalline thermoplastic produced from propene or propylene monomer. It is a linear hydrocarbon resin and also it has the lowest density among plastic [15].List of physical and chemical properties of polypropylene are listed in Table 1.

Table1. Physical Properties of Polypropylene

S.No	Parameters	Values
1.	Tenacity	3.5 – 8 gm/denier
2.	Density	0.91 gm/cubic centimetre
3.	Elasticity	Very good
4.	Moisture regain	0
5.	Melting Point	170°C
6.	Resistance to heat	Moderate
7.	Elongation at break	10 – 45 %

The density of Polypropylene ranges between 0.895 – 0.92 g/cm^3. Young's modulus lies between 1300 – 1800 N/mm^2. It has a melting point of 170°C and becomes brittle below 0°C. The glass structure of the quenched glass largely depends on the proportions of silica and aluminium along with calcium and magnesium. ASTM standard values of specific gravity of bitumen ranges were taken from codal provision. The same physical parameters of specific gravity of 0.905 and bulk density 520 - 620 kg/m^3 were observed by the mixing of polypropylene bags. The chemical properties were analyzed in the experiment and the results were observed. The waste polypropylene materials do not affect the road pavement in the future, since; it has more corrosion resistance and high acid resistance. No bleaching of the pavement due to the presence of materials within the range of temperature 65°C. Increase in temperature may lead to bleaching of road pavement. The above readings were observed within the range specified by researchers [5].

Table 2. Chemical Properties of Polypropylene

S.No	Parameters	Values
1.	Acid	Highly resistant
2.	Base	No effect
3.	Effect of Bleaching	No effect under 65°C
4.	Organic Solvent	No effect
5.	Protection against light	Looses energy
6.	Protection against Mildew	Good

When compared to bitumen the waste polypropylene cement bags can be used for the road construction in the pavement top layers. This work was initiated to improve the wastage reduction of polypropylene materials as well as to promote the cost-effective methods for the road construction work. The physical and chemical properties were satisfied and were found to be similar to conventional bituminous materials.

Properties and Function of Coarse Aggregate

In the road pavement structure, the main function of construction work is coarse aggregate. It has to withstand the stresses due to the wheel loads. The different layers of the pavement are formedusing different sizes of coarse aggregate. The different pavement layer is to distribute wheel load magnitude as the bottom-most layer. So that, the wearing surface of the pavement may use any type of materials and mixes with the stone aggregate. Hence it is an important aspect to study the strength of coarse aggregate also. Coarse aggregate testing values are shown in the below table no. 4.2.

Table 3.Properties of Coarse Aggregate Values

S.No	Test	Values
1.	Impact Value Test	17.28 %
2.	Crushing Value	28.33 %
3.	Abrasion Test	14.4 %

Table 3 shows the strength parameters studies of coarse aggregate materials. All the testing values are within the limits. In general, for the convention road pavement construction the impact Value test, Crushing and abrasion tests are less than 30%, 30% and 40% respectively. The standard values of coarse aggregates testing values are obtained in the work. Hence, this result was correlated to the present study. Herewith, further studies of the pavement construction are partial replacement of Waste Polypropylene cement bag in the conventional pavement materials.

Properties and Function of Marshall Stability:

Marshall Stability is determined to study the maximum load-carrying capacity of road pavement with standard temperature as 60 degrees centigrade when applied the load for given specified material test condition [1]. Marshall Stability is the resistance of deformation of pavement structures for applied load values. Generally, the load is applied at the constant rate of deformation

Materials Research Proceedings **23** (2022) 432-443
https://doi.org/10.21741/9781644901953-47

51mm per minute to varying from time to time. The test was done to determine the density and compactness of road pavement structures.

Properties and Function of Centrifugal Extractor:

Centrifugal Extractor is used to find the bitumen percentage in bituminous paving mixtures. The same working principles were used in the alternate materials of waste Polypropylene cement bag materials. This experiment working procedure was used to suggest the aggregate weight in the pavement construction [2][3]. The Centrifugal Extractor is used to mix two substances as Waste cements bag material and bitumen. The mix generates a continuous extraction from one liquid phase point into another liquid phase solvent. The property studies results were taken with different proportionate replacement of cement bag materials such as 5% to 11%. The optimum ratio from the graph was inferred as 9%, which indicates that the bitumen bounded more strongly with the aggregates which resulted in a heavier weight.

Properties and Function of Ductility Values:

The ductility of bitumen materials is used to find out adhesive properties. Poorly graded bitumen used in the road works forms deformation or longitudinal cracking under traffic load conditions. The ability of stretch of bitumen and sticking properties were carried out. It is expressed in the distance in centimetre to which it elongates before breaking the specimens. The pull rate of bitumen is 50mm per minute at the maintained temperature of 27 degrees centigrade. The pavement road surface melts during the summer season due to increasing temperature.Variation in temperature affected the road life and as well as the public life. In this experiment, the result was observed the maximum ductility point was 77cm for the replacement value of 5%. The permissible point of the ductility test was found to beless than 75cm.

Properties and Function of Penetration Values

The penetration test is determining the grade of bitumen in different viscosity. The penetration test of bitumen is carried out on the relationship between temperature and stiffness. Characterization of waste cement bag materials has to be carried out to perform thhepenetration test values. Penetrating Grade Bitumen was found that the standard bitumen with mixing replacement of waste bag for road construction. The optimum range of replacement used is 5%, which shows that the bitumen mix is most permeable at this percentage. For the road construction, the penetration values were taken as in the ranges of 60-70cm.

Partial Replacement of Waste Polypropylene:

Marshall Stability when observed that the optimum ratio is 8.2%, which gives the load failure at 25kN. If any ratio above 7% is selected, the load was found to be less than 22kN. The waste polypropylene materials used in Pavement construction with a load limit is less than 25K. The minimum load condition should affect the pavement structure due to incremental load conditions. Since the optimum values differ for each type, to find a common ratio suitable to meet the demands of the tests.

Table 4.Percentage of Partial replacement of Polypropylene

S.No	% of Polypropylene	Load @ failure(kN)
Marshall Stability		
1	8	25
2	8.2	25.2
3	8.4	24.3
S.No	% of Polypropylene	Weight of the aggregate
Centrifugal Extractor		
1	8	843.5
2	8.2	845.2
3	8.4	844.1
Ductility Test		
S.No	% of Polypropylene	Ductility (cm)
1	8	77
2	8.2	76.7
3	8.4	75.5
Penetration Values		
S.No	% of Polypropylene	Penetration Value(cm)
1	8	67
2	8.2	66.45
3	8.4	65.9

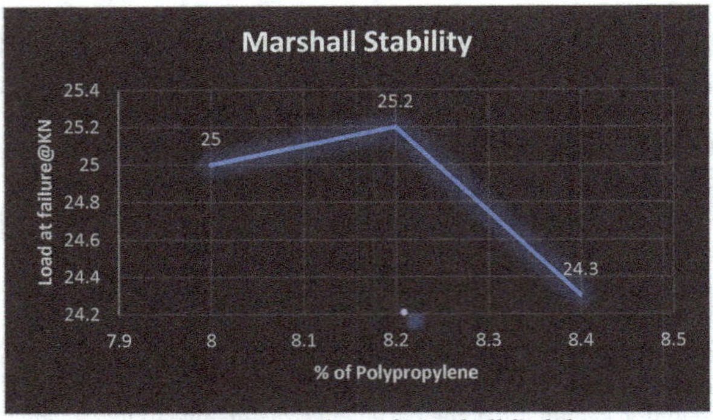

Fig. 1.Optimum Values of Marshall Stability.

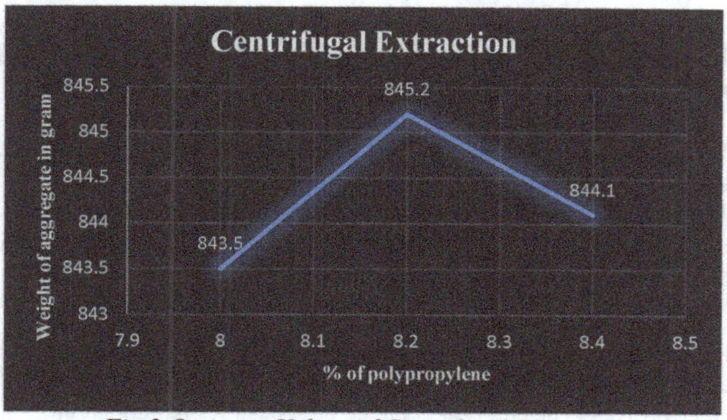

Fig.2.Optimum Values of Centrifugal Extractor.

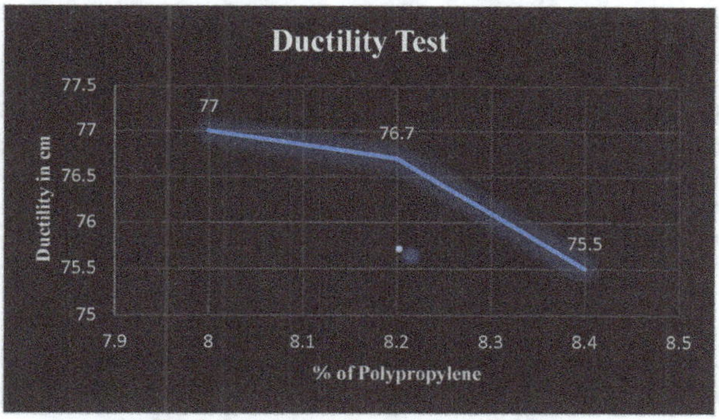

Fig.3.Optimum Values of Ductility Test

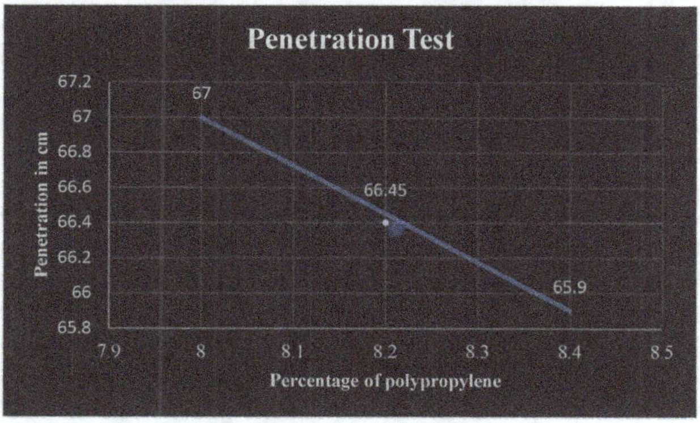

Fig.4.Optimum Values of Penetration Values

The ductility value if greater than 75cm for road construction it may harden more compared to the nature work. Penetration test values must lie between the ranges 60-70cm. The centrifugal extractor gives an optimum ratio of 9%. To satisfy all of the above criteria, the optimum temperature lies around the replacement of 8%. The result was carried out testing the values at 8 %, 8.2 % and 8.4%. The common optimum ratio was found after interpreting the data from the figures1, 2, 3 and 4.

The peak values for both Marshall Stability and Centrifugal Extraction (most relevant) @ 8.2%, within limit values for Penetration and ductility. It was inferred that the optimum value of polypropylene that can be added for road construction is 8.2%.

Conclusion

In this study, the properties of conventional bituminous mixes were analyzed. Based on that the evaluation strength of the flexible pavement was to be carried out. The waste plastic cement bag has been used in the powder. The end product of cement bags was shredded into smaller fragments and heated to 90°C for 30 minutes using a heating mantle. After heating the samples for every 20gm of plastic waste samples, 5gm of powder was obtained. The various mixes of partial replacements of waste plastic cement bag materials were analyzed for 5%, 7%, 9% and 11% in the bitumen. Based on the tests conducted for the different proportions, the optimum ratio of polypropylene was found to vary from 8% to 9% without adding any oxidant agent. Finally, the in the modified bitumen, the optimum replacement of bitumen was found to be 8.2% without the addition of binders or oxidizing agent. Results showed that Marshall Stability, centrifugal extractor, Ductility and penetration test of bitumen by using 8.2% partial replacements of waste plastic cement bag materials used in road construction was found to be most effective and efficient. This method was aimed to reduce the materials cost as well reduce pollution to the environment. The reuse and recycling of waste materials will promote sustainable development in the future.

References

[1] ASTM D 1559, Test method for resistance of plastic flow of bituminous mixtures using Marshall Apparatus, American society for testing and materials.

[2] ASTM D 6931, Indirect Tensile (IDT) Strength for bituminous mixtures, American society for testing and materials, 2007.

[3] IS: 1203, Methods for testing tar and bituminous materials: determination of penetration, Bureau of Indian Standards, New Delhi, 1978.

[4] IS: 1205, Methods for testing tar and bituminous materials: determination of softening point, Bureau of Indian Standards, New Delhi,1978

[5] IS: 2386, Methods of test for aggregates for concrete (P - I): Particle size and shape, Bureau of Indian Standards, New Delhi.1963

[6] IS: 2386, Methods of test for aggregates for concrete (P-III): Specific Gravity, Density, Voids, Absorption, Bulking, Bureau of Indian Standards.1963

[7] IS: 2386 (1963), Methods of test for aggregates for concrete (P-IV): Mechanical Properties, Bureau of Indian Standards, New Delhi.1963

[8] K. Sandhiya, L. S. Kumar, K. N. Rajkumar, R. Sandhya, and S. Sukumar, "Partial Replacement of Bitumen by Using Plastic in Bitumen Concrete," 119 (2018) 2913–2920.

[9] R. Vasudevan, Use of Plastics Waste in Construction of Tar Road.(2004)

[10] A. K. Sahu and R. K. Singh, "Application of Waste Plastic Materials in Road Construction," no. March, 2016.

[11] Amit Gawandea, G. Zamarea , V.C. Rengea , Saurabh Taydea , G. Bharsakaleb "An overview on Waste Plastic Utilization in Asphalting of Roads."Journal of Engineering Research and Studies", 3 (2012) 01-05

[12] A. Johnson, K. Appiah, and V. Nana, "Use of Waste Plastic Materials for Road Construction in Ghana," Case Studies in Construction Materials, 2016.

[13] J. Patil, A. Pune, S. D. Khandekar, S. Technical, and E. Society, "Application of Waste Plastic as an Effective Construction Application of Waste Plastic as an Effective Construction, 02 (2015)

[14] S. S.Verma., Roads from plastic waste, Science Tech Entrepreneur, The Indian Concrete Journal,(2008) 43 - 44.

[15] A. H. Mir, "Use of Plastic Waste in Pavement Construction : An Example of Creative Waste management," 05 (2015) 57–67.

[16] Ahmed Trimbakwala "Use of waste plastics in road construction" International Journal of Scientific and Research Publications, 7(2017) 137

[17] V. Swami and A. Jirge, "Use of waste plastic in construction of bituminous road," International Journal of Innovative Research in Science, Engineering and Technology" 4 (2012) 2351–2355.

[18] B. Aforla, "Assessment of Suitability of Plastic Waste in Bituminous Pavement Construction," 7 (2015) 1–6.

[19] S. Rajasekaran, Reuse of Waste Plastics Coated Aggregates-Bitumen Mix Composite for Road Application. Green Method. (2011)

[20] D. Kumar, "A Proposed Design of Flexible Pavement using Waste Plastic," (2014)21–26.

[21] V.S. Punith., Study of the Effect of Plastic Modifier on Bituminous Mix Properties.(2001)

[22] SavitaDevi. Partial replacement of bitumen by using plastic waste in bitumen concrete 3 (2016).

Investigating the Possibility of Using Clay and Fly Ash in the Production of Geopolymer Mortar

N.V. Manjunath[1,a,*], B.S. Hariprasath[1,b], S.K. Sreehaswanth[1,c], T. Gopi[1,d]

[1]Bannariamman Institute of Technology, Sathyamangalam, Erode,Tamil Nadu, India

[a]manjunathnv@bitsathy.ac.in, [b]hariprasath.ce18@bitsathy.ac.in, [c]sreehaswanth.ce19@bitsathy.ac.in [d] gopi.ce18@bitsathy.ac.in

Keywords: Geopolymer, Fly Ash, Clay, M-Sand

Abstract. The concept polymer concrete has numerous advantages over the convention concrete similarly in mortar also which uses reactive silica presented in the pozzolanic materials which is activated using alkaline materials. This paper focused on the Geopolymer mortar with fly ash and clay as raw materials and M-Sand for fine aggregate instead of river sand. Different combinations of mixtures were prepared and compared with ordinary cement mortar. The objective of this research is to identify the properties of different combinations of geopolymer mortars with conventional cement mortar, in which the fine aggregate sourced from riverbeds is replaced by Fine aggregate manufactured using Crushing stones and testing it for compressive strength, water absorption, residual alkalinity, and acid resistance. Fly ash- clay based geopolymer shows good acid resistance and shows less reduction in weight and compressive strength than ordinary cement mortar. Permeability of geopolymer mortar is less that of ordinary cement mortar. Alkalinity of geopolymer mortar is 4-20 % higher than that of ordinary cement mortar. Fly ash- clay based geopolymer has excellent compressive strength (75.7% for red clay and 52.6% for fire clay) than the conventional cement concrete and is suitable for structural applications.

Introduction

The industry that uses maximum number of natural resources is concrete industry. The carbon dioxide share of construction industry is about 7% and Class C Fly Ash utilization in Geo polymer concrete indirectly reduces the greenhouse effect [1]. Geopolymer concrete using pozzolanic materials and recycled aggregate have more advantages than other [2]. The self-cementing property of Ground Granulated Blast Furnace Slag will require less Calcium activator than the other type of pozzolanic materials [3]. Ordinary Portland Cement (OPC) has long been the dominant binder in concrete production. The environmental issues associated with FOS pollutants are well known.[4]. The amount of carbon dioxide released during the production of OPC due to limestone calcination and fossil gas burning is on the order of one tenth of a tonne of OPC produced[5]. Inorganic materials abundantsilicon (Si) and aluminum (Al) that change with alkaline activators to make cementitious geopolymers were first described by "Davidovits" as inorganic materials rich in silicon (Si) and aluminium (Al). Glassy silicates composed of sodium hydroxide (NaOH) or potassium hydroxide (KOH) and sodium or potassium silicate are the most common alkali activators used in geopolymers, while sodium hydroxide and sodium silicate are the most common due to their cost and availability.[6–8].Geopolymers are chemically like zeolites but has an amorphous microstructure. They are formed by copolymerization of individual particles of alumina and silicate by dissolving a starting material containing silicon and aluminum at a high pH of in the presence of a soluble alkali metal silicate.[9–11]. Geopolymers are well known for their exceptional mechanical properties, resistance to outbreak of fire, and acid defiance[12]. These things make geopolymers a potential construction material. In its place of OPC or other hydraulic adhesive cement paste, a low-calcium (ASTM Class F) fly ash-clay based geopolymer is used as the binder in this work to produce concrete[13, 14]. Without the presence of admixtures, the fly ash-clay based completely geopolymer paste binds the loose coarse mixture, satisfactory mixture, and various unreacted substances together to form the geopolymer concrete. Geopolymer concrete is created using conventional concrete technology methods.

Experimental procedure

The objective of this research is to experimentally observe the natural Clay and fly ash primarily based geopolymer concrete for various combos of binder and fine aggregate with exceptional proportions of alkali activator to binder ratios.

Materials:

Fly ash

M. Sand

Clay

Alkaline liquids

Fig 1: *Materials for making geopolymer mixtures*

Fly ash

Fly ash was acquired from Mettur thermal power plant. Fly ash is a thermal industrial by-product, generated in the stage of combustion of coal in the power plants. Standard consistency of the fly ash sample is 29 %.

M-Sand (manufactured Sand)
Manufactured sand conformed to zone II with the specific gravity of 2.65 is used as replacement for the fine aggregate that is sourced from Riverbeds. M-Sand is used because of the scarcity in river sand.

Clay
The studied soil sample is silty clay collected from nearby Karunya University campus which act as the binder material with Fly ash from thermal power plants. The soil contains 2%, 46% and 52% of sand, silt and clay, respectively. Its liquid limit is approximately 52 %.

Alkaline liquids
As activators, the alkaline liquid used in Geo polymerization is a mixture of sodium hydroxide and sodium silicate. A large quantity of sodium silicate solution was purchased in bulk from a local supplier. Sodium hydroxide flakes or pellets with 97-98 percent purity were also purchased in bulk from a local supplier.

Mix proportions
The manufacture of geopolymer mortar specimens was carried out using the usual concrete technology methods. The mix proportions for geopolymer mortar for different combinations are shown in table below.

Table 1: *Mix proportions for 1:1 mortar (8M 1:1 alkali activator ratios)*

Mix ID	Cement Kg/m³	Fly ash Kg/m³	Clay Kg/m³	Fine aggregate			Sodium Silicate	Sodium Hydroxide	
				River Sand	M-Sand	Clay Kg/m³		$\frac{Kg}{m^3}$	$\frac{mol}{l}$
CM1	1.166	-	-	1.166	-	-	-	-	-
GP1	-	1.166		-	1.166	-	0.19	0.19	8M
GP2	-	0.583	0.583	-	1.166	-	0.23	0.23	8M
GP3	-	-	1.166	-	1.166	-	0.26	0.26	8M
GP4	-	1.166	-	-	0.583	0.583	0.22	0.22	8M
GP5	-	1.166	-	-		1.166	0.26	0.26	8M

Geopolymer mortar combination details:
GP 1----- [100 % fly ash + 100 % M-Sand
GP 2----- [(50 % fly ash + 50 % Clay) + 100 % M-Sand
GP 3----- 100 % Clay + 100 % M-Sand
GP 4----- [100 % Fly ash + (50 % Clay+ 50% M-Sand)]
GP 5----- [100 % fly ash + 100 % Clay]

Mixing, casting, and curing:
The binders i.e., fly ash and clay and the aggregate were first mixed. The alkaline liquid prepared before one day before the casting and then added to the dry materials that is Natural Clay and Fly ash and thoroughly mixed using mixer machine.

The cube size of 70 x 70 x 70 mm and cylinder of 50 mm diameter and 100 mm height were casted for mortars. The specimens were demolded without disturbing and cured at room temperature for 1 day and the Geo polymer concrete specimens were kept in oven curing at 80°C for 24 hours.

Findings and analysis

Compressive strength

Table 2: *Compression test of mortar (8 M, 1:1 Binder to Fine aggregate ratio, 1:1 Alkali Activator ratio) by using cube of size 70x70x70 mm*

Sl. No	Mix ID	Average Ultimate Load(kN)		Compressive Strength(N/mm2)	
		3	7	3	7
1	CM1	-	106		21.63
2	GP1	205.8	215.6	42	44
3	GP2 with red clay	166.6	186.2	34	38
4	GP2 with fire clay	142.1	161.7	29	33

Table 3: *Compression test (8 M, 1:1 Binder to Fine aggregate ratio, 1:1 Alkali Activator ratio) by using cylinder of size 50 mm diameter and 100 mm height*

Sl.No	Mix ID	Average Ultimate Load(kN)			Compressive Strength(N/mm2)		
		3	7	14	3	7	14
1	CM1	-	40	73		20.38	37.20
2	GP1	76.97	84.72		39.2	43.15	
3	GP2	72.66	80.15		37.01	40.82	
4	GP3	68.96	76.67		35.12	39.05	
5	GP4	76.38	81.33		38.90	41.42	
6	GP5	74.85	78.97		38.12	40.22	

It is examined from the above tables, that the 7 days compressive strength of natural Clay and Geopolymer Mortar made with Flyash (GP2) is furthermore than ordinary Portland cement mortar by 75.7 % for red clay and 52.6 % for fire clay. Hence, it is presumed that compressive strength of geopolymer mortar is better than the OPC based concrete mixtures. It can also be observed that 3rd day strength of geopolymer concrete specimenis lesser than that of ordinary Portland cement specimen at the age of 28 days.

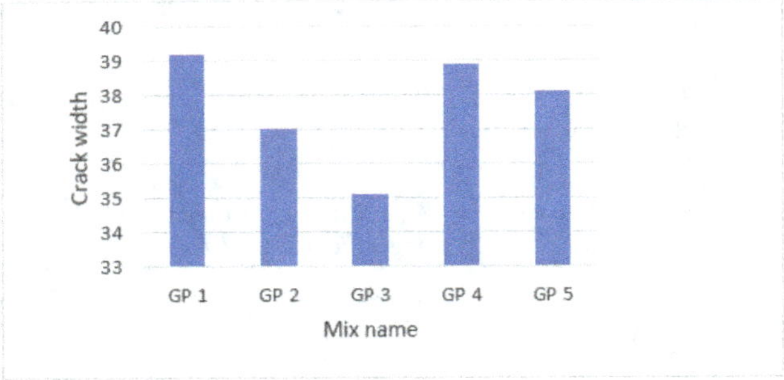

Fig 2: *Crack pattern*

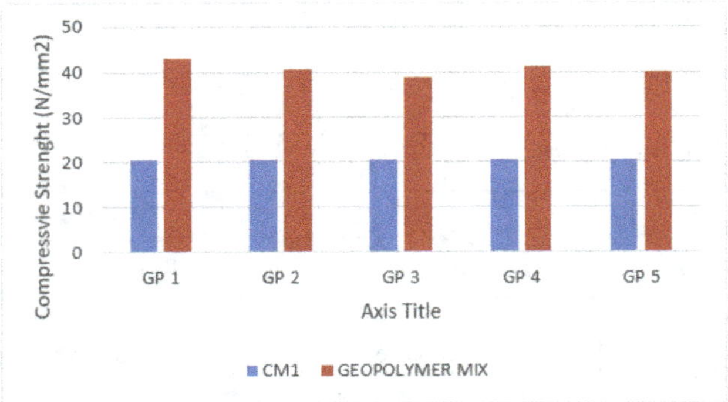

Fig 3: *3rd day compressive strength*

Fig 3 and 4 reveals the compressive strength loss of ordinary cement mortar and geopolymer mortar. It can be seen that the compressive strength of geopolymer mortar is higher than that of ordinary cement mortar.

Acid resistance test

Table 4: *weight reduction (5 % HCl)*

Mix proportion	Average reduction in weight (%)	
	7 days	14 days
CM 1	0.324	0.354
GP 1	0.112	0.121
GP 2	0.188	0.196
GP 3	0.249	0.260
GP 4	0.164	0.170
GP 5	0.152	0.161

Table 5: *Strength reduction (5 %HCl)*

Mix proportion	Average loss of compressive strength (%)	
	7 days	14 days
CM 1	68	70
GP 1	38	42.8
GP 2	41.5	46.72
GP 3	43	48.11
GP 4	41	45.9
GP 5	39.5	44.2

Table 5 and 6 shows the weight and strength reductions of ordinary cement mortar and geopolymer mortars in the presence of 5% of hydrochloric acid at 7 and 14 days. It can be seen that weight loss is lesser in the case of geopolymer mortar by 30- 70% by ordinary cement mortar at 7th day and 34-74% at 14th day. That is the performance of geopolymer mortar against acid is better than that of conventional cement mortar. In addition, the strength reduction ofgeopolymer mortar was reduced by 58-78% on the 7th day and 48-63% on the 14th daycomparedto the conventional cement mortar.

Materials Research Proceedings 23 (2022) 444-453
https://doi.org/10.21741/9781644901953-48

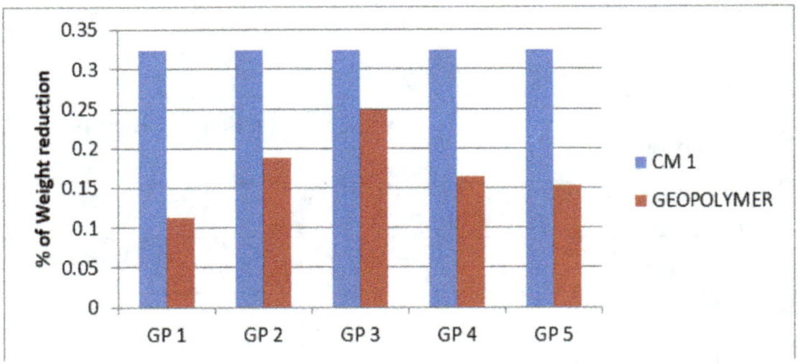

***Fig 5:** Average reduction in percentage weight after 7 days*

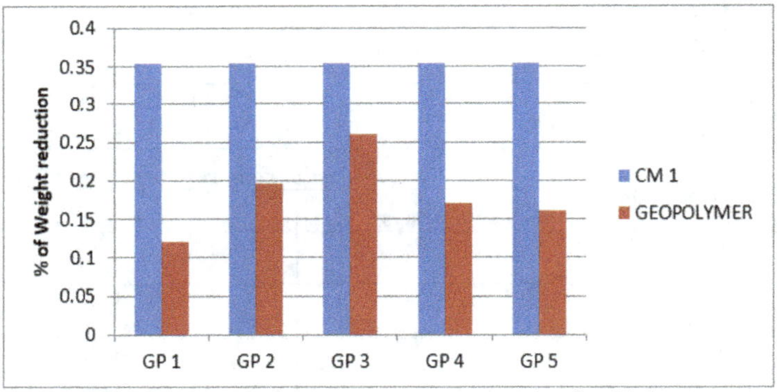

***Fig 6:**Average reduction in percentage weight after 14 days*

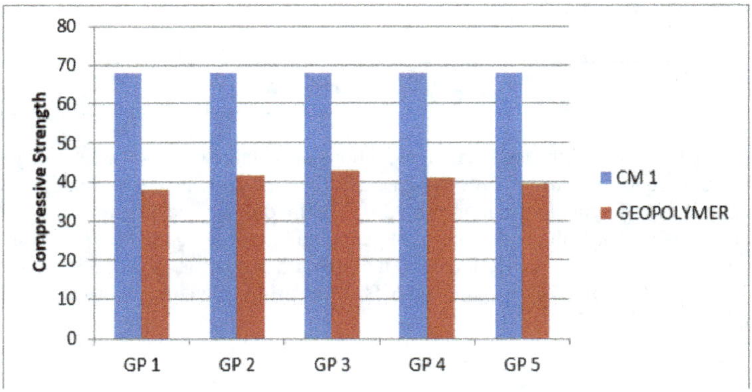

***Fig 7:** Average compressive strength loss after 7 days*

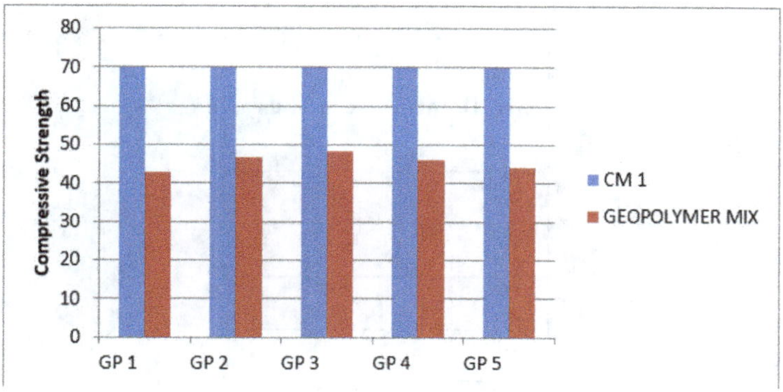

Fig 8: *Average compressive strength loss after 14 days*

Water absorption

Table 6: *Saturated water absorption*

Mix id	Weight Of saturated specimen(g)	Weight of Oven dried specimen (g)		Saturated water absorption @ 7 days (%)	Saturated water absorption @ 14 days (%)
		7 days	14 days		
CM1	458	443	444.57	3.31	3.02
GP1	515	504.85	505.15	2.01	1.95
GP2	557	544.96	545.86	2.21	2.04
GP3	568	554.42	556.15	2.45	2.13
GP4	502	490.38	491.29	2.37	2.18
GP5	535	520.38	521.85	2.81	2.52

The table shows that the saturated absorption rate of geopolymer slurries is lower than that of conventional cement slurries. The absorption rate of geopolymer mortar decreases by 15 - 40% after 7 days and 17 - 35% after 14 days. This means that geopolymer slurries are less permeable than conventional cement slurries.

Residual alkalinity

Table 8 shows the residual alkalinity. Due to the alkali activated binder, the alkalinity of geopolymer slurry is higher than that of conventional cement slurry. The alkalinity of geopolymer slurries is 420% higher than that of conventional cement slurries.

Table 7: *Residual alkalinity*

Mix ID	pH value in 14 days
CM1	10.08
GP1	12.12
GP2	11.52
GP3	10.58
GP4	11.87
GP5	10.98

Conclusion

Natural Clay and Fly Ash based geopolymer has exceptional compressive strength (75.7 % for red clay and 52.6 % for fire clay) than the conventional cement concrete and is suitable for structural applications. Natural Clay and Fly Ashmixed geopolymer shows good acid resistance and shows minimum weight reduction in compressive strength and weight than ordinary cement mortar. The permeability of geopolymer slurries is lower than that of conventional cement slurries. The alkalinity of geopolymer slurries is 4-20% higher than that of conventional cement slurries. The properties of geopolymer slurries are no worse than those of conventional cement slurries. Low-calcium, natural clay-based pozzolan ash geopolymer concrete offers a number of economic advantages over Portland cement mortar as fly ash and clay costs are lower than that of Portland cement.

References

[1] Muthadhi A, Vanjinathan J, Durai D (2016) Experimental investigations on geo polymer concrete based on class C fly ash. Indian Journal of Science and Technology 9: . https://doi.org/10.17485/ijst/2016/v9i5/87270

[2] Saravanakumar P (2018) Strength and Durability Studies on Geopolymer Recycled Aggregate Concrete

[3] Pradeep Kumar S, Kameswara Rao B, Guru Jawahar J (2018) Experimental investigation on fly ash and GGBS based geopolymer concrete incorporate black marble waste aggregate. International Journal of Engineering and Technology(UAE) 7:1233–1236 . https://doi.org/10.14419/ijet.v7i3.12486

[4] Simon S, Hemamathi A, Jenishtalouis J (2019) Strength assessment on flyash based geopolymer concrete. International Journal of Innovative Technology and Exploring Engineering 9:3956–3960 . https://doi.org/10.35940/ijitee.A5034.119119

[5] Srinivas T, Manoj Anand P (2019) Permeation properties of geopolymer concrete made with partial replacement of recycled coarse aggregates. International Journal of Innovative Technology and Exploring Engineering 8:2987–2990 . https://doi.org/10.35940/ijitee.K2317.1081219

[6] Sandeep Kumar AD, Singh D, Reddy VS, Reddy KJ (2020) Geo-polymerization mechanism and factors affecting it in Metakaolin-slag-fly ash blended concrete. In: E3S Web of Conferences. EDP Sciences

[7] Shahedan NF, Abdullah MMAB, Mahmed N, Kusbiantoro A, Tammas-Williams S, Li LY, Aziz IH, Vizureanu P, Wysłocki JJ, Błoch K, Nabiałek M (2021) Properties of a new insulation material glass bubble in geo-polymer concrete. Materials 14:1–17 . https://doi.org/10.3390/ma14040809

[8] Lakshmi Deepak A, Vara Lakshmi TVS (2019) Mechanical properties of geopolymer concrete with flyash and metakaolin. International Journal of Engineering and Advanced Technology 9:3863–3868 . https://doi.org/10.35940/ijeat.A1180.109119

[9] Srinivas T, Koushik RN (2019) Sulphate attack resistance of geo-polymer concrete made with partial replacement of coarse aggregate by recycled coarse aggregate. International Journal of Innovative Technology and Exploring Engineering 8:112–117 . https://doi.org/10.35940/ijitee.L2509.1081219

[10] Raj SR, Arulraj PG, Anand N, Balamurali K, Gokul G (2021) Influence of various design parameters on compressive strength of geopolymer concrete: A parametric study by taguchi method. International Journal of Engineering, Transactions B: Applications 34:2351–2359 . https://doi.org/10.5829/IJE.2021.34.10A.16

[11] Pu BC, Liu B, Li L, Pang W, Wan Z (2021) Influence of polypropylene fibre factor on flowability and mechanical properties of self-compacting geopolymer. Materials 14: . https://doi.org/10.3390/ma14175025

[12] Naveen Kumar G, Kailash Kumar S, Prashanth G (2019) Strengths of geo polymer concrete by adding metakaoline. International Journal of Recent Technology and Engineering 8:2682–2687 . https://doi.org/10.35940/ijrte.B2673.078219

[13] Shashikant, Prince Arulraj G (2019) A research article on “Geopolymer concrete.” International Journal of Innovative Technology and Exploring Engineering 8:499–502 . https://doi.org/10.35940/ijitee.I1106.0789S219

[14] Rambabu G, Naidu DA, Venkateswarlu D (2019) Experimental developments on corrosion resistivity of low calcium fly ash-based geopolymer concrete. International Journal of Recent Technology and Engineering 8:5320–5327 . https://doi.org/10.35940/ijrte.B1089.078219

Performance Assessment of Self Compacting Concrete Incorporating Mineral Admixtures

M. Ranjitham[1,a], A. Dinesh[2], R.J. Pragadeeshwaran[1,c], D. Sridhar[1,d]

[1]Civil Engineering, Bannari Amman Institute of Technology, Erode 638401, India

[2]Civil Engineering, Sri Ramakrishna Engineering College, Coimbatore 641042, India

[a]ranjitham@bitsathy.ac.in, [c]pragadeeshwaran.ce19@bitsathy.ac.in, [d]sridhar.ce19@bitsathy.ac.in

Keywords: Self Compacting Concrete, Mineral Admixtures, Silica Fume, Ground Granulated Blast Furnace Slag, Fly Ash

Abstract. Self-Compacting Concrete (SCC) is a newly constructed which need to be processed to be installed and assembled. It can discharge beneath its own weight, complete formwork, and achieve complete integration, at the same time in the existence of profusion reinforcement. SCC is a variety of high-strength concrete and expands to form devoid of the demand for mechanical vibrations. SCC is a non-removable concrete by its weight. The importance of concrete that you assemble is that it retains all the durability and characteristics of the concrete, meeting the expected operational requirements. Another way to diminish the expense of concrete for that is to use Mineral Admixtures (MA) such as Ground Granulated Blast furnace Slag (GGBS) Silica Fume (SF), and Fly Ash (FA), during mixing. The quantity of Portland cement was decreased by using mineral admixtures, expense of compaction will be competitive especially reason for this while using the mineral mixtures are waste or industrial product. In addition, the application of MA in the production of composite concrete not only provides economic benefits but also reduces the temperature of the hydration. The amalgamation of mineral ingredients additionally excludes the need for viscosity-improving chemical admixtures. Low water/cement (W/C) ratio which indicates to superior durability and exceptional mechanical integrity of the building. This experimental research paper familiarizes and reviewing the strength properties such as compression test, flexural strength and the split strength of SCC with different mineral compounds and compare the properties with Control Mix (CM) and workability tests of various mineral compounds (slump, L-box, U-box, and T50) also studied. From the Experimental investigation concluded that the impact of mineral mixtures on performance like compressive strength values, split tensile strength values and flexural strength values were increases as per European Federation of manufacturers for special concrete.

Introduction

Concrete is the most utilised structural material on the earth moreover its structures have been modified attributable to innovative advances materials nowadays [10]. Several varieties of concrete are generated to develop completely different concrete structures [9]. Until now, this advancement is often separated into four classifications. The primary is normal concrete fabricated from simply four materials, especially water, cement, fine aggregates(size less than 4.75mm), and coarse aggregates(size greater than 4.75mm). Due to the rapid structural development and the construction of special structures like tall structures, bridges and high-rise buildings we are in require of SCC rather than normal concrete [4]. After that, a fifth ingredient, a dehydration agent or superplasticizer was essential [8]. SCC gains good strength as well as concrete did not require any vibration. It will surge bottom along its weight, sealing the formwork and accomplishing full integration and even within the existence of closed reinforcement. The inclusion of mineral ingredients additionally dismisses the requisite for viscosity-enhancing chemical based admixtures[1]. SCC with light aggregate has self-deaeration, high fluidity and high strength, but the internal kinetic energy of light aggregate is insufficient, so it flows rather slowly and deteriorates in the densely reinforced area[3]

Experimental investigation

Materials

According to IS: 12269 1987 Portland Cement(PC) was selected of standard 43 and river sand based on IS: 383 1970 were used, corresponding to IS 383 1970 crushed stone with a particle size of 12.5 mm was collected[2] The value of 2.77 was obtained as specific gravity for coarse aggregate . To avoid the impeding effect of concrete, the maximum coarse aggregate size of SCC was limited(12.5mm)[5]. Further, the large amounts of fine fillers and/or viscosity improvers has been accomplished. Thus, the solidity of the mixture is sustained, penetration is diminished, and segregation of larger aggregates is prevented. The value of 3.15 and 2.65 was obtained as specific gravity for cement and sand.[6] In addition to these by products, FA was seized from a thermal power plant in Salem, India, SF possessed from Meridian science and technology, and GGBS collected from the Agni smelter in Herod, Tamil Nadu, India. In general, reducing quantity of water to improve stability of mixtures with low pour pressures and medium to elevated viscosity levels[7]. Low water requires a high amount compared to high pressure water filters to achieve the required defects, especially for low binder content Table 1 chemical Properties of PC and mineral admixtures[8]

Table 1 Chemical Properties of PC and mineral admixtures

Chemical proportions	Cement	FA	SF	GGBS
SiO_2	20	40	97.2	12-17
Magnesium oxide (MgO)	2.7	3	0.4	13
Al_2O_3	4.85	26	0.4	1-3
Loss On Ignition (LOI)	1.89	2.44	1.5	1.22
Fe_2O_3	0.6	5.78	0.3	22-30
Calcium Oxide (CaO)	59.76	20	0.42	30-50

Mix Proportions for calculating the strength

The various modification in the control mixtures with three types of mineral mixtures with were ready and tested to measure concrete strength. The natural aggregates are replaced by mineral admixtures by 30%, 40%, and 50% levels of FA, SF, GGBS, after a calculated the mix proportions, the water / cement(W/C) weight ratio (w / p) was designated as 0.346. Table 3, Table 4, Table 5 shows the FA, SF, GGBS of various mixes arrived to test the properties of compression test, split tensile strength and flexural strength to calculate the optimum mix. In Extra, no strategy has been identified that demonstrates all relevant presentation characteristics, in the configuration of each component it should be examined in more than a few test to determine the consequences of a different display.

Table 2: FA based SCC Concrete Mix Proportions

Materials Proportions in the mixes	Control Mix(CM) in Kg/m³	FA 30% mix in Kg/m³	FA 30% mix in Kg/m³	FA 30% mix in Kg/m³
Cement	450	300	350	200
FA	-	200	250	300
W/C	0.42	0.42	0.42	0.42
Sand	750	750	750	750
Coarse aggregate	500	500	500	500
Super plasticizer	2.5%	1.9%	2%	1.8%

Table 3 SF based SCC Concrete Mix Proportions

Materials Proportions in the mixes	Control Mix(CM) in Kg/m³	SF 30% mix in Kg/m³	SF 40% mix in Kg/m³	SF 50% mix in Kg/m³
Cement	450	300	350	200
SF	-	200	250	300
W/C	0.42	0.42	0.42	0.42
Sand	750	750	750	750
Coarse aggregate	500	500	500	500
Super plasticizer	2.5%	1.9%	2%	1.8%

Table 4 GGBS based Self-Compacting Concrete Mix Proportions

Materials Proportions in the mixes	Control Mix(CM) in Kg/m³	GGBS 30% mix in Kg/m³	GGBS 40% mix in Kg/m³	GGBS 50% mix in Kg/m³
Cement	450	300	350	200
Blast furnace slag	-	200	250	300
W/C	0.42	0.42	0.42	0.42
Sand	750	750	750	750
Coarse aggregate	500	500	500	500
Super plasticizer	2.5%	1.9%	2%	1.8%

Casting, curing at 7 and 28 days and workability and Mechanical properties of SCC

The compressive strength studies were investigated at 150mm x 150mm x 150mm (3D cube molds) while a cylinder with a blank form with a size of 150mm x 300 mm was used to determine the strength. Flexible power studies are made of 100mm x 100mm x500mm crystals. In all three FA, SF and GGBS minerals and the sample was cast and tested for pressure, separation, and flexibility in the concrete after 28 days. Prior to this flow test was performed using performance tests such as L-box (Fig.3), U-box (Fig.2), V-funnel (Fig.1), the T50 test was designed to study concrete working components that include skill achievement. of pouring and passing. Decay flow rate was used to assess the flow rate of concrete that it comprises based on the measurement of the distribution width. The base value of the SCC is 650 mm and the 800 mm limit of new concrete mixing. During the deviation of the decay flow, the time required to reach a maximum of 50 cm of decomposition flow is allocated (T50). V-funnel testing is used to determine the filling volume of the substance. The time it takes for the concrete to flow down is noted in a second. The degradation of the decay flow here determines the positive strength of the decaying concrete under its weight against the impact of the surface outside the external boundaries. Due to the sticky nature of some concrete mixtures, they form, descending flow measurements are made. At the same time, the slump-flow rate (T50) of the concrete was calculated when the concrete collapsed to a depth of 50 cm. V-funnel and L-box tests are performed through procedures announced under the European Union Committee. The maximum time when the concrete mixture can be applied to the V is 10 seconds. In the L-Box experiment, a simultaneous removal of the control gate was initiated to allow the concrete to flow to self-assembly using a horizontal block in the box and to determine the h2 / h1 ratio if the concrete flowed freely in it. Like water, at rest; it will be horizontal so the assignment will be equal to the unity. The minimum tolerable value will be 0.789 (0.8) [EFNARC, 2002]. The U-box test method is utilized to test the filling volume of the concrete that you assemble. The height of the concrete on both sides is measured and provides an unnecessary amount of filling capacity. The contrast in height is h1-h2 filling height. If the concrete flows spontaneously like water, at rest it will be horizontal. Adequate length of filling length is 30 mm wide as recommended by European standards.

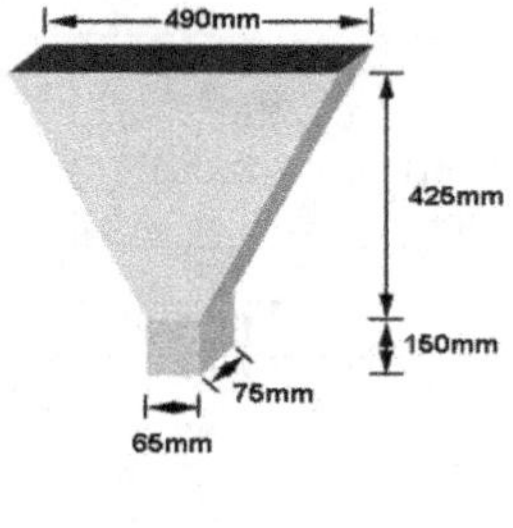

Fig.1 V-funnel test apparatus

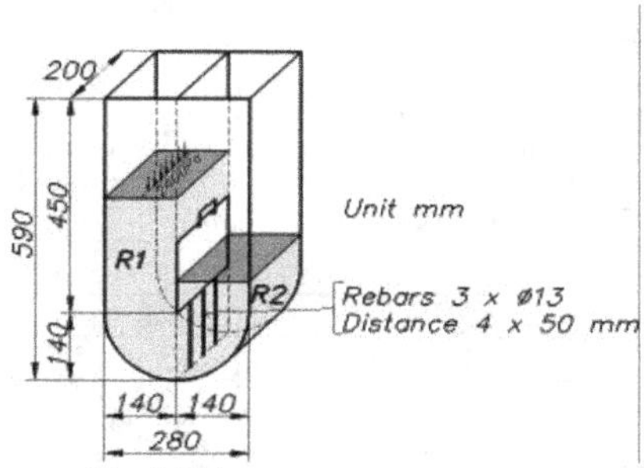

Fig. 2 U-box test apparatus

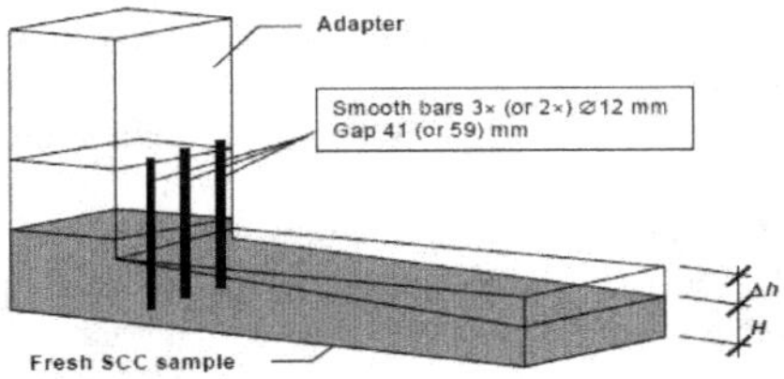

Fig.3 L – box apparatus

The Workability and Mechanical properties of various
Refer Table 5 Limitations according to European standards and Table 6 for workability values of SCC mixes

Table 5 Limitations according to European standards

Method of workability test	**Minimum range in** (mm)	**Maximum range** in (mm)
Slump flow	550	700
T_{50}	0	6
L box	1.0	1.2
V funnel	0	14
U box	0	25

Table 6 workability ranges of SCC mixes

Mixtures with mineral admixtures	**W/C ratio**	**Slump value (mm)**	**V-funnelvalue (Sec)**	**U-Box value (h_2-h_1) mm**	**L-Box value (h_2/h_1)**	**T_{50} value (Sec)**
FA 30% mix	0.42	650	11	23.5	0.9	6.1
FA 40% mix	0.42	655	9	24	0.95	6.64
FA 50% mix	0.42	640	8.95	23.5	0.96	7
SF 30% mix	0.42	670	8	25	0.87	6
SF 40% mix	0.42	630	7	27	0.96	5.56
SF 50% mix	0.42	690	6	25	0.93	7
GGBS 30% mix	0.42	660	12	24	0.91	5.2
GGBS 40% mix	0.42	665	7	28	0.97	5.58
GGBS 5 0% mix	0.42	680	6	24	1.0	7

Refer Fig 4,5,6 for compressive, flexural and split tensile values for mixes containing mineral admixtures like FA, SF and GGBS

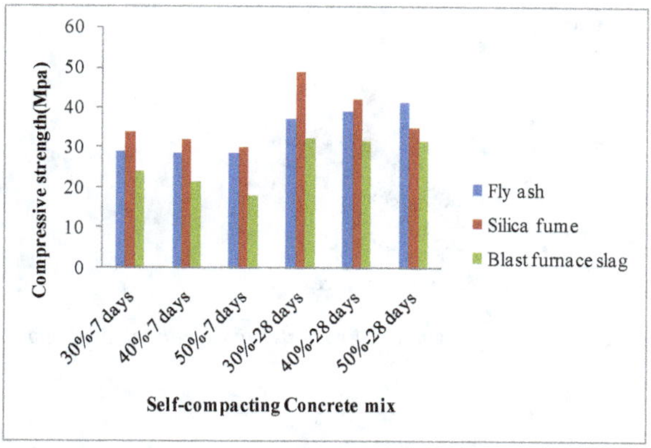

Fig 4. Compressive Strength values of various mixes including mineral admixtures

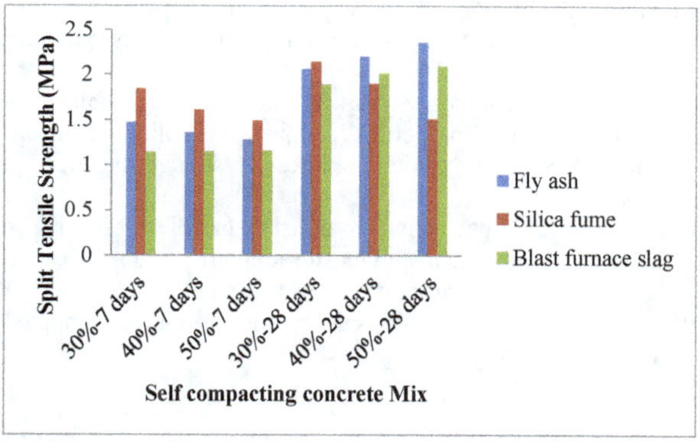

Fig 5. Split Tensile Strength values of various mixes with mineral admixtures

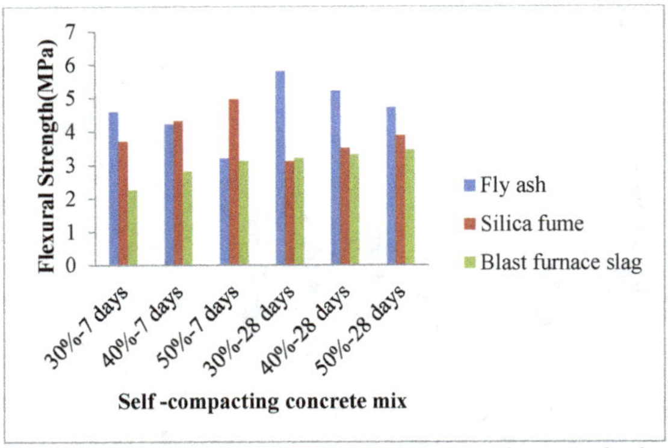

Fig 6. Flexural Strength values of various mixes with mineral admixtures

Summary

The mechanical properties workability test of SCC combinations and the effects of the tests as per the following.

1. All the SCC substantial blends had a good result in the workability. Compare with all the mineral admixtures, the GGBS series attained slump property parallel with FA and SF series.
2. The utilization of mineral admixtures worked on the SCC and achieved good strength
3. The effects of the mechanical properties (compressive, split, and flexure) had shown the huge execution contrasts and the high compressive strength has been gotten for the FA Specimen series.
4. The 30% FA based SCC specimen accomplishes excellent Flexural strength and 50% GGBS specimens achieved decent split tensile strength
5. Optimum water/cement (W/C) proportion was picked as 0.376 by weight, the proportion significantly past or not exactly this might cause isolation combined with hindering inclination in self-compacting concrete mixtures.

References

[1] M. Ranjitham, B Piranesh, A.Vennila, "Experimental investigation on high-performance concrete with partial replacement of fine aggregate by foundry sand with cement by mineral admixture: International Journal of Advanced Structures and Geotechnical Engineering 03: 2319-5347,2014.

[2] Bonen, D. Shah, S.P.(2005) "Fresh and hardened properties of self-consolidating concrete construction". *Progress in Structural Engineering Materials* Vol. 7, No.1, pp. 14–26. https://doi.org/10.1002/pse.186

[3] Bouzoubaa, N. Lachemi, M.(2001) " Self-compacting concrete incorporating high volumes of class F fly ash preliminary results" *Cement and Concrete Research* Vol. 31, No. 3, pp. 413–420. https://doi.org/10.1016/S0008-8846(00)00504-4

[4] Burak Felekoglu, Selcuk Turkel, Bulent Baradan (2007) " Effect of water/cement ratio of the fresh and hardened properties of self-compacting concrete" *Building and Environment* Vol. 42, No. 4, pp.1795-1802. https://doi.org/10.1016/j.buildenv.2006.01.012

[5] Dinakar, P. Babu, K.G. Santhanam, M.(2008) "Durability properties of high volume fly ash self compacting concretes". *Cement Concrete Composites* Vol. 30 No. 10, pp. 880–886. https://doi.org/10.1016/j.cemconcomp.2008.06.011

[6] EFNARC, Specification and guidelines for self-compacting concrete. UK, 2002. pp.32, ISBN 0953973344.

[7] Gesoglu, M. Güneyisi, E. Özbay, E. (2009) "Properties of self-compacting concretes made with binary, ternary, and quaternary cementitious blends of fly ash, blast furnace slag, and silica fume" *Construction and Building Materials* Vol. 23, No. 5, pp.1847–1854. https://doi.org/10.1016/j.conbuildmat.2008.09.015

[8] Gesoglu, M. Ozbay, E.(2007) "Effects of mineral admixtures on fresh and hardened properties of self-compacting concretes: binary, ternary and quaternary systems" *Materials and Structures* Vol. 40, No. 9, pp.923–937. https://doi.org/10.1617/s11527-007-9242-0

[9] Karjinni, V. Shrishail, B. Anandini, and Dada Patil.S (2009) "An Investigation on the characteristic properties of high performance SCC with mineral admixtures" *The Indian Concrete Journal* Vol. 83, No. 9, pp. 15-19.

[10] Khatib, J.M. (2008) "Performance of self-compacting concrete containing fly ash" *Construction and Building Materials* Vol. 22 No. 9, pp. 1963–1971. https://doi.org/10.1016/j.conbuildmat.2007.07.011

Laboratory Studies on the Effectiveness of Quarry Dust and Bottom Ash with Treated Marine Clay for Adaptable Flexible Pavement Sub-Grade

Surya Teja Veeranki, Abhilash Nadakuditi*

Civil Department, Andhra Loyola Institute of Engineering and Technology, Andhra Pradesh, India

veerankisuryateja@gmail.com, abhiced@aliet.ac.in*

Keywords: Stabilization, Marine Clay (MC), Bottom Ash (BA), Quarry Dust (QD), OMC, MDD, CBR

Abstract. The soil found on the seabed is classed as Marine clay. The major proportion of marine soil is clay so it is generally referred as marine clay (MC). The MC is often weak and lacks stability in heavy loads. This research is concerned with the potential of BA (Bottom ash) and QD (Quarry dust) as soil stabilizers based on resistance enhancement. Soil stabilization points to make strides the geotechnical features of the MC. The engineering properties of MC have been built-up, such as grain size distribution, particle density and soil plasticity. The soil sample was blended and compacted with various quantities of the BA and QD i.e. 2.5%, 5%, 7.5%, 10% 12.5% and 15% for compaction and strength test. The dry-weight method was utilized to prepare the samples. A standard Proctor test was run to determine the stabilized floor OMC and MDD. In the interim, the CBR was conducted to obtain the strength of the stabilized soil. Test results indicate that the MDD of the MC has been improved by 0.19 (g/cc) on addition of 10% BA and it has been improved by 0.246 (g/cc) when 10% QD is added when compared with untreated MC. Laboratory analyses of the cyclic plate load test revealed the ultimate load carrying capacity of the treated MC model flexible pavement has been increased by 349.9% at OMC when compared with untreated MC model flexible pavement.

Introduction

Many parts of the world, such as India, the United States, Egypt, etc., are struggling with construction issues due to clay soil.

The study is focused on the following points:

1. Advancement of locally accessible soil utilizing a few eco-friendly and cheap squander materials.
2. Assessment of quality characteristics of virgin as well as mixed soil utilizing diverse extents of QD and BA.
3. Assurance of fitting QD and BA substance proportion can accomplish the greatest pick up in quality of soil.

Wide run of soil alteration strategies are accessible based on the sort of soil and its characteristics, sort of the development, time accessible, related taken a toll. It has been watched that mechanical by-products can cause exceptional alter within the soil properties in terms of quality characteristics, thickness, sharpness etc., parched too serves rural benefits by increasing crop yield. It may be a well-established truth that the stack coming from the superstructure is eventually borne by the soil. The sandy soil has the property of liquefaction while sweeping soil imbibes part of water posturing danger to little structures, canal-linings, asphalts. Subsequently when an inadmissible condition is met, conceivable elective arrangement can be either of • Forsake the site • Expel and supplant the soil • Overhaul the arranged structure accordingly • Treat the soil to alter its properties or Ground Improvement The final strategy recorded over is known as Ground Improvement Technique or Geotechnical Engineering Practice which is used in this work.

Objectives of the study

- To decide the properties of MC, BA and QD,
- To assess the performance of Raw MC with the addition of BA and its suitability for sub-grades of flexible pavements.
- To assess the performance of BA stabilized MC with the addition of QD and its suitability for sub-grades of flexible pavements.

Materials and Methods

Marine clay (MC). The MCs largely found within the states of West Bengal, Orissa, Andhra Pradesh, Tamilnadu, Kerala, Karnataka, Maharashtra and some parts of Gujarat.
The MC used in this work has been collected near sea coast, Kakinada, Andhra Pradesh
The volume changes of soils are occurs due to variations in moisture content present in the soil. Amid dry-season, a shrinkage crack may occurs close the surface, expanding to profundity of around 1.5m, demonstrating for compressibility.

Furnace BA (FBA). Furnace BA (FBA) is composed of agglomerated fly fiery debris collected at the furnace bottom or combustion chamber of the power station boiler. The grain size of furnace BA ranges from fine sand to fine gravel (more prominent than 5 mm size).
There are two types of BA
A) Dry BA: The ash is in a solid state at the furnace bottom
B) Wet BA: The ash refers to the molten state of the ash which leaves the furnace as a liquid.

BA (Bottom ash) Properties. The BA used in this work has been collected form VTPC (Vijayawada Thermal Power Corporation). Fly fiery debris is, therefore, ideal for backfilling retaining walls or for building dikes over delicate soils since of its:
High internal angle of friction, Low unit mass, Low compressibility, diminished settlement when utilized as fill fabric, Ease of compaction and self-hardening properties, coming about in a conceivable decrease in fill pressures on structures.
To use BA in embankment fill, attention should be paid to the control of:

- Erosion caused by surface runoff and internal piping, dust and leaching,
- Care should be taken with the use of acidic BA in backfill against concrete structures to ensure that they do not attack the concrete.

QD (Quarry dust). The QD brought from Padmavathi crusher present at Yeleswaram from 40 Km away Kakinada.

Remedial measures to overcome issues of marine clay soils. If soil incorporates high deformation, the preventive measures are required. These measures can be broadly classified into the following categories:

- Avoiding highly compressible soils
- Alterations to these soils

In case of Asphalt sub-grade, stabilization procedures can be adopted using different mechanical squander considering the economy additionally chemical added substances for simple blending and early comes about. The support strategies moreover play imperative part in progressing the stack carrying capacity of the marine clay beds.

Soil Replacement. It consists of replacing the weak marine clay by using the rich soils available within the vicinity. So that the allowable bearing pressure of the foundation bed or sub-grade of the flexible pavement can be improved and also the deformations can be controlled up to some extent.

Sand Cushion Method. In this method, the whole profundity of the marine clay stratum or a portion thereof is expelled and supplanted with a sand pad, compacted to the desired density and thickness. The deformation varies inversely as the thickness of the sand layer and directly as the density of the sand. Arrangement of sand pad is likely based on the suspicion that it would retain upward and lateral deformations. The sand pad strategy in this way bristles with serious restrictions especially when it is embraced in profound strata. Foundation engineers regularly recommend a few subjective thicknesses for sand cushion.

Stabilization of Marine Clay. Soil stabilization could be a strategy where natural or manmade additives or folios are utilized to make strides the properties of soils. Chemical added substances, such as lime, cement, Rice Husk Fiery debris, Fly fiery debris and other chemical compounds have been utilized in marine clays stabilization for numerous a long time with different degrees of success.

Compositional variety through ionic or isomorphism substitution inside the clay mineral crystal lattice can take off the structural unit with a net negative charge. The presence of this net negative charge implies that soluble cations can be pulled in or adsorbed on to the surface of the clay mineral auxiliary units without modifying the essential structure of the clay mineral. The most common soluble cations are Na+, K+, Ca2+, Mg2+, H+, and NH4+.

Cation exchange capacity (C.E.C.) is measured in terms of milli-equivalent of the atomic weight of solvent/100 gram of the dry solid, which varies widely for various types of clay minerals.

Laboratory test results & Discussions

The impact of diverse materials on the compaction and CBR properties were discussed in this article. In the laboratory, index tests, swell tests, strength tests were conducted by supplanting MC partially by BA and QD at distinctive rates with a view to determine the supplanting extents. The CBR values of the untreated and treated MC were determined at the respective OMC of the various mixes obtained from the IS modified compaction test.

The impact on the addition of BA and QD on compaction, CBR, swelling, strength properties and Atterberg limits of MC were discussed in detail in the following sections.

Effect of BA on Compaction properties of MC. Individual influence of BA on the compaction properties of MC as observed in the laboratory testing by replacing MC with BA partially by 2.5%, 5%, 7.5%, 10% and 12.5% respectively.

The ideal percentages of distinct individual additives observed during the laboratory experimentation appeared below.

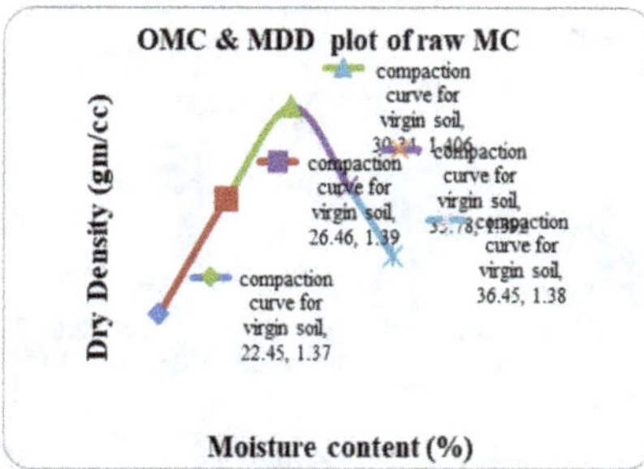

Fig 1. Compaction properties of raw MC

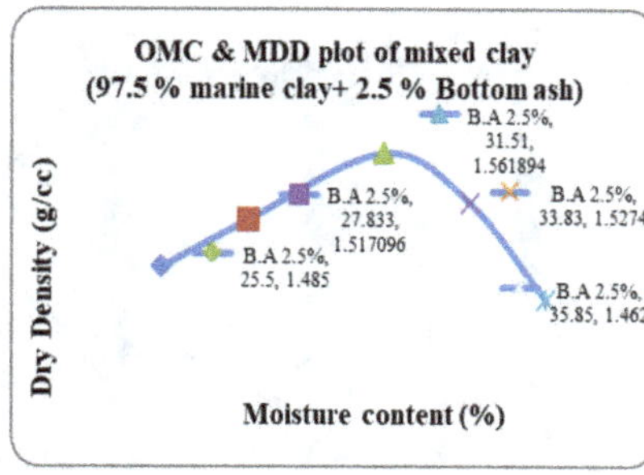

Fig 2. Compaction properties of 2.5% BA treated MC

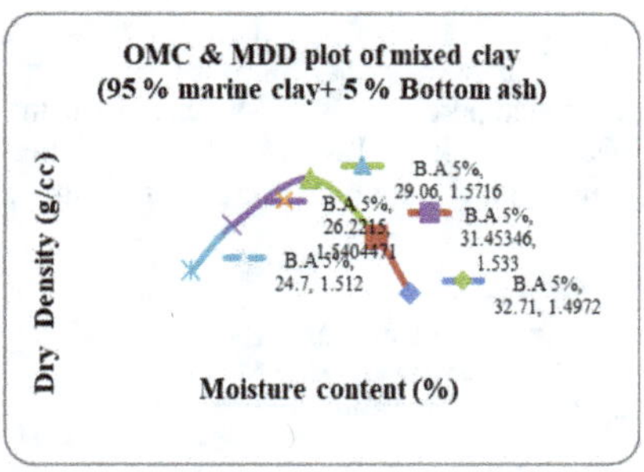

Fig 3. Compaction properties of 5% BA treated MC

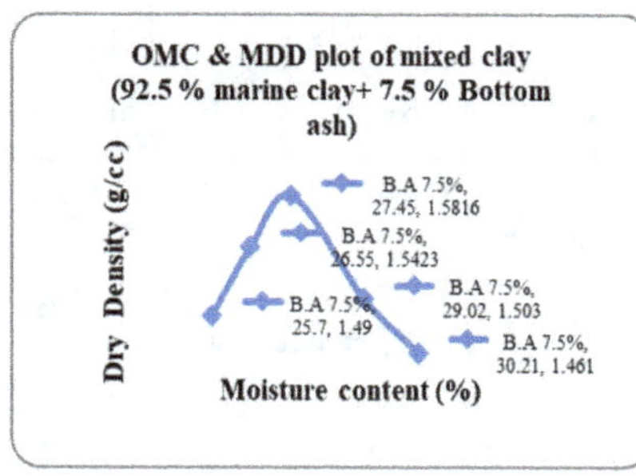

Fig 4. Compaction properties of 7.5% BA treated MC

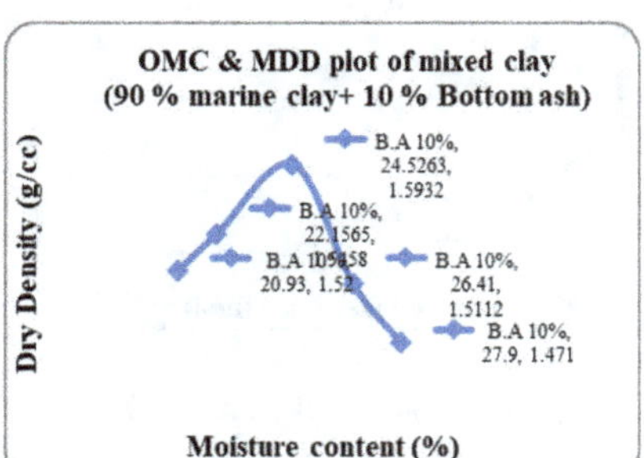

Fig 5. Compaction properties of 10% BA treated MC

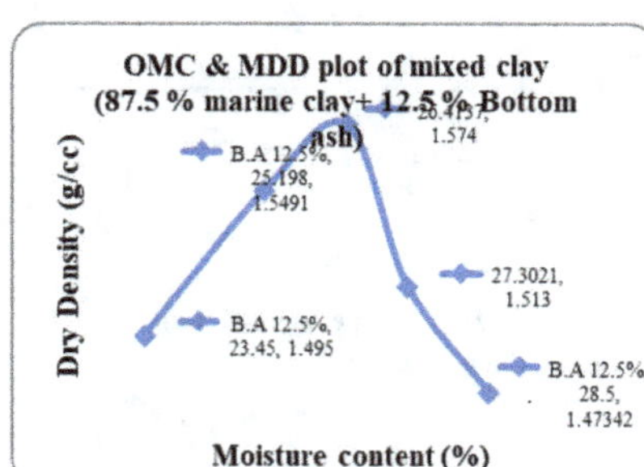

Fig 6. Compaction properties of 12.5% BA treated MC

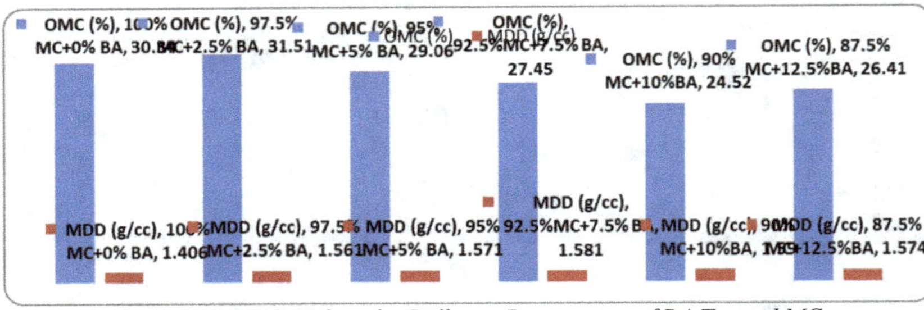

Fig 7. OMC & MDD Values for Different Percentages of BA Treated MC

Of all the different combinations of BA tried in this investigation, it was observed that optimum is found to be obtained at 10% BA. However, more than the adding of 10%, there is lowering the MDD values of treated MC.

Effect of BA on CBR of MC. CBR values of distinct blended percentages of MC and BA using OMC obtained from standard proctor tests are determined. The soaked CBR test is done. Variation in CBR of raw MC and MC with 7.5%, 10%, 12.5% partial replacement by BA is presented in the following figures.

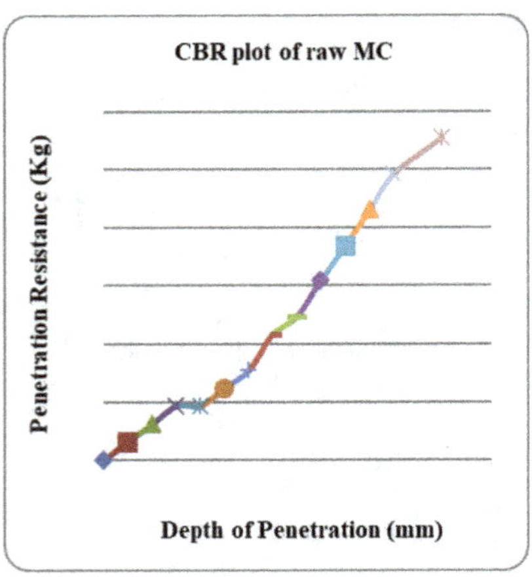

Fig 8. CBR for untreated Marine Clay

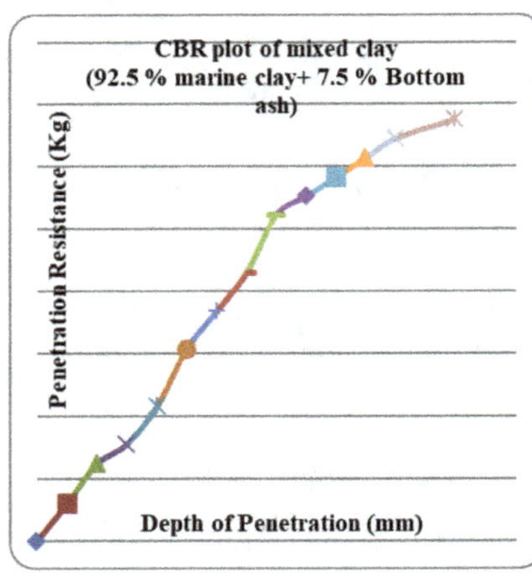

Fig 9. CBR of 7.5% BA treated MC

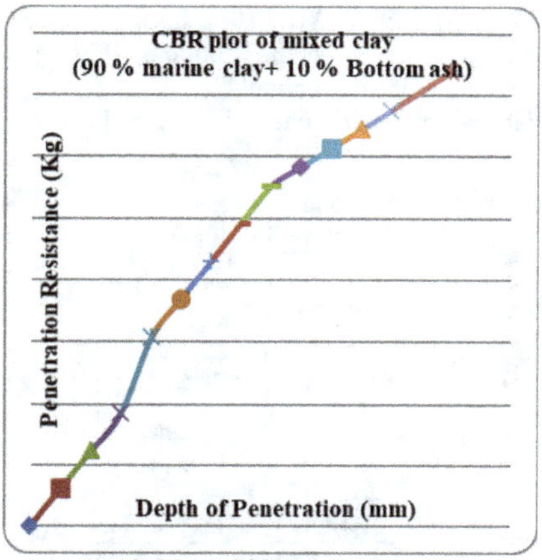

Fig10.CBR of 10% BA treated MC

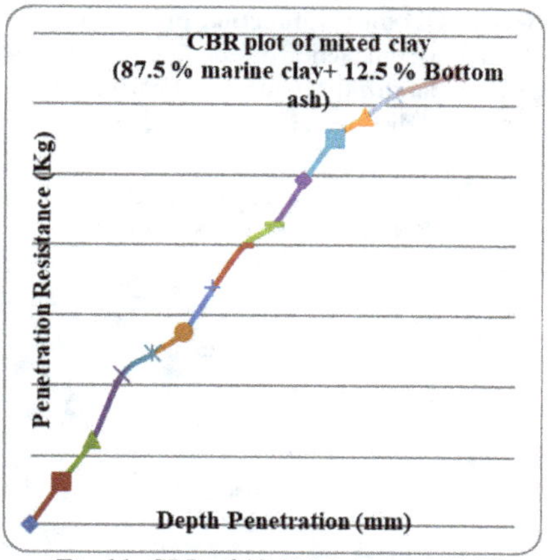

Fig 11. CBR of 12.5% BA treated MC

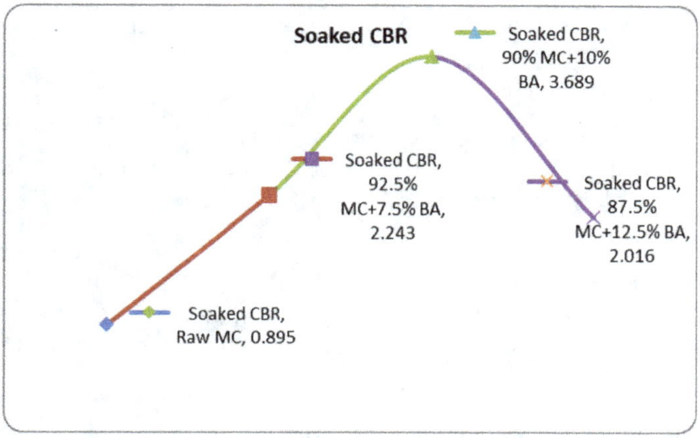

Fig 12. CBR Values for Different Percentages of BA Treated MC

It was observed that optimum CBR is found to be obtained at 10% BA. However, other combinations of the adding of BA, there is a decrease in CBR values of treated MC. The stabilized MC i.e., raw MC with 10% replacement with BA is further evaluated for the compaction properties and CBR by the addition of QD at various proportions.

Effect of QD on Compaction properties of stabilized MC (90% MC+10% BA). nfluence of QD on the compaction properties of stabilized MC as observed in the laboratory testing by replacing stabilized MC with QD partially by 5%, 10% and 15% respectively.

The ideal distinct additives observed during the laboratory experimentation appeared below.

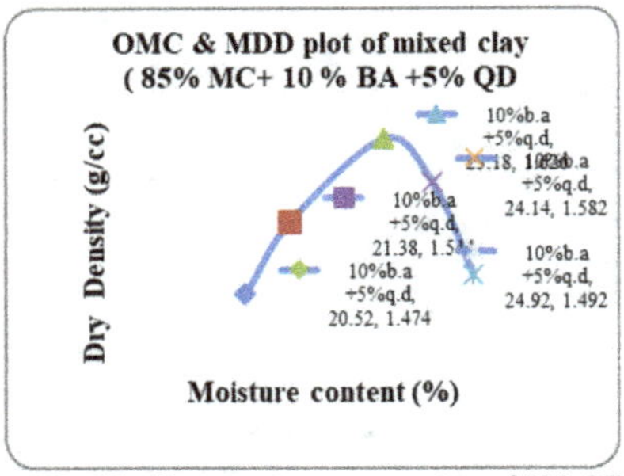

Fig 13. Compaction properties of 5% QD treated stabilized MC

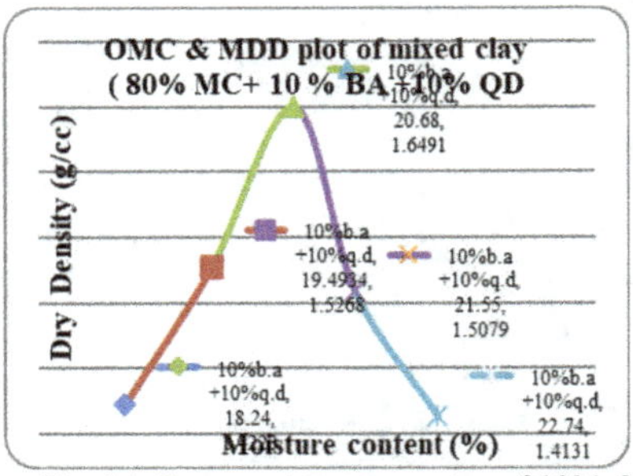

Fig 14. Compaction properties of 10% QD treated stabilized MC

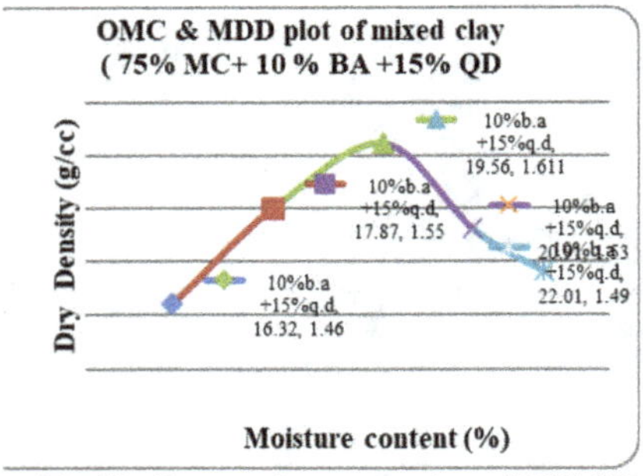

Fig 15. Compaction properties of 15% QD treated stabilized MC

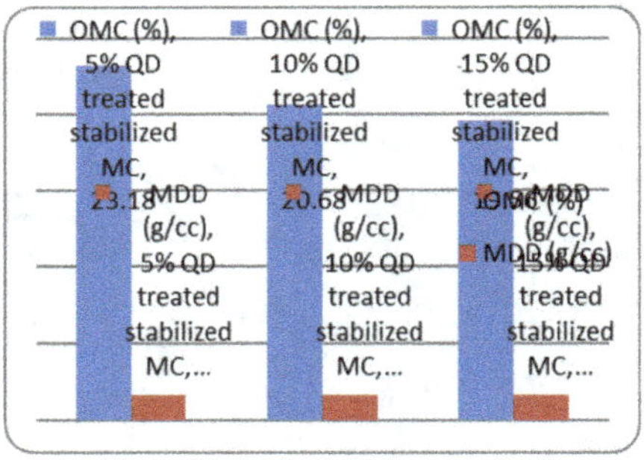

Fig 16. OMC & MDD Values for Different Percentages of QD Treated stabilized MC

Effect of QD on CBR of stabilized MC (90% MC+10% BA). CBR values of various mixes of 10% QD treated stabilized MC are calculated. The soaked CBR value is determined. Variation in CBR of stabilized MC and MC with 7.5%, 10%, 12.5% partial replacement by QD is presented in the following figures.

Materials Research Forum LLC
https://doi.org/10.21741/9781644901953-50

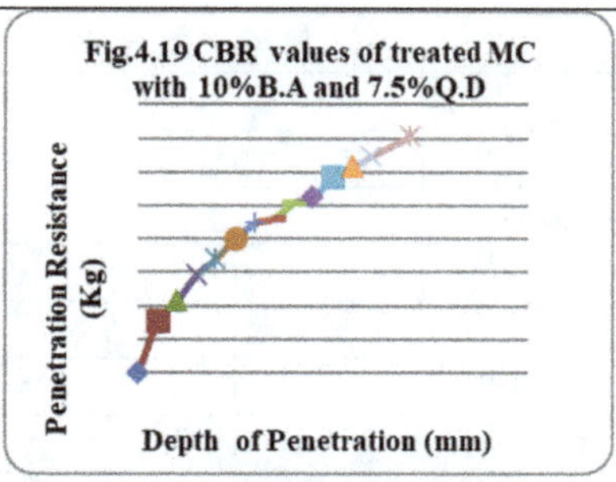

Fig 17. CBR of 7.5% QD treated stabilized MC

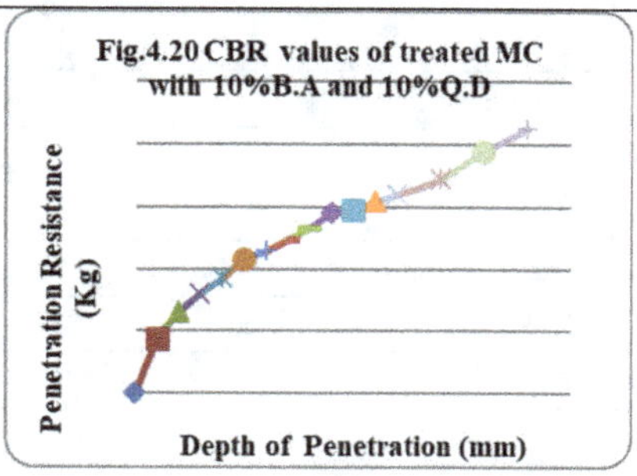

Fig 18. CBR of 10% QD treated stabilized MC

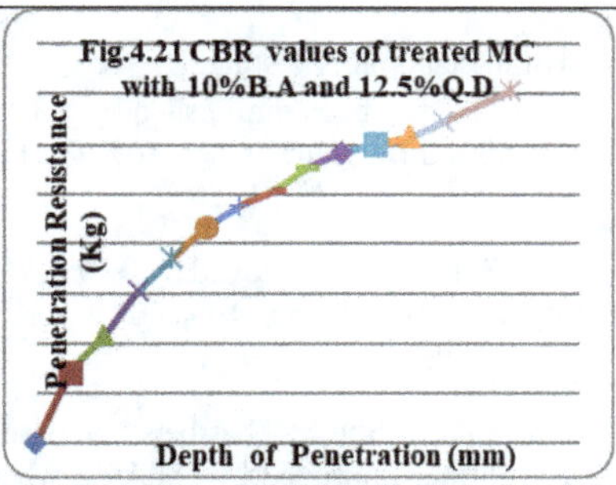

Fig 20. CBR of 12.5% QD treated stabilized MC

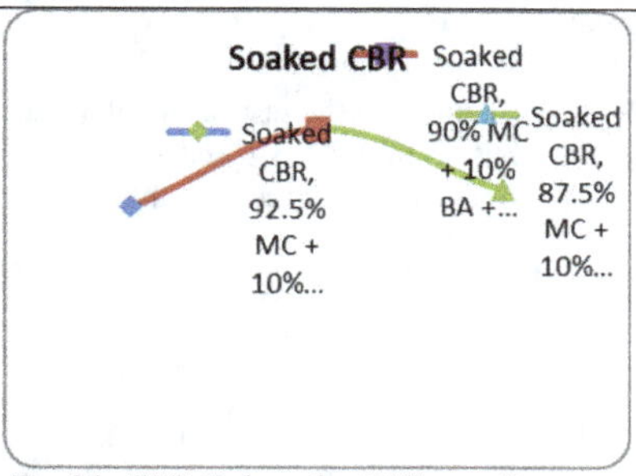

Fig 21. CBR Values for Different Percentages of QD Treated stabilized MC

Properties of BA, QD, Untreated and Treated MC

Table.1 Comparison of physical properties of MC, BA and QD before and after treatment

Sl.No	Property		BA	QD	Untreated (Raw) MC	Treated MC (10% BA)	Treated MC (10% BA+ 10% QD)
1	Grain size distribution						
	Gravel (%)		-	16.36	0	-	3.2
	Sand (%)		86	83.18	1.2	9.6	17.8
	Fines	Silt (%)	10	0.46	4	4.3	4.1
		Clay (%)	4	0	94.8	86.1	74.9
2	AtterBerg limits						
	Liquid Limit (%)		NON PLASIC	NON PLASIC	64.52	44.52	36
	Plastic Limit (%)				30.12	27.96	24.09

Materials Research Proceedings **23** (2022) 462-471 https://doi.org/10.21741/9781644901953-50

	Plastic Index (%)			34.40	16.58	11.91
3	Soil Classification	SW	SW	CH	CI	CI
4	Specific Gravity	2.21	2.87	2.4	2.58	2.62
5	**Compaction properties**					
	Optimum Moisture Content (%)	23	13.4	30.34	24.52	20.68
	Maximum Dry Density (g/cc)	1.53	1.94	1.4	1.59	1.64
6	Cohesion (kN/m^2)	7.2	4.8	120	65	25
7	Angle of Internal Friction (°)	9	16	3.5	6	8
8	CBR Value (soaked) (%)	3.59	8.96	0.89	3.68	7.84

Summary

The following conclusions are drawn based on the results of the laboratory testing.

- It is found from the test results that the O.M.C of the M.C has been diminished by 5.82% in addition to 10% BA and it has been further diminished by 9.66% when 10% Q D is added when compared with untreated MC.
- It is found from the test results that the M.D.D of the MC has been improved by 0.19 (g/cc) in addition to 10% BA and it has been improved by 0.246 (g/cc) when 10% QD is added when compared with untreated MC.
- It is observed from the test results that the C.B.R. the value of the MC has been increased by 412.17% in addition to 10% BA and it has been further improved by 880.89% when 10% QD is added when compared with untreated MC.

References

[1] Al Quadi, I.L (1994), Laboratory Evaluation of Geo-synthetics Reinforced Pavement Sections, TRR-1739, TRB, 1994, pp. 25-31.

[2] Al-Omari, R.R and Oraibi, W.K (2000), Cyclic behavior of reinforced expansive clay, Jr. of the Japanese Geotechnical Society of Soils and Foundations, Vol. 40, No.2; 2000,pp1-8. https://doi.org/10.3208/sandf.40.2_1

[3] Al-Rawas, N.M (2000), Effect of curing and temperature on lime stabilization, Proc. of Second Australian Conf. on Engineering Materials, Sydney, 1981, pp.611-662.

[4] Ambily, A.P and Gandhi, S.R (2004), Experimental and Theoretical Evaluation of Stone Column in Soft Clay, ICGGE, pp 201-206.

[5] Ambily, A.P and Gandhi, S.R (2004), Experimental and Theoretical Evaluation of Stone Column in Soft Clay, ICGGE, pp 201-206.

[6] AnandJ.Puppala, EkarinWattanasanticharoen and LaureanoR.Hoyos (2003), Ranking of Four Chemical and Mechanical Stabilization Methods to Treat Low-Volume Road Subgrades in Texas, Jr.-Transportation Research Record, Vol. 1819B, 2003, pp. 63-71. https://doi.org/10.3141/1819b-09

[7] Anandarajah. A and Chu. J (1997), Laboratory Determination of shear strength parameters for marine clay, Journal of the Institution of Engineers, Singapore, Vol.14, No.3, pp 39-46.

[8] Arvind Kumar, Baljit Singh Walia and Asheet Bajaj (2007), Influence of Flyash, Lime and Polyester Fibers on Compaction and Strength Properties of Expansive Soil, J.Mat in Civil Engineering, ASCE, Vol. 19, Issue. 3, 2007, pp. 242-248. https://doi.org/10.1061/(ASCE)0899-1561(2007)19:3(242)

[9] Balasubramaniam, A.S., Bergado, D.T., Buensuceso, B.R. and Yang, W.C (1989), Strength and deformation characteristics of lime treated soft clays, Geotechnical Engineering (AIT), 20, 1989, pp. 49-65.

Keyword Index

About the Editors

Dr. M. Vasudevan is currently working as Assistant Professor and Head of the Department of Civil Engineering at Bannari Amman Institute of Technology Sathyamangalam, Tamil Nadu, India with a teaching experience of 6 years. He obtained his bachelor's degree in Agricultural Engineering from Kerala Agricultural University, India in 2007, followed by master's degree (M.Tech.) in Environmental Engineering from Motilal Nehru National Institute of Technology Allahabad, India in 2009. He completed his Ph.D. from Indian Institute of Technology Madras in 2015 in the area of Environmental Engineering. He has published about 45 research papers in peer-reviewed international journals and 70 international/national conferences. He has also authored 8 book chapters and co-edited 2 books. He is a professional member of Indian Geotechnical Society and Indian Society for Technical Education and Indian Water Works Association.

Mrs. V. Preetha is Assistant Professor in the Department of Civil Engineering, Bannari Amman Institute of Technology, Tamilnadu, INDIA. She obtained BE in Civil Engineering from Kongu Engineering College, Erode, INDIA in 2006, Masters in Structural Engineering from Bannari Amman Institute of technology, Erode, India in 2009 and Pursuing her Ph.D in Structural Engineering from Anna university. Having worked for about 1 year in Jain Housing and Construction as Quantity surveyor, she joined Larsen and Toubro Pvt Limited, ECC division, Headquarters, Manapakkam Campus as Senior Design Engineer during 2009 to 2011 and then joined BIT in 2012. Her research interests include Earthquake Engineering, Steel concrete Composite Structures and structural design. She has published around 15 peer-reviewed journal papers, 20 conference papers, 1 book chapters and filed 2 patents. Her service includes field studies in the area of concrete technology and non-destructive testing methods as consultancy projects. She is a professional member of Institution of Engineers India, Indian Geotechnical Society and Indian Society for Technical Education.

Mr. B. Jeyanth is Assistant Professor in the Department of Civil Engineering, Bannari Amman Institute of Technology, Tamilnadu, INDIA. He has obtained BE in Civil Engineering from Adhiyamaan Engineering College, INDIA in 2012, Masters in structural Engineering from Gnanamani College of Engineering, INDIA in 2014. He started his Professional carrier as Assistant Professor and presently with 7 years of Experience in Teaching. His research interest includes study on the behavior of fiber reinforced concrete. He has published around 5 peer-reviewed journal papers, 9 conference papers. He is a professional member of International association of engineers (IAENG).

www.ingramcontent.com/pod-product-compliance
Lightning Source LLC
LaVergne TN
LVHW052216200926
848481LV00001B/120
9781644901946